Experimental
Organic and Medicinal Chemistry
Principles & Practice

Experimental
Organic and Medicinal Chemistry
Principles & Practice

T. Durai Ananda Kumar
M.Pharm.,
Sr. Assistant Professor,
Dept of Pharmaceutical Chemistry,
Gokaraju Rangaraju College of Pharmacy,
Hyderabad-500 090. India.

N. Swathi
M.Pharm.,
Sr. Assistant Professor,
Dept of Pharmaceutical Chemistry,
Gokaraju Rangaraju College of Pharmacy,
Hyderabad-500 090. India.

PharmaMed Press
An imprint of Pharma Book Syndicate
A Unit of BSP Books Pvt. Ltd.
4-4-309/316, Giriraj Lane,
Sultan Bazar, Hyderabad - 500 095.

Experimental **Organic and Medicinal Chemistry:** ***Principles & Practice*** **by** ***T. Durai Ananda Kumar and N. Swathi***

Published by

PharmaMed Press

An imprint of Pharma Book Syndicate

A unit of BSP Books Pvt. Ltd.

4-4-309/316, Giriraj Lane, Sultan Bazar, Hyderabad - 500 095.
Phone: 040-23445605, 23445688; Fax: 91+40-23445611
E-mail: info@pharmamedpress.com

ISBN: 978-93-52300-79-2 (HB)

Foreword

Organic-medicinal chemistry deals with the synthesis of molecules having diverse pharmacological actions. Adequate training with guidelines on chemical reactions, product characterization and estimations is very crucial for the graduate pharmacy students. There are a few books available particularly related to medicinal chemistry.

Synthesizing a drug molecule is only a part of the story, but understanding the process and reaction mechanisms are the needs of the hour, because the Drug Regulatory Authorities insist on product knowledge and process understanding, as a part of the QbD approach. Further process must be laboratory friendly and innoxious to environment. Thus drug synthesis is a major challenge to supply in tons to meet the, ever growing health care needs. In this context, the efforts made by Mr. T. Durai Ananda Kumar and Mrs. N. Swathi, to bring the title ***Experimental* Organic and Medicinal Chemistry: *Principles and Practice*** are very praise-worthy and helpful for laboratory instructions and practice. Both the authors had spent more than 9 years on teaching and conducting practicals of organic-medicinal chemistry.

This book consists of several topics covering both theory and practical concepts. A few topics to mention are: safety measurements, general laboratory practices, synthesis of drugs and drug intermediates, green synthetic strategy, estimations of drugs, qualitative analysis of inorganic, organic and natural products, isolation and determination of active principles from natural sources. In addition, preparation of a few solutions and reagents are included as a ready reference.

This book is a good companion for students of B. Pharmacy and a source book for M. Pharmacy (Pharmaceutical chemistry, Medicinal Chemistry). I recommend this book to the students.

Hyderabad
25-02-2015

C. V. S. Subrahmanyam,
Principal,
Gokaraju Rangaraju College of Pharmacy,
Hyderabad-500 090.

PREFACE

தொட்டனைத் தூறு மணற்கேணி மாந்தர்க்குக்
கற்றனைத் தூறும் அறிவு - குறள் 396

In sandy soil, when deep you delve, you reach the springs below;
The more you learn, the freer streams of wisdom flow – **Couplet 396**

With the blessing of our parents we present this book entitled "***Experimental* Organic and Medicinal Chemistry: *Principles and Practice***" to the active academic community.

The modern medicinal chemistry utilizes several novel drug discovery tools to identify the drug-like molecules (lead) and to convert them into therapeutically potential molecules. The advanced and adequate practice in synthetic medicinal chemistry is essential for pharmacy graduates (B. Pharmacy and M. Pharmacy) to receive recognition in academia and industry sectors.

The material available for practice is complex in nature or inadequate. This stimulated us to design and develop this book with copious information. The basis of research is dependent on the verification of facts. Extensive review of literature and the large number of experiments conducted in our research laboratories for the pharmacy students enabled us to load the information in this material. This book serve as a primer for academic community.

The experiments described are very simple to carry out and does not require any special equipments. The entire concepts are described in seven parts spreading through 60 chapters and the salient features of this book are:

- Safety measurements and general laboratory practices are outlined in simple language.
- The role of reagents, alternative reagents and hazards associated are highlighted.
- Pharmaceutical relevance of chemical reactions are illustrated.
- Neat and self explanatory chemical reaction mechanisms involved in the synthesis of drugs and drug intermediates are discussed.
- The significance of green synthesis strategy with few experiments are outlined.
- Limit tests, qualitative analysis of inorganic, natural and synthetic organic compounds are illustrated in a lucid manner.
- Estimations of natural and organic-medicinal compounds along with isolation of active principles are described.
- Preparation of important solutions and reagents are given.

This book is intended for the students of B. Pharmacy, M. Pharmacy (Pharmaceutical Chemistry, Medicinal Chemistry) and other organic-medicinal chemistry discipline. Adequate measures have been taken to remove errors, if any remains kindly light upon it.

The suggestions and constant encouragement we received from Prof. Dr. C.V.S. Subrahmanyam, Principal, Gokaraju Rangaraju College of Pharmacy, Hyderabad, is the driving force for the completion of this book. Our sincere gratitude to Dr. G. Gangaraju, President, and Mr. G.V.K. Rangaraju, Vice-President, Gokaraju Rangaraju Educational Society for the facilities provided.

We also thank our colleagues and students for their constructive criticism. Our special thanks to our professional friends Mr. P. Mohan, Ms. N. Yamini, Ms. Y. Eswaramma, and Ms. P. Vaishnavi for their valuable contributions in different phases of this project.

We gratefully acknowledge the timely help extended by Mr. P. Gopi, Lab technician and Mr. I. Laxman Rao, Librarian. We would like to thank Mr. Anil Shah, Director, Mr. A. Santhosh, Mr. Naresh Daveragave and his team member Mrs. Sandhya, BSP Books (P) Ltd, Hyderabad for their support and co-operation extended.

-Authors

In theory, theory and practice are the same.
In practice, they are not. - Albert Einstein.

CONTENTS

GENERAL LABORATORY GUIDELINES AND PRACTICES

CHAPTER 1

GENERAL LABORATORY SAFETY

1.1 General Tips

Do's

- Ensure good general ventilation.
- Wear laboratory coats (fastened), safety goggles and gloves.
- Legs should be covered with closed-toe shoes.
- Confine long hair and loose clothing.
- Use fume hood for cautious reactions.
- Replace all caps and lids on reagent/chemical containers.
- Return reagent/chemical containers to proper locations.
- Wash off any chemicals spilled on skin or clothes immediately.
- Clean all spills in and around balances.
- Wash hands before [especially before eating], and after handling contaminated material.
- Use bulb pipette/micro pipette for measuring harmful liquid chemicals.

Don'ts

- Don't wear contact lenses during laboratory work.
- Don't perform unauthorized reaction, and work alone if procedures are hazardous.
- Don't eat anything during the lab work inside the laboratory.
- Don't store food or beverages in the lab or in chemical refrigerator.

- Don't put face inside the fume hood and ensure the minimum storage of chemicals in the fume hood.
- Don't bring contaminated materials out of the cabinet until they have been decontaminated.
- Don't pipette harmful/evaporable substances.

1.2 Specific Rules

1. ***Safety Awareness:***
    - Develop and implement preventative maintenance program and safety awareness for chemical operations.
    - *Personal protective equipments*: Personal protective equipments (PPEs) such as respirators, goggles, gloves and lab coats should be available in work place and use them whenever necessary.
    - *Safety equipments:* Gain knowledge on the location and operation of safety equipments, such as safety showers, eyewashes, fire extinguishers, first-aid kits and explosion-proof refrigerators.
2. ***Reactive Safety:***
    - Use personal protective equipments (PPEs) and keep appropriate fire suppression equipment in laboratory.
    - Never use any laboratory equipment without proper training.
    - Always wear lab coats (apron), safety goggles and gloves in the laboratory.
3. ***Chemical safety:***
    - Chemical exposure results in burns, inflammation, allergic responses, eye irritation and dizziness. Hence minimize all chemical exposures, observe exposure limits and institute chemical hygiene plan.
    - Use prescribed chemicals only and do not substitute.
    - *Chemical disposal:* Equip the laboratory with waste solvent container for collecting waste organic solvents and solutions. Neutralize any acids and alkalis before pouring them down the drain.
    - *Decontamination:* Decontamination is the removal or neutralization of toxic agents, it includes both sterilization and disinfection.
    - Use proper shields when handling with peroxides. Ethers and alkenes should be tested for peroxides before using them.

4. ***Compressed Gas Safety:***
   - Read label and tags to identify gaseous chemicals and do not use unknown gases.
5. ***Flammable Safety:***
   - Keep containers closed, away from heat, sparks and sources of ignition.
   - Limit the quantity of flammable liquids storage in the laboratory.
   - Use sparkless "flammable storage" refrigerator for flammables.
   - Be aware of the fire extinguishers location and its operation.

   Easy acronym for the use of fire extinguisher

   PASS: **P**ull **A**im **S**queeze **S**weep
   - *Pull:* Pull the pin at the top of the extinguisher.
   - *Aim:* Aim the nozzle toward the base of the fire, stand approximately 8 feet away from the fire.
   - *Squeeze:* Squeeze the handle in short intervals to discharge the extinguisher.
   - *Sweep:* Sweep the nozzle back and forth at the base of the fire.
6. ***Lab Safety:***
   - Never under-estimate the risk and don't proceed with unfamiliar procedure.
   - Acquire complete training and understanding for safe working.
   - Use fume hood for conducting reactions with hazardous and toxic substances and also in case where flammable gases and vapours are produced.
   - Ensure that the exhaust blower is operating and air is entering the fume hood.
   - Ensure that gas supply, water supply, power supply, vacuum lines, compression lines and heating apparatus are turnoff properly.
7. ***Corrosive Safety:***
   - Prevent exposure to corrosive fumes and vapours.
   - Use proper personal protective equipments (PPEs) for eyes, face, hands and body.
8. ***Toxic Safety:***
   - Provide routine medical monitoring.
9. ***Glassware Safety:***
   - Borosilicate glassware is recommended for all laboratory glassware and never use cracked or chipped glassware.
   - Pyrex and shatterproof glassware can be used for special applications.

10. ***Ultraviolet radiation safety:***

- Ultraviolet (UV) lamps and arcs should not be viewed directly. Exposure of eye and skin to UV radiation cause eye damage and skin burns.

11. ***Environmental Protection:***

- Aware of possible environmental threats and institute environmental protection measures.

11. ***Labeling Containers:***

- Label all the reagent/chemical containers to avoid confusion and chemical accidents.

1.3 First Aid

First aid is the immediate help given to an ill or injured person using readily available materials. It include cleaning minor cuts, scrapes and scratches, treating minor burns, applying bandages, dressings, cold compress, cold pack ice bag and splint. It also includes calling emergency medical services.

The very first step in the first aid is not to create any panic condition.

All injuries that are a result of a spill must be attended immediately.

- ***Bleeding:*** Flush the wound thoroughly with water, apply antiseptic and bandage to prevent contamination. If the bleeding is uncontrollable then rise the bleeding part and apply pressure to the wound with sterile gauze.
- ***Inhalation:*** Move to fresh air.
- ***Swallowing:*** Get emergency medical assistance.
- ***Thermal burns:*** Apply cold water and or ice immediately to the burned area until the pain subsides. Wrap to protect the area from infection.
- ***Chemical burns:*** Flush the affected area with plenty of water for several minutes (at least 15 minutes). Acid or minor bromine burns may be treated with 5% sodium carbonate solution. Alkali burns may be washed with 5% acetic acid or saturated boric acid. Wrap the burned area loosely.
- ***Eyes:*** Flush with water in the eye wash immediately for about 15 minutes.
- ***Skin:*** Flush the skin with water first then with soap. Neutralize an acid with *baking soda* and neutralize a base with boric acid. Use safety shower for large amount of chemical slipped on the body.
- ***Fainting and shock:*** Place the victim in a prone position with head lower than the feet.

CHAPTER 2

GENERAL LABORATORY PRACTICES

2.1 Knowledge about Chemicals

Class 1: Explosives

- Background color: Orange
- Symbol: Exploding ball

Class 2: Gases: Non-flammable gases, flammable gases, poisonous gases

- Background color: Varies
- Symbol: Varies

Class 3: Flammable and Combustible Liquids: Store flammable liquids inside cabinets or in storage rooms designed for flammables.

- Background color: Red
- Symbol: Flame

Class 4: Flammable Solids: Flammable solids and spontaneously combustibles are dangerous when they are wet.

- Background color: Red and White candy striped, Blue if reacts with water

Class 5: Oxidizers and Organic Peroxides: Oxidizers are incompatible with organics, hence separate oxidizers from organics. Oxidizers promote combustion and produce fire (explosion) by giving off electrons and undergoing reduction. Never open a bottle that has solid formation around the lid (Friction caused by unscrewing the cap leads to an explosion).

- Background color: Yellow
- Symbol: Flaming ball

Class 6: Poisons and Etiologic Materials: Inhalation hazard

- Background color: White with black letters

Class 7: Radioactive Materials

- Background color: White on the bottom, yellow on the top
- Symbol: Trefoil

Class 8: Corrosives: Acids and bases.

- Background color: Black on the bottom and white on the top
- Symbol: Piece of metal and a hand damaged by liquid

Class 9: Miscellaneous: Hazardous materials not specified in other classes.

Safety Precautions

Laboratory should be well ventilated and Personal Protective Equipments (PPEs) should be worn (especially during distillation and operations under reduced pressure/elevated temperature). Toxicity and flammability of solvents are major concern in laboratories. Most chemicals are carcinogenic and cause deleterious effect to health. Some of the important hazards are:

- ***Explosions:*** Autoxidation of ethers, tetrahydrofuran (THF) and related compounds forms explosive peroxides.
- ***Flammables:*** Low boiling point compounds such as acetaldehyde, acetone, benzene, carbon disulphide, diethyl ether and ethyl acetate are highly inflammable.
- ***Oxidising agents:*** Chromic acid and perchloric acid are known violent explosives.
- ***Tissue toxics:***
    1. Both vapour and liquid dimethyl sulphate are highly poisonous, hence it should be used in fume cupboard and rubber gloves should be worn.
    2. Inhalation leads to giddiness and more toxic effects.
    3. If the liquid is accidently splashed on the skin, wash immediately with ammonium hydroxide and then rub gently by wad of cotton wool soaked in ammonia solution.
    4. Liquid and vapours of benzene are potentially carcinogenic and can be replaced with less toxic toluene.
- ***Acids:***
    1. Concentrated acids should be added slowly to the flask containing water under cooling for dilution purpose.
    2. Chlorosulphonic acid is more dangerous than sulphuric acid.
    3. Picric acid stain on the skin can not be removed readily, hence use rubber gloves.
- ***Reactive halides and anhydrides:*** Acid chlorides, inorganic halides and anhydrides are highly toxic and lachrymatory, cause damage to mucous membranes and lungs.
- Organic azides, nitrates and picrates are highly hazardous.

2.2 Glasswares

2.2.1 Reaction Vessels

Reaction vessels are generally made up of glass. In general vessels can be used for storing, collecting, mixing, heating, boiling, distilling and cooling chemicals. It can be conveniently grouped into:

- ***Reaction flasks:*** They are made up of glass, which tolerate large pressure difference due to vacuum and pressure. The necks of the flasks are connected to the condenser for refluxing or heating the reaction mixture.
    - *Flat bottom flask:* These flasks are useful when stirring is required.
    - *Round bottom flask (single-neck and multi-neck):* These are useful in distillation and heating the reaction mixture and are also known as *Florence flasks*.
- ***Erlenmeyer flask:*** Cone shaped, flat bottomed cylindrical flask useful in titration, heating, mixing and cooling.
- ***Volumetric flasks:*** These are used in the preparation of liquids with volumes of high precision.
- ***Evaporatory flasks:*** Centered, pear shaped flasks with socket for the use in rotary evaporator.

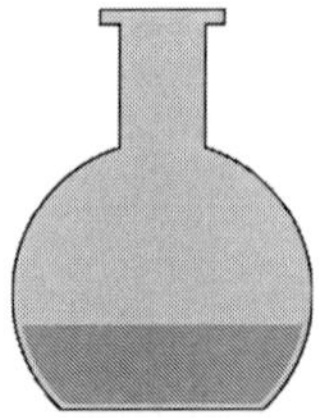

Flat bottom flask

Round bottom flask

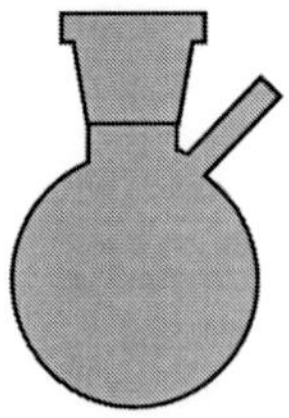

Round Bottom Flask with septum inlet

Two-Neck Angled Round Bottom Flask

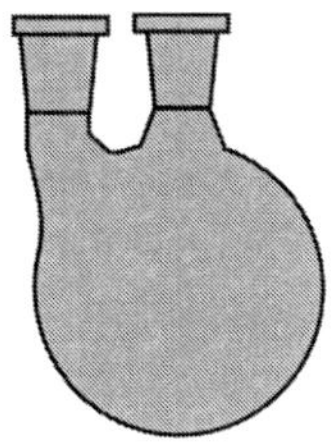

Two-Neck Vertical Round Bottom Flask

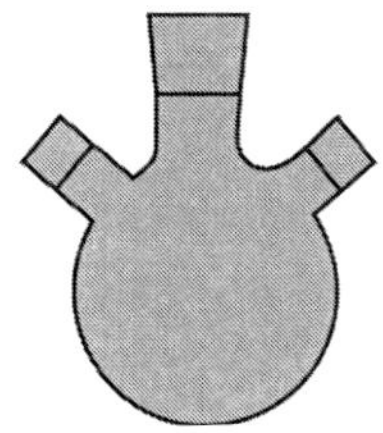

Three-Neck Angled Round Bottom Flask

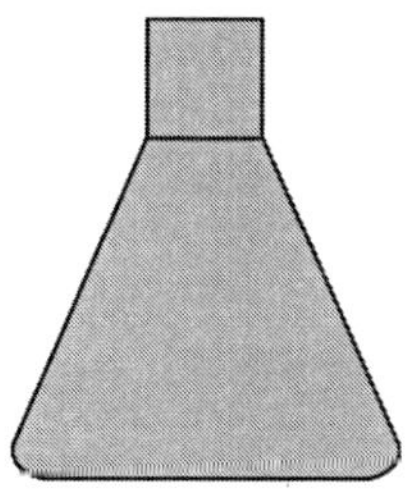

Erlenmeyer Flask

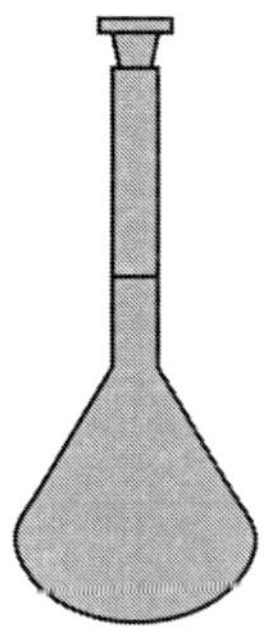

Volumetric Flask

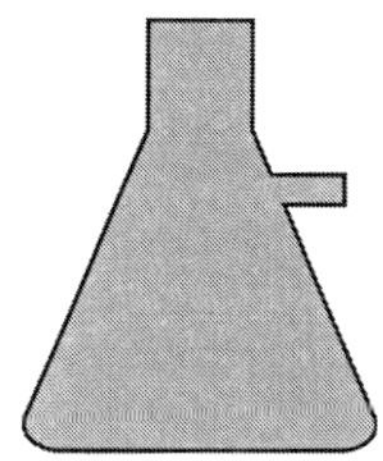

Filtering Flask

2.2.2 Condensers

Condensers are used in refluxing the liquids and distillation process. It condenses the solvent vapours and sends them back to flask thereby reduce the solvent loss. In fractional distillation use of air condenser slows the rate of hot vapour rise and provides better separation between the mixture of components.

Types of condensers

- ***Air condenser:*** A glass tube as the inner part of a *Liebig condenser*, useful for liquids with boiling point above 90 °C. The heat of the liquid will be conducted to the glass, which is cooled by air.
- ***Coil condenser:*** An open tube sealed by glass coil/spiral (water circulates in it).
- ***Double surface condenser:*** A tube in which the vapour is condensed between an inner and outer water cooled jacket, which is useful for liquids having boiling point below 40 °C. *Davies type* and *Double coil type* are the two types of double surface condenser known.
- ***Liebig condenser (west condenser):*** An inner glass tube surrounded by a glass jacket through which water is circulated. It allows rapid heat transfer and hence rapid condensation.

- ***Allihn condenser:*** Long glass tube with water jacket and series of bulbs. The bulbs on the tube increases the surface area upon which the vapour constituents may condense.
- ***Grahams condenser (Inland revenue condenser):*** It contains a coolant spiral coil jacket (condensate path) for the circulation of vapours to get condensed.

Air condenser

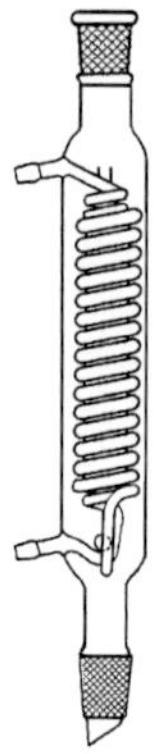

Coil condenser

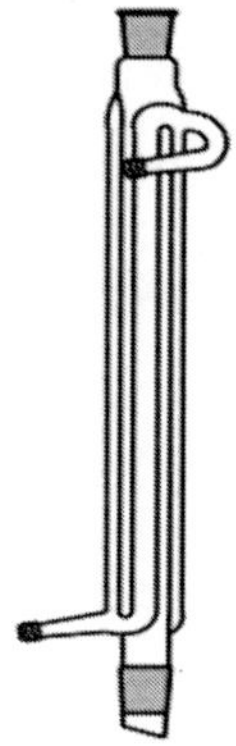

Double surface condenser

Liebig condenser

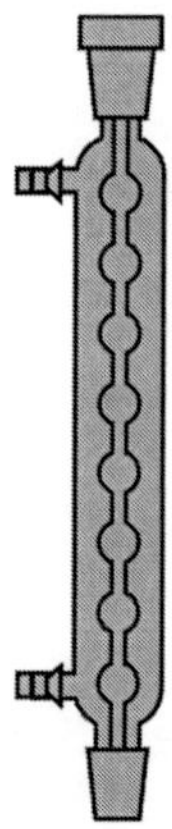

Allihn condenser

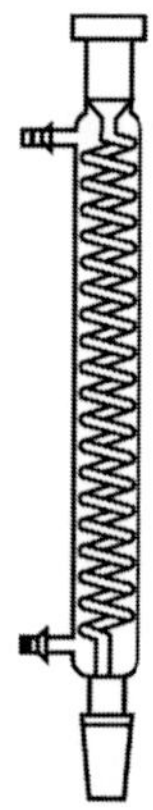

Grahams condenser

2.2.3 Distillation Heads

- ***Knee tube:*** Mostly utilised in the removal of solvents.
- ***Claisen distillation head:*** The left-hand socket accommodates the capillary tube for use in distillation under vacuum. The right-hand socket is for placing suitable thermometer.
- ***Splash-Guard:*** Employed in rotary evaporators and can be used as splash protection (bubble inhibitor).

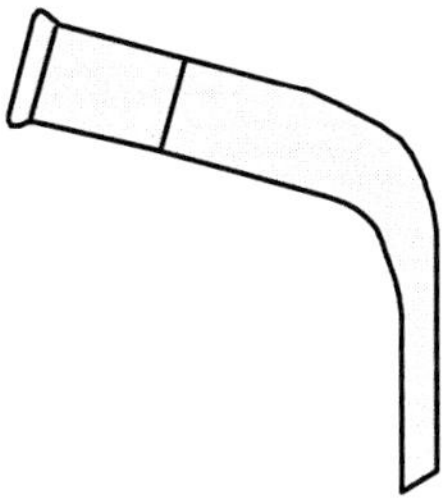

Knee tube

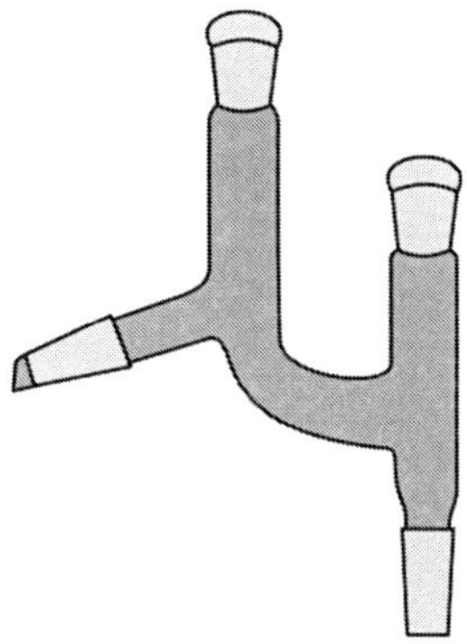

Claisen distillation head

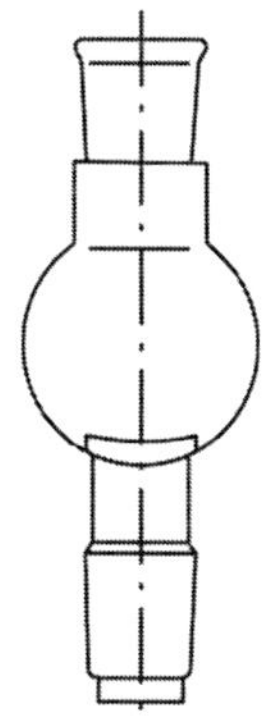

Splash Guard

2.2.4 Adapters/Connectors

- ***Claisen adapter:*** It is used when the reaction requires both reflux and addition of reactants at the same time. One joint connects to condenser for reflux and the other gets connected with dropping funnel (addition of reactants).
- ***Claisen distillation adapter:*** This adapter has two connections, one is socket for thermometer and the other one connects to flask.
- ***Vacuum distillation adapter:*** This adapter provides connection to vacuum pump to facilitate the distillation under reduced pressure (more suitable for thermolabile substances).

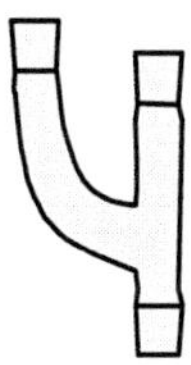

Claisen adapter

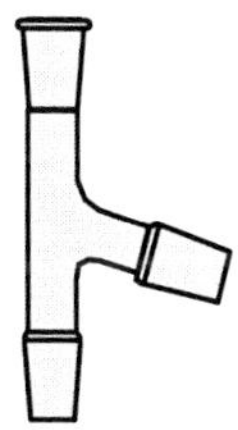

Claisen distillation adapter

Vacuum distillation adapter

- ***Calcium guard tube (drying tube):*** It is used to exclude moisture from experimental setup during reflux/distillation. A tube like structure with one end bend, which connects the tube to reaction vessel. The end contains bulb to pack solid desiccant materials (granular alumina, silica gel, calcium sulphite and calcium chloride). A cotton plug (glass wool plug) to the bottom opening prevents the contact of these desiccants to the reaction vessel.

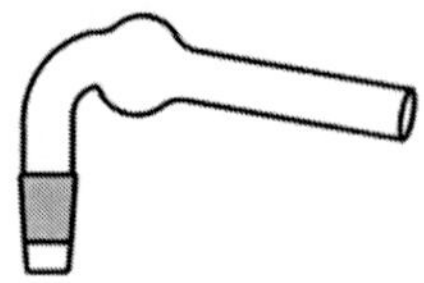

Calcium guard tube

2.2.5 Dean-Stark Apparatus

It is also known as distilling trap used in the continuous removal of liberated water from the reflux mixture (azeotropic distillations). Solvent vapours along with water escapes in to the condenser and then drip into the trap. The two immiscible liquids gets separated into two layers, the high density water flow to the collector through the side arm and low density liquid flows back into flask.

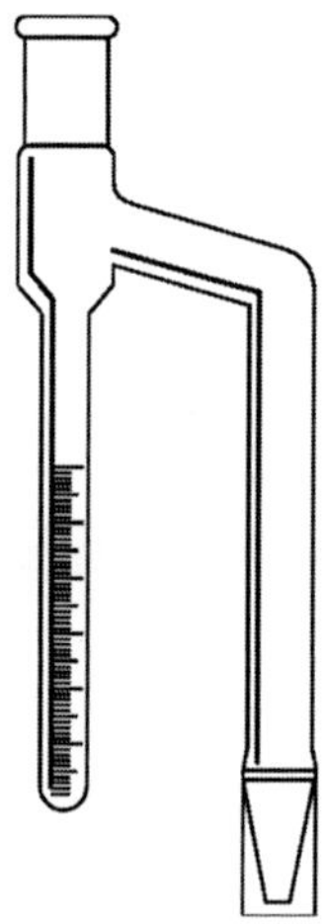

Dean-Stark apparatus

2.2.6 Beakers

Beakers are commonly made up of borosilicate glass, but plastic [polythene, polystyrene, polytetrafluoroethylene (PTFE)] beakers are also available for gamma spectral analysis. They are generally graduated, cylindrical with flat bottom and a spout (beak), which assists in pouring. Griffin with sprout and conical beaker are the two common types of beakers in use, for stirring, mixing, and heating of liquids.

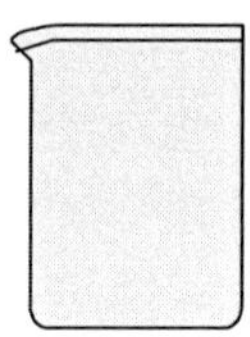

Beaker

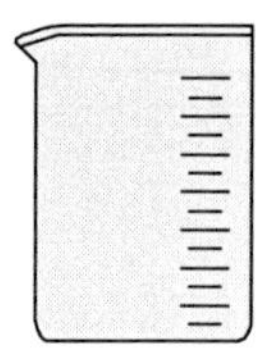

Beaker (graduated)

2.2.7 Funnels

Funnels are made up of either glass or plastics and are useful for the filtration.

Dropping funnel: It is useful in transferring/adding the liquids and reagents in a controlled manner (dropwise). Because quick addition of liquids and reagents may result in vigorous side reactions.

Separating funnel: It is useful in separation of mixture of components by distributing (partition) them into two immiscible solvents (liquid-liquid extraction) with different densities.

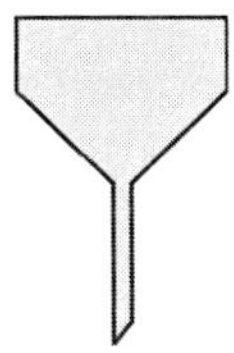

Funnel

Dropping Funnel

Separatory Funnel

Buchner funnel: It is useful in filtering the mixture/suspension to collect recrystallized compounds and generally made up of porcelain. Pouring the suspension to be filtered onto the filter paper placed on the surface of perforated plate of funnel and applying the vacuum through the side arm results in effective filtration. If the same funnel contains smaller plate then it is called as *Hirsch funnel.* Usually the filter paper is moistened with solvent to prevent the leakage. It provides several fold efficient filtration than normal filtration.

Buchner Funnel

2.2.8 Soxhlet Extractor

Soxhlet extraction is effective in the separation of desired compound with limited solubility than the impurity. The simple filtration is suitable for the separation, if the desired compound has significant solubility in the solvent.

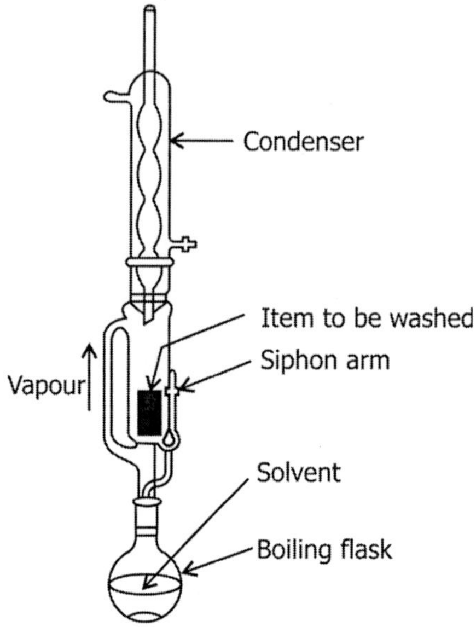

Soxhlet extractor

2.3 Boiling Chips

Super heating is a condition in which heat will be trapped around the gas surface and there is no release of energy. This is responsible for the formation of bubbles, which converts into gas phase and can be explosive. Super heated solution results in bumping and causes loss of liquid. Stirring and agitation ensures that the heat is properly distributed in the reaction mixture. Use of porous chips, sticks and wooden splints prevents solvent overheating. Air trapped within the porous chip/stick/wooden splint initiates the bubble formation to the top of the liquid and allows solution to boil evenly.

2.4 Magnetic Stirrer

It is a laboratory device that employs rotating magnetic field to cause a stirrer bar immersed in a liquid to spin very quickly.

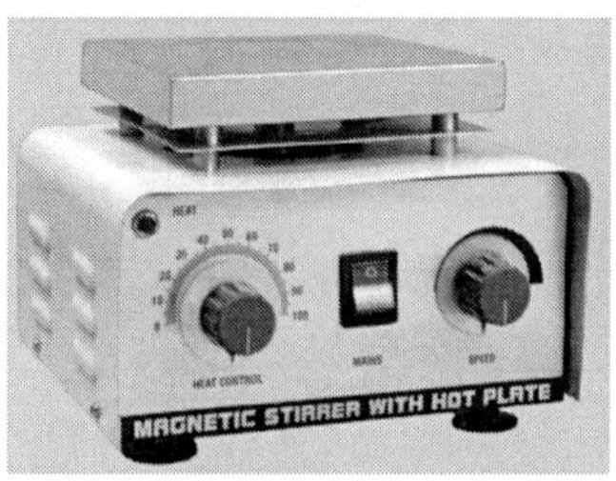

Magnetic stirrer

2.5 Stirring Rod (Stir Bar/Flea)

It is a piece of solid glass - magnetic bar generally made up of borosilicate (Pyrex), used for the purpose of mixing chemicals and liquids. The stir bar motion is driven by electromagnets present in the stirrer. They will be coated with Teflon or glass. Glass coating is suitable for the mixing of liquid alkali metals solutions.

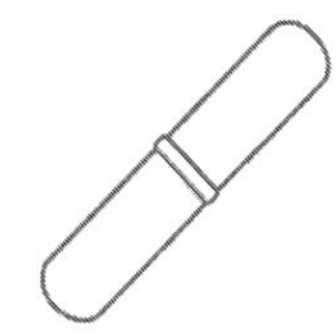
Stirring rod

2.6 Kipp's Apparatus

It is also known as Kipp's generator and useful in the preparation of small volume of gases. In general they are made up of glass and polyethylene. The apparatus is made up of three vertically stacked cylinders and the solid material which generates gas will be placed in the middle cylinder. The top cylinder will be loaded with acid which in turn connected to the middle and bottom cylinder through the tube. The middle cylinder has separate tube with stopcork to collect the evolved gas. The closure of the stopcock will raise the pressure of the gas which expels the acid back to the top of the cylinder.

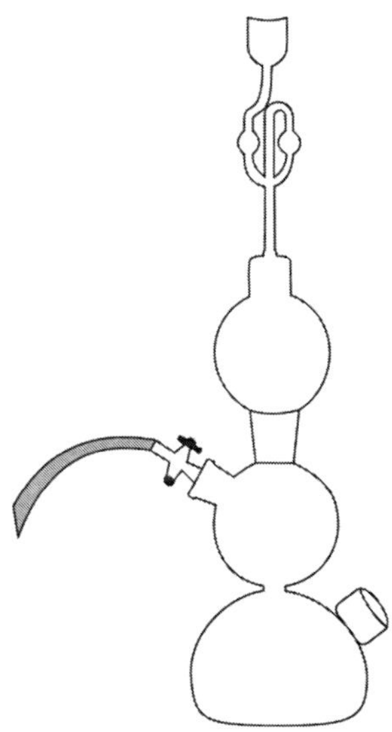
Kipp's apparatus

2.7 Glassware Cleaning

- Glassware can be cleaned by immersing them in dichromate solution (cleaning solution), followed by draining and rinsing with distilled water.

Preparation of dichromate solution: Dissolve 5 gm of potassium/sodium dichromate in 5 ml of water and cool. Add 100 ml of sulphuric acid slowly with stirring (exothermic reaction, hence use fume cupboard).

- Glasswares can be rinsed with acetone and then with petroleum ether or diethyl ether to remove traces of impurities prior to use.
- Alcoholic solution of sodium hydroxide (120 gm of sodium hydroxide in 120 ml of water, dilute to 1000 ml with ethanol) can be used for cleaning polyethylene ware.

2.8 Solvents

Saturated aliphatic hydrocarbons

- Light petroleum

Aromatic hydrocarbons

- ***Benzene:*** Carcinogen, b.p 84 °C. Contains thiophen as impurity and can not be removed by simple distillation. But can be removed by shaking with sulphuric acid, neutralizing with 10% sodium carbonate and with water followed by drying with anhydrous calcium chloride.
- ***Toluene:*** b.p 112-113 °C. Contains methyl thiophen and can be removed by shaking with sulphuric acid.

Halogenated hydrocarbons

- ***Dichloro methane (CH_2Cl_2, DCM, methylene chloride):*** b.p 40-41 °C. It should be stored in amber coloured bottle and protected from light. It is better substituent for diethyl ether.
- ***Chloroform ($CHCl_3$):*** Carcinogen, b.p 61 °C. It can be replaced with dichloro methane (DCM).
- ***Carbon tertachloride (CCl_4):*** Carcinogen, b.p 76.5 °C. The vapour breathing, contact with skin and eye should be strictly restricted.

Aliphatic alcohols

- ***Methanol:*** b.p 64.7 °C
- ***Ethanol:*** b.p 78.37 °C
    - Dehydrated ethanol
    - Super dry ethanol
- ***Propan-1-ol:*** b.p 97.5 °C
- ***Propan-2-ol:*** b.p 82.5 °C

Ethers

- ***Diethyl ether (ether):*** b.p 34.6 °C
- ***Tetrahydrofuran (THF):*** b.p 66 °C
- ***1,4-Dioxane:*** b.p 101 °C

Ketones

- ***Acetone:*** b.p 56 °C
- ***Butan-2-one:*** b.p 79.64 °C

Esters

- ***Methyl acetate:*** b.p 56.9 °C
- ***Ethyl acetate:*** b.p 77.1 °C
- ***Dimethyl Formamide (DMF):*** b.p 153 °C
- ***Dimethyl Sulphoxide (DMSO):*** b.p 189 °C

2.9 Heating Reaction Mixtures

- ***Bunsen burners:*** Supported with tripods and ceramic centred gauze

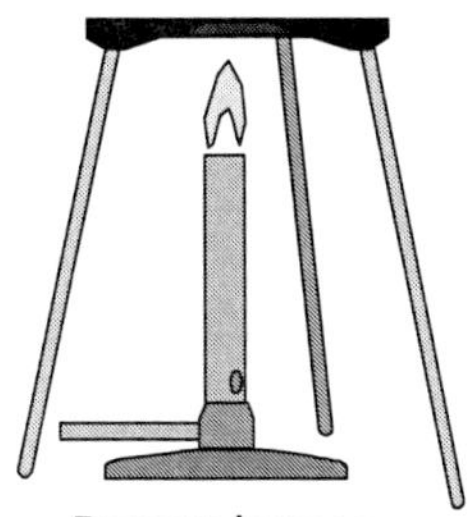

Bunsen burner

- ***Electrically heated water bath:*** For liquids with boiling point less than 100 °C.

Electrically heated water bath

- ***Steam bath:*** A useful device for heating flammable substances at below 100 °C.

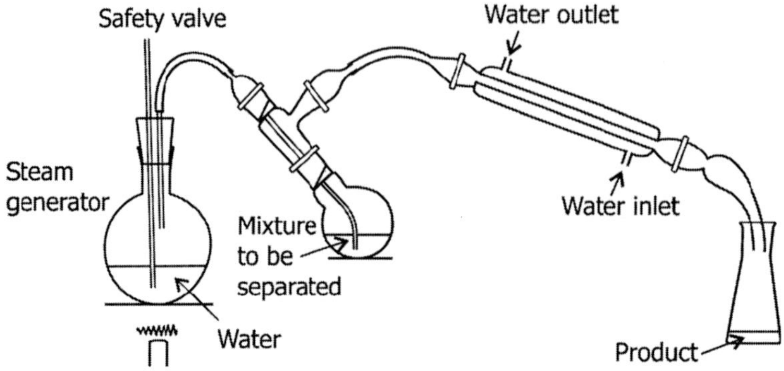

Steam bath

- ***Heating mantle:*** It is also known as isomantle, which supplies heat to the containers and flasks. The heating element is insulated and hence the containers can be directly placed on the mantle. Heating mantle with Rheostat is useful for maintaining the intended temperature. It distributes the heat evenly over the surface of the flask and produces no/less harmful hotspots.

Heating mantle

- ***Oil baths:*** useful in temperature above 100 °C.
    - *Hard hydrogenated cotton seed oil:* For temperature upto 250 °C, more advantageous because it is clear, non sticky liquid and solidifies on cooling.
    - *Medicinal paraffin:* For temperature upto 220 °C.
    - *Glycerol and dibutyl phthalate:* For temperature upto 140-150 °C.
    - *Silicone fluids:* Silicone fluids are polymers of inorganic compounds and not derived from petroleum and organics, hence posses unique properties. (e.g., polydimethylsiloxanes, methylphenyl silicone fluid).

2.10 Purification Methods

Common methods employed in the purifications are:

1. Solvent extraction

2. Washing
3. Distillation
4. Recrystallization
5. Sublimation
6. Chromatography
7. Drying

1. **Solvent Extraction:** Organic compounds can be separated from inorganic impurities by shaking the reaction suspension with suitable immiscible solvents (benzene, carbon tetra chloride, diethyl ether, etc). Evaporation of organic layer (containing organic compound) leaves the residue. Addition of electrolytes (ammonium sulphate, calcium chloride and sodium chloride) ensures the organic layer separation and increases organic extraction.
2. **Washing:** Phenol impurities can be removed by adding (washing with) sodium hydroxide. Organic bases can be removed by adding dilute hydrochloric acid/sulphuric acid. Unsaturated hydrocarbons of alcohols and ethers can be removed by adding concentrated sulphuric acid.
3. **Distillation:** Distillation is the method of separating mixtures based on thc differences in volatility of components in a boiling liquid mixture.
   - ***Simple and fractional distillation:*** Non-volatile compounds can be separated by simple distillation. Liquids and low melting point solids (volatile compounds) can be purified by fractional distillation. Distillation apparatus consists of

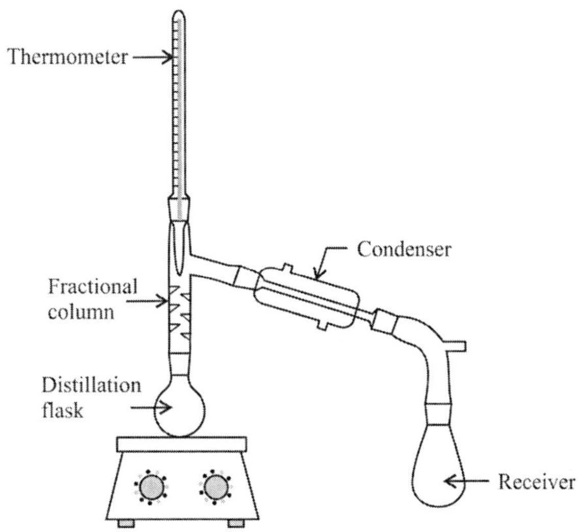

Distillation unit

distillation flask fitted with fractional column which in turn attached to condenser and thermometer. In case of simple distillation, column is not necessary and flask can be attached directly to condenser. Adding boiling chips to the flask reduces the super heating.

Upon heating, the contents of the flask vaporises, passes through the column and condenses back to flask. The vapour passing into the condenser is lowest boiling component of mixture. In fractional distillation the initial and final fractions (lower and higher boiling point components respectively) has to be rejected. Stir the mixture mechanically, in case of suspended insoluble materials present in the mixture to be distilled.

- ***Vacuum distillation:*** Distillation under reduced pressure (lower than atmosphere) is denoted as vacuum distillation. Thermolabile, sensitive and high boiling point liquids can be purified by vacuum distillation.
- ***Steam distillation:*** Distillation of temperature sensitive materials (natural aromatics) can be purified by this method.
- ***Isopiestic/isothermal distillation:*** Metal free solutions of volatile acids and bases can be obtained by this method. This can be effected by placing distilled water and solution to be purified in separate beaker in sealed desiccator at room temperature for several days.

4. **Recrystallization:** Recrystallization is the most suitable and effective method for the purification of organic solid compounds. A solid product prepared may contain impurities, and recrystallization separates impurity from product.

Simplified general procedure for the purification of impure solids

1. Select the right solvent based on the solubility of sample (trial and error method).
2. Place beaker containing known amount of solvent on hot plate.
3. Add small amounts of boiling solvent to the impure solid (partially soluble in solvent) taken in the beaker.
4. Dissolve the solid by heating and stirring if necessary (add more solvent, if necessary to obtain saturated solution).
5. Warm the stemless funnel with the vapours of solvent and filter the hot solution through fluted filter paper placed in funnel.
6. Dissolve the crystals if any formed with adding additional amount of solvent.
7. Cool the clear liquid (collected in the beaker) to room temperature.
8. Place the beaker in an ice-bath (do not disturb) and filter the crystals formed through Buchner funnel (previously wet with solvent).

Steps involved in recrystallization process

1. Selection of solvent
2. Solubilization of the compound
3. Decolorization
4. Impurities filtration
5. Crystallization
6. Filtration
7. Drying

Selection of solvent: Solvent can be one or mixture of two liquids. In general solvent with partial solubility will be selected for recrystallization. Solute should be insoluble at room temperature and soluble in high temperature (boiling point). A solvent that can completely dissolve a solid in hot condition gives crystals when the temperature is low. Soluble impurities stay in solution and some of the pure compound may also remain dissolved.

Mixture of solvents also can be used when single solvent is not applicable. Second solvent can be added to reduce the strength of first solvent. Most common mixture includes alcohol (strong solvent) and water (weak solvent). For polar molecules (containing oxygen and/nitrogen atom) alcohol-water is more suitable.

Solubilization of the compound: Sample should be dissolved with the aid of solvent by the slow addition of solvent.

Decolourization: The coloured impurities present along with the product can be removed by the treatment with charcoal and filtration. Coloured impurities will adhere to the large surface of charcoal and can be separated by filtration. This step is not necessary, when there is no coloured impurities present.

Impurities filtration: Insoluble impurities will be removed by the filtration through fluted filter paper (provides larger surface area and increase rate of filtration) and stem less funnel (avoids crystallization in the stem during crystallization).

A simple filtration in hot condition separates the product from insoluble impurities. Filtrate upon cooling crystallizes out the pure product and the soluble impurities remain in the solvent. Thus both soluble and insoluble impurities can be removed to obtain the pure solid product.

Crystallization: Solvent upon cooling and/or evaporation initiates crystallization of the solid product.

Filtration: Solid product which crystallizes out can be collected by simple filtration.

Drying: Separated solid may contain some residual solvent in it, so it requires drying to remove solvent completely.

Recrystallization never offer 100% product due to loss while physical handling, impurities and some amount will remain in saturation with solvent. Excess of solvent, inadequate warming and selection of wrong solvent also reduces the yield (crystallization).

Procedure for the purification of benzoic acid

1. Heat the benzoic acid with enough water to dissolve all soluble material.
2. Filter the hot solution to remove insoluble material (if any).
3. Cool the solution to effect crystallization.
4. Filter, wash with water and dry the product.

Procedure for the purification of benzamide

1. Dissolve finely powdered benzamide in boiling water.
2. Filter the insoluble impurities through filtration.
3. Allow it to cool to room temperature and collect the crystals through filtration.
4. Wash the crystals with cold water.

5. **Sublimation:** The chemical transformation of solids into vapours is known as sublimation. Volatile solids can be converted into vapours on heating and vapours condense as purified solid on cooling. It helps in the purification of volatile solids like naphthalene, iodine, camphor, etc.

Purification of iodine

1. Cover the impure iodine placed in the beaker with glass dish.
2. Place ice cubes on the glass dish (cover).
3. Heat the beaker gently by keeping over the wire gauze.
4. The impure iodine vapourizes leaving the impurities.
5. Iodine vapours on cooling sublimes into solid.
6. Isolate the pure iodine from glass dish by scratching.

6. **Chromatography:** Chromatography is the collective term for the set of techniques applicable for the separation of mixtures. The mixture to be separated dissolve in mobile phase which carries it through the stationary phase. Column chromatography and preparative thin-layer chromatography (TLC) are the two important separation techniques useful in organic chemistry laboratories.

Column chromatography: Useful in the separation of large quantities of mixtures.

Thin-layer chromatography (TLC): Thin-layer chromatography is a simple, inexpensive analytical and purification technique.

Thin-layer chromatography for salicylic acid and acetyl salicylic acid (aspirin)

- *Preparation of sample:* Dissolve the specified quantity of salicylic acid and acetyl salicylic acid in ethanol separately.

  [Use these solutions for spotting on the TLC plate]
- *Preparation of developing chamber:* Prepare mobile phase (hexane: ethyl acetate: acetic acid, 6.5:3:0.5) and load into developing chamber (solvent tank). Cover the chamber with lid to ensure the saturation of chamber with solvent vapour. Line the inside of the beaker with filter paper wet with the mobile phase; this also saturates the chamber with solvent vapour.
- *Preparation of TLC plate:* Prepare the silica gel paste [5 gm of silica gel with 30 ml of water]. Pour this paste onto the glass plate and spread uniformly. [Alternatively aluminium backed precoated plates can be used.] Handle TLC plates by edges only and avoid touching the surface. Draw a line (light) about 1-2 cm from the bottom of the TLC plate.
- *Loading the TLC plates:* Dip the capillary into the sample solution and spot on the plate. Touch the tip of the capillary onto the plate, and allow it to dry. Spot the sample for three times and the size should be in between 0.5-1 mm. Use different capillary tube for each sample.
- *Developing the TLC plates:* Place the TLC plate into the chamber and make sure that the mobile phase level is below the baseline present in plate. Solvent rises by capillary action and component present in the sample will rise up. Allow the plate to elute until the solvent front is about 1 cm from the top. The rate at which sample travels is depends upon their partitioning with silica gel. A more polar component spends more time on the polar silica gel surface, move up more slowly than the non-polar component.
- *Visualising the TLC plates:* Remove the plate, mark the solvent front and allow it to dry. Visualize the spots of TLC plate by placing in a UV chamber and keeping in iodine chamber. Different components will appear at distinct locations, and pure sample produces single spot.

  Note: UV light is harmful to eyes and skin, so don't expose hands or eyes to UV light.
- *Calculation of Retention factor (R_f):* The ratio of the distance travelled by the sample and distance travelled by the mobile phase (solvent front) is known as R_f.

$$\text{Retention factor } (R_f) = \frac{\text{Distance travelled by the solute}}{\text{Distance travelled by the solvent}}$$

7. Drying

Removal of Solvents: Removal of solvent from recrystallized compounds can be effected by heating them in an oven above the boiling point of the solvent followed by cooling in a desiccator.

Removal of water: Thermal stability of the compounds determines the method of choice. Direct distillation is suitable for removing organic liquid from solids and liquids. During distillation, guard tube with calcium chloride/silica gel prevents the entry of moisture. Substances with melting point above 100 °C and stable in air can be dried using hot air oven. Drying compounds in vacuum desiccator over sulphuric acid, phosphorous pentoxide, silica gel, calcium chloride and similar desiccants is the safest method of drying water. Water in gases can be removed by adsorption on to a drying agent.

Desiccator

Most common desiccants

- *Calcium sulphate (anhydrous):* Very efficient and suitable for most of the organic compounds.
- *Magnesium sulphate (anhydrous)*
- *Sodium sulphate (anhydrous)*
- *Calcium chloride:* Absorbs water and becomes hexahydrate below 30 °C (slow in action and less efficient). It is suitable for drying aryl halides, esters, ethers and hydrocarbons. It is unsuitable for drying alcohols, amines, amides, amino acids, ketones and phenols.
- *Calcium oxide (anhydrous):* It is suitable for drying alcohols and amines, unsuitable for acids and esters.
- *Copper (II) chloride (anhydrous):* It is suitable for esters and alcohols.
- *Potassium carbonate (anhydrous):* It is suitable for salting out water soluble alcohols, amines and ketones and unsuitable for phenols, thiols and related compounds.
- *Granulated silica gel:* Physical adsorption is the mechanism, which is efficient in room temperature.

2.11 Melting Point

Melting point is the important physico-chemical characteristic of molecules, which can be used for identification purpose. The solubility profile and hydrogen bonding nature are the two important properties, which influence the melting point of the molecules. Melting occurs over a temperature range and temperature at which first tiny drop of liquid appears is termed as melting point.

Determination of melting point for pure/impure organic compounds

Sample filling:

1. Grind the sample, whose melting point has to be determined into fine powder.
2. Fill 2-3 mm level of capillary tube with fine powder of sample. [use sealed capillary tube or seal the one end of capillary tube].
3. Drop the capillary tube through 1 meter length glass tube (broken burette or condenser can be used) several times.

Determination of melting point:

1. Place the capillary tube filled with sample in the melting point apparatus.
2. Switch on the instrument to heat the sample until the contents of the tube dissolves.
3. Use the knob to set rise of temperature per minute (best set at 2 °C per minute).
4. Appearance of first tiny drop indicates the melting of sample and the temperature range should be noted as melting range.
5. Change in crystalline structure, wet appearance and shrinkage of the sample should not interpreted as melting point.

Determination of melting point of mixture of compounds

Sample filling 1: For pure product

1. Grind the pure (recrystallized) sample, whose melting point has to be determined in to fine powder.
2. Fill 2-3 mm level of capillary tube with fine powder of sample, use sealed capillary tube or seal the one end of capillary tube.
3. Drop the capillary tube through 1 meter length glass tube (broken burette or condenser can be used) several times.

Sample filling 2: **For mixture of pure and impure products**

1. Grind the pure (recrystallized) and impure products, whose melting point has to be determined in to fine powder.
2. Fill 2-3 mm level of capillary tube with fine powder of sample, use sealed capillary tube or seal the one end of capillary tube.
3. Drop the capillary tube through 1 meter length glass tube (broken burette or condenser can be used) several times.

Determination of melting point

1. Place the capillary tubes filled with pure sample and mixture separately in the melting point apparatus.
2. Switch on the instrument to heat the sample until the contents of the tube dissolves.
3. Use the knob to set rise of temperature per minute (best set at 2 °C per minute).

4. Appearance first tiny drop indicates the melting of sample and the temperature range should be noted as melting range.
5. Change in crystalline structure, wet appearance and shrinkage of the sample should not interpreted as melting point.

Note:

- The determined melting range may differ (small extend) from literature value, if the solid is not completely pure.
- Rapid heating of melting point apparatus and error in the thermometer may also cause this.

2.12 Theoretical Yield

The maximum amount of product that results from a reaction is called as theoretical yield. It is based upon the molar quantities of the reactant/reagent used in least amount. It can be calculated from stoichiometric factors derived from the coefficient of the products and reactants. E.g., Reaction between benzaldehyde (2 moles, 5 gm) and acetone (1 mole, 2 ml) to form dibenzalacetone.

5 gm of benzaldehyde produces 0.0236M of dibenzalacetone

2 ml of acetone produces 0.0273M of dibenzalacetone

It is impossible to generate more than specified quantity of reactants (i.e., excess of acetone will remain unreacted). Limiting reagent is the one that produces the least amount of product (not necessarily the one present in the lowest mass). In this case benzaldehyde is limiting reagent and hence yield calculations must based on the quantity of benzaldehyde.

The theoretical yield of dibenzalacetone

$$= \frac{0.0236 \times 234.3 \text{ gm (dibenzalacetone)}}{1\text{M dibenzalacetone}} = 5.53 \text{ gm dibenzalacetone}$$

Relative yield

Relative yield is also known as fractional yield, which serve to measure the effectiveness of the synthetic procedure. It can be calculated by the following formula.

$$\text{Relative yied} = \frac{\text{Practical yield}}{\text{Theoretical yield}}$$

$$\text{Percentage yied} = \frac{\text{Practical yield}}{\text{Theoretical yield}} \times 100$$

The ideal or theoretical yield of a chemical reaction would be 100%, but it is impossible to achieve due to limitations. Yields can be grouped as following:

- Quantitative yield: 100%
- Excellent yield: > 90%
- Very good yield: > 80%
- Good yield: > 70%
- Fair yield: > 50%
- Poor yield: 40%

PREPARATION OF DRUG INTERMEDIATES AND DRUGS

CHAPTER 3

ACYLATION

Acylation is the process of addition of acyl group to organic compounds. The compounds which provide acyl groups are known as acylating agents (e.g.: acyl halides). The most important acylation reactions are listed below.

1. **Friedel-Crafts acylation:** It involves the reaction between arenes and acylchlorides/anhydrides in presence of Lewis acid catalyst ($AlCl_3$). This reaction produces ketones and can be converted into corresponding alkanes by reduction (Clemmenson/Wolff-Kishner reduction).
2. **C-Acylation:** Attachment of acyl group (substitution) to aromatic rings forms ketone.
3. **N-Acylation:** It involves the reaction between amines (ammonia, primary and secondary amines) and acylating agents to form amides (primary, secondary and tertiary amides).
4. **O-Acylation:** It involves the reaction between phenolic–OH group and acylating agents to form esters.
5. **S-Acylation:** It involves the reaction between thiol group and acylating agents to form thiol esters.

General reaction mechanism

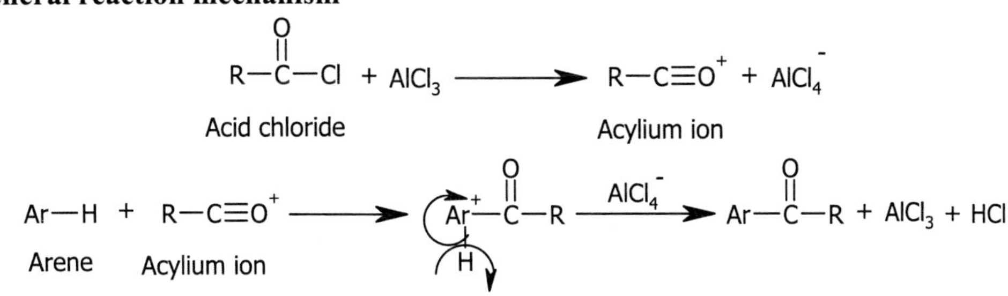

Important acylation reactions

Schotten-Baumann reaction: It involves the reaction between aliphatic amines/alcohols and acid chlorides in aqueous alkali solution to generate amides/esters.

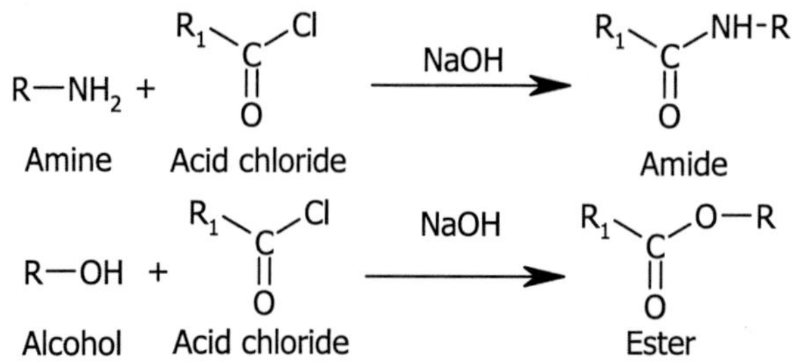

Kostanecki reaction: It involves the acylation of *ortho*-hydroxyaryl ketones with aliphatic acid anhydrides and subsequent cyclization to form chromones/coumarins.

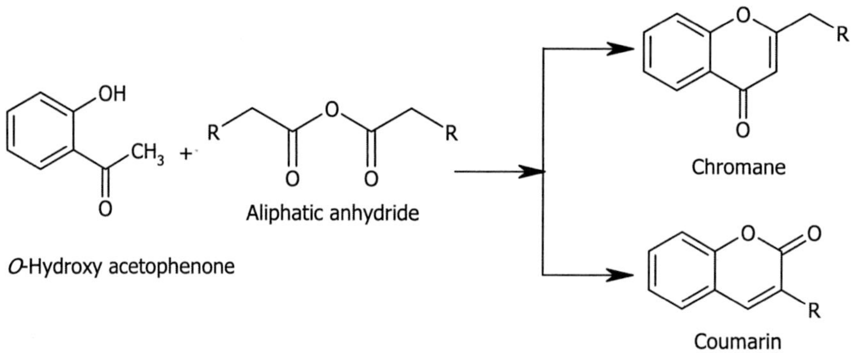

Houben Hoesch reaction: In this reaction nitrile group reacts with an arene (e.g., resorcinol) compound and forms an aryl ketone using hydrochloric acid and aluminium chloride as catalyst.

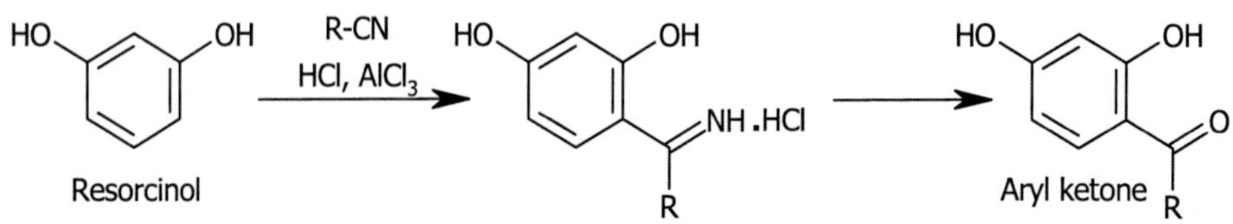

Acetylation: Acylation with acetyl chloride and/acetic anhydride is known as acetylation. Acetylation is also known as ethanoylation. It involves the replacement of active hydrogen of hydroxyl and or amino (1° & 2°) groups by an acetyl group of acetylating agent. The alcohols (R-OH), phenols (Ar-OH), primary amines (R-NH_2) and secondary amines (R-NH-R) can be directly acetylated. O-Acetylation and N-acetylation are the two types of acetylation reactions known.

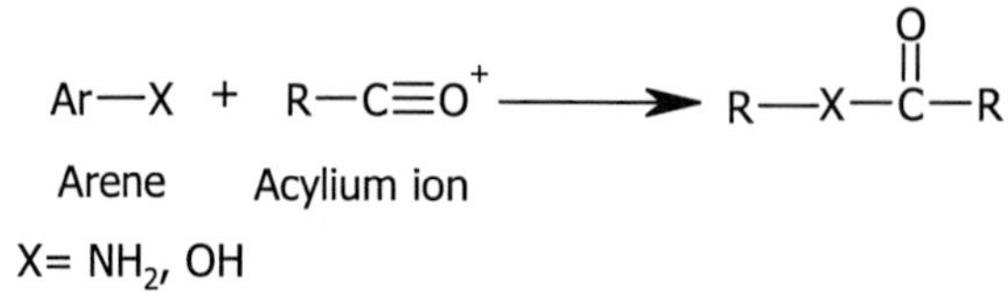

- *O-Acetylation:* Acetylation of alcohol and phenols offer acetyl esters.
- *N-Acetylation:* It is an efficient method for the characterization and protection of amines. Acetylation of primary and secondary amines offer amides.

Acylating agents

- **Acetyl chloride:** Acetyl chloride in pyridine is the most common acetylating agent, combinations with $ZrOCl_2.8H_2O$, zinc dust and thermally decomposed Ni-Fe-hydrotalcite also can be used. The high reactivity and water instability (explosive) limits the use of acetyl chloride.
- **Acetic anhydride:** In view of the above mentioned limitations, acetic anhydride is preferred for acetylation. Generally acetylation with acetic anhydride will be carried out in presence of bases, such as triethylamine, pyridine and 4-dimethyl amino pyridine (DMAP). Acetylation by acetic anhydride proceeds very well in aqueous medium, such as acetic acid and sodium bicarbonate solution.
- Aliphatic carboxylic acids
- Benzoyl chloride

Catalysts in acylation

1. *Acid halides*: Aluminium chloride ($AlCl_3$), ferric chloride ($FeCl_3$), boron trifluoride (BF_3), cadmium chloride ($CdCl_3$), zinc chloride ($ZnCl_2$) and stannous chloride ($SnCl_2$).
2. *Metal alkyls and alkoxides*: Aluminium phenoxide.
3. *Protonic acids*: Sulphuric acid (H_2SO_4), hydrogen chloride (HCl) and phosphoric acid (H_3PO_4).
4. *Combination of proton acids and Lewis acids:* Hydrochloric acid-aluminium chloride (HCl-$AlCl_3$) Hydrochloric acid-boron trifluoride (HCl-BF_3), acid-borontrifluoride (H_2SO_4- BF_3).
5. *Liquid super acid*: Sulphuric acid (100%), antimony pentafluoride (SbF_5), niobium (v) fluoride (NbF_5), arsenic pentafluoride (AsF_5) and tantalum pentafluoride (TaF_5).
6. Sodium hydride (NaH) and N,N-dimethyl amino pyridine (DMAP) in N-acylation.
7. Rutherium (III) chloride ($RuCl_3$), Lanthanide nitrate ($LaNO_3$), niobium (v) chloride ($NbCl_5$), sodium bisulphate ($NaHSO_4$), silicon di-oxide (SiO_2), imidazoles and ionic liquids promotes the acetylation reaction.
8. Ethyl acetate, ammonium acetate in acetic acid and acetic acid with ferric-monomorillonite also can be used.

Pharmaceutical Applications

1. Acetophenone, an acetylation product of benzene is useful in the synthesis of perfumes and pharmaceuticals.

2. Aspirin, an anti-inflammatory and antiplatelet drug can be prepared by the acetylation (O-acetylation) of salicylic acid.

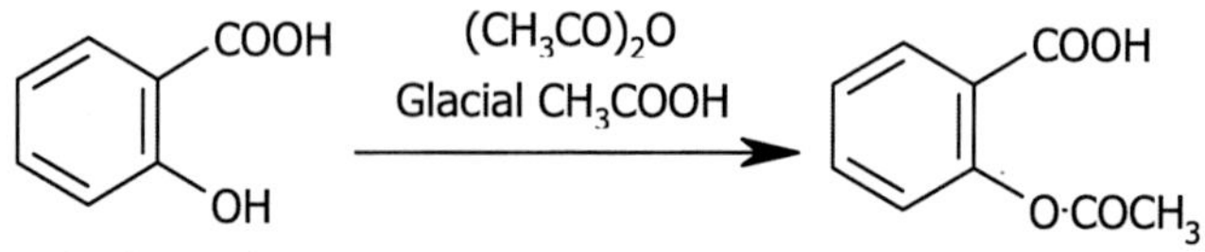

Salicylic acid — Aspirin

3. Acetylation of 4-amino phenol (N-acetylation) offers 4-hydroxy acetanilide (paracetamol), which is the intermediate in the preparation of phenacetin.

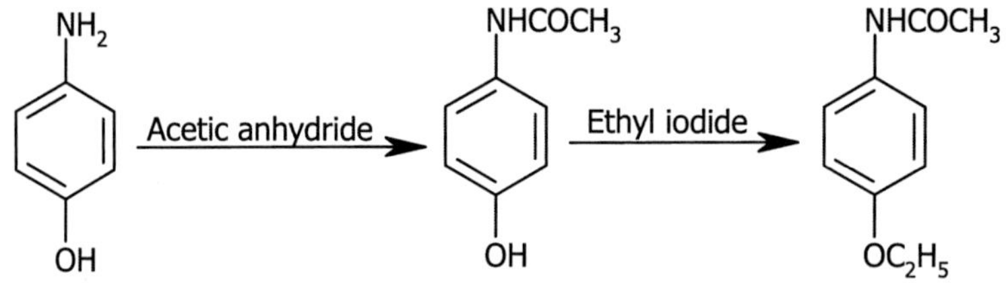

Para-Amino phenol — Paracetamol — Phenacetin

4. Metoclopramide, an anti-histaminergic and anti-emetic drug can be prepared by the acetylation (N-acetylation) of 4-amino-5-chloro-2-methoxy benzoic acid.

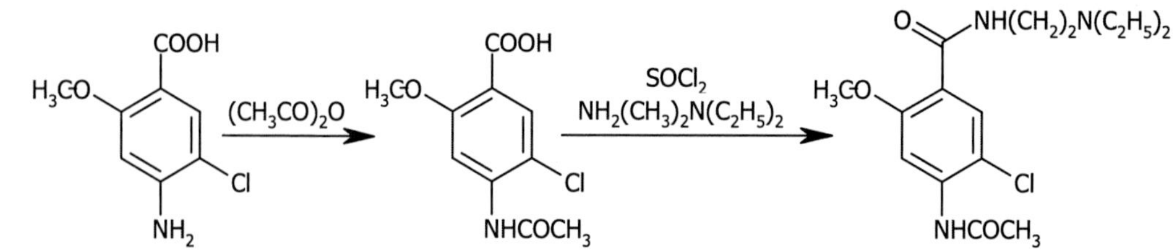

4-Amino-5-chloro-2-methoxy benzoic acid — Metoclopramide

5. Preparation of salacetamide (anti-pyretic, anti-inflammatory).

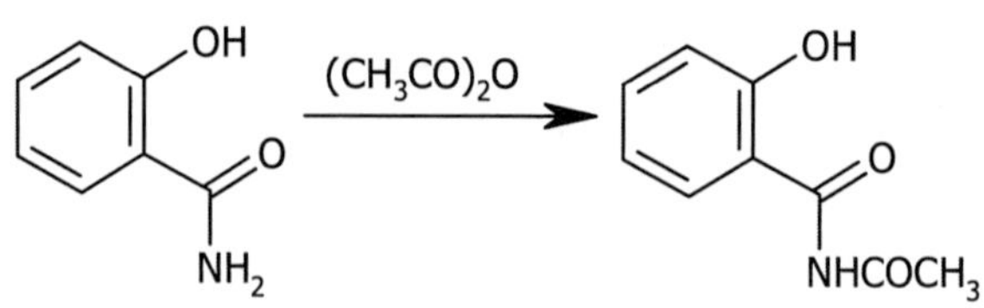

Salicylamide — Salacetamide

6. Preparation of indomethacin (anti-inflammatory agent).

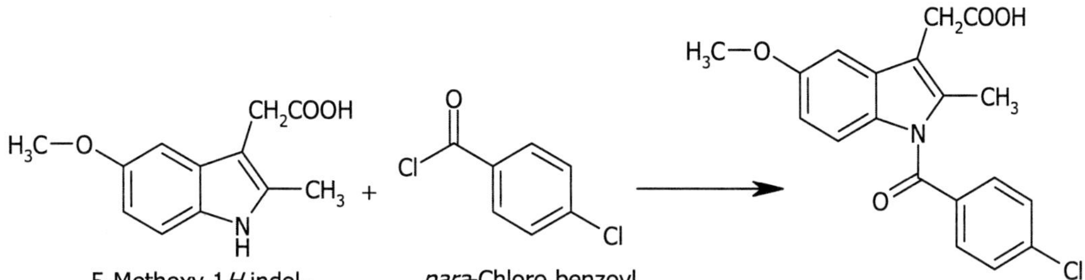

5-Methoxy-1*H*-indol-3-yl-acetic acid *para*-Chloro benzoyl chloride Indomethacin

7. Preparation of amidotrizoic acid (contrast agent).

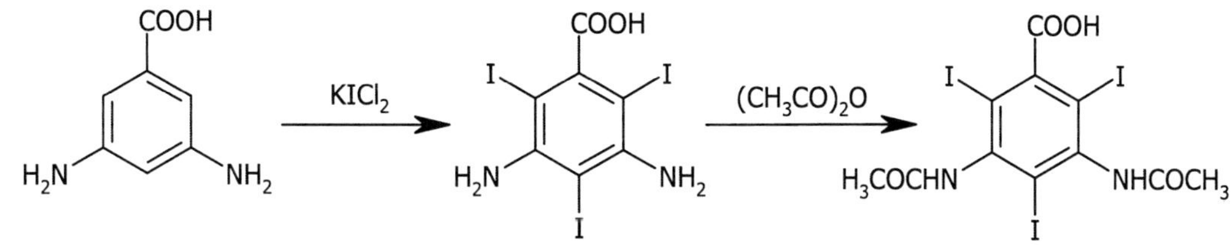

3,5-Diamino benzoic acid 3,5-Diamino benzoic acid Amidotrizoic acid

8. 2-Acyl-6-methoxynaphthalene, the precursor of anti-inflammatory drug naproxen can be prepared through C-acylation.

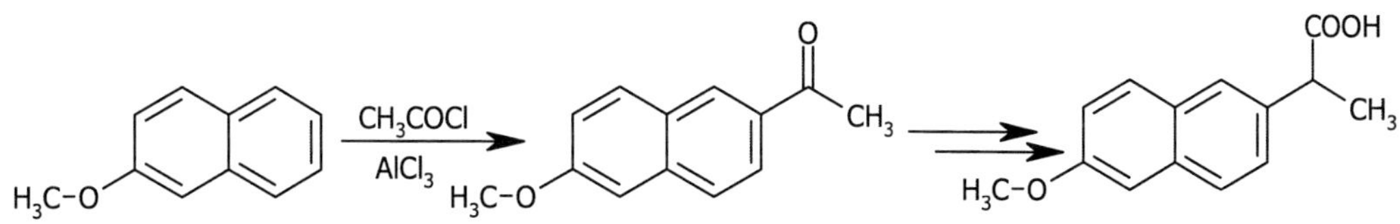

6-Methoxy naphthalene 2-Acyl-6-methoxy naphthalene Naproxen

9. *Para* acylated anisole, an intermediate in the preparation of sweetening agent 2-(4-methoxy benzoyl) benzoic acid can be prepared through C-acylation of anisole.

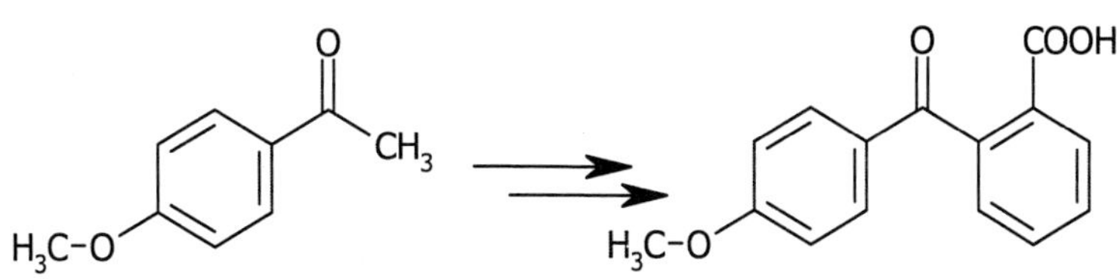

para-Acylated anisole 2(4-Methoxy-benzoyl)benzoic acid

10. *Para* acylated veratrole, an intermediate for the synthesis of cardiotonic drug vesarinone can be synthesized through C-acylation of veratrole.

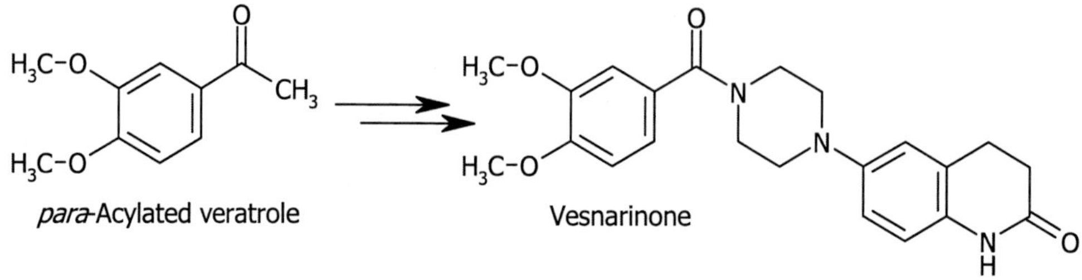

3.1 Preparation of Acetanilide

Traditional Name: Antifebrin *IUPAC Name:* N-phenyl ethanamide.

History of acetanilide: The acetanilide was discovered serendipitously and introduced as medicine in the name of antifebrin for the analgesic and antipyretic effects. The toxic effects cyanosis and methemoglobinemia made their use obsolete and brought paracatamol and phenacetin as alternatives (acetanilide derivatives).

Method-1

Chemicals: Aniline, Acetic anhydride, Glacial acetic acid and Charcoal.

Principle: Acetanilide is also known as N-phenylacetamide, antifebrin, acetyl aniline and acetyl amino benzene. Acetanilide preparation involves the replacement of active hydrogen of amino group by an acetyl group with the help of acetylating agent (N-acetylation). Aniline reacts readily with any of the acetylating reagents (acetic anhydride/acetyl chloride) in presence of acetic acid and undergoes N-acetylation.

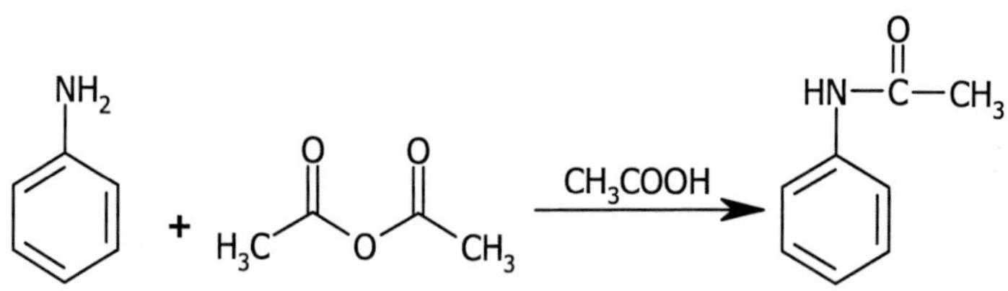

Movement of π electrons towards carbonyl oxygen makes carbon positive and attracts the unpaired electrons in amine nitrogen. Elimination of hydrogen atom from the amino group removes the positive charge present on it. Movement of electrons from carbonyl oxygen to form π bond, facilitates the bond breaking between carbonyl carbon and acetoxy oxygen to form acetanilide.

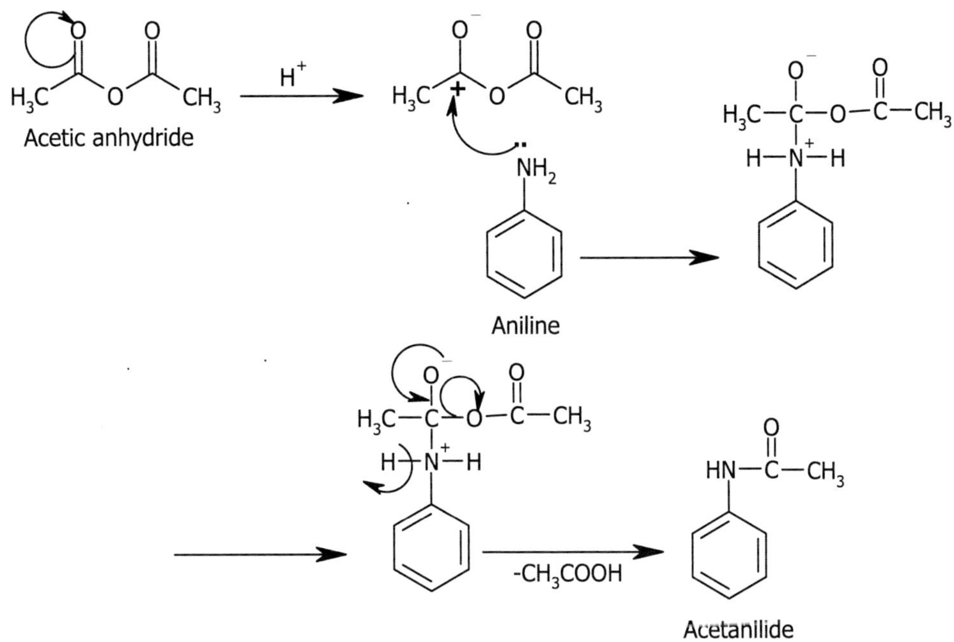

Procedure:

1. Place acetic anhydride (10 ml, 105 mmol) and glacial acetic acid (10 ml, 175 mmol) to the flask containing aniline (10 ml, 110 mmol).
2. Swirl the flask to dissolve the contents completely.
   [Decolorize the solution by boiling with charcoal and filter, if it is coloured].
3. Attach condenser to the flask and boil it gently for 10 minutes.
4. Pour the hot liquid into beaker containing cold water (200 ml) with stirring.
5. Collect the crystallized acetanilide by filtration under suction and wash with water.
6. Dry and recrystallize the product from hot water.
7. The melting range of pure acetanilide is 113-115 °C.

Method-2

Chemicals: Aniline, Acetic anhydride, Hydrochloric acid, Charcoal and Sodium acetate.

Principle: Aniline undergoes acetylation readily with acetylating reagent (acetic anhydride/acetyl chloride) in the presence of hydrochloric acid. Aniline solubility will be enhanced by the hydrochloric acid by getting converted into anilinium ion. Anhydrous sodium acetate converts the anilinium ion into aniline *in situ*, which then undergoes acetylation reaction. This improves reaction rate and product yield.

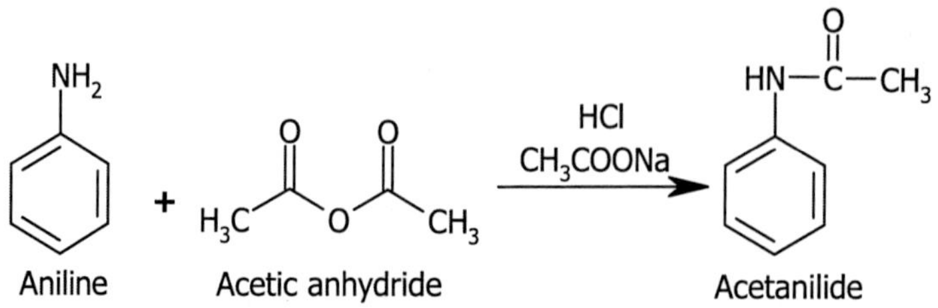

Procedure:

1. Place hydrochloric acid (4.6 ml) to the beaker containing water (125 ml), and swirl the flask to effect solution.
2. To the resulting dilute acid solution, add aniline (5 ml, 55 mmol).
3. Swirl the flask to dissolve the contents completely.

Reaction Mechanism

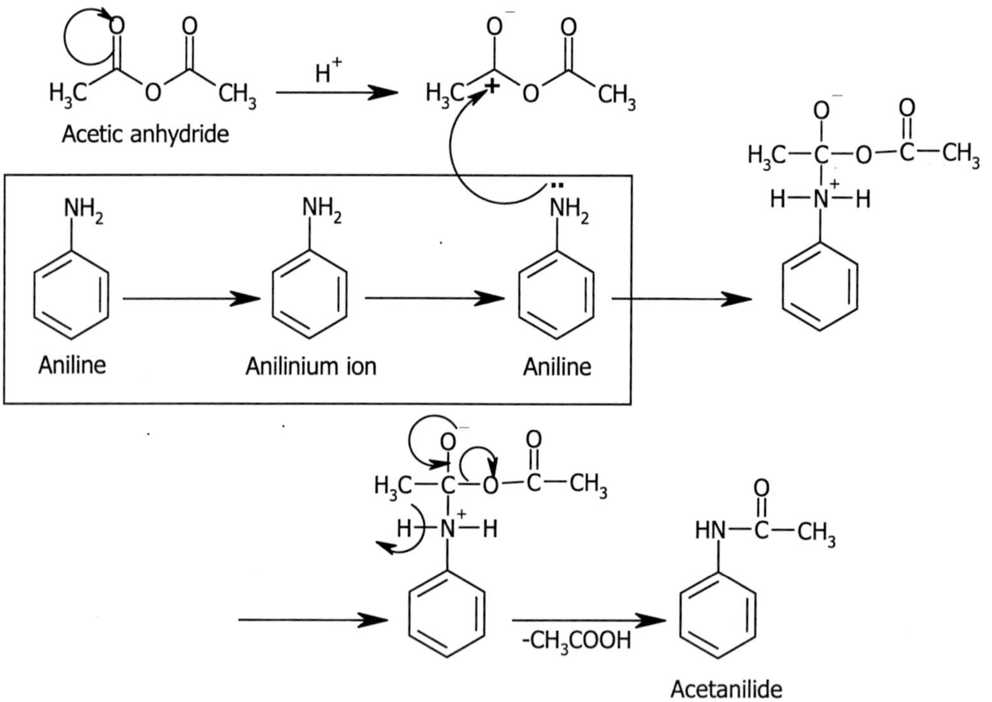

[Decolorize the solution by boiling with charcoal for 2 minutes and filter, if it is coloured].

4. Warm the solution containing the dissolved aniline hydrochloride to about 50 °C, and add redistilled acetic anhydride (6.4 ml, 68 mmol).

5. Swirl the flask vigorously and pour it immediately to the solution of sodium acetate (8.3 gm of crystallized sodium acetate in 25 ml of water).
6. Cool the reaction mixture to 5 °C in an ice- bath and continue to swirl the mixture until crystalline product separates out.
7. Filter under suction, wash the residue with water (10 ml) and drain well.
8. Dry and recrystallize the product from hot water.
9. The melting range of pure acetanilide is 113-115 °C.

3.2 Preparation of Paracetamol

Traditional Name: Acetaminophen ***IUPAC Name:*** *para*-Hydroxy acetanilide.

Chemicals: 4-Amino phenol (4-hydroxy aniline), Acetic anhydride, Hydrochloric acid, Sodium bicarbonate and Charcoal.

Principle: Preparation of paracetamol involves acetylation of 4-amino phenol with acetic anhydride in the presence of hydrochloric acid. The reaction takes place by the replacement of active hydrogen of amino group (N-acetylation) by an acetyl group (electrophile) of acetic anhydride.

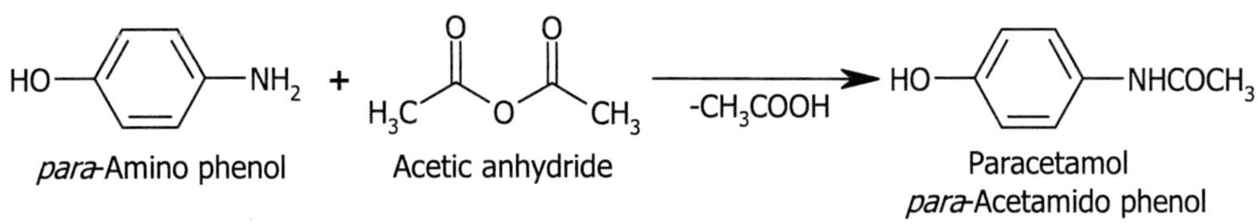

The electropositive carbonyl carbon (due to indutive effect) will be attacked by π electrons of amino group of 4-amino phenol. Deprotonation of the amino group removes the positive charge present in it. Movement of electrons from carbonyl oxygen to form π bond with carbon, breaks the bond between carbonyl carbon and electro negative group of acetylating reagent to generate paracetamol.

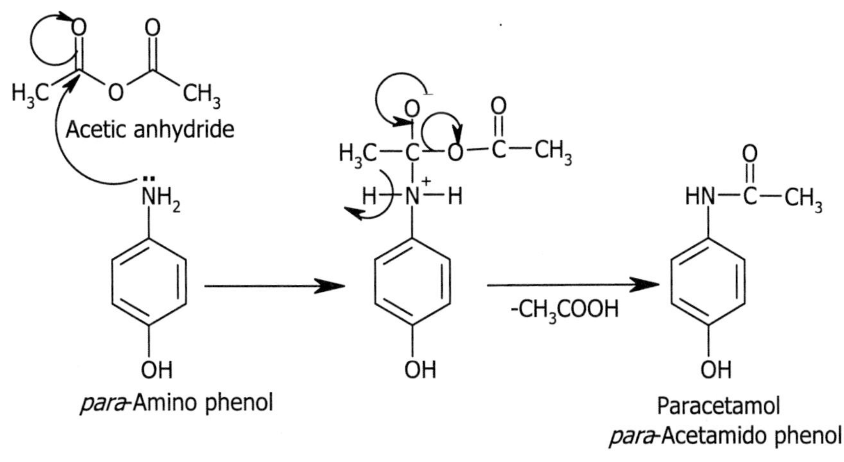

Procedure:

1. Place *para*-amino phenol (1.1 gm, 10 mmol) and water (3 ml) in a conical flask and stir it vigorously at room temperature, in order to suspend the solid in the water.
2. Add acetic anhydride (1.2 ml, 13 mmol) to the stirred suspension shake gently to effect mixing and warm on water bath (perform in fume hood).
3. Stir the flask to dissolve the solid and continue until it reprecipitate.
4. Filter the solid under vacuum after 10 minutes, wash with cold water and dry it.
5. Recrystallize the product from hot water (10 ml).
6. Allow the clear solution to cool to room temperature and isolate the recrystallized product by suction filtration.
7. The melting range of pure paracetamol is 168-170 °C.

Note:

- Though $-NH_2$ is more nucleophilic than –OH, excess acetic anhydride should be avoided as there is chance of double acetylation of 4-aminophenol.
- High temperature may also facilitate the double acetylation, hence it should be avoided.

3.3 Preparation of Aspirin

Traditional Name: Acetyl salicylic acid, spiraeic acid. ***IUPAC Name:*** Ethanoyloxy benzene carboxylic acid.

History of aspirin: Aspirin was developed from the observations that willow bark could cure malarial fever and its extracts act as a remedy for acute rheumatism. Later it is identified that the salicin (β-glycoside), an active constituent present in the bark is responsible for the therapeutic effect. In the biological system this salicin transforms into salicylic acid and produce effect. This made medicinal chemists to prescribe salicylic acid for the relief of fever and pain. Severe gastric irritations produced by salicylic acid prompted the need for the alternative. Thus aspirin, a prodrug of salicylic acid was developed. The name ASPIRIN is derived from Acetylated SPIRaeic acid.

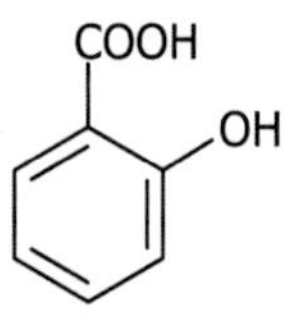

Salicylic acid

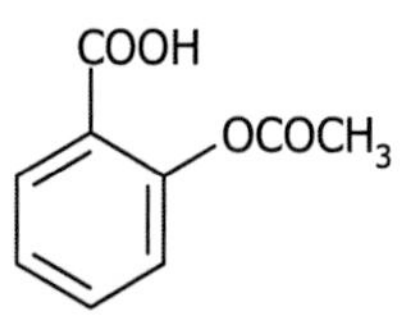

Aspirin
Acetyl salicylic acid

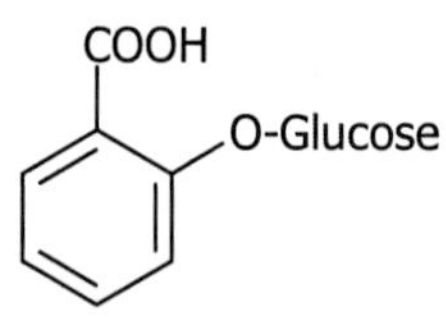

Salicin

Principle: Preparation of aspirin (acetyl salicylic acid) involves the acetylation of salicyclic acid with acetic anhydride in presence of glacial acetic acid. In place of glacial acetic acid, sulphuric acid or phosphoric acid also can be used. The reaction takes place by the replacement of active hydrogen of hydroxyl group (O-acetylation) by an acetyl group of acetic anhydride. The acetylation of alcohols and phenols are special case of esterification.

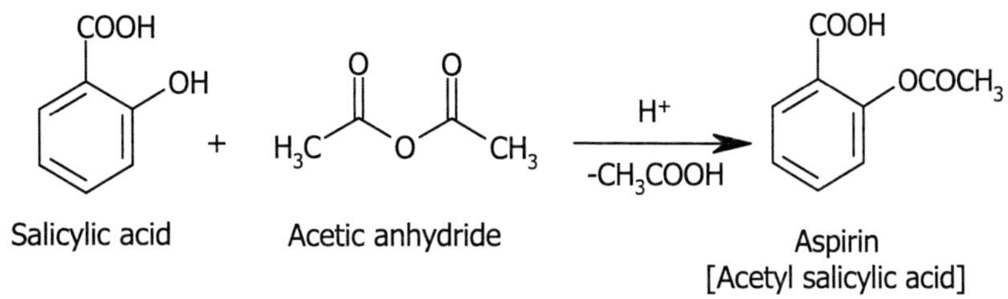

The electropositive carbonyl carbon (due to inductive effect) will be attacked by π electrons of oxygen of hydroxy group of salicyclic acid. Deprotonation removes the positive charge present in oxygen. Movement of electrons from carbonyl oxygen to form π bond with carbon, breaks the bond between carbonyl carbon and electro negative group of acetylating reagent to generate aspirin.

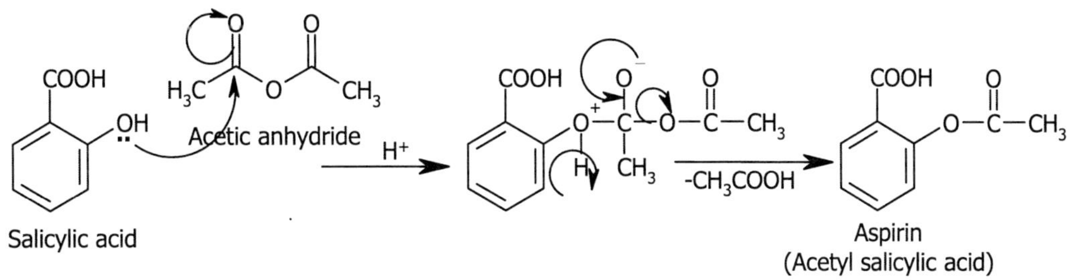

The acetyl salicylic acid can be prepared using acetyl chloride in pyridine. Pyridine absorbs the formed hydrogen chloride.

Method 1:

1. Place distilled acetic anhydride (10 ml, 105 mmol) and glacial acetic acid (10 ml, 174 mmol) in a dry beaker.
2. Add salicyclic acid (1 gm, 7 mmol) to the mixture and condense for 45 min.
3. Pour the hot mixture into cold water (100 ml).
4. Stir vigourously until shiny white crystals of acetyl salicylic acid (aspirin) separates out.
5. Filter and wash the product with cold water for several times.
6. Purify the product by recrystallization with ethanol.
7. The melting range of pure aspirin is 135-137 °C.

Method 2:

1. Dissolve salicylic acid (1.3 g, 9.4 mmol) in dry pyridine (1.5 ml) contained in a conical flask.
2. Add acetyl chloride (1 ml) immediately to the mixture with shaking. [Maintain the temperature between 50-60 °C]
3. Heat the mixture by placing on a water bath for 5 minutes.
4. Cool the reaction mixture and pour into beaker containing cold water with vigorous stirring. [Aspirin may separate as oil, but continuous and rapid stirring converts oil into crystals]
5. Filter the solid and wash with water.
6. Purify the acetyl salicylic acid isolated using the mixture of water and acetic acid (1:1).
7. The melting point of pure aspirin is 135-137 °C.

Method 3:

1. Place salicylic acid (1 gm, 7 mmol) in a beaker and add 4 drops of sulphuric acid (alternatively phosphoric acid 85% can also be used).
2. Add acetic anhydride (1 ml, 1 mmol) and mix thoroughly.
3. Heat the mixture gently by placing on a water bath for 15 minutes. [Maintain the temperature between 50-60 °C].
4. Cool the reaction mixture and pour into beaker containing cold water (10 ml) to decompose excess of acetic anhydride.
5. Stir the mixture and collect the crystals by vacuum filtration.
6. Purify the product by recrystallization with ethanol.
7. The melting point of pure aspirin is 135-137 °C.

3.4 Preparation of N-Acetyl Glycine

Traditional Name: Acetyl glycocoll, Aceturic acid, Acetamido acetic acid

IUPAC Name: 2-Ethanoylaminoethanoic acid.

Chemicals: Glycine, Acetic anhydride and Distilled water.

Principle: Amino acid glycine reacts readily with acetic anhydride and undergoes N-acetylation to form N-acetyl glycine.

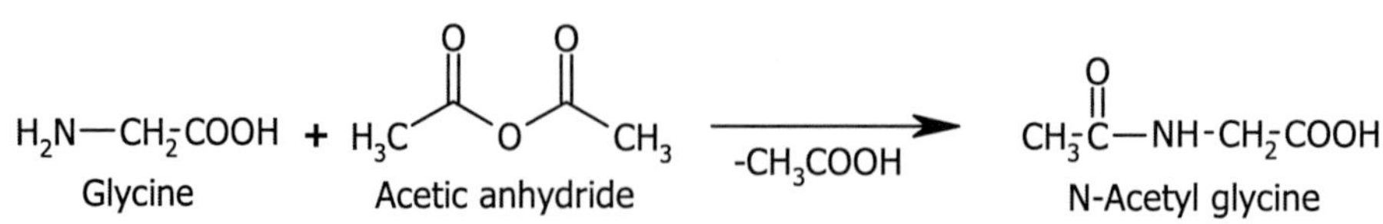

Movement of π electrons towards carbonyl oxygen (negative inductive effect) of acetic anhydride makes carbon positive. The unpaired electrons of nitrogen of glycine are attracted by positively charged carbon of carbonyl group. Amino group bears positive charge and removal of hydrogen from primary amine removes the charge. Movement of electrons from carbonyl oxygen to form π bond with carbon breaks the bond between carbonyl carbon and electronegative group of acetylating reagent (leaving group) and generates N-acetyl glycine.

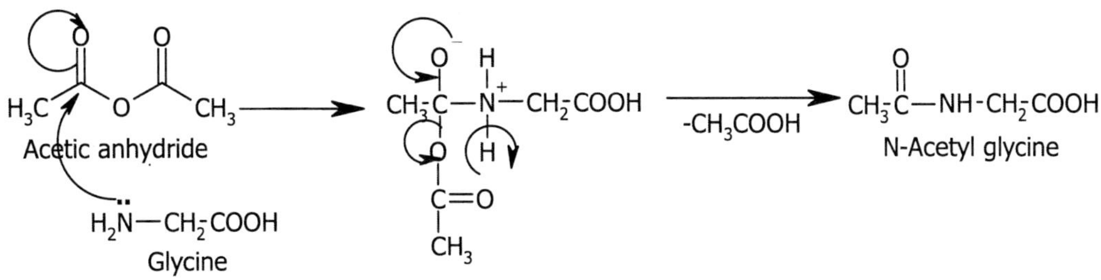

Procedure:

1. Place glycine (1 gm, 13 mmol) and water (4 ml) in a 50 ml conical flask.
2. Stir vigorously using mechanical stirrer until solid dissolves completely.
3. Add acetic anhydride (2.5 ml, 26 mmol) in one portion and continue stirring for 20 min.
4. Cool the reaction mixture in a refrigerator overnight, N-acetylglycine gets crystallizes out.
5. Filter, wash the residue with ice-cold water and dry at 100 °C.
6. Evaporate the combined filtrate and washings to dryness under reduced pressure on a water bath at 50-60 °C.
7. Purify the N-acetylglycine by recrystallization with boiling water.
8. The melting range of pure N-acetylglycine is 207-208 °C.

3.5 Preparation of Hippuric Acid

Traditional Name: Benzoyl glycocoll. ***IUPAC Name*:** Benzoyl amino ethanoic acid.

Chemicals: Benzoyl chloride, Glycine, Sodium hydroxide, Hydrochloric acid, Carbon tetrachloride and Charcoal.

Principle: Acylation of glycine with benzoyl chloride in basic medium generates hippuric acid (benzoyl glycine).

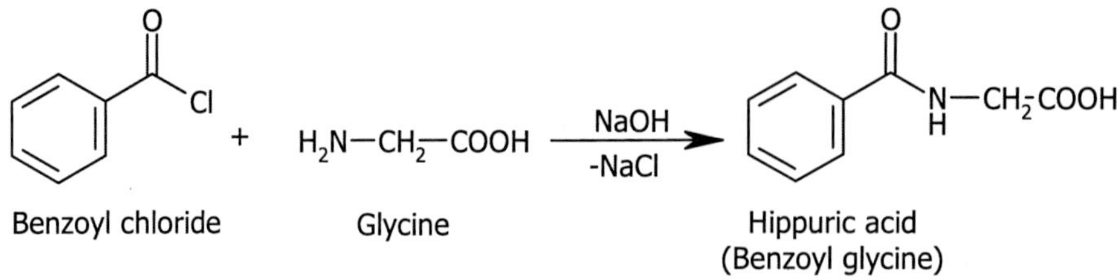

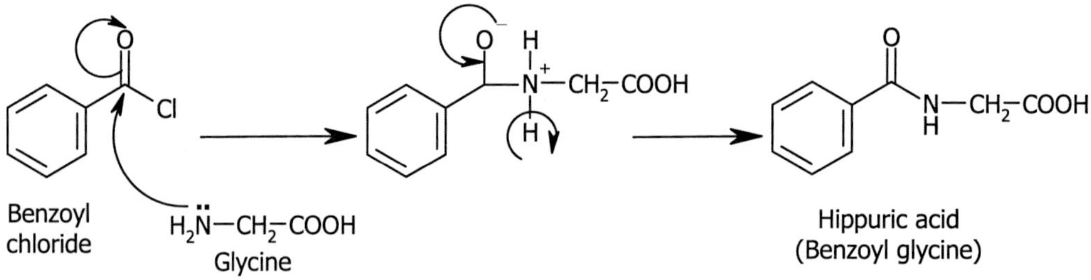

Procedure:

1. To the flask containing glycine (1 gm, 13 mmol) in cold sodium hydroxide (10 ml, 10%) add benzoyl chloride (2 ml, 17 mmol) in portions.
2. Shake the flask to dissolve the contents after each addition.
3. Cool the mixture and acidify with hydrochloric acid (10%).
4. Hippuric acid will separate upon acidification (benzoic acid may present in small amounts).
5. Filter the separated hippuric acid, wash with cold water and dry.
6. Heat the precipitate with carbon tetrachloride (5 ml) for 10 minutes and filter.
7. Purify the hippuric acid by recrystallization with boiling water.
   (Use charcoal to remove if any colouring impurities are present).
8. The melting range of pure hippuric acid is 187-188 °C.

CHAPTER 4

OXIDATION

The process of oxidation involves loss of electrons/loss of hydrogens/addition of oxygen. Loss of electrons in an atom increases its valence. After oxidation carbon atom gains bonds to more electronegative elements (commonly with oxygen) and transforms to more oxidized state.

Types of oxidation reactions

- *Simple composition reactions:* It involves the addition of oxygen to the molecule.

$$CH_3CHO \xrightarrow{(O)} CH_3COOH$$

- *Dehydrogenation:* It involves removal of hydrogens from the molecule.

$$CH_3CH_2OH \xrightarrow{(O)} CH_3CHO + H_2O$$

- *Addition of oxygen and dehydrogenation*: It involves both the addition of oxygen to the molecule and removal of hydrogen from the molecule.

$$CH_4 + O_2 \longrightarrow CH_2O + H_2O$$

Common Oxidants

- ***Chromium oxidants***

  Pyridinium chlorchromate (PCC): It oxidizes primary alcohols to aldehydes, and secondary alcohols to ketones. More acidic than pyridinium dichloride and collin. This reagent is not suitable for molecules containing acid labile groups. Buffering the reagent with ethyl acetate avoids this problem.

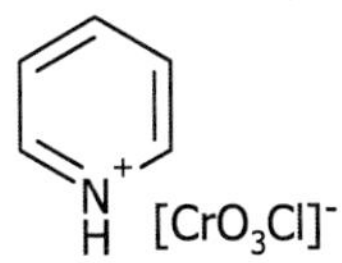

Collin's reagent: The solution of Sarett reagent (CrO_3.2 P_y) in methylene chloride is called as Collin's reagent. Non acidic reagent oxidizes primary alcohols to aldehydes and secondary alcohols to ketones. Requires excess of reagent (stoichiometric quantities) and will not cause lysis of labile group.

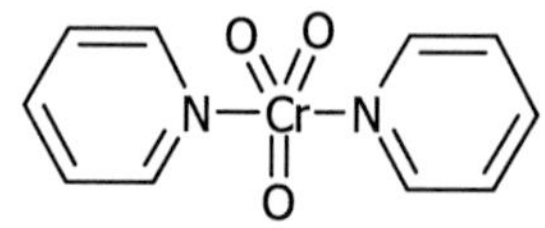

Pyridinium dichromate: It oxidizes primary alcohols to either aldehydes or carboxylic acids and secondary alcohols to ketones. Less acidic than pyridinium chlor chromate (PCC) and more acidic than Collin' reagent.

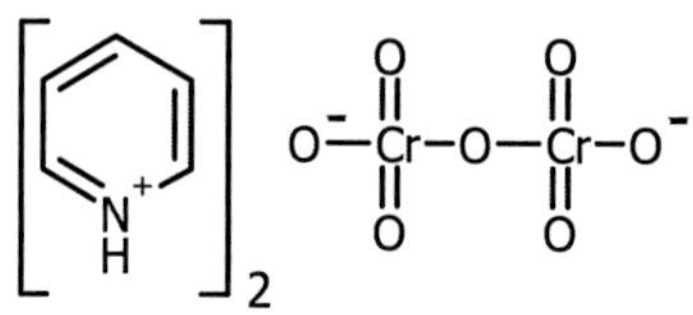

Jones reagent: The reagent is a solution of Chromium trioxide (CrO_3) in dilute sulphuric acid. It can be prepared by dissolving potassium/sodium dichromate in sulphuric acid. It oxidizes primary alcohols to carboxylic acids and secondary alcohols to ketones. This reagent is not suitable for the oxidation of molecules with acid labile groups.

$$CrO_3^-,\ H_2SO_4$$

- ***Ruthenium oxidants***

Tetrapropylammonium perruthenate (TPAP): It oxidizes primary alcohols to aldehydes (oxidation to carboxylic acid is rare) more rapidly and secondary alcohols to ketones. N-Methylmorpholine-N-oxide is useful in recycling this reagent.

$$(C_3H_7)_4N^+RuO_4$$

- ***Manganese oxidants***

Manganese dioxide (MnO_2): Mild oxidant, oxidizes allylic, propargylic and benzylic alcohols to aldehydes and ketones.

$$MnO_2$$

Potassium permanganate ($KMnO_4$): Powerful oxidant, aqueous medium promotes the reaction, where as organic solvents tampers the oxidation potential. It oxidizes benzylic position of aromatic system to carboxylic acid. Non-chemoselectiveness of the regent limits its use.

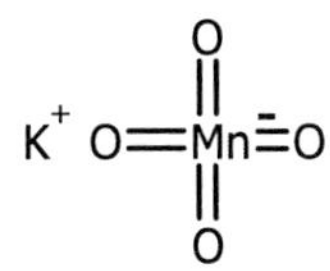

- ***Dimethyl sulphoxide:*** It initiates very mild oxidation and is used in Swern oxidation and in Nicolau's synthesis of rapamycin.

$$(CH_3)_2SO_2$$

- ***Iodine:*** Hypervalent iodine reagent (+3 and +5 oxidation state of iodine), [Dess-Martin Periodinane (DMP)] prepared from 2-iodobenzoic acid oxidizes the sensitive functional groups. It oxidizes primary alcohol to aldehyde (not to carboxylic acid) and secondary alcohol to ketones.

$$I_2$$

- ***Sodium chlorite:*** It oxidizes alcohol to carboxylic acid more efficiently. Use of resorcinol (electrophile scavenger) along with the sodium chlorite prevents the oxidation of olefin functionalities.

$$NaClO_2$$

- ***Lead tetraacetate:*** Lead tetraacetate is a strong oxidizing agent and also a source for acetyloxy groups.

$$Pb(OCOCH_3)_4$$

- ***Sodium periodate ($NaIO_4$):*** Sodium periodate is useful in the conversion of 1,2-diols into two corresponding aldehydes and further into carboxylic acid.

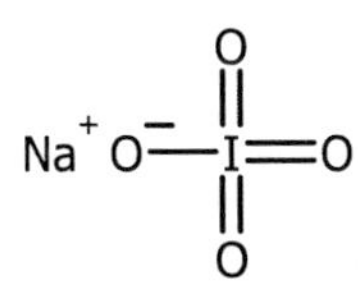

- ***Epoxidation oxidants***

  Perbenzoic acid: It is a simple peroxy acid, useful in the generation of epoxides.

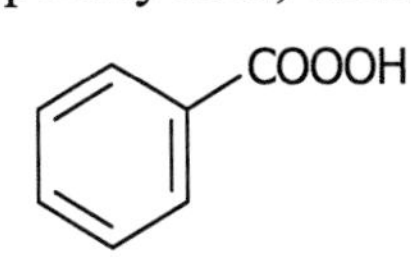

  meta-Chloro/nitro perbenzoic acid: Electrophilic oxidant, oxidizes olefins and reacts preferentially with electron rich carbon – carbon double bond (C=C).

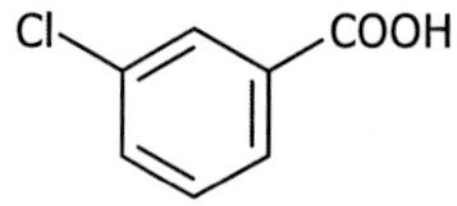

meta-Chloro perbenzoic acid (*m*-CPBA)

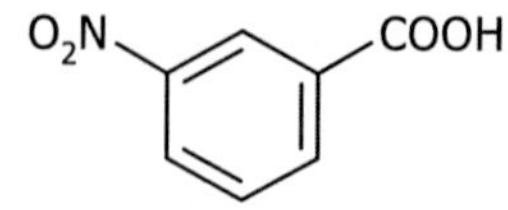

meta-Nitro perbenzoic acid (*m*-NPBA)

- ***Osmium tetroxide:*** Osmium tetroxide oxidizes olefins into 1,2-diol in presence of pyridine through 1,3-cycloaddition reaction. It forms highly unstable intermediate osmic ester (osmylation), which upon hydrolysis forms 1,2-diol.

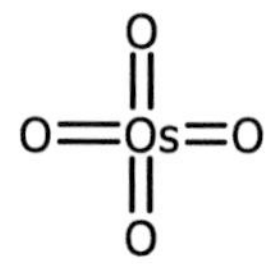

- ***Hydrogen peroxide:*** Hydrogen peroxide (simple peroxide) is a strong oxidizer. Higher oxidation capacity of hydrogen peroxide provides bleaching property.

$$H_2O_2$$

- ***Fenton reagent:*** The mixture of hydrogen peroxide and ferrous sulphate/ferrous acetate is known as Fenton reagent. It can be used for the hydroxylation of arenes and oxidation of barbituric acid to alloxan.

- ***Selenium dioxide:*** Selenium dioxide commonly oxidizes C-H bonds attached to various activating groups. Aldehydes, ketones and olefins are oxidized with high yields under mild conditions. Alcohols, amines, phenols and mercaptans are oxidized with poor yields under vigorous conditions. Alkanes, ethers and allyl halides undergo oxidation only in severe reaction conditions.

$$SeO_2$$

- ***Ozone (O_3):*** Ozone is a triatomic molecule contains three oxygen atoms and is a powerful oxidizing agent. The reaction proceeds more rapid with increased temperature and pressure.

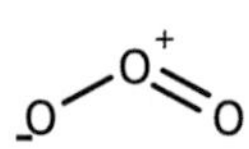

- ***Dimethyldioxirane:*** Powerful, selective and electrophilic oxidant, which oxidizes unreactive olefins under mild conditions.

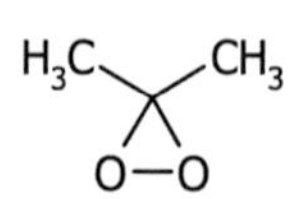

- ***Perphthalic acid:*** It is useful in the epoxidation and formation of alkene oxide.

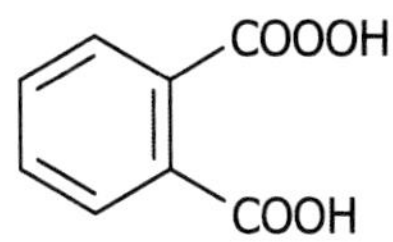

- ***Per trifluroacetic acid:*** It is useful in the synthesis of peptides.

$$CF_3COOOH$$

- ***N-Bromo succinamide:*** *N*-Bromo succinamide mediates oxidations only in the presence of UV light and peroxides. This oxidative reaction takes place through free radical mechanism. It oxidizes primary alcohols into aldehydes, and secondary alcohols into ketones. Secondary alcohols can be selectively oxidized using *N*-bromo succinamide in aqueous dimethoxyethane without affecting the primary alcohol. Through oxidative decarboxylation and deamination, amino acids can be converted into aldehydes.

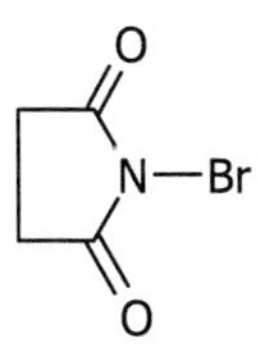

Important oxidation reactions

Osmylation: The oxidation of olefins by osmium tetroxide through 1,3-cycloaddition forms osmic ester (unstable intermediate), which upon hydrolysis generates 1,2-diol. This reaction is known as Osmylation.

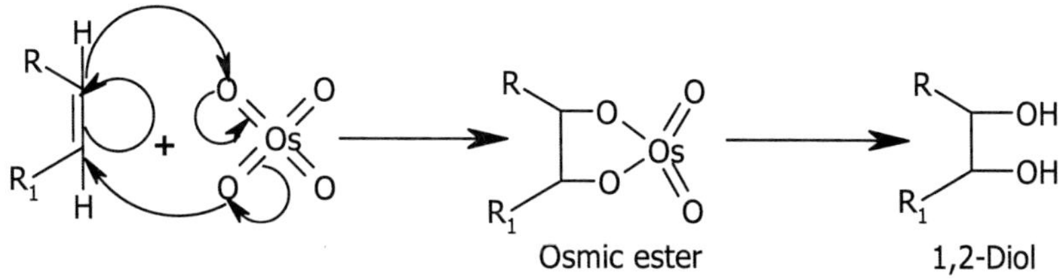

Syn-selective preparation of 1,2-diol from alkenes by the use of osmium tetroxide along with *N*-methyl morpholine-*N*-oxide is known as Upjohn dihydroxylation.

Swern oxidation: The dimethyl sulfoxide and oxalyl chloride generates dimethylchlorosulphonium *in situ*. It oxidizes primary and secondary alcohols to aldehydes and ketones, respectively. Aldehydes will not undergo any oxidation. It avoids the use of toxic metals (chromium) and proceeds even under very mild conditions. The side product dimethyl sulphide is the major drawback.

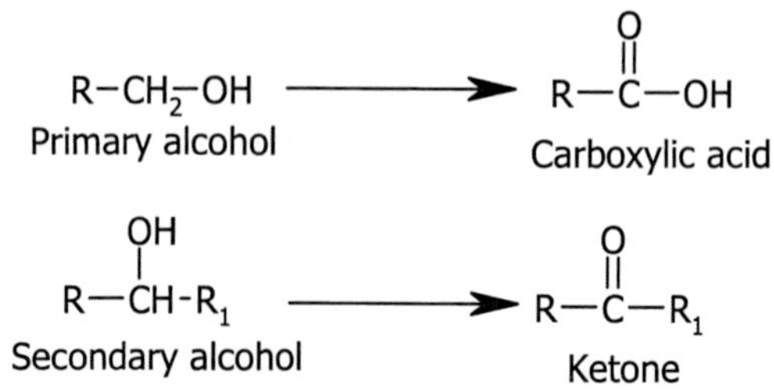

Rubottom oxidation: It involves the oxidation of silyl enol ethers by *m*-chloroperbenzoic acid (mCPBA) and gives α-hydroxy ketones.

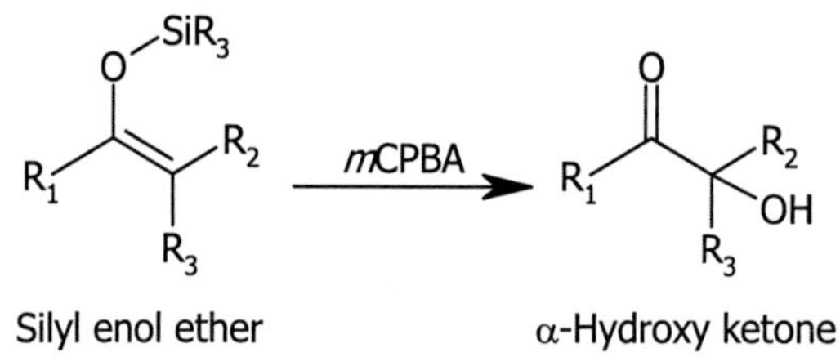

Ozonolysis: Oxidative cleavage of alkenes into aldehydes, ketones and carboxylic acid by ozone is called as ozonolysis. Generally this reaction will be carried out with dichloro methane at –78 °C. Reductive workup with zinc in acetic acid/dimethyl sulphide produces ketones and aldehydes. Oxidative workup with hydrogen peroxide forms carboxylic acid.

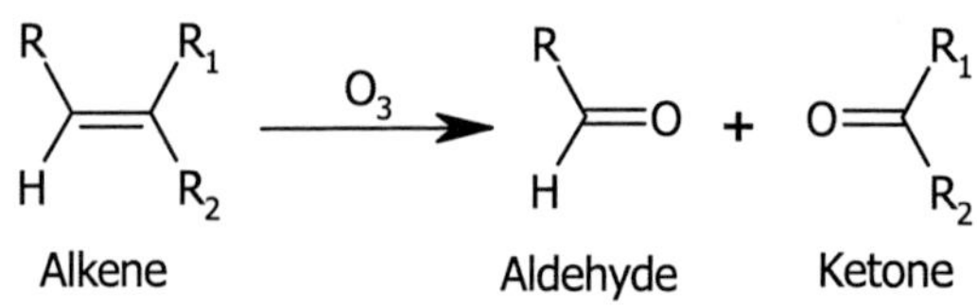

Oppenauer oxidation: The aluminium-catalyzed hydride shift from the α-carbon of an alcohol to the carbonyl carbon of ketone. It proceeds over a six membered transition state and is known as Oppenauer oxidation.

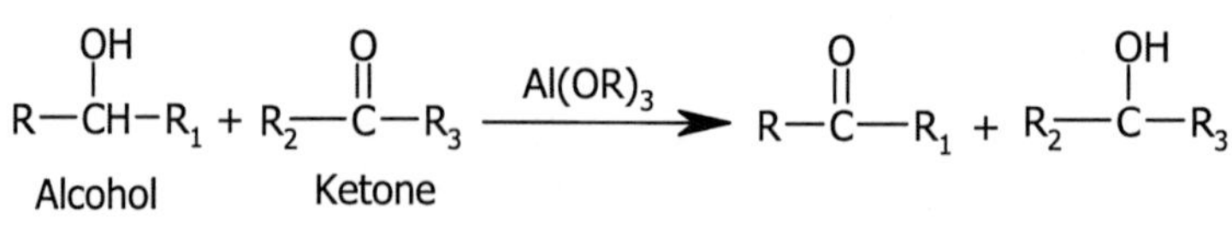

Jones oxidation: Oxidation of primary alcohols to carboxylic acid and secondary alcohols to ketones by cromic acid is known as Jones oxidation reaction.

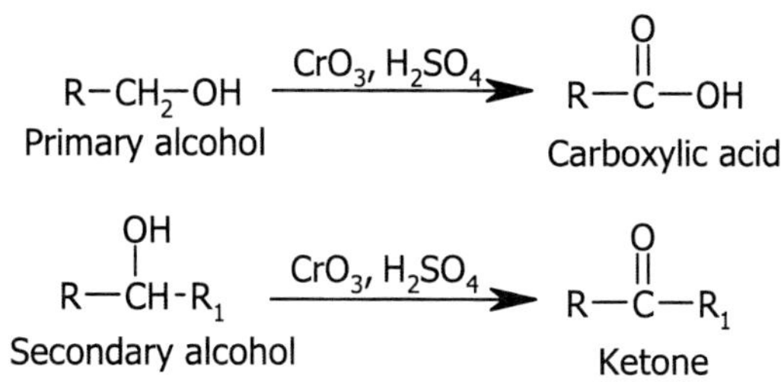

Jacobse-Katsuki epoxidation: Enantioselective formation of epoxides from substituted olefins (prochiral allylic alcohols) by Mn-Salen catalyst along with sodium hypochlorite is called as Jacobe-Katsuki epoxidation.

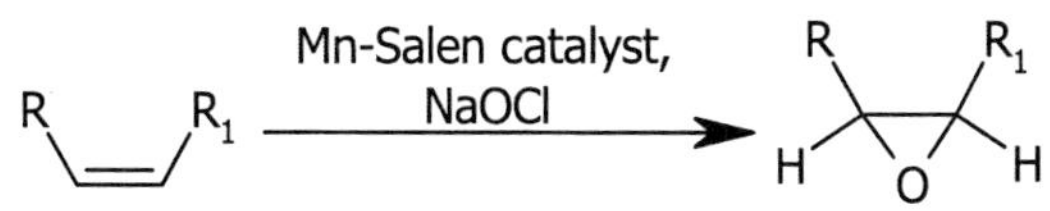

Prilezhaev reaction: Epoxidation of alkene by peracid (*meta*-chloroperbenzoic acid, peracetic acid, magnesium mono-perphthalate) to form oxirane is called as Prilezhaev reaction.

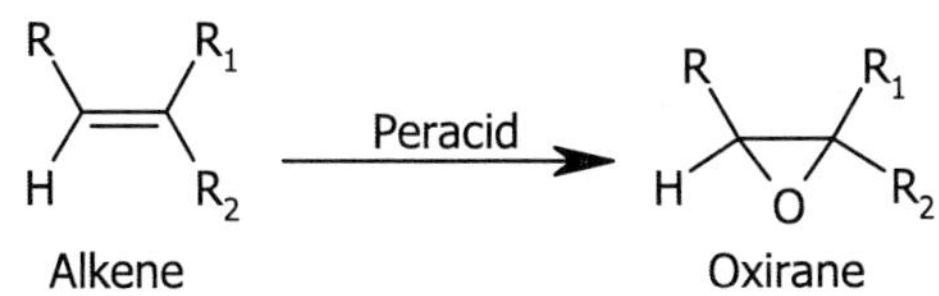

Dess-Martin oxidation: A selective and very mild oxidation of alcohols to aldehydes or ketones by Dess-Martin periodinane (DMP, hyper valent iodine) called as Dess-Martin oxidation.

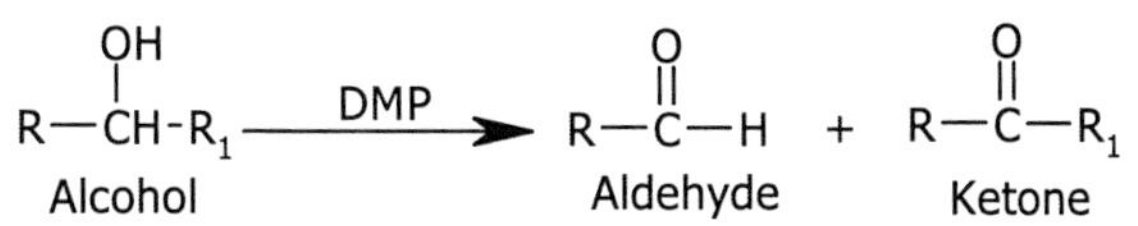

Baeyer-Villiger oxidation: Conversion of ketones into esters and cyclic ketones into lactones by oxidative cleavage mediated through *m*CPBA/hydrogen peroxide of carbon-carbon bond adjacent to carbonyl group.

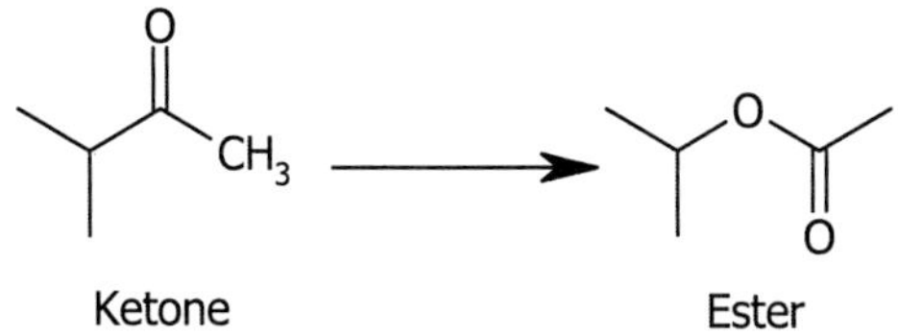

Sommet reaction: Arylalkyl halides upon reaction with hexamethylene tetramine (hexamine) form an intermediate of hexamethylene tetramine complex with arylalkyl halides. This complex formed undergoes hydrolysis to yield quaternary salt.

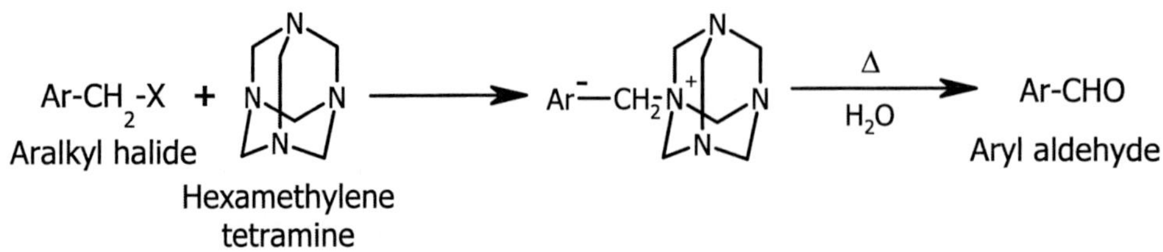

Sarett oxidation: Oxidation of primary and secondary alcohols to aldehydes and ketones, respectively by Collin's reagent (CrO_3-pyridine) is known as Sarett oxidation.

Pharmaceutical Applications

- Preparation of cortisone and hydrocortisone using osmium tetroxide.

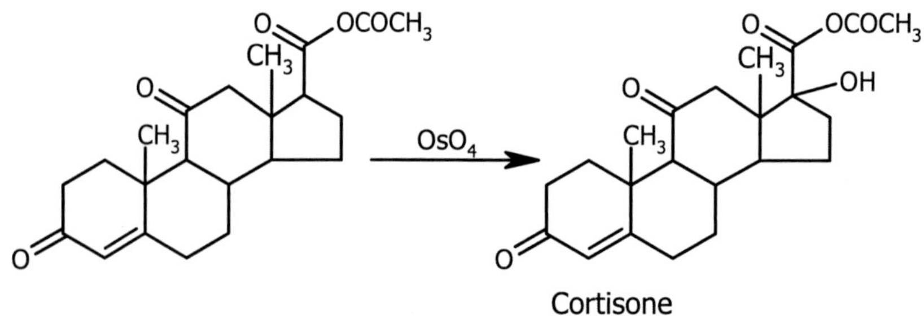

- Preparation of sulindac using sodium metaperiodate ($NaIO_4$).

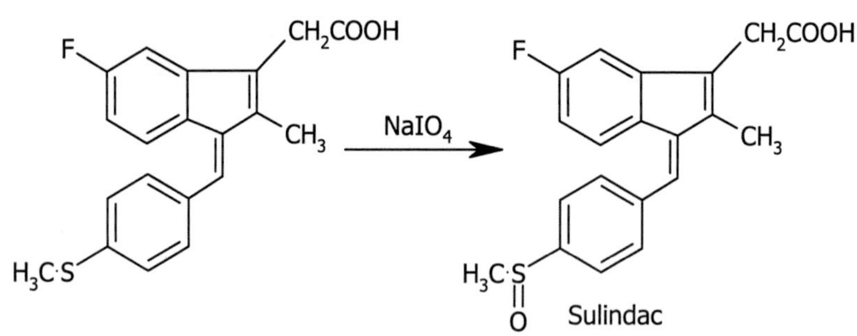

- Preparation of carbamazepine by N-bromo succinamide.

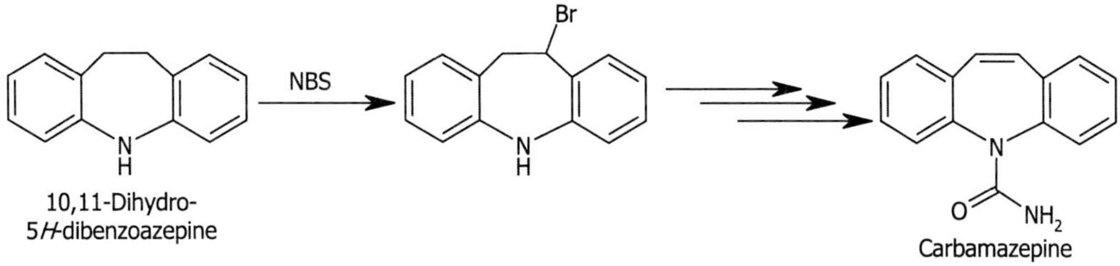

- Preparation of Vitamin A_1 and A_2 by potassium permanganate, N-bromo succinamide and lithium aluminium hydride.

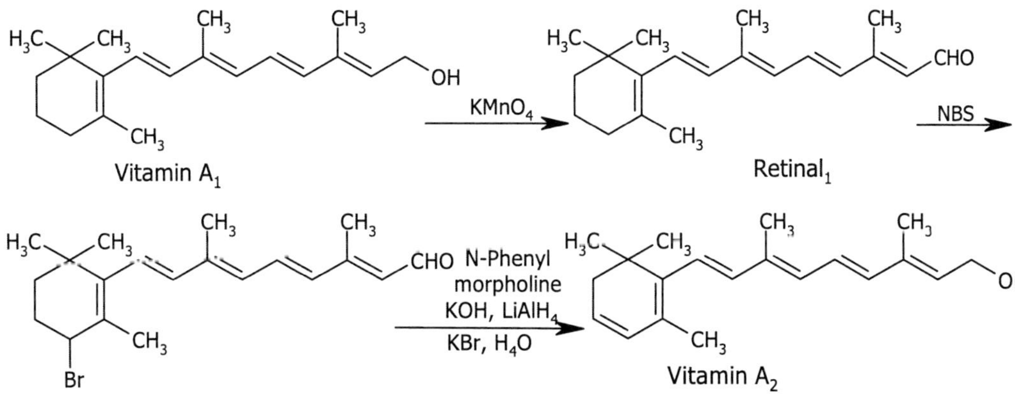

- Preparation of azatadine (antihistamine) using peracid.
- Preparation of PGE_1 (vasodilator) using osmium tetraoxide.

4.1 Preparation of Benzil

IUPAC Name: 1,2-Diphenylethane-1,2-dione

Chemicals: Benzoin, Nitric acid and Ethanol.

Principle: Benzil can be prepared by the oxidising benzoin using nitric acid as oxidizing agent.

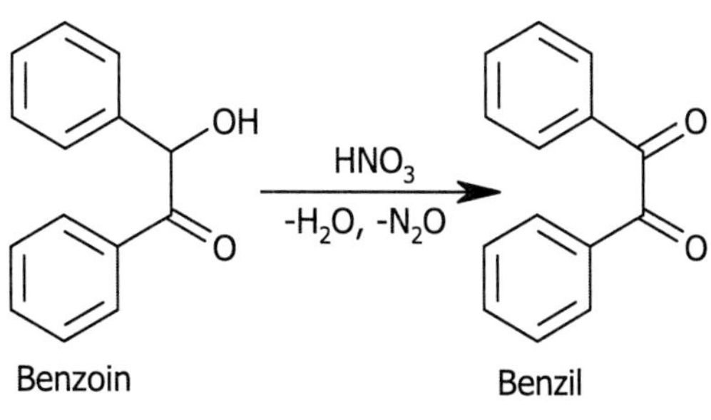

Reaction mechanism:

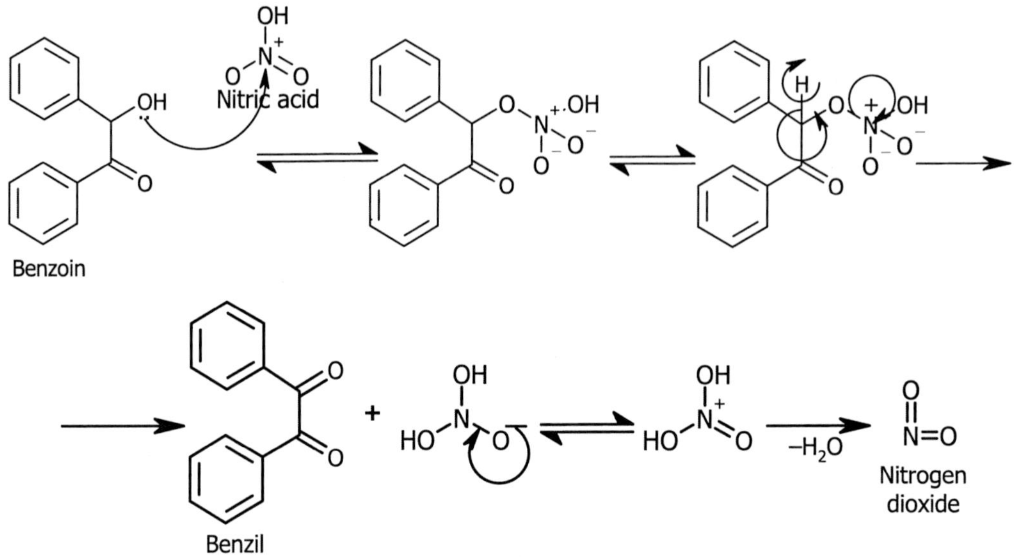

Procedure:

1. Place benzoin (1 gm, 4.7 mmol) and nitric acid (5 ml) in a flask.
2. Heat the reaction mixture for 1½ hr by placing in a water bath (until the evolution of oxides of nitrogen ceases).
3. Pour the reaction mixture into cold water (20 ml) contained in a beaker.
4. Stir well until the oil crystallizes completely as a yellow solid.
5. Filter the crude benzil by vacuum filtration and wash with water.
6. Dry and recrystallize the benzil with ethanol.
7. The melting range of pure benzil is 94-96 °C.

Note: Carry out the reaction in fume hood to avoid the breathing of evolved nitrogen oxides.

4.2 Preparation of Benzoic Acid

Traditional Name: Dracyclic acid, Carboxy benzene

Chemicals: Benzyl chloride, Sodium carbonate (anhydrous), Potassium permanganate, Hydrochloric acid and Sodium sulphite.

Principle: Benzoic acid can be prepared by the oxidation of benzyl chloride using potassium permanganate in basic medium. In alkaline medium, permanganate functions as powerful oxidizing agent. Alternatively, oxidation of benzyl chloride can be done with dilute nitric acid, but reaction proceeds very slow. Chlorinated alkyl group due to the positive inductive effect, increases the susceptibility of oxidation.

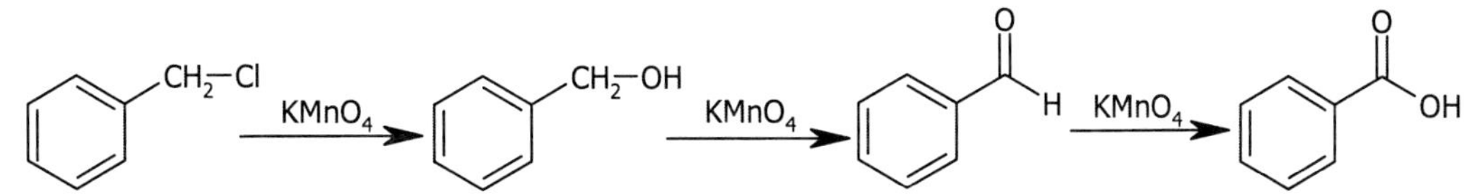

Benzyl chloride Benzyl alcohol Benzaldehyde Benzoic acid

Potassium permanganate dissociates and generates manganate ion (MnO_4^-), which attacks benzyl chloride and forms benzyl alcohol transition complex, which consequently gives benzaldehyde and manganic acid. Concentrated hydrochloric acid converts manganic acid ester into benzoic acid and manganese dioxide. Addition of sodium sulphite solution dissolves brownish manganese dioxide and gives white crystals of benzoic acid.

Reaction mechanism

Conversion of benzyl chloride into benzyl alcohol: Benzyl chloride reacts with manganate ion, undergoes nucleophilic substitution reaction and forms benzyl alcohol.

$$KMnO_4 \longrightarrow K^+ + MnO_4^-$$

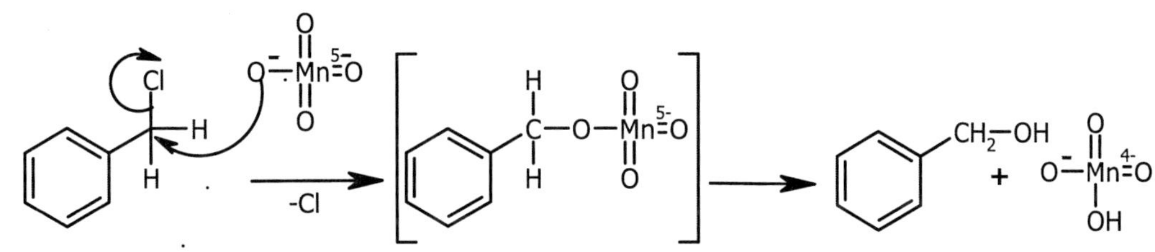

Benzyl chloride Benzyl alcohol

Oxidation of benzyl alcohol into benzaldehyde: Manganate ion oxidizes benzyl alcohol and produces benzaldehyde.

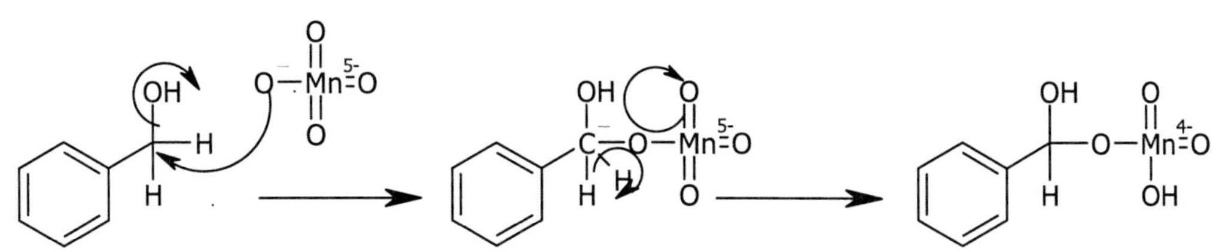

Benzyl alcohol

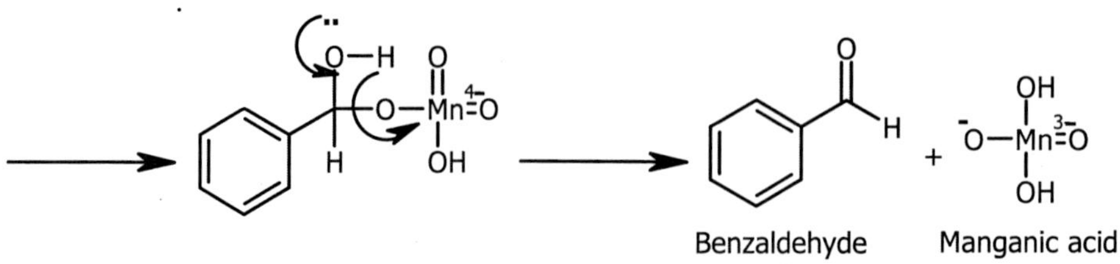

Oxidation of benzaldehyde into benzoic acid: Benzaldehyde then oxidizes to give benzoic acid.

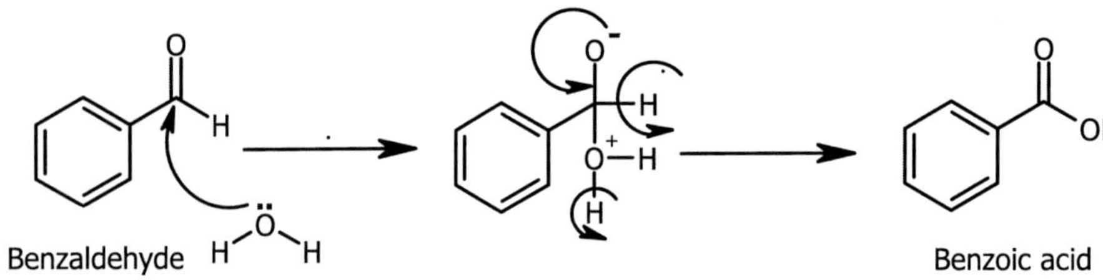

Procedure:

1. Place anhydrous sodium carbonate (2.5 gm, 23.59 mmol) and potassium permanganate (5 gm, 31.64 mmol) in a round bottom flask containing cold water (100 ml) and attached with a reflux condenser.
2. Add benzyl chloride (2.5 ml, 22.7 mmol) slowly with stirring to get clear solution and boil the mixture, until the liquid running from the condenser contains no oily drops of benzyl chloride (~1½ hr).
   (Violet colour of permanganate changes to brown precipitate of manganese dioxide)
3. Cool the reaction mixture under running water, shake vigorously to avoid the formation of large lumps.
4. Add concentrated hydrochloric acid (25 ml) slowly to precipitate benzoic acid.
5. Add sodium sulphite solution (50 ml, 20%) to dissolve brown precipitate of manganese dioxide and to get white precipitate of benzoic acid.
6. Filter under suction and wash the residue with cold water.
7. Dry and recrystallize the product from hot water.
8. The melting point of pure benzoic acid is 121 °C .

CHAPTER 5

REDUCTION

Reduction is a chemical transformation in which electrons are gained. Reduction can be defined as process, where addition of hydrogen to a compound or subtraction of oxygen from a compound results.

Types of reductions

1. Catalytic reduction.
2. Metal hydride reduction.
3. Dissolving metal reduction.

Hydrogenation (reduction) differs from protonation or hydride addition. In hydrogenation, products have same charge, whereas in protonation products carries charge. Hydrogenation is of two types

1. Heterogeneous hydrogenation
2. Homogeneous hydrogenation

Hydrogen Sources

1. Hydrogen gas.
2. Hydrazine.
3. Dihydronaphthalene.
4. Dihydroanthracene.
5. Isopropanol.
6. Formic acid.

Reducing Agents

- Alkali metals [Zinc (Zn), platinum (pt)].
- Metal hydrides [Lithium aluminium hydride ($LiAlH_4$), sodium boro hydride ($NaBH_4$)].

1. Catalytic Reduction

Many organic compounds are reduced by passing their vapour mixed with hydrogen over nickel heated at 200-300 °C.

Catalysts used in Reduction

- ***Lindlar's catalyst:*** Palladium-Barium sulphate (Pd-$BaSO_4$) poisoned with quinoline or Palladium-Calcium carbonate (Pd-$CaCO_3$) poisoned with lead acetate.
- ***Wilkinson's catalyst:*** Chlorotris(triphenylphosphine)rhodium (coordination compound) is known as Wilkinson's catalyst, catalyzes the hydrogenation of alkenes.

$$RhCl(PPh_3)_3$$

- ***Adam catalyst:*** Platinum dioxide (Platinum (IV) oxide hydrate) is known as Adam's catalyst and is useful in hydrogenation reactions.

$$PtO_2.H_2O$$

- ***Nickel:*** Powerful catalyst for the reduction of multiple bonds present in aliphatic, alicyclic and aromatic compounds.

Important Catalytic Reduction

Rosenmund reduction: Reduction of acid chlorides into aldehydes by palladium supported barium sulphate.

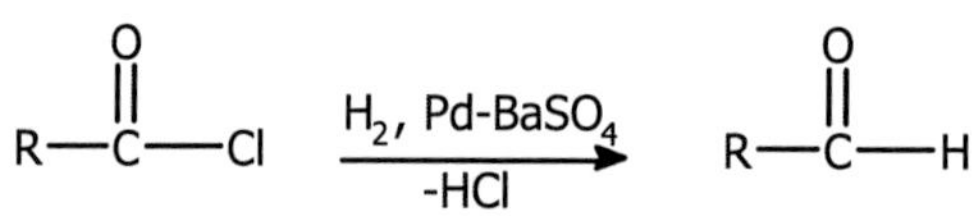

5.1 Preparation of Benzhydrol

Traditional Name: Hydroxy diphenyl methane

IUPAC Name: Diphenylmethanol

Chemicals: Benzophenone, Zinc powder, Sodium hydroxide and Ethanol.

Principle: Benzophenone upon reduction by zinc dust and alcoholic sodium hydroxide forms benzhydrol (secondary alcohol). Metallic zinc serves as the source of electrons (electron donor). Alcoholic sodium hydroxide (reducing agent) liberates nascent hydrogen, thus acts as source of hydrogen. Both benzophenone and nascent hydrogen gets adsorbed on the surface of zinc. The high reactive nascent hydrogen thus reduces benzophenone into benzhydrol.

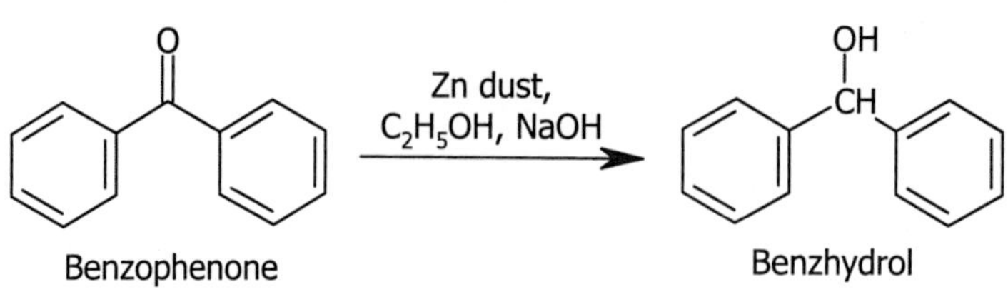

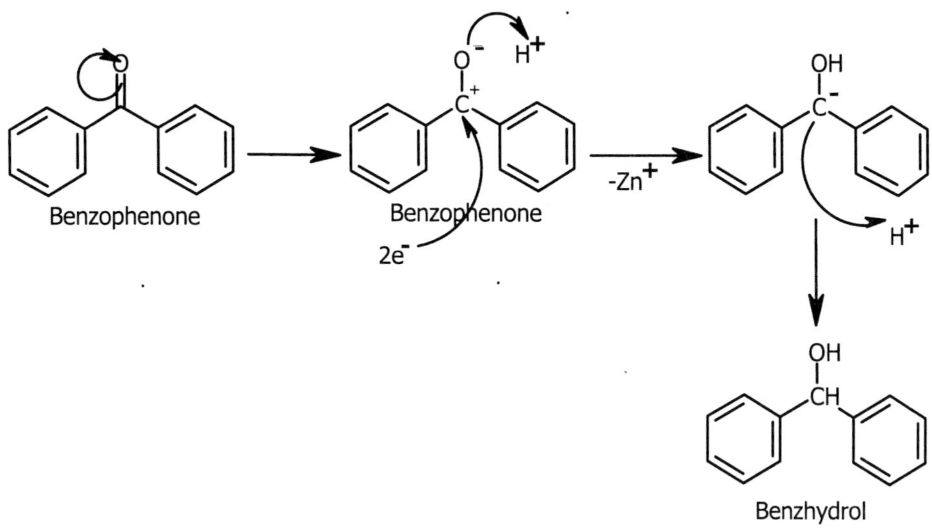

Procedure:

1. Place benzophenone (1 gm, 5.5 mmol), sodium hydroxide (1 gm, 25 mmol), in ethanol (10 mL) and zinc powder (1.5 gm) in a round bottom flask.
2. Mix the reactants well, attach reflux condenser to the flask and warm gently (on water bath) for 2 hr.
3. Cool the flask to 60 °C and then filter the reaction mixture under suction.
4. Wash the residue twice with 2 mL portions of hot ethanol.
5. Pour the clear filtrate into ice-cold water (50 ml) and acidify with hydrochloric acid.
6. Viscous oil separates out and solidifies on cooling (after standing overnight).
7. Filter the white crystalline mass separated under suction and dry.
8. Recrystallize the solid from hot ethanol.
9. The melting point of the pure benzhydrol is 68 °C.

Note: The filtered zinc dust is inflammable, therefore do not allow it to contact with combustible material, destroy it by treating with mineral acid.

2. Metal Hydride Reduction

There are many metal hydrides available for the reduction of molecules and they include:

(a) ***Lithium aluminum hydride (LAH):*** Lithium aluminium hydride ($Li^+[AlH_4]^-$) is the most versatile metal hydride used as a reducing agent in organic synthesis. It is widely used in fine chemical industries, especially for the reduction of carbonyl compounds (esters, carboxylic acids and amides). LAH is white solid, but appear grey-white after contamination.

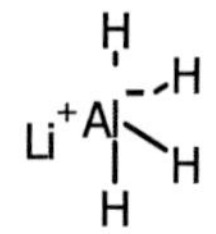

Reduction with LAH is usually carried out in etheral solutions. LAH violently react with water and atmospheric moisture and releases inflammable gaseous hydrogen (frothing of hydrogen). This reaction is exothermic, hydrogen can ignite spontaneously and form explosive mixture with air. LAH on reaction with water releases lithium hydroxide, which is corrosive to skin and eye.

$$LiAlH_4 + 4\,H_2O \longrightarrow LiOH + Al(OH)_3 + 4\,H_2$$

LAH is most powerful than sodium borohydride because of weaker Al-H bond than B-H bond in sodium borohydride. In lithium aluminium hydride, hydrogens bear electron pair and a negative charge. In case of sodium borohydride, a bare proton with no surrounding electrons and positive charge is possible. Hydrogen is considered to be a nucleophile, when hydrogen is attached to an atom that is less electronegative than itself and becomes electron rich.

Special warnings

1. All apparatus and reagents should be perfectly dry.
2. Reactions should be run rigourously under nitrogen, with reaction temperature below 70 °C.
3. Hydride should be added first to the solvent in the nitrogen purged apparatus.
4. Gentle efficient stirring is essential to prevent the crust above the level of the liquid or at the bottom of the flask.
5. Use oil or graphite bath to avoid local overheating.
6. Destroy the unreacted and excess of LAH by slow careful addition of ethyl acetate (diluted with dry THF) after reaction under nitrogen.

Pharmaceutical applications

1. Chemical conversion of Vitamin A_1 into Vitamin A_2.

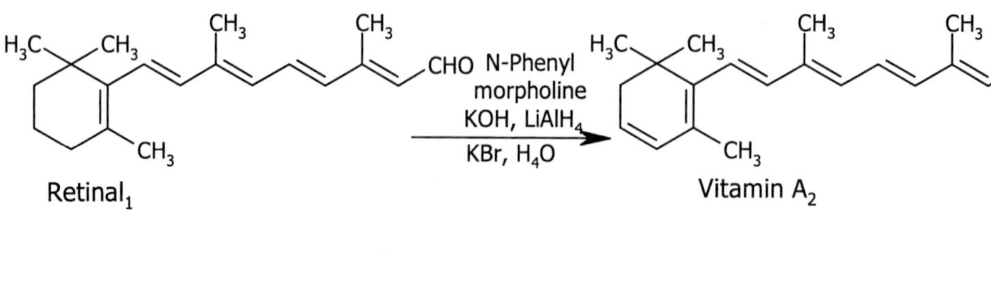

Retinal$_1$ Vitamin A_2

2. Preparation of Thiamphenicol.

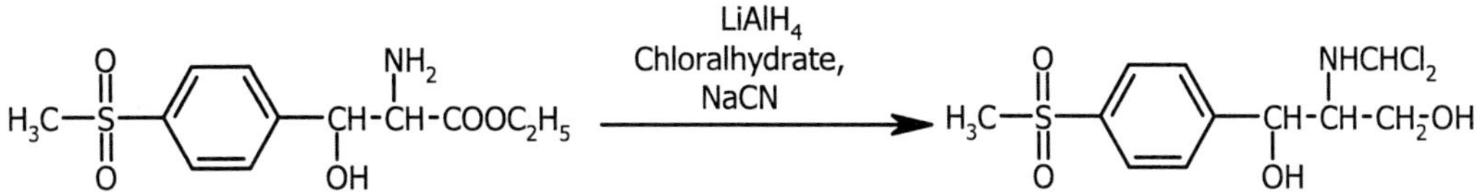

2-Amino-1-[4-(methylsulfonyl)phenyl] propane-1,3-diol — Thiamphenicol

3. Reduction of lactum for preparing a cyclic amine.

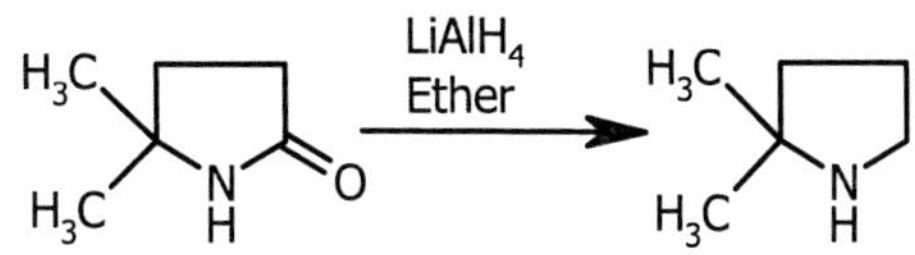

Lactum — Cyclic amine

(b) ***Alkoxyaluminate reagents:*** Lithium aluminium hydride can be converted into lithium tetramethoxyaluminium [$LiAl(OCH_3)_4$] with 4 equivalents of methanol, lithium trimethoxyaluminium [$LiAl(OCH_3)_3$] with 3 equivalents of either ether, THF or glyme and lithium triethoxyaluminium [$LiAl(OC_2H_5)_3$] with 3 equivalents of ethanol.

(c) ***Sodium borohydride:*** Sodium borohydride ($NaBH_4$) is an important hydride used in the separation of uranium salts. It is used in the reduction of glucose to D-glucitol (alditol, D-sorbitol). D-sorbitol is an artificial sweetener and used as a sugar subtituent in foods.

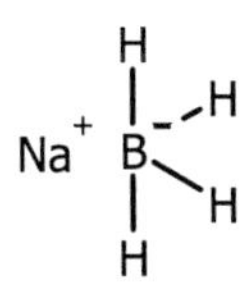

Pharmaceutical applications

1. Preparation of D-Sorbitol (artificial sweetener).

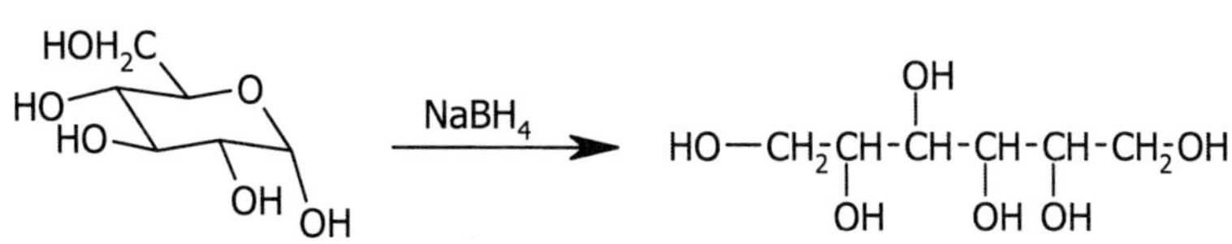

D-Glucose — D-Sorbitol (alditol)

2. Preparation of Ibuprofen (anti-inflammatory drug).

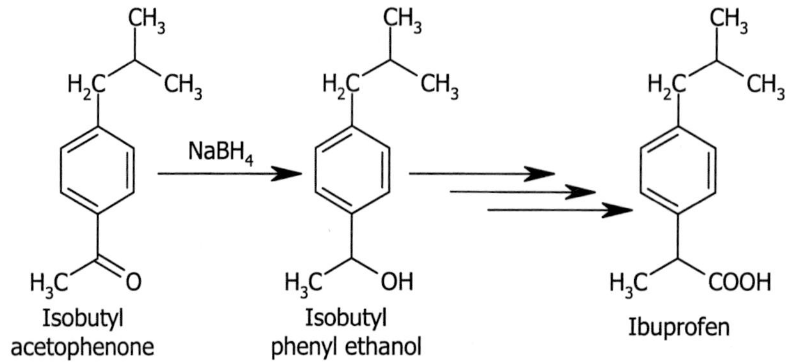

3. Preparation of Fenoprofen (anti-inflammatory drug).

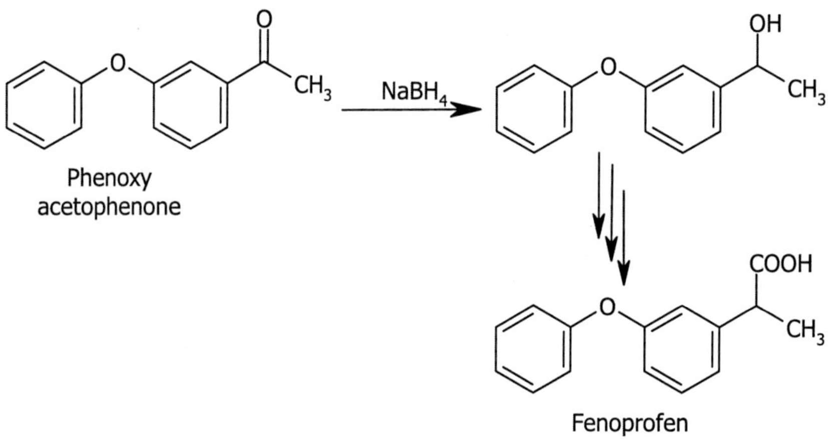

4. Preparation of Roxaltidine (antiulcer drug).

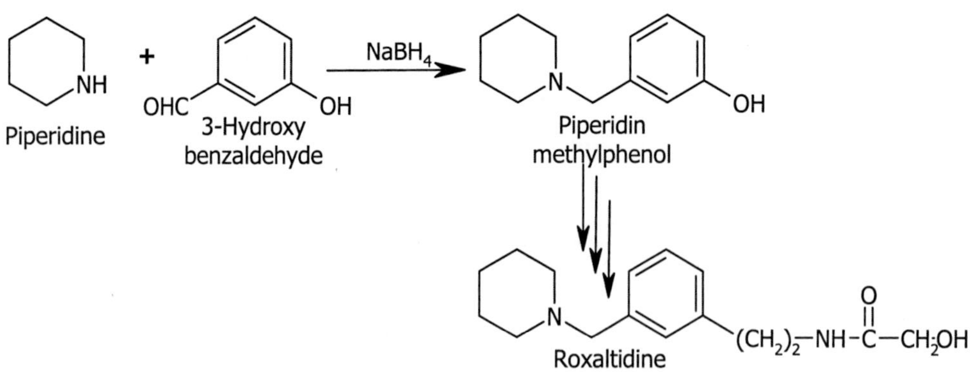

(d) ***Lithium borohydride:*** Lithium borohydride ($LiBH_4$) reduces lactones into diols. It also reduces aldehydes, ketones, acid chlorides, epoxides and esters, but not acids, nitriles and nitro compounds.

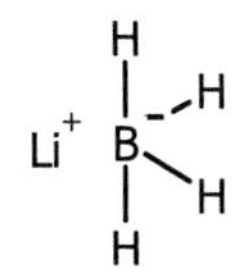

(e) ***Magnesium borohydride:*** Magnesium borohydride ($MgBH_4$) readily reduces esters.

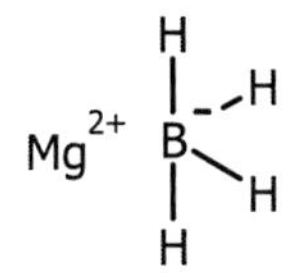

(f) ***Aluminium borohydride:*** Aluminium chloride with 3 equivalents of sodium borohydride generates aluminium borohydride [$Al(BH_4)_3$]. It is a very powerful reagent for the reduction of the lactones, epoxides, carboxylic acids, tertiary amides, nitriles, ketones, aldehydes and esters.

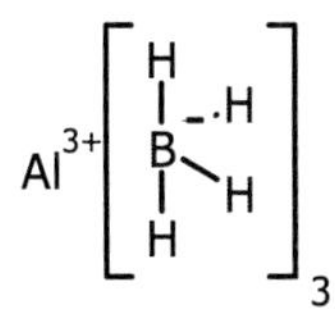

(g) ***Zinc borohydride:*** Zinc chloride reacts with sodium borohydride and generates zinc borohydride [$Zn(BH_4)_2$]. It is a most useful reagent for the reduction of α,β-unsaturated aldehydes and ketones.

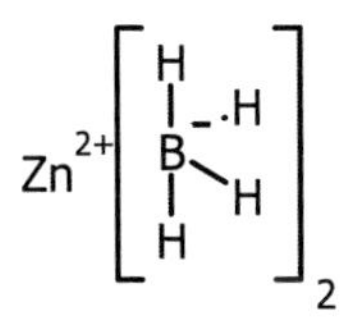

(h) ***Cerium borohydride:*** Addition of cerium salts to sodium borohydride generates cerium borohydride [$Ce(BH_4)_2$], also known as Luche reagent. It is very selective reagent for the reduction of conjucated aldehydes and ketones.

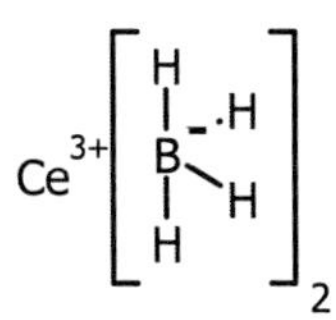

(i) ***Sodium cyanoborohydride:*** Reaction of sodium borohydride (SBH) with hydrogen cyanide gives sodium cyanoborohydride. It readily reduces iodides, bromides and tosylates to hydrocarbons.

$$NaBH_4 + HCN \xrightarrow{THF} NaBH_3CN$$

(j) ***Aluminium hydride:*** Reaction of lithium aluminium hydride (LAH) with aluminium chloride gives aluminium hydride [AlH_3].

(k) ***Diisobutylaluminium hydride:*** This is also known as Dibal/Dibal-H ([$(CH_3)_2CHCH_2]_2$-AlH$)_2$, most widely useful in reducing esters, lactones and nitriles.

$$LiAlH_4 + AlCl_3 \xrightarrow{THF} 4AlH_3 + LiCl$$

(l) ***Alkoxy and alkyl borohydride:*** Alkylborohydrides are more reactive than sodium borohydride (SBH) but less reactive than lithium aluminium hydride (LAH). Acyloxy borohydrides [$NaBH_4$-*n*-(O_2CR)n] selectively reduces aldehydes and ketones.

(m) ***Super-hydride:*** Lithium hydride reacts with triethylborane and generates super-hydride. Alkyl groups attaches to the boron of borohydride reagent, which are powerful hydride nucleophile than $LiBH_4$. They are useful in the reductive dehalogenation of alkyl halides.

(n) ***Selectrides:*** Potassium tri-*sec*-butyl borohydride is known as *K*-Selectride and lithium tri-*sec*-butyl borohydride is known as *L*-Selectride and both reduce aldehydes and ketones into corresponding alcohols.

$$B(C_2H_5)_3 + LAH \xrightarrow{THF,\ 65\ °C} LiBH(C_2H_5)_3$$

(o) ***Diborane:*** The stable dimer of borane is a potent reducing agent for aldhydes, ketones, lactones, epoxides, acids, tertiary amides and nitriles.

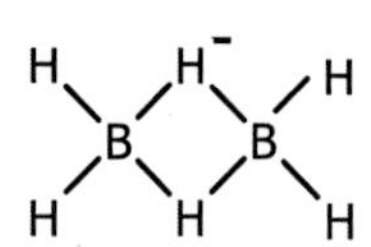

5.2 Preparation of *meta*-Nitro Benzyl Alcohol

Chemicals: Benzaldehyde, Sodium Borohydride, Methanol and Sodium hydroride.

Principle: *Meta*-Nitrobenzaldehyde on reduction with sodium borohydride gives *Meta*-nitro benzylalcohol. Sodium borohydride is chemoselective for the carbonyl groups and doesn't react with other reducible functional groups. Hence nitro group present in 4-nitro benzaldehyde will not undergo reduction with sodium borohydride. Sodium borohydride acts as donar of hydride ion, which attacks sp^2 carbon of carbonyl group and forms tetrahedral intermediate (sp^3 carbon). The formed alkoxide on protonation affords alcohol. Sodium borohydride is relatively stable in water, whereas lithium aluminium hydride generates lithium hydroxide by the reaction with water or trace of oxygen, which is more corrosive.

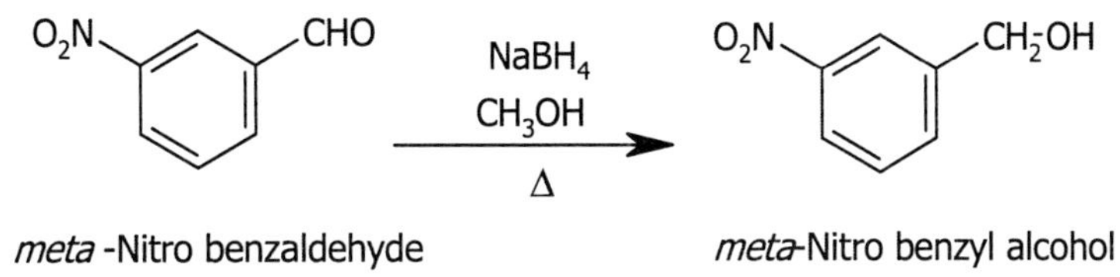

Reaction mechanism:

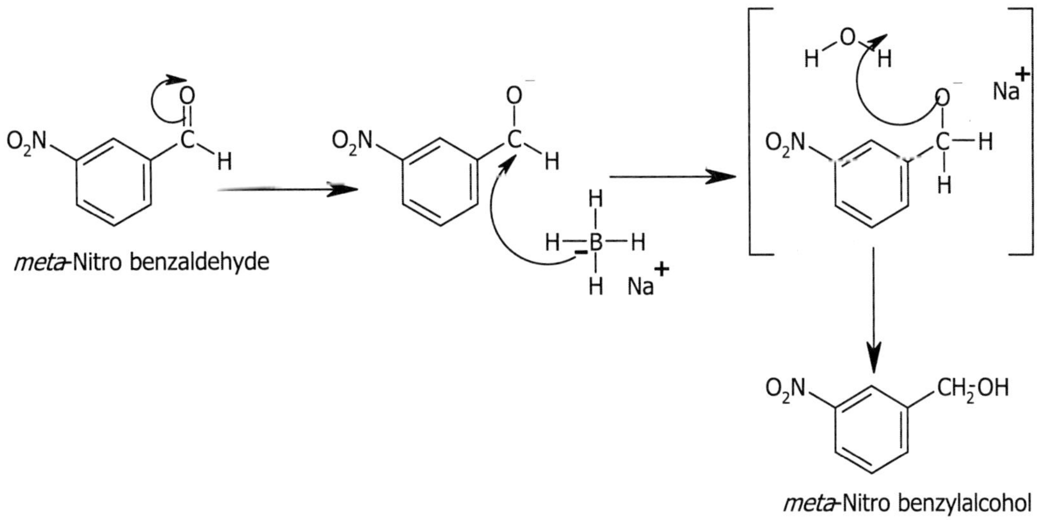

Procedure:

1. Dissolve *meta*-nitrobenzaldehyde (1.5 gm, 10 mmol) in methanol (10 ml) contained in a flask.
2. Prepare sodium borohydride solution by dissolving sodium borohydride (140 mg, 3.7 mmol) in sodium hydroxide (2ml, 2 M)] to the flask with stirring.
3. Dilute the sodium borohydride solution with 2 ml of water, 2 ml of 2 M sodium hydroxide and finally with water (18 ml).
4. Add the diluted sodium borohydride solution dropwise to the flask and maintain the temperature in between 18 to 25 °C.
5. Treat small portion of reaction mixture with dilute sulphuric acid. [after the addition of 3/4th of the dilute sodium borohydride solution].

6. Evolution of hydrogen indicates the completion of reduction.
7. Dilute the reaction mixture with ice cold water (10 ml).
8. Extract the aqueous mixture with ether (10 ml) and wash the collected ether layer with water.
9. Dry the organic layer passing through anhydrous magnesium sulphate.
10. Evaporate the organic layer to collect yellowish oil.
11. Cool the oil by placing in an ice bath and isolate the pale yellow solid of *meta*-nitrobenzyl alcohol.
12. The melting point of pure *meta*-nitrobenzyl alcohol is 30 °C.

Note:

- Ethanol vapours forms irritant orange smoke.

5.3 Preparation of N-(*m*-Nitrobenzyl) Aniline

Chemicals: N-(*m*-Nitrobenzylidene) aniline, Methanol and Sodium borohydride.

Principle: N-(*m*-Nitrobenzylidene) aniline on reduction with sodium borohydride gives N-(*m*-nitrobenzyl) aniline. Sodium borohydride is chemoselective for the carbonyl and olefin groups and doesn't react with other reducible functional groups such as nitro group. Hence nitro group present in N-(*m*-nitrobenzylidene) aniline will not undergo reduction with sodium borohydride. Sodium borohydride is stable and safer than lithium aluminium hydride. Lithium aluminium hydride generates lithium hydroxide by the reaction with water or trace of oxygen, which is more corrosive, where as sodium borohydride will not. Sodium borohydride acts as donar of hydride ion, which attacks sp^2 carbon of imine present in N-(*m*-nitrobenzylidene) aniline. The formed tetrahedral intermediate (sp^3 carbon), affords N-(*m*-nitrobenzyl) aniline.

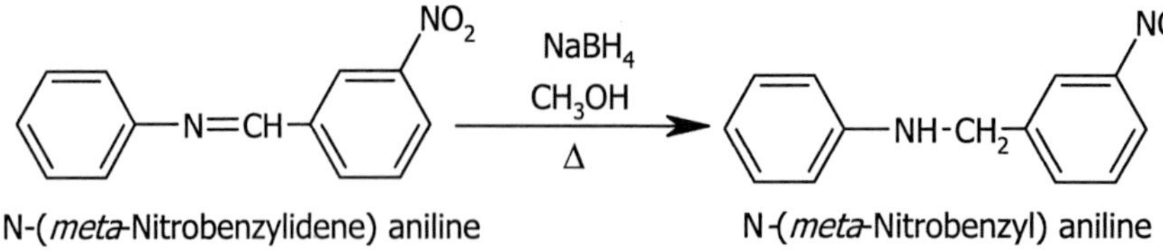

Reduction of N-(*m*-nitrobenzylidene) aniline

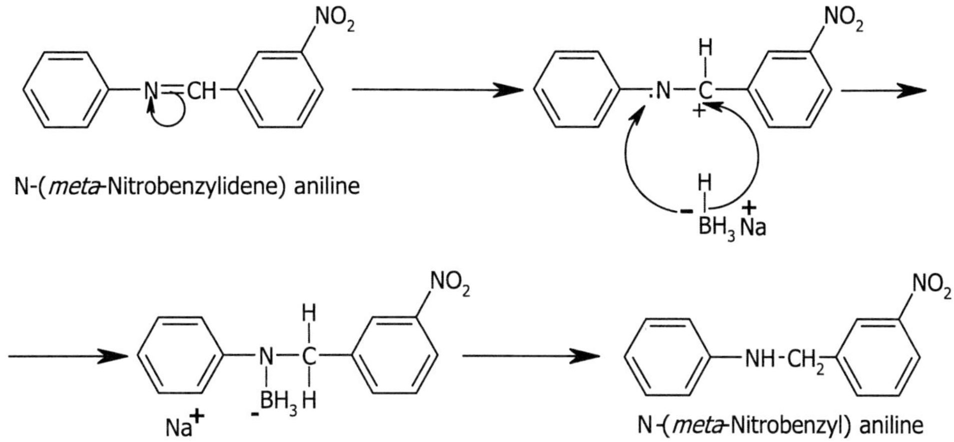

Procedure:

1. Place N-(*m*-nitrobenzylidene) aniline (schiff base; 1 gm, 4.4 mmol) and methanol (10 ml) into two necked round bottom flask.
2. Place the two necked round bottom flask attached with reflux condenser loaded with magnetic bead on a magnetic stirrer unit.
3. Warm the solution to about 40 °C and add sodium borohydride (170 mg, 4.4 mmol) with stirring over a period of 30 minutes.
4. Reflux the mixture for 15 minutes, add water (10 ml) and cool.
5. Collect the solid product separated by filtration and dry.
6. Purify the product by recrystallization with mixture of methanol and water.
7. The melting range of pure N-(*m*-nitrobenzyl) aniline is 84-85 °C.

Note: Refer chapter 22 page no. 160 for the preparation of Schiff base.

Alkali Metal Reduction

Hydrogenation is a chemical reaction between molecular hydrogen and olifinic compounds usually in presence of a catalyst. Hydrogenation has three components, the unsaturated substrate, the molecular hydrogen and catalyst. It reduces or saturates olifinic compounds. Normally non-catalytic reaction takes place in very higher temperature. Platinum, palladium, rhodium and ruthenium are highly active metal catalysts. Tin also acts as a reducing agent (sparingly).

$$R{-}CH{=}CH_2 + H_2 \xrightarrow[\text{R= alkyl, aryl}]{\text{Catalyst}} R{-}CH_2{-}CH_3$$

Zinc is important metal for the reduction of organic molecules, in the following combinations.

- Zinc and hydrochloric acid
- Zinc and acetic acid (for the reduction of α-halo ketones)
- Zinc and sodium hydroxide
- Zinc-copper couple and ethanol

Important alkali metal reductions:

Clemmensen reduction: Deoxygenation of aldehydes and ketones into alkanes by amalgamated zinc in presence of hydrochloric acid.

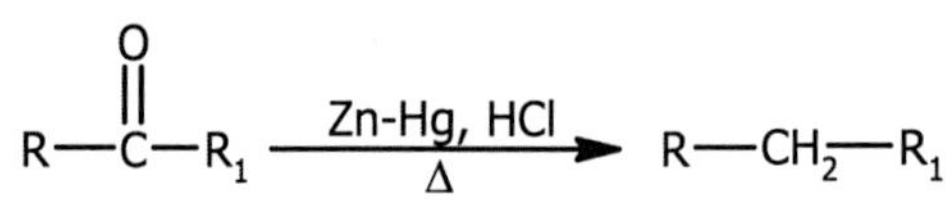

Sabatier-Senderens reduction: Catalytic hydrogenation of organic compounds in the vapour phase by passage over hot, finely divided nickel.

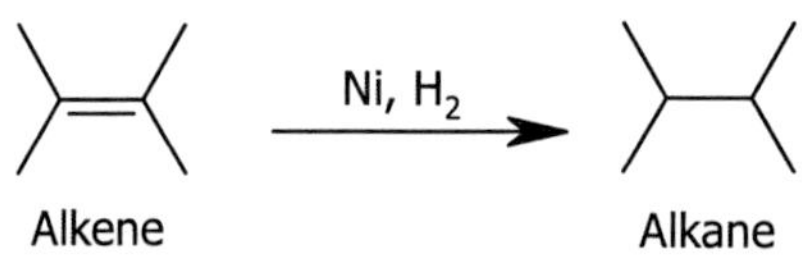

Staudinger reduction: Conversion of azides into amines by hydrogenation (mild reduction), using triphenyl phosphine. (PPh_3)

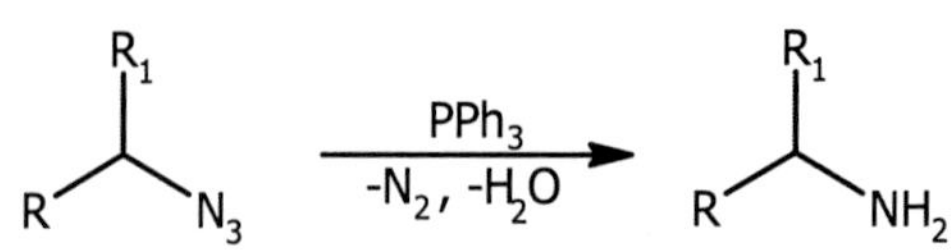

Corey-Bakshi-Shibata reduction: Enantioselective reduction of ketones into alcohols by borane and chiral oxazaborolidine catalyst.

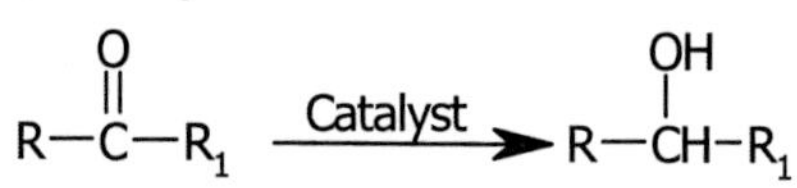

3. Dissolving metal reduction

The reduction of aromatic compounds with alkali metal in suitable solvent is known as dissolving metal reduction. Electron is transferred from alkali metal to the organic molecule and acidic solvent supplies protons to reactive intermediates. Water, alcohol and ammonia (also primary and secondary amines) act as a acidic solvent and quenches the radical anion. This type of reduction is applicable for the reduction of molecules with polarised groups such as $-C=O$, $-CN=N-$ and $-CN$.

Important dissolving metal Reduction Reactions

Benkeser reduction: Reduction of naphthalene by calcium in diethylamine-dimethyl amine is known as Benkeser reduction. It is more suitable for large scale production as it gives better yields.

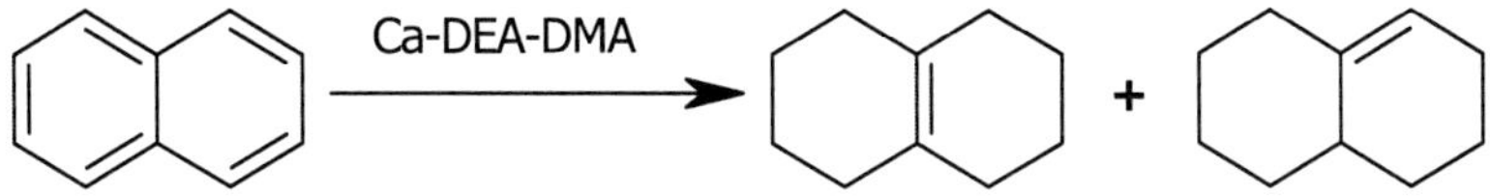

Bouveault-Blanc reduction: Reduction of esters into primary alcohols using absolute ethanol and sodium metal.

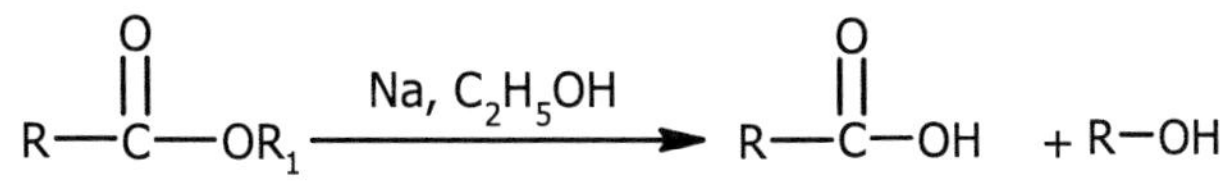

Meerwein-Ponndorf-Verley (MPV) reduction: Reduction of ketones into corresponding alcohols by aluminium isopropoxide in isopropanol is known as Meerwein-Ponndorf Verley (MPV) reduction.

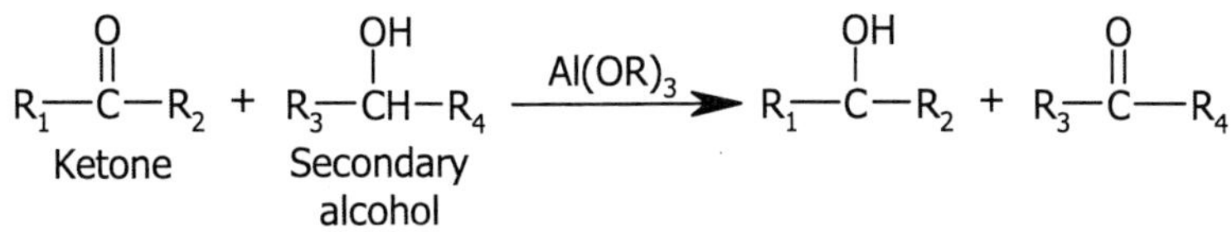

Pinacol reduction: Reduction of ketones into corresponding alcohols (pinacol) by acidic solvent is known as pinacol reduction.

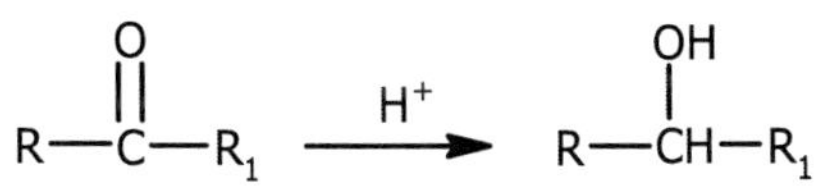

Wolff-Kishner reduction: Reduction of aldehydes and ketones (carbonyl compounds) using hydrazines and potassium hydroxide into alkanes. Hydrazine forms hydrazone with carbonyl compounds, then oxidizes into gaseous nitrogen and yields alkanes (reduction product).

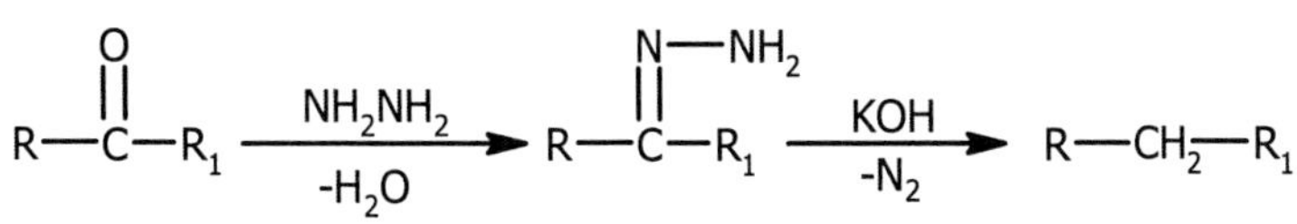

Birch Reduction

Reduction using alkali metals as the electron source are the most important synthetic transformations in organic chemistry. The dissolving metal reductions of aromatic compounds with alkali metal (sodium/lithium/potassium), alcohol (ethanol/*t*-butanol) in liquid ammonia is known as Birch reduction. Reduction of aromatic compounds (having benzenoid ring) into 1,4-cyclohexadiene (unconjucated diene) by lithium in liquid ammonia is the typical example of Birch reduction. This reaction is distinguished from catalytic hydrogenation, where aromatic ring is completely reduced into cyclohexane. These reactions are generally carried out by adding the metal to a solution of the substrate in liquid ammonia and dry methanol or ethanol or *t*-butanol (proton source). These alcohols supply a proton to the radical anion and also to the penultimate carbanion. The initial process is usually an electron transfer to form a radical anion and then second electron transfer forms dianion.

Selective Reduction of Aromatic Rings

Group 1 metals Li and Na (dissolving in liquid ammonia) ionize to give Li^+ or Na^+ and e^- $(NH_3)_n$ and produces blue colour to the solution. Blue solution with their solvated electrons is a reducing agent in Birch reduction. The reduction of NH_3 to NH_2^- and H_2 is slow and better electron acceptor will get reduced in preference (blue colour gets fade). The electrons go into lowest lying antibonding orbital of benzene, and generates radical anion. The radical anion is very basic and picks up a proton from the ethanol. The molecule turns into non-anionic, remains radical and takes up another electron, which pairs with radical to give an anion.

$$Na \longrightarrow Na^+ + e^-$$

$$\underset{\text{Blue solution}}{e^-[NH_3]_n} \longrightarrow \underset{\text{Colourless solution}}{^-NH_2 + \frac{1}{2}H_2}$$

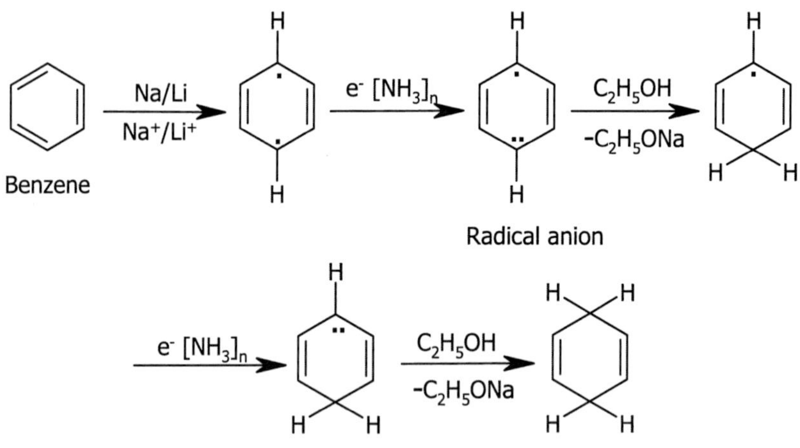

Cyclohexadiene

Regiochemistry

It is determined at protonation step, the anion delocalizes and reacts at either end to give a conjucated diene (more stable). The electron withdrawing groups promote *ipso* (site of substitution) and/*para* direction (1 and/4). The electron donating groups promote *ortho and meta* reduction. In case of anisole (having electron donating group) gives small amount (about 20%) of the conjucated product. But in aniline it is impossible to stop the isomerization and it gives conjucated enamines. Birch reduction also useful in reducing alkynes, it reduces them to *trans*-alkenes. But lindlar catalyst reduces them to *cis*-alkenes.

Pharmaceutical Applications

1. Dissolving metal reductions are used in the addition of acetyl group on carbonyl groups in case of the synthesis of citral, vitamin A and 17 α-ethinyl estradiol.

Vitamin A1

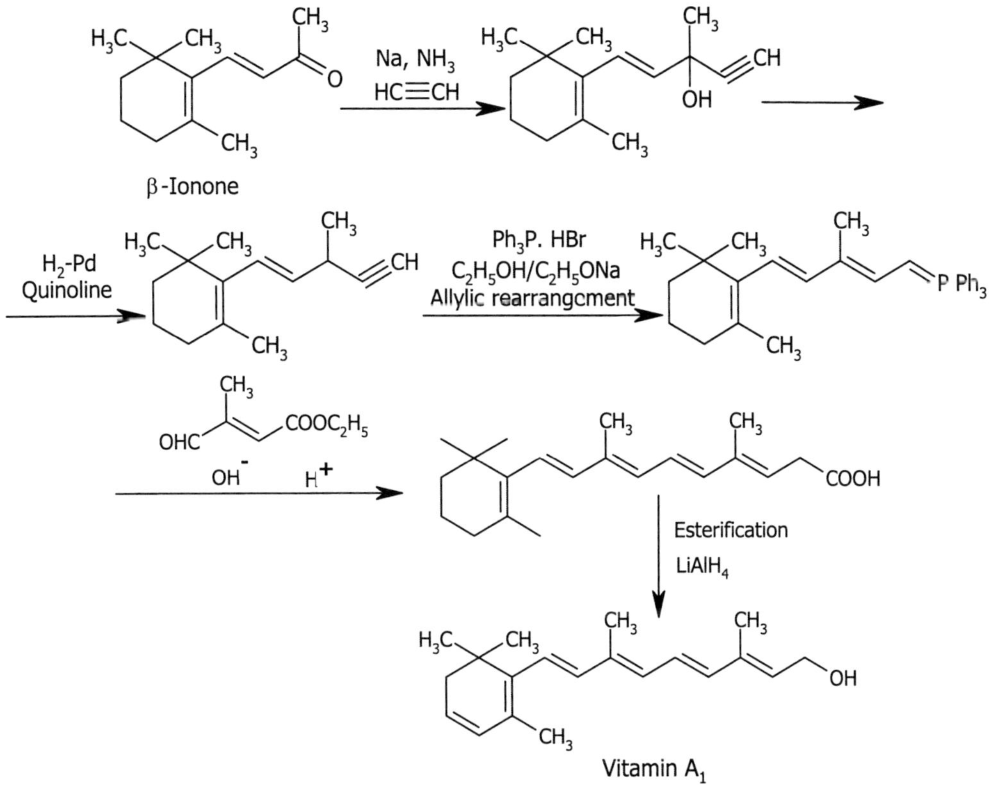

17-α-Ethinyl estradiol

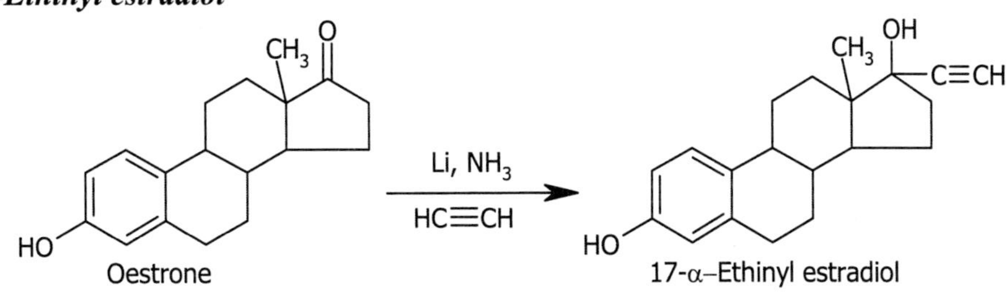

2. The most important active pharmaceutical ingredients (APIs) allylstrenol, tibolone, spironolactone, medrogestrone, lynestrenol and epicillin also can be prepared through Birch reduction.

Epicillin

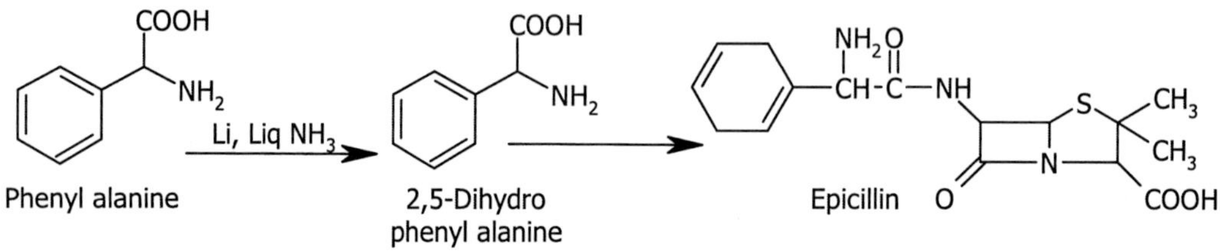

3. Useful in the reduction of 1-naphthol into 1,4-dihydro naphthol (intermediate in the preparation of β-blocker Nadolol.

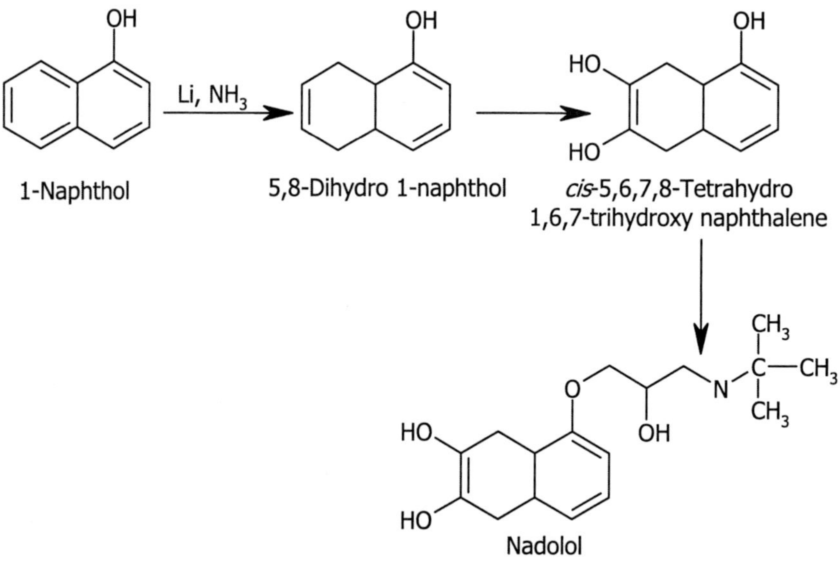

4. Used in the conversion of ephedrine into methamphetamine.
5. Birch reduction is also used for the removal of protecting groups during the synthesis of peptides. (oxytocin and glutathione).
6. Birch reductions are utilized for the reduction of terminal double bonds in the synthesis of steroids [cortisone and terpenoids (zingiberene)].
7. Preparation of cimetidine [anti-ulcer drug].

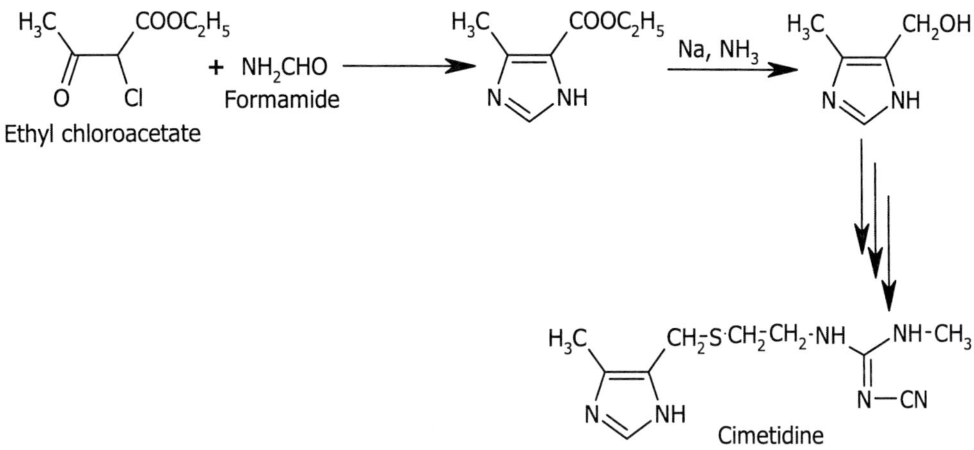

5.4 Preparation of Cyclohexa-2,5-Diene Carboxylic Acid

Chemicals: Benzoic acid, Sodium metal, Ammonia, Ethanol, Hydrochloric acid, Magnesium sulphate and Sodium chloride.

Principle: Benzoic acid upon reduction with sodium metal in presence of ammonia and ethanol produces cyclohexa-2,5-diene carboxylic acid (1,4-dihydro benzoic acid). This reaction is a typical example of Birch reduction.

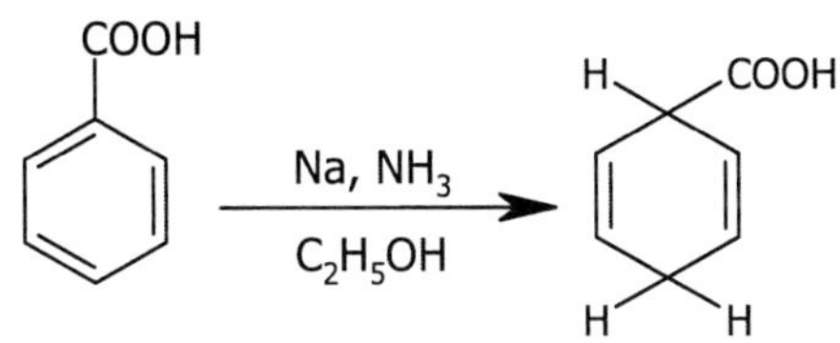

Benzoic acid — Cyclohexa-2,5-diene carboxylic acid (1,4-Dihydro benzoic acid)

Sodium metals (dissolving in liquid ammonia) ionize to Na^+ and $e^-(NH_3)_n$ and produces blue colour in the solution. Blue solution with their solvated electrons is reducing agent.

$$Na \longrightarrow Na^+ + e^-$$

$$e^- [NH_3]_n \longrightarrow {}^-NH_2 + \frac{1}{2} H_2$$

Blue solution — Colourless solution

The electrons of aromatic ring reduces ammonia to NH_2^- and hydrogen gas (blue colour dissappears), and forms sodium amide. The reduction of NH_3 to NH_2^- and H_2 is slow and better electron acceptor will get reduced in preference. The electrons go into lowest lying antibonding orbital of benzoic acid, and generates radical anion. The radical anion is very basic and picks up a proton from the ethanol. The molecule turns into non-anionic, remains radical and takes up another electron, which pairs with radical and forms pentadienyl anion. Pentadienyl anion takes up another proton from ethanol and forms 1,4-dihydro benzoic acid.

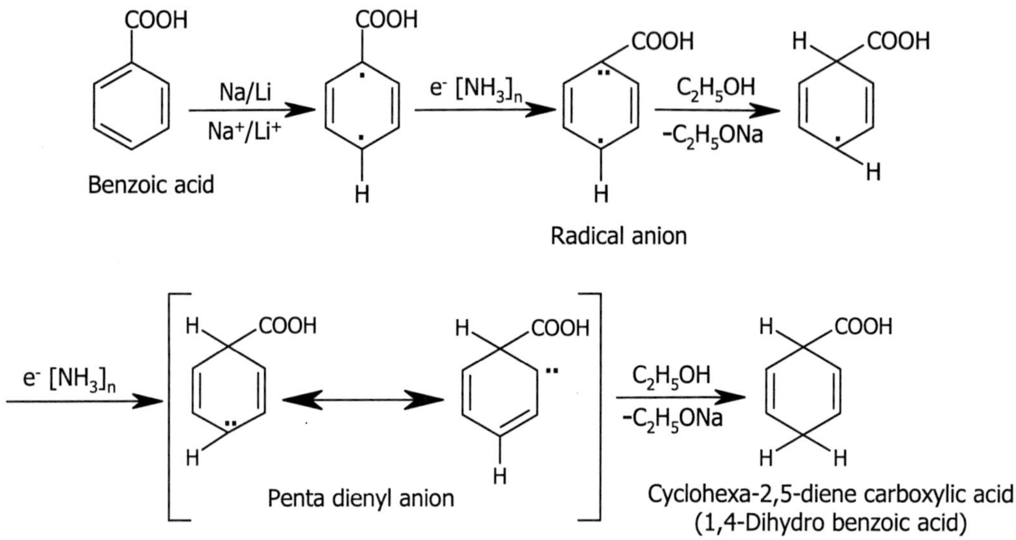

Procedure:

Preparation of cyclohexa-2,5-diene carboxylic acid (1,4-dihydrobenzoic acid)

1. Dissolve benzoic acid (1 gm, 8 mmol) in ethanol (10 ml) and liquid ammonia (60 ml).
2. Stir the solution, add sodium metal (620 mg) in small pieces, followed by of and then ammonium chloride (1.46 gm, 27 mmol).
3. Evaporate the ammonia and dissolve the residual material in ice water (50 ml).
4. Acidify the solution with 10% hydrochloric acid.
5. Extract with four 10 ml portions of ether and wash the ether layer with saturated sodium chloride solution.
6. Dry the ether layer by passing through anhydrous magnesium sulphate and concentrate.
7. Distill the pale yellow oil obtained at 96-98 °C to collect cyclohexa-2,5-diene carboxylic acid.

CHAPTER 6

HYDROLYSIS

Ester Hydrolysis

The splitting of an ester into acid and alcohol is known as ester hydrolysis (reverse of esterification). Both esterification and hydrolysis can be catalyzed by acids/bases. Generally hydrolysis is favoured by base catalysis. Base converts the acid component of ester into salt and favours the forward reaction (irreversible). Base catalyzed hydrolysis of triglycerides (esters of long chain fatty acids) are known as saponification.

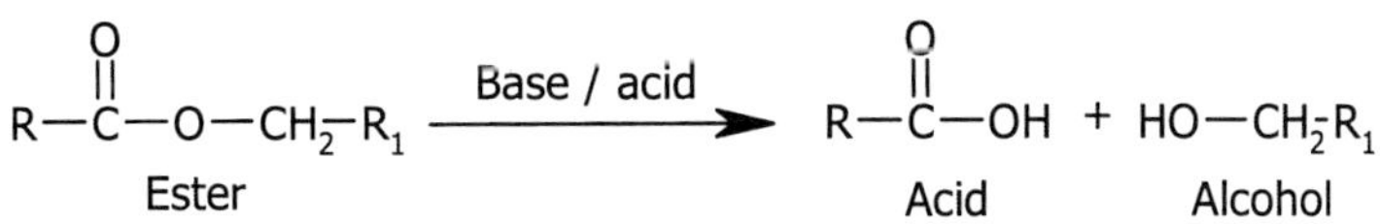

Ester hydrolysis reactions are classified into eight classes based on their reaction mechanisms.

1. AAC1: Acid catalyzed acyl cleavage unimolecular hydrolysis.
2. AAC2: Acid catalyzed acyl cleavage bimolecular hydrolysis.
3. AAL1: Acid catalyzed alkyl cleavage unimolecular hydrolysis.
4. AAL2: Acid catalyzed alkyl cleavage bimolecular hydrolysis.
5. BAC1: Base catalyzed acyl cleavage unimolecular hydrolysis.
6. BAC2: Base catalyzed acyl cleavage bimolecular hydrolysis.
7. BAL1: Base catalyzed alkyl cleavage unimolecular hydrolysis.
8. BAL2: Base catalyzed alkyl cleavage bimolecular hydrolysis.

   'A' represents acid catalysed, 'B' represents base catalysed.

   'AC' represents breaking of the bond at 'acyl' carbon (acid carbon).

   'AL' represents breaking of the bond at 'alkyl' carbon (alcohol carbon).

   '1' represents 'unimolecular mechanism', '2' represents 'bimolecular mechanism.

Acyl cleavage (AC) Alkyl cleavage (AL)

Acyl bond breaking is easier than the alkyl bond breaking. Protonation takes place on acyl oxygen atom only. Sterically hidered esters will undergo hydrolysis with difficulty.

AAC1 and AAC2: The most common type of ester hydrolysis, proceeds through the tetrahedral intermediate and acyl cleavage. Water molecule acts as proton donor (for another water molecule to form hydronium ion) as well as nucleophile.

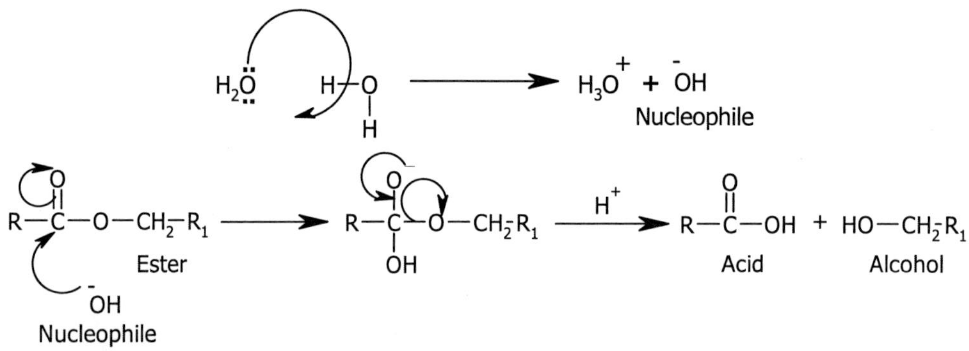

AAL1: This is the common mechanism for the hydrolysis of esters of tertiary alcohols (very rare). Tertiary alkyl, allyl and benzyl groups appears as stable carbonium ion and facilitates the reaction.

AAL2: Unlikely mechanism

BAC1: Unlikely mechanism

BAC2: This is the common mechanism for the base catalysed hydrolysis of esters, proceeds with tetrahedral intemediate with acyl cleavage.

BAL1: This occurs only in neutral/basic solutions and when 'R' comes off as a stable carbonium ion.

BAL2: This occurs in case of hydrolysis of β-lactones under neutral conditions. Cleavage of C-O opens the four membered ring and releives the strain.

Amide Hydrolysis

The splitting of an amide into acid and amine is known as amide hydrolysis.

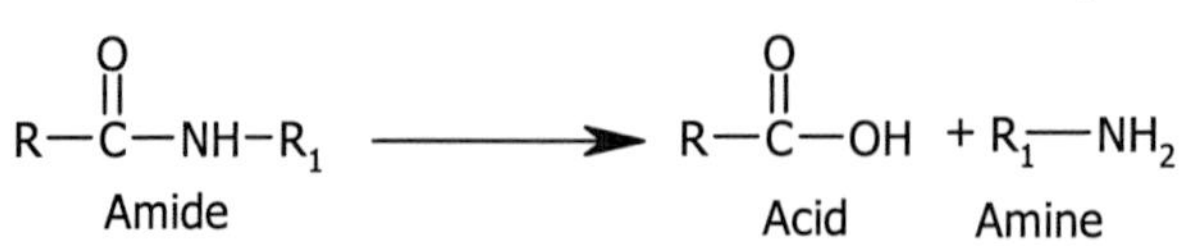

6.1 Preparation of Benzoic Acid

Traditional Name: Dracyclic acid, Carboxy benzene

Chemicals: Benzamide, Sodium hydroxide (10%) and Hydrochloric acid.

Principle: Carboxylic acids can be prepared by hydrolysis of acid derivatives (esters/ amides). Amides can be hydrolysed to carboxylic acids and ammonia or amines by heating with acid (acid hydrolysis) or base solution (base hydrolysis). Base hydrolysis of benzamide produces benzoic acid. This base hydrolysis is analogous to saponification of esters. In basic medium reaction is driven to completion by formation of the carboxylic acid salt (sodium benzoate), which upon hydrolysis gives benzoic acid. The conditions for both acid and base promoted amide hydrolysis are considerably more severe than corresponding reactions of esters, because amides are less reactive than esters.

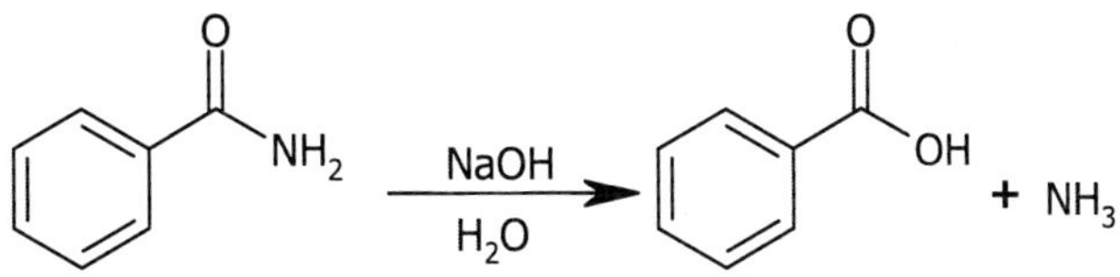

Benzamide Benzoic acid

The reaction proceeds in the following way

Nucleophilic attack on carbonyl carbon of benzamide

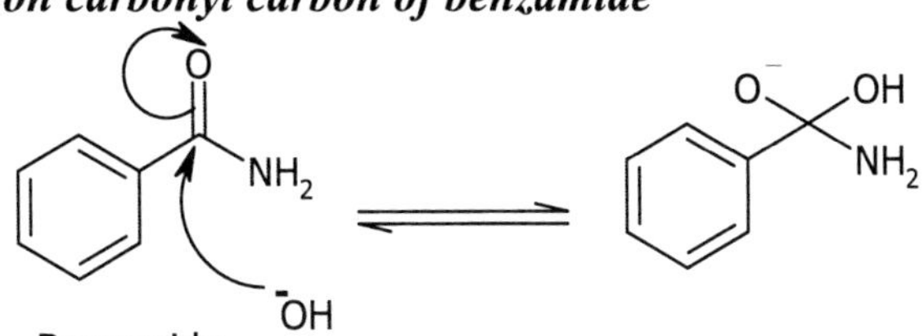

Benzamide

Formation of benzoic acid anion

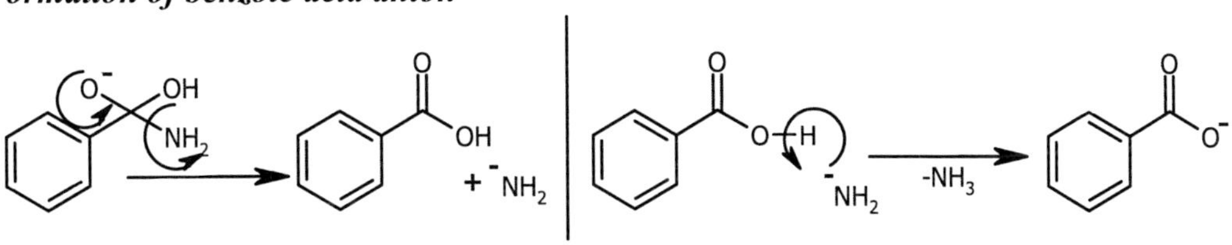

Formation of benzoic acid

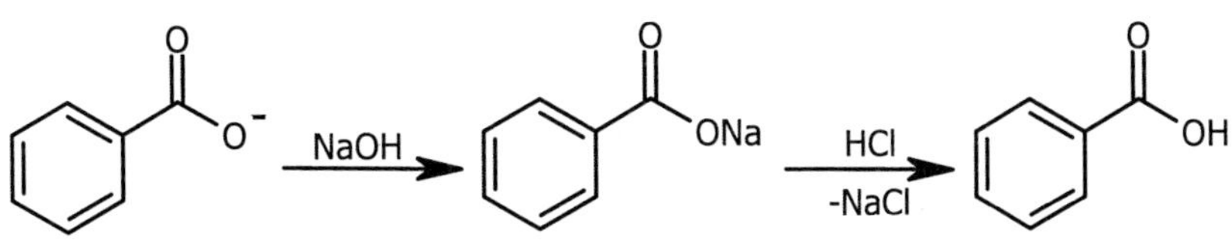

Sodium benzoate Benzoic acid

Procedure:

1. Place benzamide (1 gm, 8.3 mmol) into a round bottom flask and add aqueous sodium hydroxide (15 ml, 10%).
2. Attach reflux condenser to the flask and boil the mixture for 30 minutes, until no ammonia gas is coming out (use litmus paper to check the evolution of ammonia).
3. Cool the flask in ice-bath and acidify the mixture by the addition of hydrochloric acid with stirring.
4. Benzoic acid immediately separates out in acidic condition (as it converts soluble sodium benzoate into insoluble benzoic acid).
5. Filter under suction and wash the residue with ice-cold water.
6. Dry the residue and recrystallize it from hot water.
7. The melting point of pure benzoic acid is 121 °C.

CHAPTER 7

AROMATIC ELECTROPHILIC SUBSTITUTION

Aromatic substitution refers to the replacement of an atom (generally hydrogen) or a group attached to the benzene by another atom or group. This includes electrophilic, nucleophilic and free radical substitution reactions. Aromatic compounds most often undergo electrophilic substitution reaction. In general these reactions occurs through two steps

1. *Addition*: The π electrons of the aromatic ring system attack the electrophile and form addition product called as cationic intermediate (Meisenheimer complex).
2. *Elimination*: The complex eliminates proton and form substituted product.

Nitration, halogenation, sulphonation, Friedel-Crafts alkylation and Friedel-Crafts acylation reaction are the typical examples of aromatic electrophilic substitution reaction. Halogenation of aromatic compounds can be effected by the reaction with either of chlorine (Cl_2), bromine (Br_2) and iodine (I_2).

The five-membered heterocyclic systems, such as pyrrole, furan and thiophene are highly activated towards electrophilic attack compared to benzene. The hetero atom (nitrogen, oxygen, sulphur) with lone pair of electrons increases the stability of the cationic intermediate.

General reaction mechanism

Benzene ring with its delocalization of six π electrons in a cyclic conjugated system (above and below the plane) provides site for the attack of electrophiles. The benzene ring acts as an electron donor (Lewis base, Nucleophile) and reacts with electron acceptors (Lewis acid, Electrophile). The ability of the compound to undergo electrophilic substitution is a good test for aromaticity (bromine test). This reaction can be explained through the following steps:

***Step* 1:** *Generation of an electrophile (E^+):* Dissociation of reagent under suitable reaction conditions, generates an electrophile.

$$E—X \longrightarrow E^+ + X^-$$

***Step* 2:** *Attack of electrophile on aromatic ring:* It involves the temporary disruption of the aromatic π system to form resonance-stabilized carbocation (rate determining step).

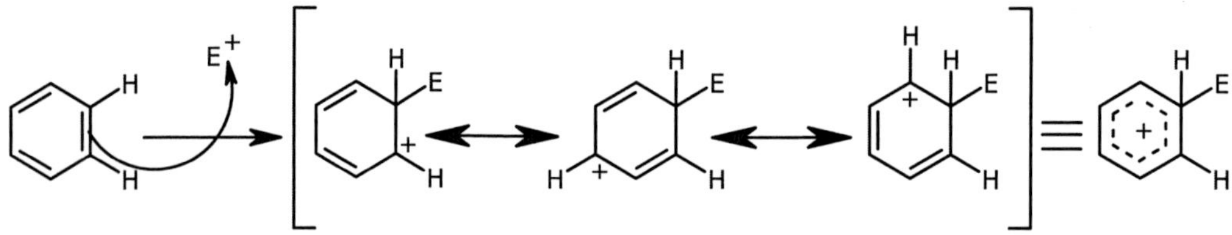

***Step* 3:** Deprotonation by the action of base to give substitution product.

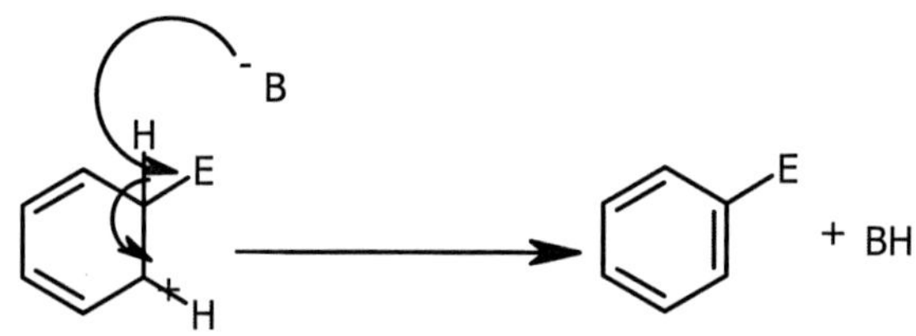

IPSO substitution: A special kind of electrophilic aromatic substitution where the leaving group is not hydrogen is called *ipso substitution.* Synthesis of picric acid by the reaction of salicylic acid with nitric acid and sulphuric acid is an example for *ipso substitution* reaction.

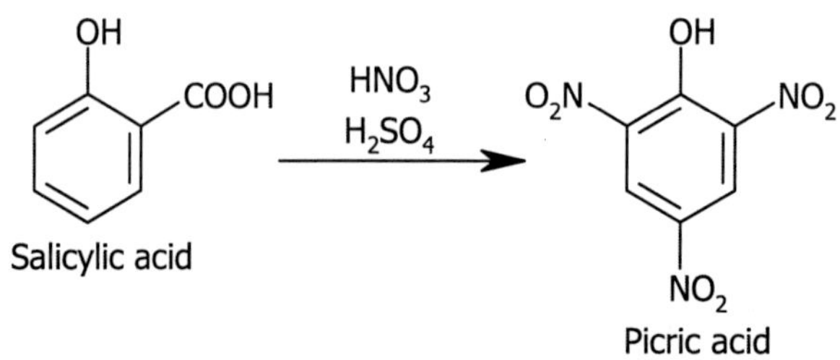

Effect of groups on electrophilic aromatic substitution

The nature of substituent present in the aromatic ring system decides the position of electrophilic attack. These groups are broadly divided into:

- Ring activating groups: *Ortho, para* director
- Ring deactivating groups
  - *Meta* directors.
  - *Ortho, para* directors.

Ortho/para directors: A group that causes attack to occur chiefly at *ortho* and *para* positions is called as *ortho*, *para* director. The presence of an electron-releasing (donating) group (x) makes *ortho*- and *para*-positions of the ring electron rich and makes it reactive site for attack of electrophile. The resonance stabilization of positive charge will be optimized at *ortho and para* position.

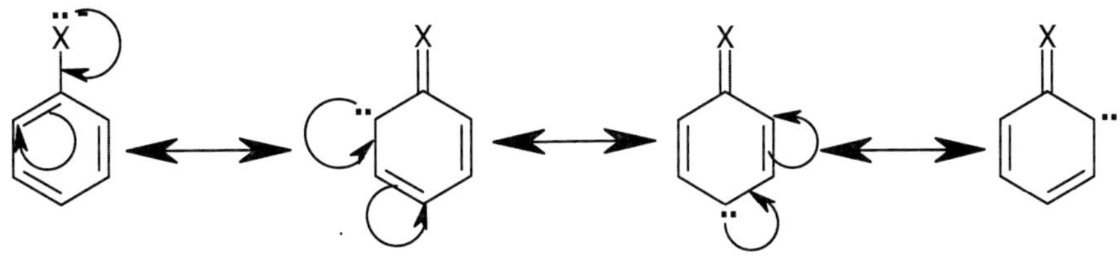

X= Elelctron releasing (donating) groups

Meta directors: A group that causes attack to occur chiefly at *meta* position is called as *meta* director. The electron withdrawing substituent (Y) on the ring favours the electrophilic attack prefentially to *meta* position. These groups produce *meta* position electron rich than the *ortho* and *para* positions (electron deficient) and stabilize resonance charge at *meta* position.

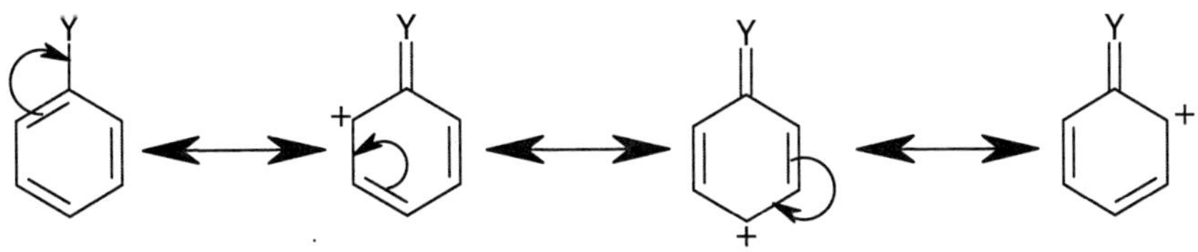

Y= Elelctron withdrawing (accepting) groups

Ring activating groups (*Ortho and para* directors)

Strongly activating: $-NH_2$, -NHR, $-NR_2$, -OH

Moderately activating: $-OCH_3$, $-OC_2H_5$, $-NHCOCH_3$

Weakly activating: $-C_6H_5$, $-CH_3$, $-C_2H_5$, etc

Ring deactivating groups (*Meta* directors)

$-NO_2$, $-N(CH_3)_3^+$, -CN, -COOH, -COOR, $-SO_3H$, -CHO, -COR

Ring deactivating groups (*Ortho and para* directors)

-F, -Cl, -Br, -I

Role of catalysts in aromatic electrophilic substitution

Lewis acids [anhydrous $AlCl_3$ and $FeCl_3$] and mineral acids (HCl, HNO_3, H_2SO_4) are the catalysts used in aromatic electrophilic substitution reactions. They help in the generation of electrophile. The catalyst should be electron deficient to cause polarization of the attacking molecule. Anhydrous $FeCl_3$, and $FeBr_3$ acts as a catalyst for halogenations reactions. Anhydrous $FeCl_3$ (or $FeBr_3$) takes up Cl^- ion from the Cl_2 and gives Cl^+ (electrophile). Anhydrous $AlCl_3$ is suitable for Friedel-Crafts alkylation and Friedel-Crafts acylation.

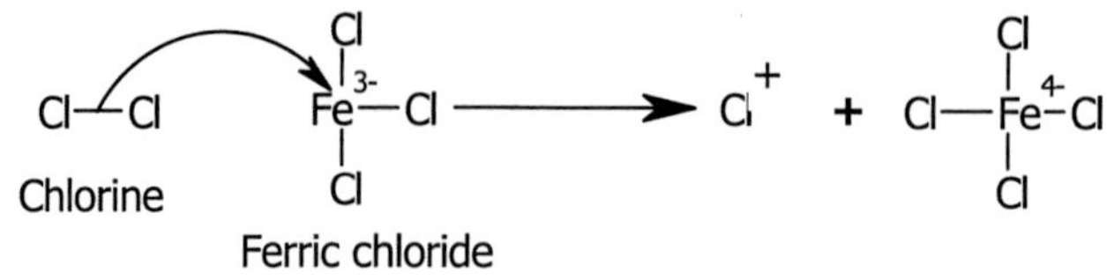

7.1 Preparation of Nitrobenzene

Traditional Name: Oil of mirbane, Essence of mirbane

Chemicals: Benzene, Nitric acid, Sulphuric acid, Sodium hydroxide (1N) and Calcium Chloride.

Principle: Preparation of nitrobenzene involves aromatic electrophilic substitution reaction. Mixture of concentrated nitric acid and sulphuric acid called as mixed acid reagent. Nitric acid generates nitronium ion, NO_2^+ (an effective nitrating agent) using sulphuric acid as a catalyst. Nitronium ion (electrophile) introduced into the benzene by electrophilic substitution forms nitrobenzene. Nitrobenzene is an important drug intermediate used in the manufacture of paracetamol, phenacetin and sulphonamides. This nitration reaction is also involved in the preparation of drugs, such as mebendazole, primaquine, pamaquine and metronidazole.

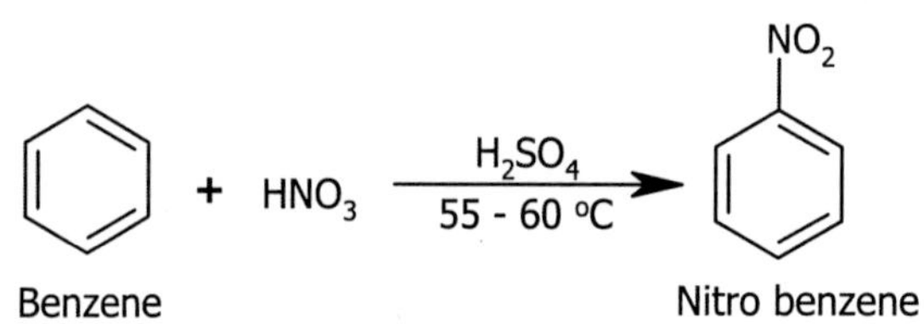

This reaction proceeds in the following way

Reaction Mechanism

Generation of the electrophile: Sulphuric acid protonates nitric acid and generates electrophile (nitronium ion [NO_2^+]) with the elimination of water molecule.

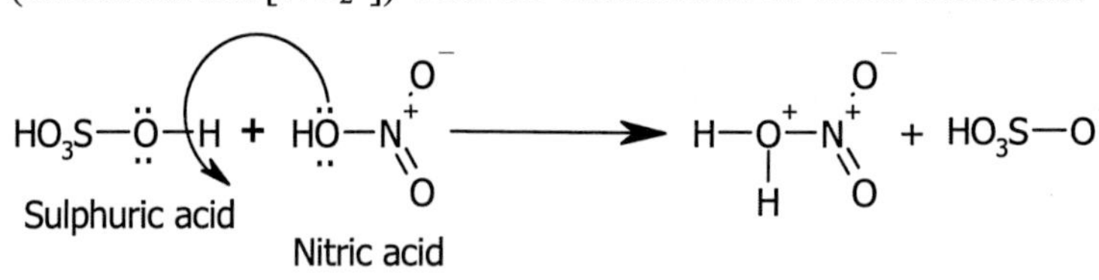

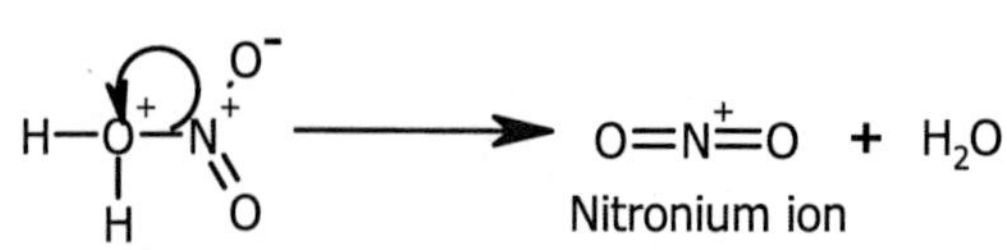

Electrophilic attack on π electrons of benzene: Nitronium ion (electrophile) attacks π electrons of benzene and forms a carbocation through resonance.

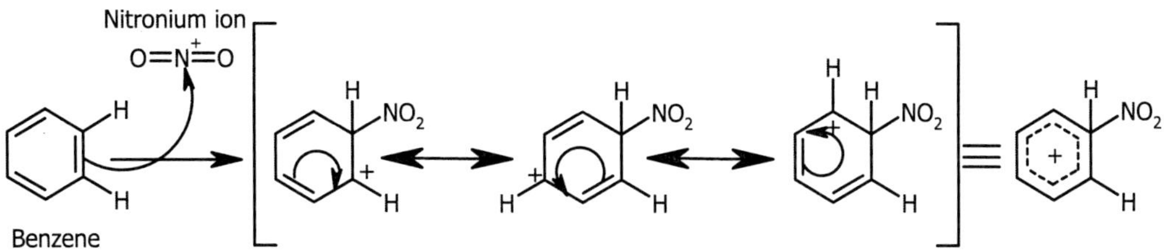

Deprotonation of carbocation: Loss of proton (H^+) from the carbocation generates nitrobenzene.

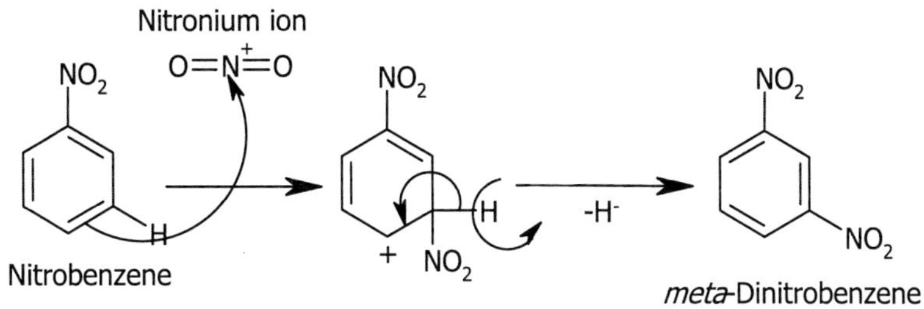

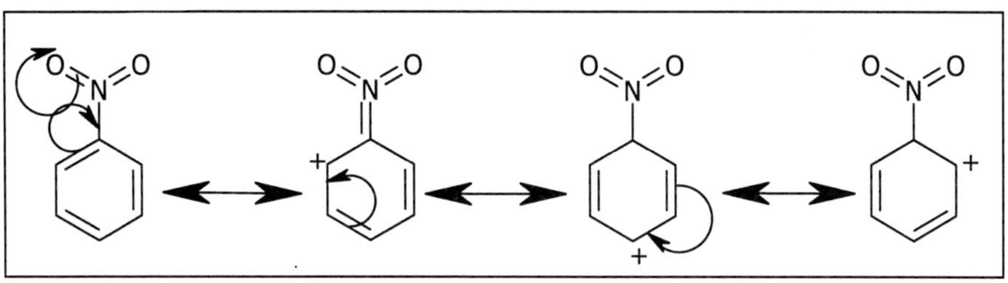

$$HO_3S-O^- + H^+ \longrightarrow HO_3S-\ddot{O}-H$$

Sulphuric acid

Procedure:

1. Place nitric acid (3.5 ml) in a round bottom flask and add sulphuric acid (4 ml) to the flask drop wise with shaking.
2. Keep the mixture in cool condition during the addition by immersing the flask in an ice bath (to ensure the temperature below 55 °C).
3. Add benzene (3 ml, 33 mmol) in small portions with shaking to ensure thorough mixing after each addition and maintain the temperature below 55 °C.
4. Attach the reflux condenser to the flask and reflux the mixture in a water bath maintaining the temperature at 55-60 °C for 45 minutes.

5. Shake the flask occasionally to ensure good mixing of two immiscible (benzene and acid mixture) layers.
6. Pour the contents into cold water (50 ml) containing beaker, stir the mixture well in order to remove the acid and then decant the aqueous layer.
7. Add sodium hydroxide (3 ml, 1N) to the mixture and then collect the organic layer with the help of separating funnel.
8. Pour the organic layer into a conical flask containing anhydrous calcium chloride (500 mg) to dry the organic layer.

   (*If necessary warm the mixture to get clear solution*).
9. Filter and distill at 210 °C to collect the pure form of nitro benzene.

Precautions: The temperature should not exceed 55 °C, as it favors the polynitration (formation of dinitro benzene) and oxidative breakdown of aromatic ring.

7.2 Preparation of *meta* Dinitrobenzene

Chemicals: Nitrobenzene, Nitric acid, Sulphuric acid.

Principle: *meta* Dinitrobenzene preparation involves aromatic electrophilic substitution of nitro benzene with nitronium ion (electrophile). Nitric acid component of mixed acid generates nitronium ion, NO_2^+ (an effective nitrating agent) and sulphuric acid acts as a catalyst. Nitronium ion (electrophile), introduced into the nitrobenzene by electrophilic substitution and forms di-nitrobenzene. Nitro group (*meta*-directing group) of nitro benzene directs the incoming electrophile (NO_2^+) into *meta* position and generates *meta*-dinitrobenzene.

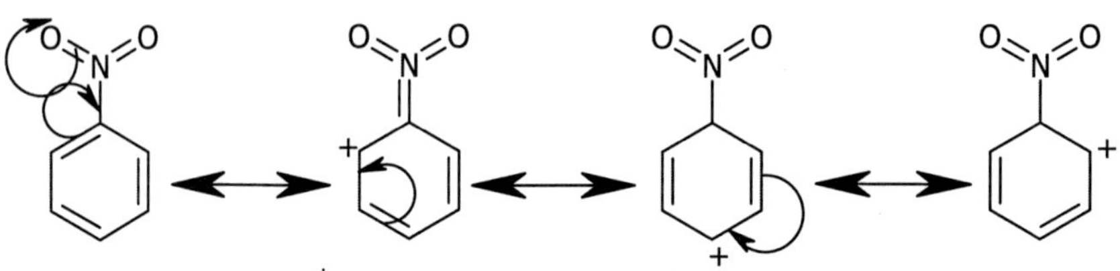

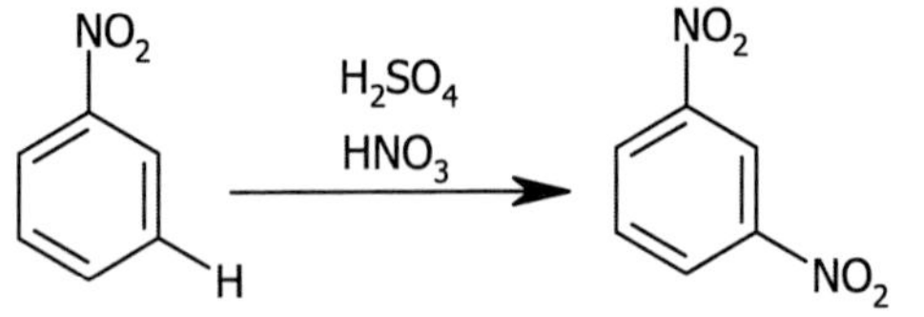

Nitrobenzene *meta*-Dinitrobenzene

$$HO_3S{-}O^- + H^+ \longrightarrow HO_3S{-}\ddot{\underset{..}{O}}{-}H$$

Sulphuric acid

Procedure:

1. Place sulphuric acid (7 ml) and fuming nitric acid (5 ml) into round bottom flask.
2. Add few boiling chips to the flask and place it in a fume cupboard.
3. Add nitrobenzene (4.3 ml) an small portions to ensure thorough mixing.
4. Attach reflux condenser and heat the mixture [with occasional shaking] by placing on a water bath for 30 minutes.

   [*heat until the evolution of brown nitric oxide gas ceases*]
5. Allow the mixture to cool to room temperature and pour into beaker containing cold water (170 ml) with vigorous shaking.
6. Collect the solid product by vacuum filtration and wash with cold water.
7. Transfer the isolated solid into flask and add rectified spirit (30 ml).
8. Attach reflux condenser and heat the contents to dissolve the solid product.

   [*filter if necessary through preheated Buchner funnel to obtain clear solution*].
9. The colourless crystals of *meta*-dinitrobenzene deposits upon cooling.
10. The melting range of pure *meta* dinitrobenzene is 89-90 °C.

7.3 Preparation of *Para* Bromoacetanilide

IUPAC Name: N-Acetyl *p*-bromoaniline.

Chemicals: Acetanilide, Bromine, Glacial acetic acid and Sodium metabisulphite.

Principle: Acetanilide reacts with bromine in presence of glacial acetic acid and forms *para*-bromoacetanilide. It is very difficult to obtain mono bromo substitution product in case of aniline. High reactivity of aniline leads to (2,4,6-tribromo aniline). Slow addition of phenol in carbon disulfide and maintaining the temperature below 5 °C produces mainly *para* bromo phenol. Converting aniline into less reactive by N-acetylation (simple acylation of amine to amide) favours selective aromatic electrophilic substitution.

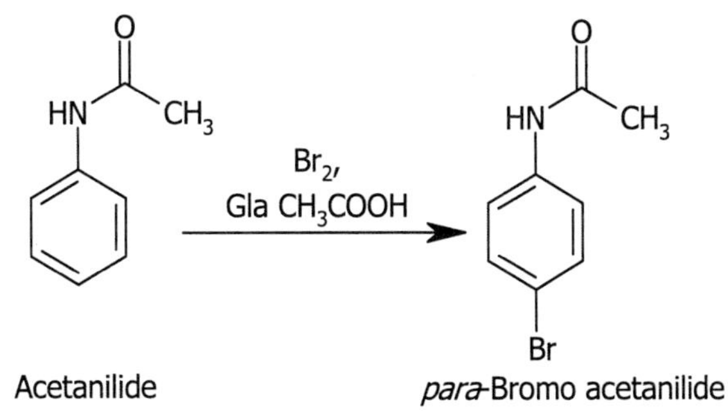

The lone pair of electrons on the nitrogen atom of the amide is conjugated with carbonyl group. But the electrons are delocalized into the benzene ring, more weakly than in amine. This resonance controls the aromatic electrophilic substitution. Reaction still occurs in the *ortho* and *para* postions (mainly *para*), but it occurs chiefly in one position only. Theoretically the possibility of *ortho* product is twise as much as *para* product because there are two ortho positions. But steric hindrance produced by acetamido group inhibits *ortho* substitution and favours *para* substitution.

$$Br_2 \longrightarrow \overset{+}{Br} + \overset{-}{Br}$$

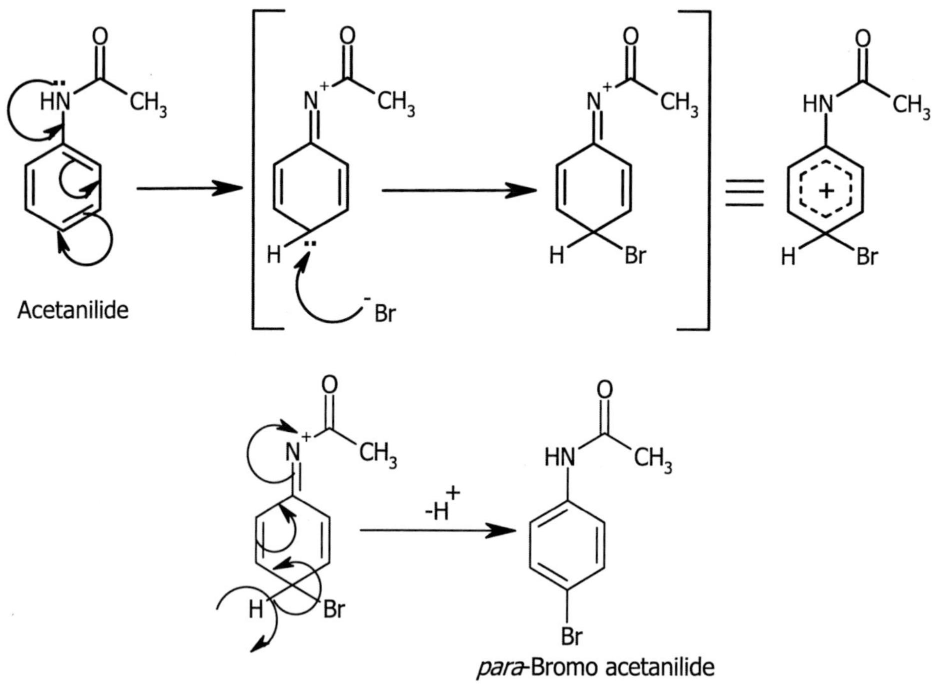

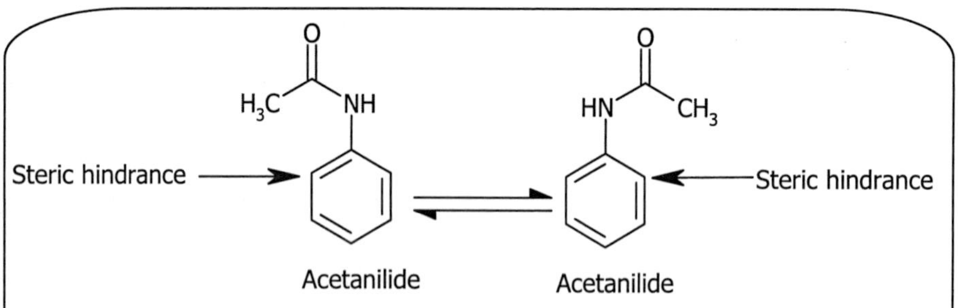

Steric hindrance favours substitution at *para* position

Acetanilide undergoes bromination and produces mixture of *ortho* and *para* bromo acetanilides. The *ortho* bromoacetanilide forms only small amount and is more soluble in ethanol than *para* bromoacetanilide. Hence *ortho* bromoacetanilide can be readily eliminated by recrystallization with ethanol.

Procedure:

1. Place acetanilide (1.35 gm, 10 mmol) in a flask and dissolve with cold glacial acetic acid (4.5 ml).
2. Prepare the solution of bromine (0.53 ml, 11 mmol) in glacial acetic acid (25 ml) and transfer it into a dropping funnel or burette placed.
3. Add the bromine solution slowly to the flask (place in cold water) with constant shaking to ensure thorough mixing.
4. After complete addition, an orange yellow colour will appear, due to excess of bromine and part of product may crystallize out.
   (If the orange yellow colour produced is strong, add sodium metabisulphite to remove the colour)
5. Allow the reaction mixture to stand at room temperature for about 30 minutes and pour the mixture into cold water (40 ml) containing beaker.
6. Filter the precipitate under suction and wash with cold water.
7. Dry and recrystallize the product from ethanol.
8. The melting point of pure *para*-bromo acetanilide is 167 °C.

Precautions:

1. Bromine produces painful burns and the vapors are unpleasant. Hence it should be handled in the fume cupboard with great care.
2. Bromine burns should be treated immediately with glycerin.
3. If the vapour is inhaled relief may be obtained by holding hand kerchief soaked in alcohol near to the nose.

CHAPTER 8

BECKMANN OXIME REARRANGEMENT

Migration of chemical unit (an atom, an ion, a group of atoms) from one atom to another atom (same/different molecule) is known as rearrangement reactions. Most commonly known rearrangement is Whitmore 1,2-shift, where one atom migrates to adjacent atom.

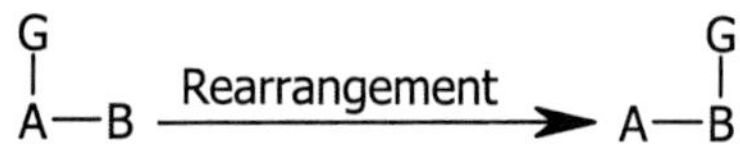

A = Migration origin,
B= Migration terminus,
C= Migrating group.

An acid catalyzed transformation of ketoxime (cyclic and acyclic) into corresponding amide is termed as Beckmann oxime rearrangement. Acidic catalysts namely sulphuric acid, formic acid, liquid sulphur dioxide, hexa methyl phosphoramide (HMPA), thionyl chloride, Phosphorous pentaoxide (P_2O_5), methane sulphonic acid and hydrochloric acid, acetic acid and acetic anhydride promotes this reaction.

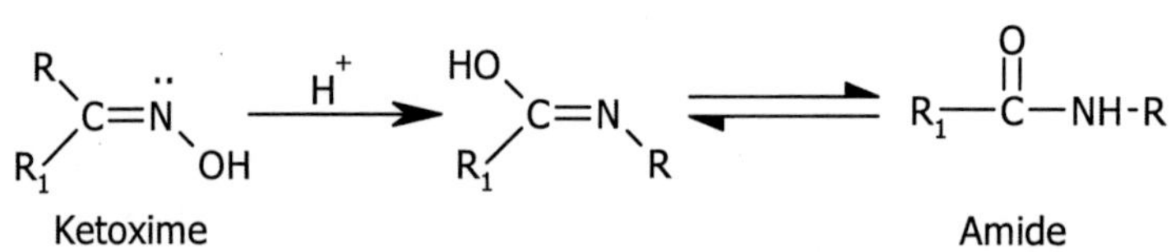

Reaction mechanism

Acid converts -OH group into better leaving group (proton acid protonates to form H_2O^+, Phosphorous pentachloride (PCl_5) forms ester leaving group-$OPCl_4$, Sulphuric acid (H_2SO_4) forms ester leaving group-OSO_2OH. Detachment of leaving group generates electron deficient nitrogen species (loss of water or loss of anion results). Migration of the group –R to the electron deficient nitrogen occurs and forms carbocation. Attack of water on carbocation, loss of proton from the intermediate and tautomerisation yields amide compound.

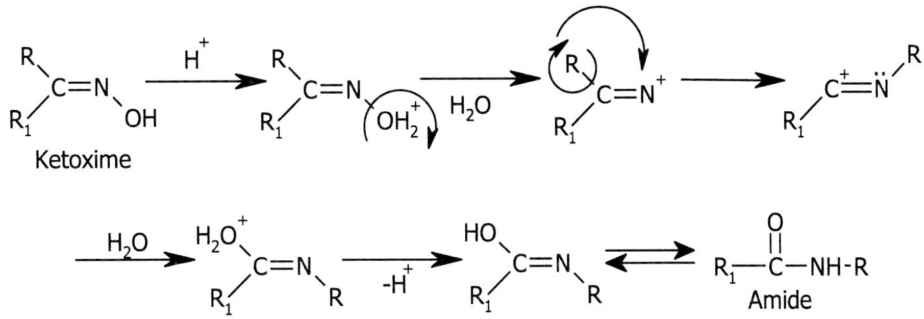

Stereospecificity

The migration of the group depends upon the orientation of the group. The migrating group is always *anti* (i.e., *trans*) to the hydroxyl group (ketoxime) and this reaction is streospecific in nature. In the *isomer*-1 of 2-bromo-5-nitroacetophenone oxime, -OH and -Br groups are close enough for cyclization reaction to occur in cold sodium hydroxide. Due to the migration of *anti*-CH_3 group to the nitrogen atom (Beckmann oxime rearrangement) it gives the product 2.

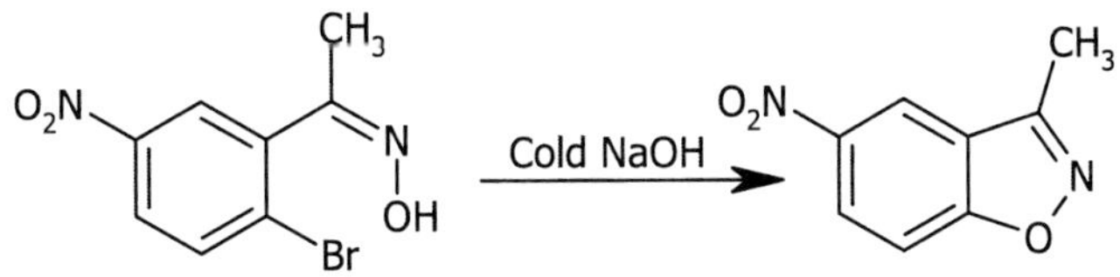

Isomer 1 Product 1

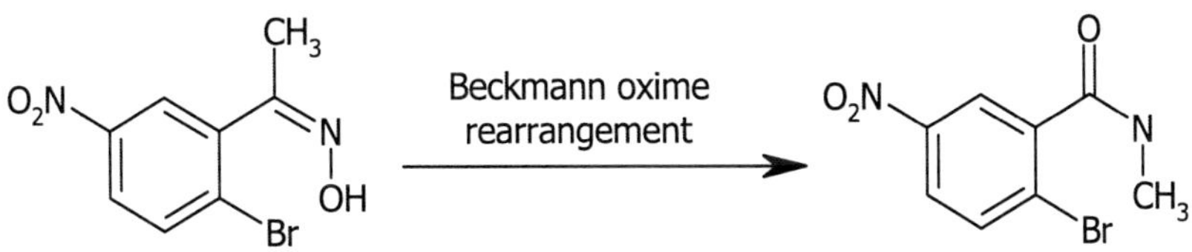

Isomer 1 Product 2

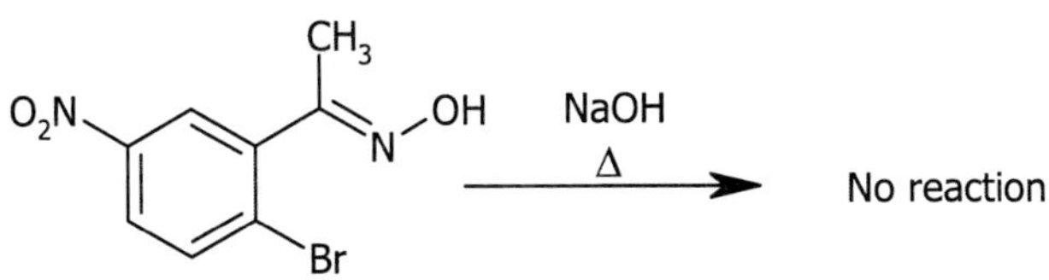

Isomer 2

In the *isomer* 2, -OH and -Br groups are faraway and hence no cyclization reaction even heating with sodium hydroxide. Due to the migration of *anti* aryl group to the nitrogen atom it gives the product 3.

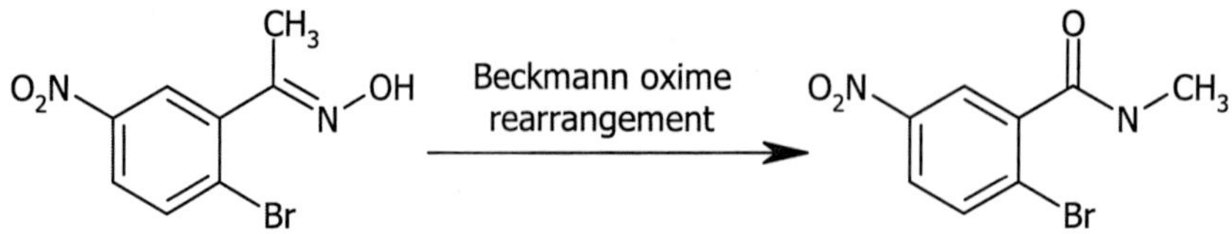

Isomer 2 Product 3

Oximes where both R and R_1 are alkyl groups, mixture of two amides will form, due to the isomerization (before migration). In the Beckmann oxime rearrangement of unsymmetrical ketones, there are two groups that could migrate. Hence there are two possible *geometrical isomers* for the unsymmetrical oxime (due to *cis/trans* isomers of C=N bond), this leads to the generation of mixture of products. But the ratio of the products is exactly the ratio of geometrical isomers in the starting materials. The group migrated in each case is the group *trans* to the -OH in the starting materials. Some aldoximes undergo rearrangement in the presence of poly phosphoric acid (PPA), but the reaction is not a general rule.

Pharmaceutical applications

1. *Anti* (*trans*) group migration strategy is used in the determination of the configuration of organic compounds.
2. Azalides are semi synthetic macrolides in which nitrogen is introduced into a macrolactone ring *via* Beckmann oxime rearrangement.
3. The semisynthesis of hormones such as oestrone and testosterone involves Beckmann oxime rearrangement strategy.

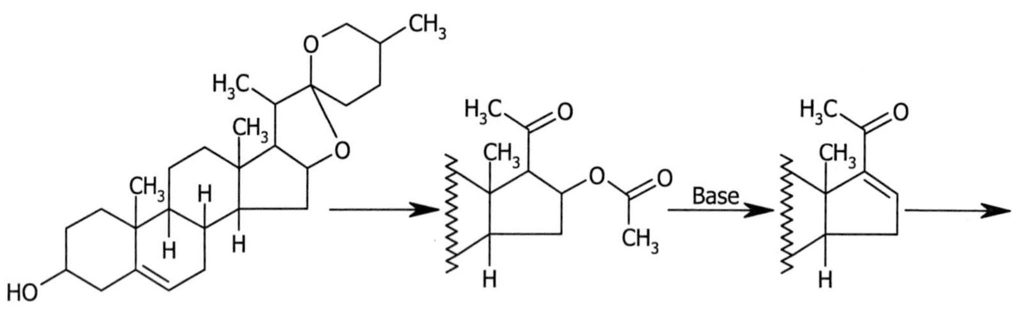

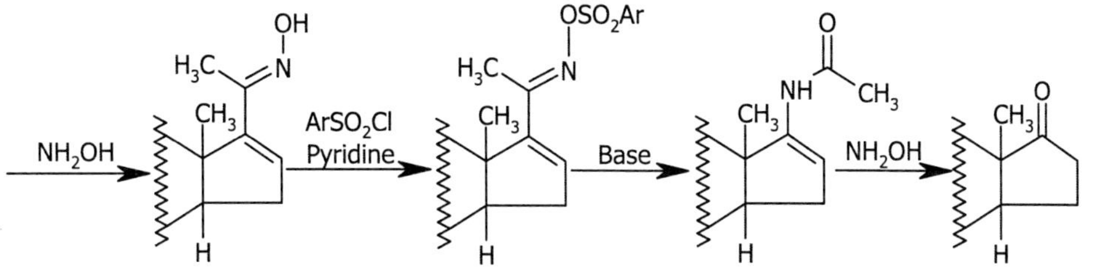

D-Ring of testosterone and oestrone

4. Beckmann oxime rearrangement strategy is involved in the synthesis of isoquinoline

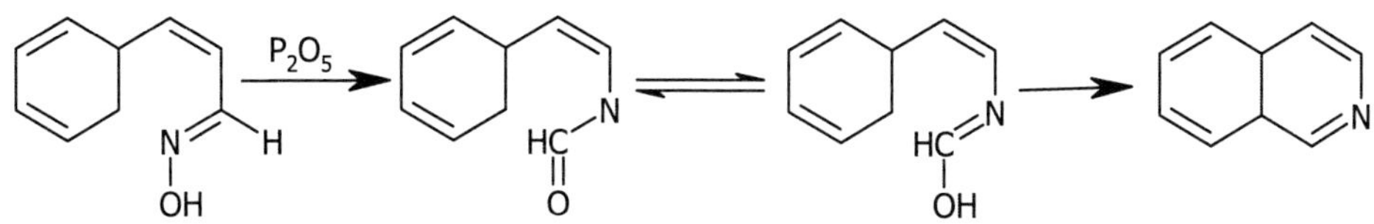

8.1 Preparation of Benzanilide

Traditional Name: N-Benzoyl phenylamine, N-Benzoyl aniline

***IUPAC Name*:** N-Phenyl benzamide

Chemicals: Benzophenone, Hydroxyl amine hydrochloride, Rectified spirit, Sodium hydroxide pellets, Hydrochloric acid (dilute) and Methanol.

Principle: Benzanilide can be prepared from benzophenone oxime. Benzophenone on condensation with hydroxyl amine hydrochloride in the presence of excess of sodium hydroxide yields benzophenone oxime. Due to the polarization, carbonyl oxygen of benzophenone will bear partial negative charge, and allows protonation to occur. The positively charged carbonyl carbon (electrophile) will be attacked by electron rich nitrogen of hydroxyl amine (nucleophile) and forms an intermediate. Removal of hydrogen from nitrogen followed by the dehydration generates benzophenone oxime. The optimum pH for this reaction is 7. Strong acidic condition converts hydroxyl amine into conjugate acid and reduces the concentration of nucleophilic hydroxyl amine and results in no reaction.

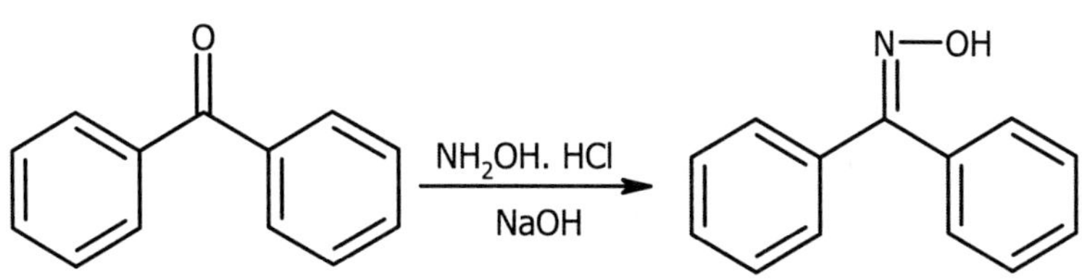

An acid catalyzed transformation of benzophenone oxime into corresponding amide (benzanilide) occurs by Beckmann oxime rearrangement. Benzophenone oxime on treatment with phosphorous penta chloride (PCl_5) in anhydrous ether forms benzanilide. Phosphorous penta chloride (PCl_5) converts -OH group into better leaving group -OPCl_4 (ester). Detachment of leaving group generates electron deficient nitrogen species (loss of water or anion). Migration of the phenyl group to the electron deficient nitrogen occurs and forms carbocation. Addition of water on carbocation, loss of proton from the intermediate and tautomerisation yields benzanilide.

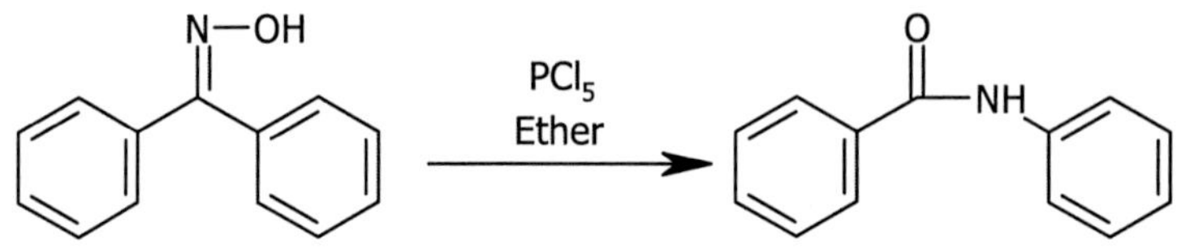

Benzophenone oxime Benzanilide

This reaction proceeds in the following way

Formation of benzophenone oxime

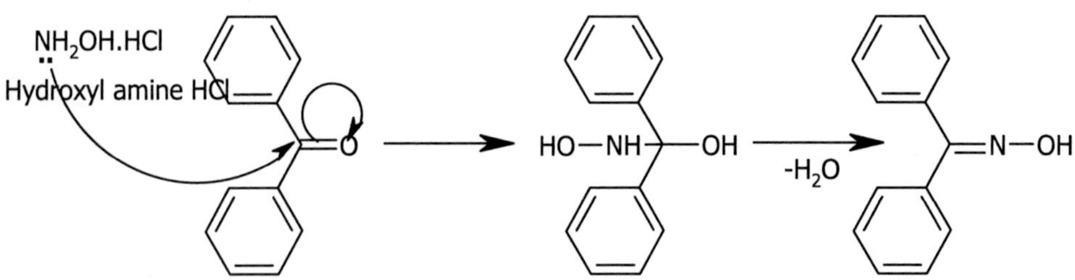

Benzophenone Benzophenone oxime

Formation of benzanilide

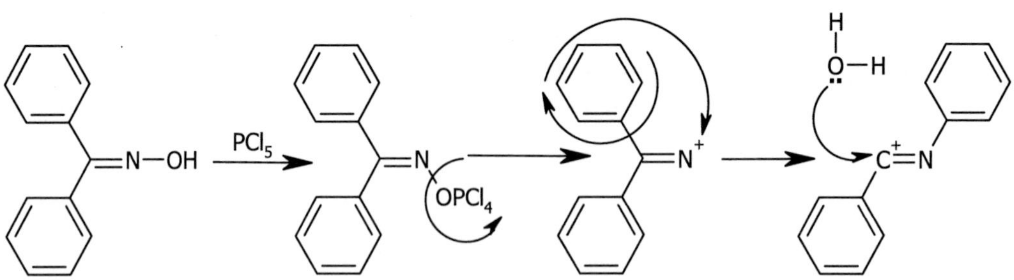

Benzophenone oxime Carbocation

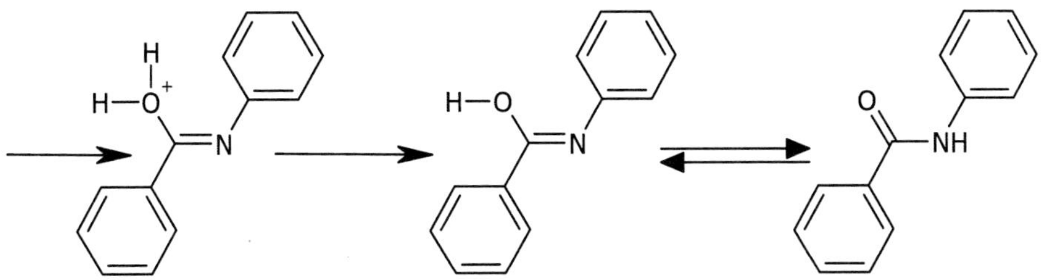

Procedure:

***Step* 1: Preparation of Benzophenone oxime**

1. Place benzophenone (2.5 gm, 13.7 mmol), hydroxyl amine hydrochloride (1.5 mg, 22 mmol), rectified spirit (4 ml) and water (5 ml) in a round bottom flask.
2. Add sodium hydroxide pellets (2.8 g, 70 mmol) in small portions with shaking. (If the reaction becomes too vigorous, cool the flask).
3. Attach the reflux condenser and reflux for 5 to 10 minutes.
4. Cool and pour the contents into beaker containing dilute hydrochloric acid (7.5 ml of hydrochloric acid in 50 ml of water).
5. Filter the precipitate, wash with cold water and dry in an oven at 40 °C (or in a desiccator).
6. Purify the product by recrystallization with methanol.
7. The melting point of pure benzophenone oxime is 142 °C.

Precautions:

- Hydroxyl amine hydrochloride causes skin irritation.
- Traces of oxygen or moisture convert oxime into benzophenone and nitric acid.
- Hence, it should be stored in vacuum desiccator filled with dry carbon dioxide or nitrogen.

***Step* 2: Preparation of Benzanilide**

1. Dissolve benzophenone oxime (1 gm, 5 mmol) in anhydrous diethyl ether (10 ml) in a small conical flask and add powdered phosphorous pentachloride (1.5 gm, 7.2 mmol) (*alternatively thionyl chloride (1.5 ml) can be used*).
2. Distil off the solvent and other volatile products (using water bath on a rotary evaporator).
3. Add water (15 ml) and boil to break up any lumps.
4. Decant the supernatant liquid and purify the product by recrystallization with boiling ethanol.
5. The melting point of pure benzanilide is 163 °C.

8.2 Preparation of Acetanilide

Traditional Name: Antifebrin. ***IUPAC Name*:** N-phenyl ethanamide.

Chemicals: Acetophenone, Methanol, Sodium acetate dihydrate, Hydroxyl amine hydrochloride and Sulphuric acid.

Principle: Acetanilide can be prepared from acetophenone oxime. Acetophenone on condensation with hydroxyl amine hydrochloride in presence of sodium acetate yields acetophenone oxime. Polarization of carbonyl group of acetophenone allows protonation. The positively charged carbonyl carbon (electrophile) will be attacked by electron rich nitrogen of hydroxyl amine (nucleophile) and forms an intermediate. Removal of hydrogen from nitrogen followed by the dehydration generates acetophenone oxime.

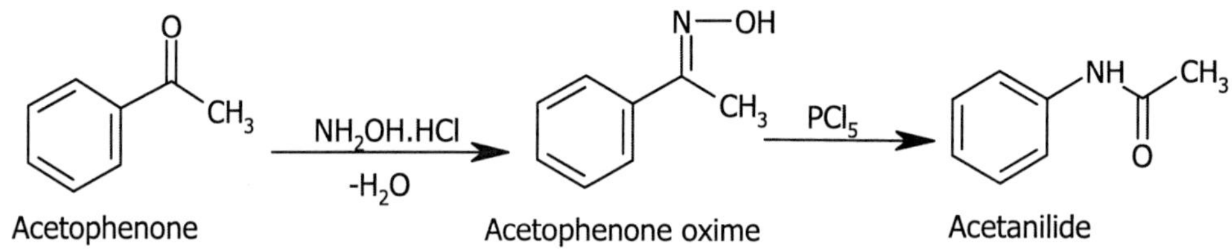

An acid catalyzed transformation of acetophenone oxime into corresponding amide (acetanilide) occurs by Beckmann oxime rearrangement. Acetophenone oxime on treatment with phosphorous penta chloride (PCl_5) in anhydrous ether forms benzanilide. Phosphorous penta chloride (PCl_5) converts -OH group into better leaving group $OPCl_4$ (ester). Detachment of leaving group generates electron deficient nitrogen species (loss of water or anion). Migration of the phenyl group to the electron deficient nitrogen occurs and forms carbocation. Addition of water on carbocation, loss of proton from the intermediate and tautomerization yield acetanilide. This reaction proceeds in the following way.

Formation of acetophenone oxime

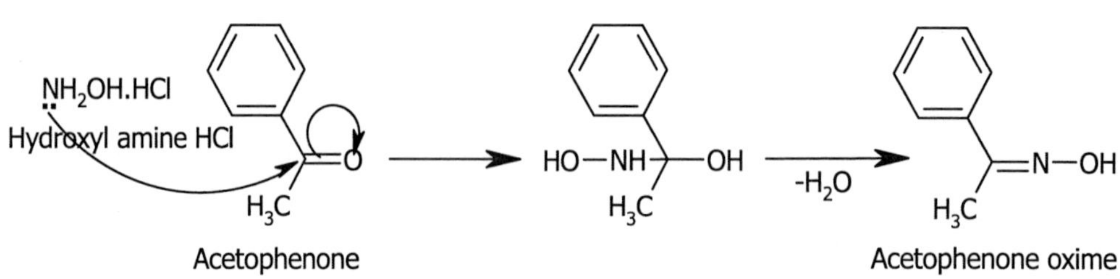

Formation of acetanilide

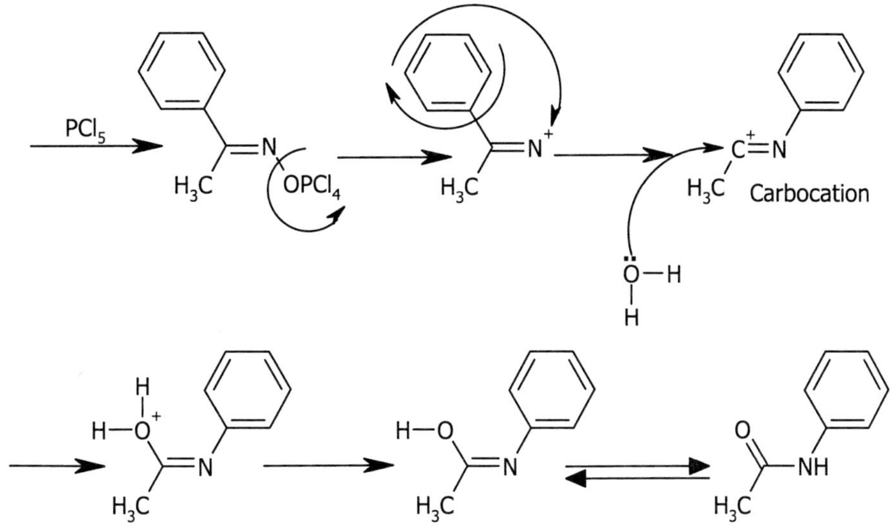

Procedure:

***Step* 1: Preparation of Acetophenone oxime**

1. Dissolve acetophenone (3 ml, 26 mmol) in the mixture of methanol (8 ml) and water (28 ml) loaded in the flask.
2. Add sodium acetate dihydrate (3 gm, 37 mmol) and hydroxyl amine hydrochloride (3 gm, 44 mmol) to the flask.
3. Heat the reaction mixture by placing in a water bath for 10 minutes and cool in salt-ice bath.
4. Filter to isolate the precipitated acetophenone oxime and purify by recrystallization with water.
5. The melting point of the pure acetophenone oxime is 59 °C.

***Step* 2: Preparation of Acetanilide**

1. Heat the sulphuric acid (6 ml) by placing in a water bath. (*thionyl chloride also can be used, but it requires heating upto 70-80 °C*).
2. Add acetophenone oxime (3 gm, 22 mmol) slowly with stirring to the hot sulphuric acid.
3. Continue the stirring for 15 minutes, cool and pour the mixture into crushed ice with stirring.
4. Isolate the solid product by filtration and wash with water.
5. Purify the acetanilide by recrystallization with boiling water.
6. The melting range of the pure acetanilide is 113-115 °C.

CHAPTER 9

BILTZ SYNTHESIS

Phenytoin, an anti-convulsant agent belongs to imidazolidine-2,4-dione (hydantoin) family. It is used in the treatment of grandmal epilepsy, psychomotor epilepsy ulcers, epidermolysis bullosa and inflammatory diseases. Phenytoin also possesses neuroprotective and cardio protective properties. Conventionally phenytoin and its derivatives are synthesized by Bucherer-Bergs reaction. Synthesis of phenytoin by base catalyzed condensation of benzil with urea is known as Biltz synthesis. This reaction occurs through Benzilic acid rearrangement.

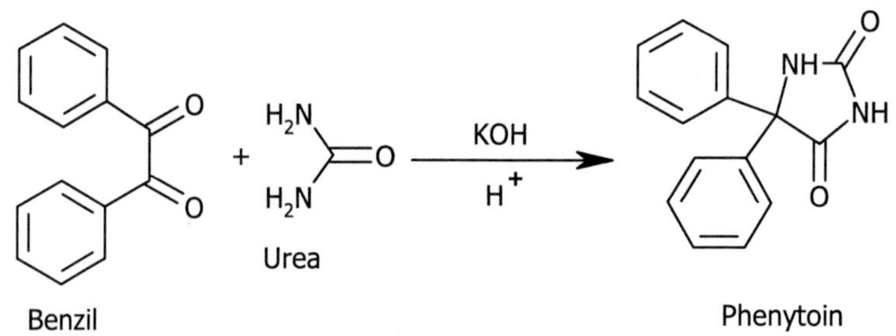

9.1 Preparation of Phenytoin

Traditional Name: Diphenyl hydantoin. ***IUPAC Name*:** 5,5-Diphenyl imidazolidine-2,4-dione.

Chemicals: Benzil, Urea, Ethanol, Sodium hydroxide (30%) and Hydrochloric acid (dilute).

Principle: Synthesis of phenytoin by base catalyzed condensation of benzil with urea is known as Biltz synthesis. Benzil condenses with urea in the presence of sodium ethoxide and produces phenytoin. Phenytoin preparation from benzil involves the reaction mechanism (chemical transformation) known as Benzilic acid rearrangement. The reaction progress through the Pinacol-Pinacolone rearrangement also described below.

Benzilic acid rearrangement: Sodium Ethoxide base [C_2H_5ONa] induced rearrangement of benzil to benzilic acid ester *via* phenyl group migration is known as Benzilic acid rearrangement and is a representative of 1,2-rearrangements. Ethoxide anion attacks the one of the carbonyl carbon of benzil and gets converted into Benzilic acid ester. The ester carbonyl carbon will be attacked by lone pair electrons of the urea nitrogen and eliminates ethanol. The acid proton will add to hydroxyl group and removes it as water. The intermediate carbocation formed will undergo cyclization by the attack of lone pair of electron on the urea nitrogen and generates phenytoin.

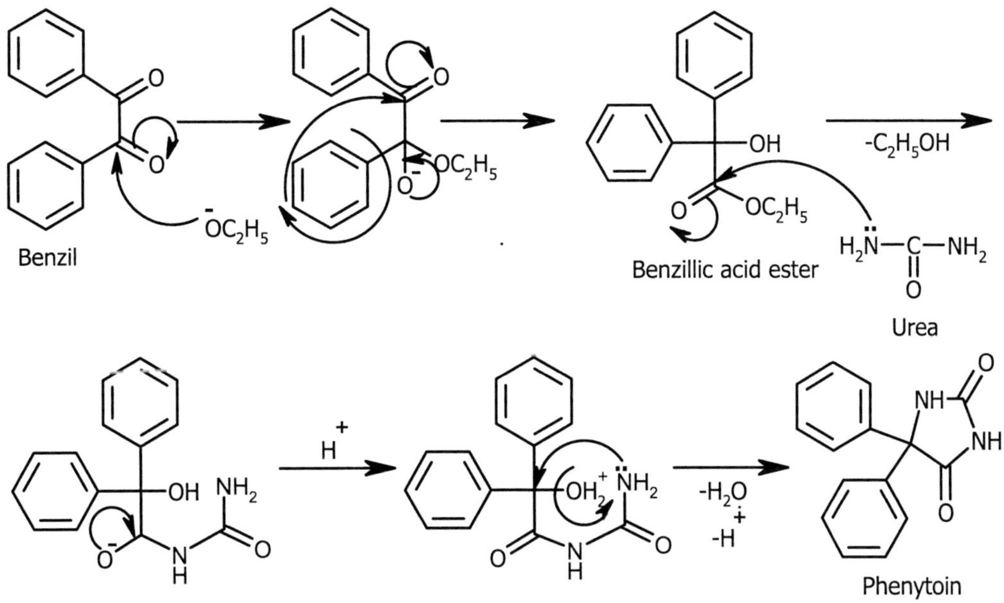

Benzillic acid rearrangement

Pinacol-Pinacolone rearrangement

Completely substituted 1,2-diols (pinacols) on treatment with mineral acids, undergoes dehydration and rearrangement to form ketones is called as Pinacol-Pinacolone rearrangement reaction. Base abstracts proton (H^+) from urea and generates nucleophile. Nucleophile attacks the electrophilic center in carbonyl carbon of benzil and upon cyclization forms 1,2-diol (1,2-glycols/pinacols). One of the hydroxyl group will be eliminated as water molecule after protonation by an acid. Removal of water molecule generates electron deficient carbon atom. Phenyl group shift (1,2-shift) to electron deficient carbon and oxidation of another hydroxyl group generates carbonyl group (pinacolone). [Shifting of phenyl group and generation of carbonyl group is characteristic of Pinacol-Pinacolone rearrangement reaction].

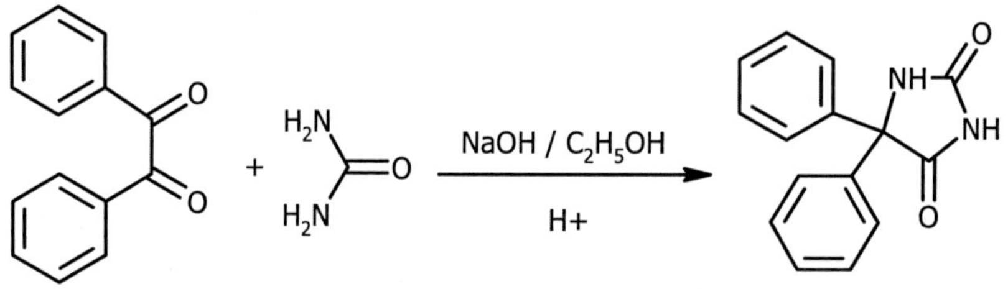

Benzil Urea Phenytoin

Generation of nucleophile

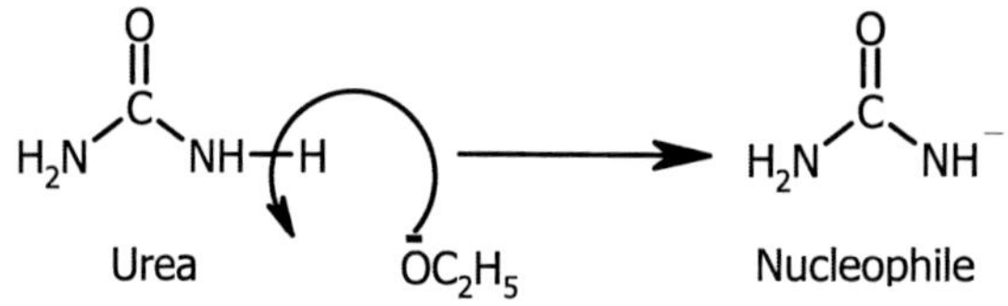

Nucleophilic attack and pinacol formation

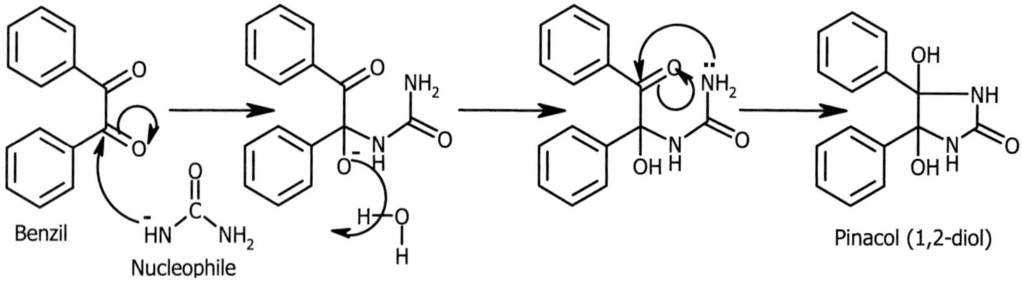

Rearrangement and pinacolone formation

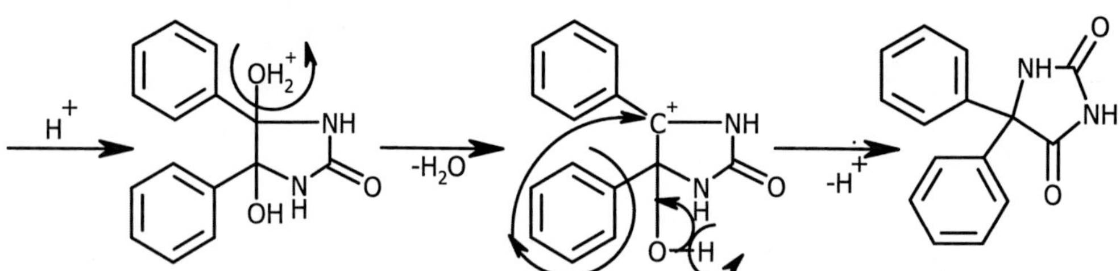

Phenytoin

Procedure:

1. Place benzil (1.1 gm, 5 mmol), urea (0.6 gm, 10 mmol), sodium hydroxide (3 ml, 30%) and ethanol (15 ml) in a round bottom flask.

Procedure:

1. Place benzil (1.1 gm, 5 mmol), urea (0.6 gm, 10 mmol), sodium hydroxide (3 ml, 30%) and ethanol (15 ml) in a round bottom flask.
2. Attach reflux condenser and reflux the mixture for 2 hr.
3. Cool the reaction mixture to room temperature and pour into water (25 ml) contained in beaker.
4. Mix thoroughly and allow it to stand for 15 minutes.
5. Filter to remove any insoluble by product.
6. Acidify the filtrate with hydrochloric acid, cool in an ice bath and immediately filter the precipitate under vacuum.
7. Dry and recrystallize the product from rectified sprit.
8. The melting range of pure phenytoin is 297-298 °C.

Note: Refer 4.1 (Refer Page 55) for the preparation of benzil from benzoin

CHAPTER 10

ERLENMEYER-PLOCHL AZLACTONE SYNTHESIS

The transformation of amino acid into other amino acids through the formation of oxazolone (azlactone) is named as Erlenmeyer-Plochl azlactone amino acid synthesis. It is mainly restricted to the synthesis of aromatic amino acids, because the unsaturated azlactones (intermediates) are prepared using aromatic aldehydes.

Acylation of glycine with acetic anhydride/benzoyl chloride forms acylglycine (acetyl/benzoyl glycine). Intramolecular condensation of acylglycines (by dehydration of acylglycines) in the presence of acetic anhydride and sodium acetate leads to the cyclization of *N*-acyl amino acid. It produces five membered heterocycle (azlactone) containing both nitrogen and oxygen with the oxygen adjacent to a carbonyl group. Aromatic aldehyde condenses (Knoevenagel condensation) with the reactive methylene group of the azlactone under the influence of base. Hydrolysis of azlactone moiety on boiling with water yields unsaturated α-acylamino acids. α-Acylamino acid upon reduction and hydrolysis forms amino acid. Unsaturated α-acylamino acids on drastic hydrolysis gives α-oxo acid.

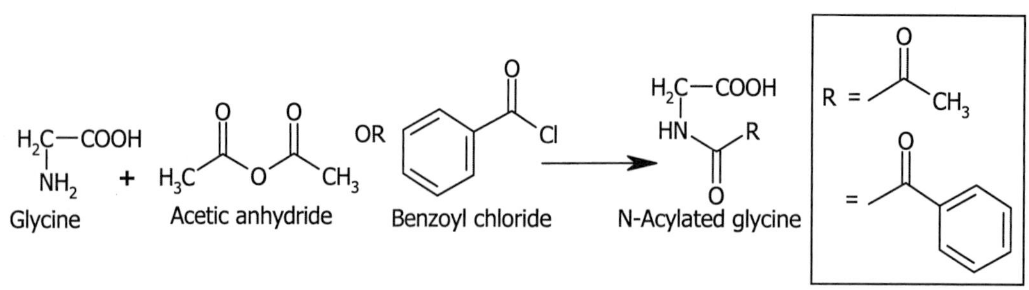

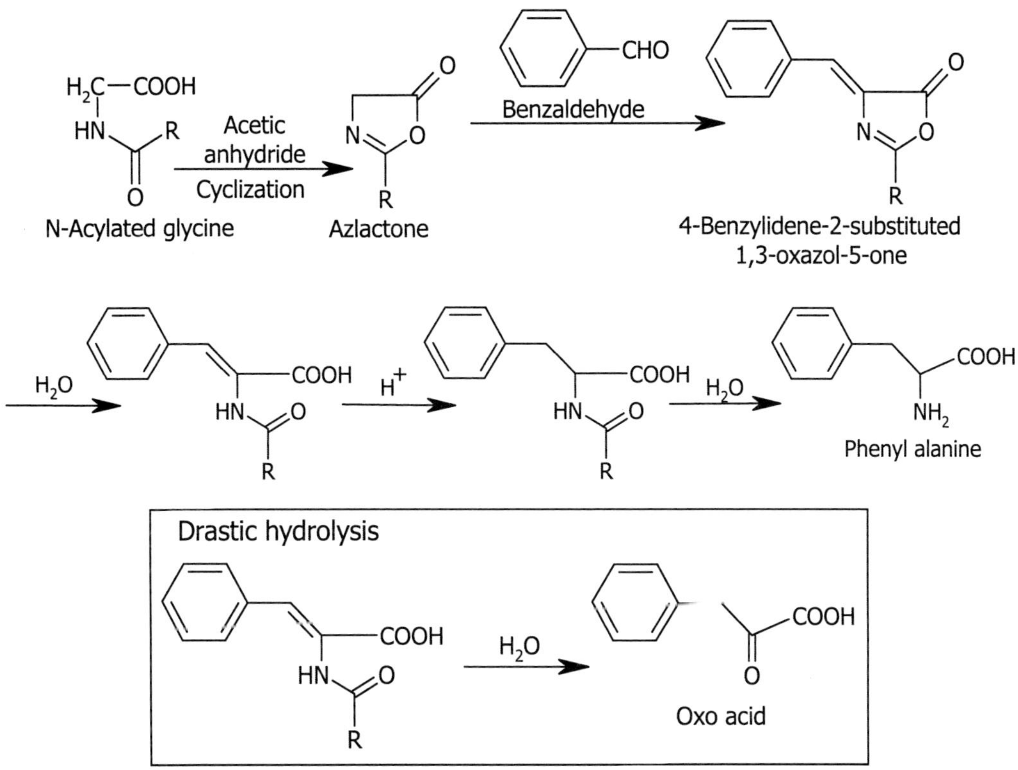

Heating azlactone in acetic anhydride solution with red phosphorous and hydroiodic acid also yields acylaminoacrylic acid (single step).

Pharmaceutical importance

1. Azlactone moiety provides a basic skeleton with drug-like characteristics.
2. Azlactone exhibits antifungal, antibacterial and anti-inflammatory activities.
3. Azlactone generates imidazole and penicillin type drug intermediates (pharmaceutical synthons).
4. Useful in the synthesis of hormonal compounds.

10.1 Preparation of 4-Benzylidene-2-methyloxazol-5-one

Chemicals: *N*-Acetyl glycine, Acetic anhydride, Benzaldehyde, Sodium acetate, Hydrochloric acid and Methanol.

Principle: Preparation of 4-benzylidene-2-methyloxazol-5-one is an intermediate step in the Erlenmeyer Plochl azlactone synthesis. Intramolecular condensation of acetylglycines (dehydration of acetylglycine) in the presence of acetic anhydride and sodium acetate produces five membered heterocycle azlactone moiety. Benzaldehyde condenses with the reactive methylene group of the azlactone under the influence of sodium acetate. Hydrolysis of azlactone moiety with boiling water yields unsaturated α-acylamino acid. The α-acylamino acid upon reduction and further hydrolysis produces phenyl alanine.

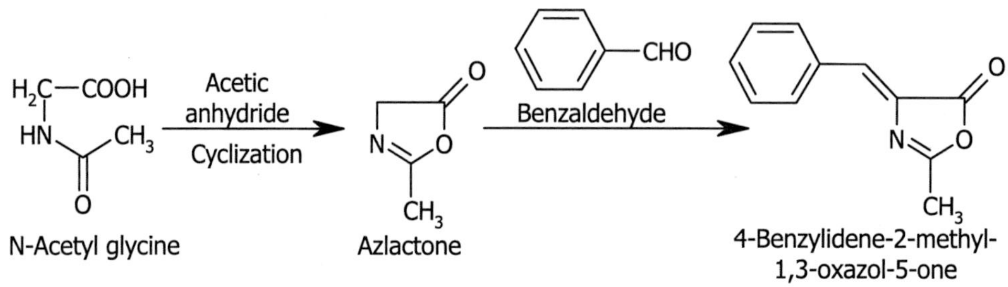

Reaction mechanism

Formation of azlactone

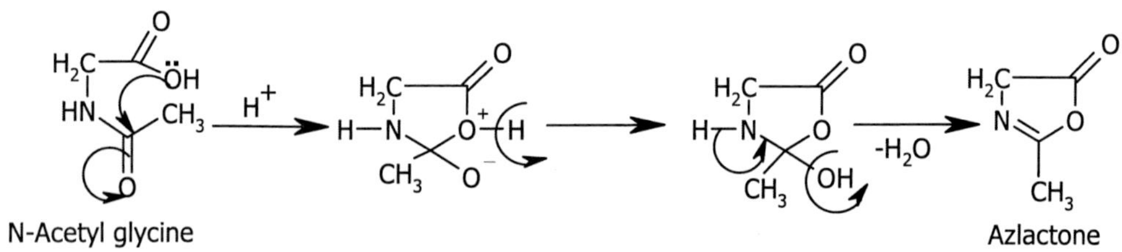

Formation of 4-benzylidene-2-methyloxazol-5-one

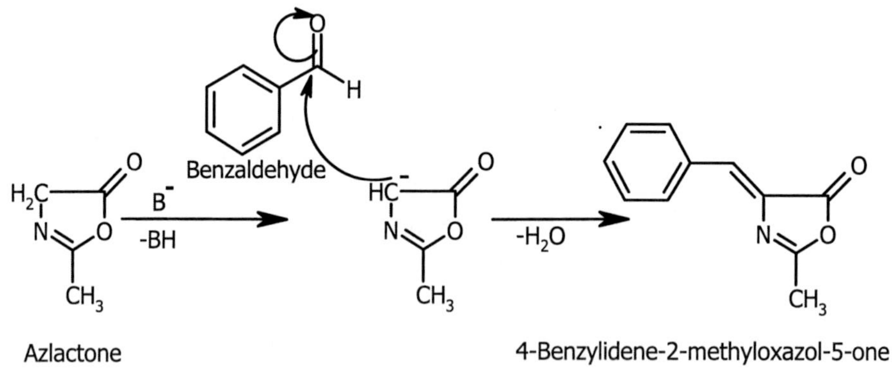

Procedure:

1. Warm the mixture of acetylglycine (1 gm, 8.6 mmol), distilled benzaldehyde (1.3 ml, 13 mmol), anhydrous sodium acetate (516 gm, 6.3 mmol) and acetic anhydride (2 ml, 21 mmol) loaded in the flask.

2. Attach reflux condenser to the flask and stir occasionally to effect the solution.
3. Boil the reaction mixture for 1 hr by placing flask on the water bath, cool to room temperature and place in a refrigerator overnight.
4. Stir the resulting yellow crystalline solid with cold water (5 ml).
5. Filter and wash with cold water.
6. Purify the product by recrystallization with carbon tetrachloride. (alternatively ethyl acetate-light petroleum can be used).
7. The melting point of 4-benzylidene-2-methoxyoxazol-5-one is 150 °C.

Note: *Refer 3.4 (page 44) for the preparation of acetyl glycine.

CHAPTER 11

FISCHER-SPEIER ESTERIFICATION

Carboxylic acids on reaction with large excess of alcohol and in the presence of strong acid catalyst, forms an ester. The interaction between an acid and an alcohol is a reversible process and proceeds very slowly. Equilibrium will be attained after refluxing for several days. But strong acid catalyst, either sulphuric acid or dry hydrogen chloride helps to reach the equilibrium within few hours. This reaction is called as Fischer-speier esterification (acid-catalyzed esterification).

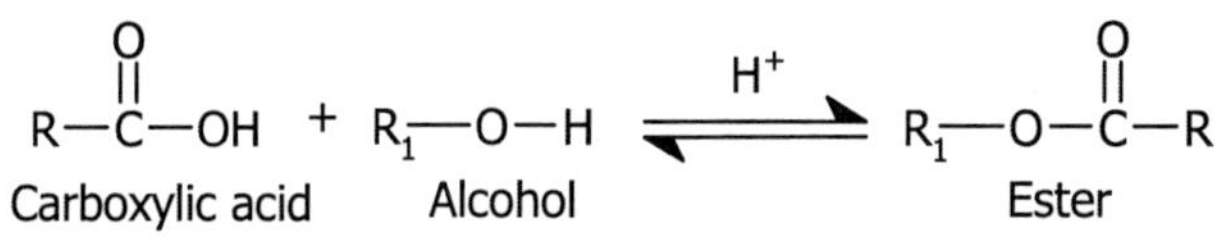

Protonation of the carbonyl oxygen of carboxylic acid forms oxonium ion and addition of alcohol on the carbonyl carbon (oxonium ion) forms addition product. Transfer of proton followed by loss of water molecule generates the ester. All these reactions are reversible and an excess of one reactant, usually the alcohol favours ester formation.

Formation of oxonium ion, addition product and transfer of proton

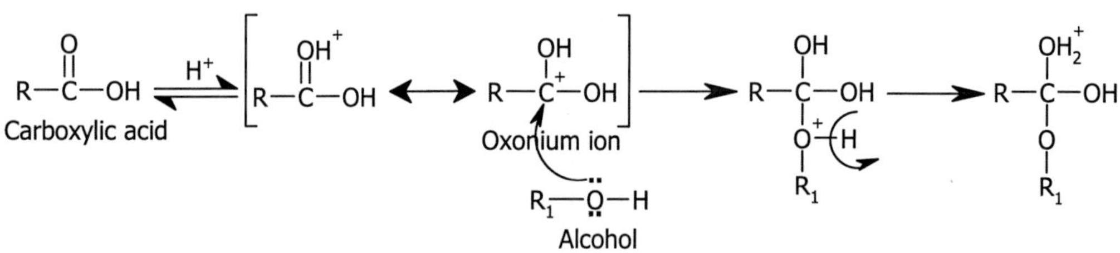

Loss of water molecule and formation of ester

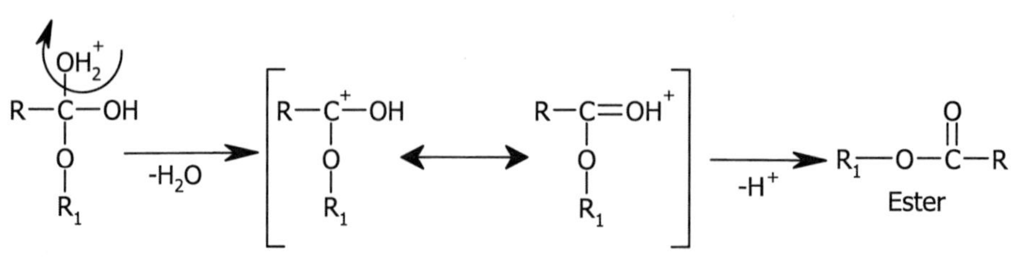

11.1 Preparation of *n*-Butyl Acetate

***IUPAC Name*:** Butyl ethanoate.

Chemicals: Glacial acetic acid, *n*-Butyl alcohol, Sulphuric acid, Sodium bicarbonate and Magnesium sulphate (anhydrous).

Principle: The esterification of *n*-butyl alcohol with acetic acid forms *n*-butyl acetate in presence of sulphuric acid. Sulphuric acid acts as a catalyst and also as a dehydrating agent. The large excess of *n*-butyl alcohol promotes the reversible reaction, hence should be removed by distillation.

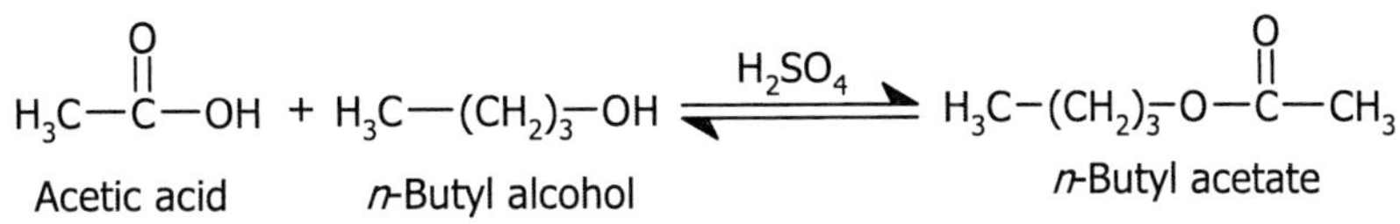

The reaction involves following steps:

Protonation of the carbonyl oxygen forms oxonium ion and addition of *n*-butyl alcohol on the carbonyl carbon (oxonium ion) forms addition product. Transfer of proton followed by loss of water molecule generates the ester. All these reactions are reversible and an excess of one reactant, usually the alcohol favours ester formation.

Formation of oxonium ion, addition product and transfer of proton

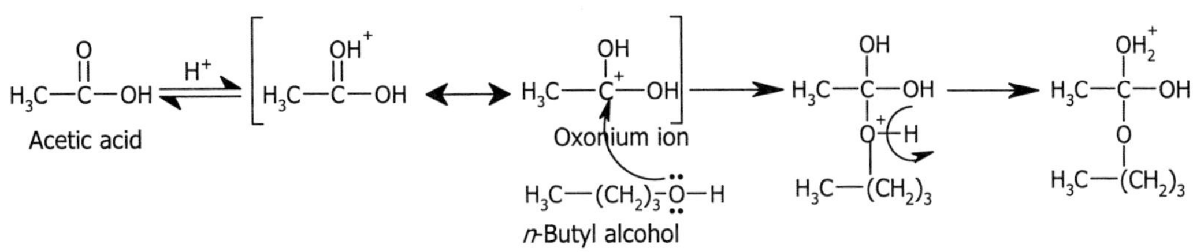

Loss of water molecule and formation of butyl acetate

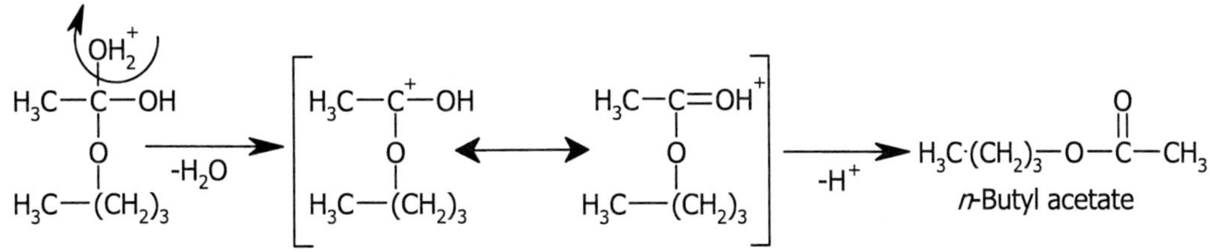

Procedure:

1. Mix *n*-butyl alcohol (10 ml, 109 mmol) and glacial acetic acid (13 ml, 229 mmol) in a round bottom flask.

2. Add few drops of sulphuric acid cautiously, attach reflux condenser and reflux gently for about 1½ hr.
3. Pour the reaction mixture into separating funnel containing water (50 ml) shake and remove the upper organic layer.
4. Wash the organic layer with water (20 ml), followed by saturated sodium bicarbonate solution (5 ml) and water (10 ml).
5. Dry the organic liquid by passing through anhydrous magnesium sulphate (1.3 gm).
6. Filter and distil the organic liquid at 124-125 °C to collect n-butyl acetate.

11.2 Preparation of Methyl Benzoate

Chemicals: Benzoic acid, Methyl alcohol, Sulphuric acid, Carbon tetrachloride, Magnesium sulphate (anhydrous) and Sodium bicarbonate.

Principle: Benzoic acid esterifies with methyl alcohol in the presence of sulphuric acid and forms methyl benzoate. Sulphuric acid acts as a catalyst and dehydrating agent. The large excess of methyl alcohol promotes the reversible reaction and can be removed by distillation once the reaction is completed.

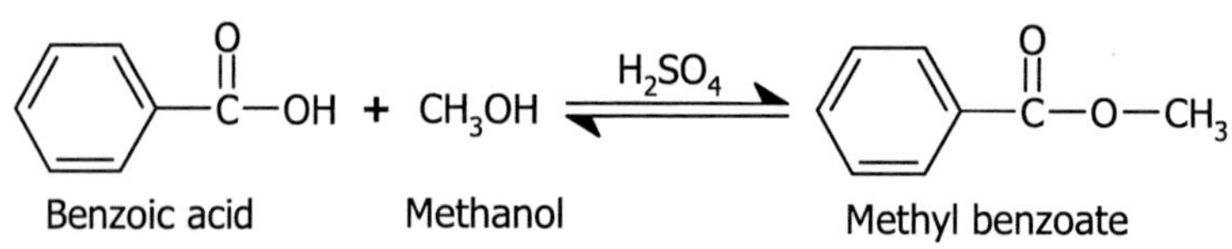

Protonation of the carbonyl oxygen forms oxonium ion and addition of methyl alcohol on the carbonyl carbon (oxonium ion) forms addition product. Transfer of proton followed by loss of water molecule generates the ester. All these reactions are reversible and an excess of one reactant, usually the alcohol favours ester formation.

The reaction involves following steps:

Formation of oxonium ion, addition product and transfer of proton

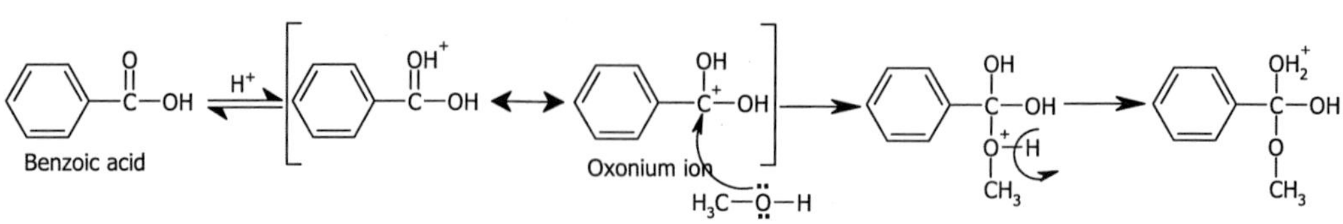

Loss of water molecule and formation of methyl benzoate

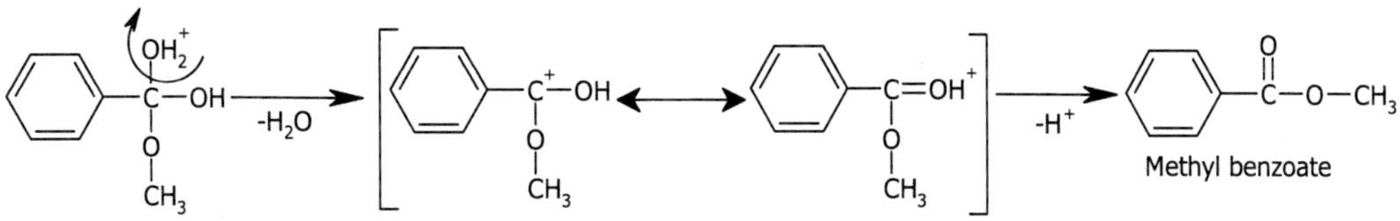

All these reactions are reversible and an excess of one reactant, usually the alcohol favours ester formation.

Procedure:

1. Place benzoic acid (1 gm, 8.2 mmol), absolute methanol (4 ml) and 2 drops of sulphuric acid into a round bottom flask.
2. Add boiling chips to the flask, attach reflux condenser and boil gently for 4 hr (longer heating improves yield).
3. Allow the mixture to cool.
4. Pour the reaction mixture into separating funnel containing water (8 ml) and rinse the flask with water and return to separating funnel.
5. Add carbon tetrachloride (1 ml), which enhances the separation of 2 distinct layers (alternatively diethyl ether can be used) and shake vigorously.
6. The heavy solution of methyl benzoate in carbon tetrachloride separates upon standing at the bottom.
7. Collect the lower carbon tetrachloride layer and reject the upper aqueous layer.
8. Transfer the carbon tetrachloride layer back to separating funnel and shake it with sodium bicarbonate solution until effervescence ceases.
9. Wash the organic layer with excess of water and dry it by pouring into dry conical flask containing anhydrous magnesium sulphate (170 mg) or anhydrous sodium sulphate (170 mg).
10. Stopper the flask, shake it for 5 minutes and allow it to stand for 30 min with occasional shaking.
11. Filter the resulting solution and distil at 198-200 °C.
12. Collect the colourless liquid of methyl benzoate.

11.3 Preparation of Benzocaine

Traditional Name: Anaesthesin. ***IUPAC Name***: Ethyl-4-amino benzoate.

Chemicals: *para*-Amino benzoic acid, Absolute ethanol, Sulphuric acid and Sodium carbonate.

Principle: Preparation of benzocaine involves esterification reaction. *para*-Amino benzoic acid (PABA) esterifies with ethyl alcohol in the presence of sulphuric acid and forms benzocaine. Sulphuric acid acts as a catalyst and dehydrating agent.

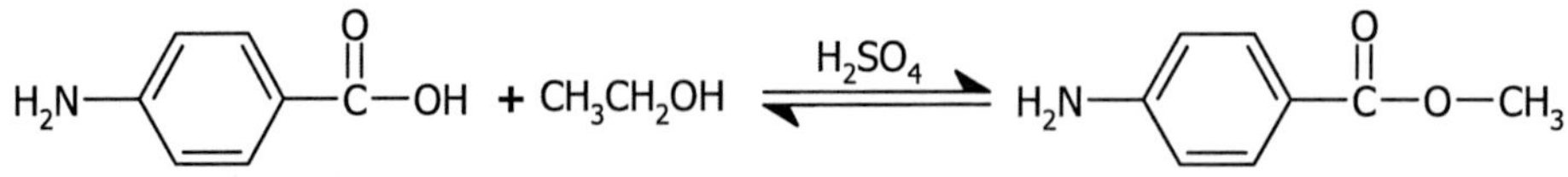

para-Amino benzoic acid   Ethanol   Benzocaine

Protonation of the carbonyl oxygen forms oxonium ion and addition of ethyl alcohol on the carbonyl carbon (oxonium ion) forms addition product. Transfer of proton followed by loss of water molecule generates the ester. All these reactions are reversible and an excess of one reactant, usually the alcohol favours ester formation.

Reaction mechanism

Formation of oxonium ion, addition product and transfer of proton

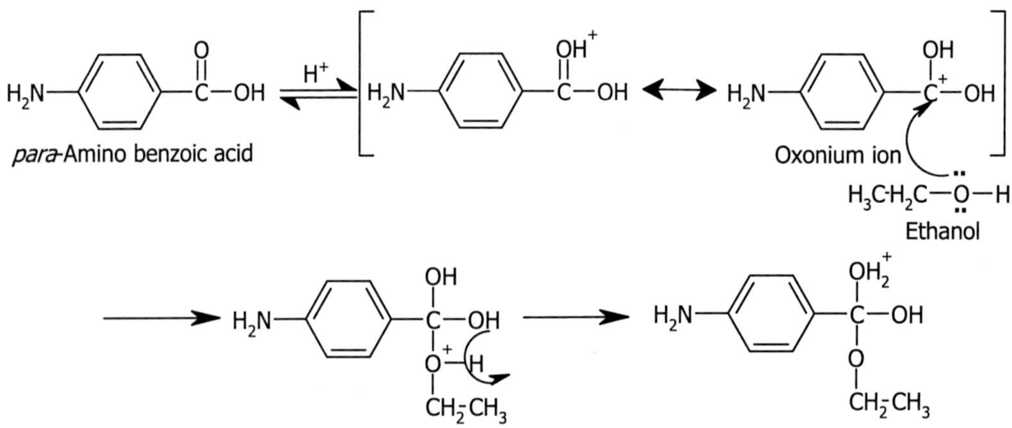

Loss of water molecule and formation of benzocaine

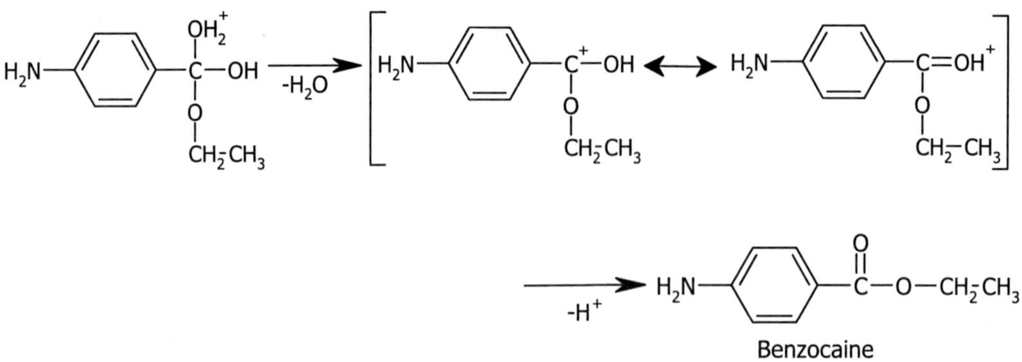

Procedure:

1. Saturate absolute ethanol (8 ml) with sulphuric acid (3 ml) in a flask.
2. Add 4-amino benzoic acid (1.2 g, 9 mmol) to the flask.
3. Add boiling chips and reflux the reaction mixture (sand bath is preferable) for 2 hours.
4. Allow the reaction mixture to cool to room temperature.
5. Transfer the reaction mixture into beaker containing distilled water and neutralize with solid sodium carbonate (ensure thorough mixing)
6. White precipitate of benzocaine separates after neutralization.
7. Collect the benzocaine by vacuum filtration and wash with water.
8. Purify the benzocaine by recrystallization with ethanol.
9. The melting point of pure benzocaine is 91 °C.

11.4 Preparation of Isoniazid

Traditional Name: Isonicotinic acid hydrazide *IUPAC Name*: Pyridine-4-hydrazide

Chemicals: Isonicotinic acid, Ethanol, Sulphuric acid, Sodium bicarbonate, Ether and Hydrazinc hydrate.

Principle: The preparation of isoniazid involves esterification of nicotinic acid followed by hydrazide formation. Isonicotinic acid undergoes esterification catalyzed by sulphuric acid and forms ethy.l isonicotinate. Ethyl isonicotinate reacts with hydrazine hydrate and gives isoniazid.

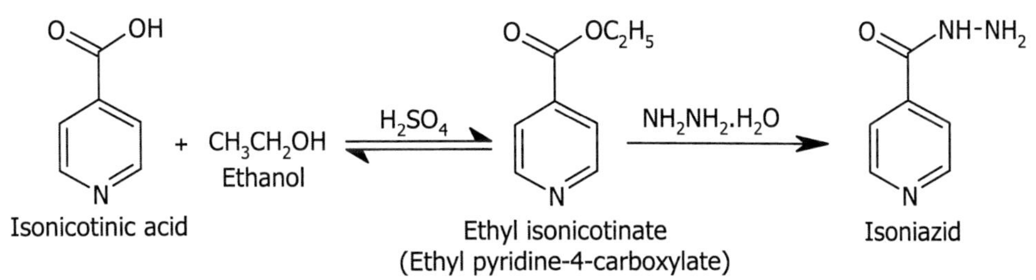

Reaction mechanism

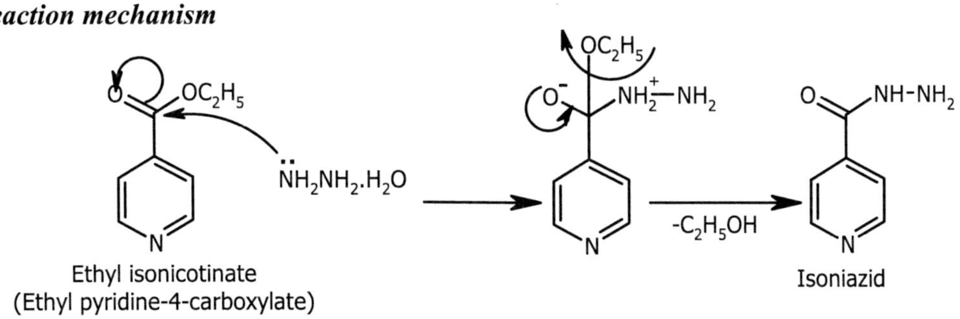

Procedure:

***Step* 1*:* Preparation of ethyl isonicotinate**

1. Place isonicotinic acid (1.23 gm, 10 mmol), ethanol (10 ml) and sulphuric acid (2 ml) into round bottom flask.
2. Reflux the mixture for 3 hr maintaining the temperature between 60-70 °C. (*Conduct the reaction in fume hood, since it produces irritating and toxic vapours*).
3. Add the reaction mixture to ice-cold water (200 ml) and neutralize the excess of acid with sodium bicarbonate.
4. Extract the aqueous layer with diethyl ether and evaporate the ether layer to obtain the residue of ethyl isonicotinate.

***Step* 2*:* Preparation of isoniazid**

1. Dissolve ethyl isonicotinate (1.6 ml, 10 mmol) in ethanol (10 ml) placed in a round bottom flask.
2. Add hydrazine hydrate (0.6 ml, 12 mmol) to the flask.
3. Reflux the reaction mixture for 2 hr and then cool to room temperature.
4. Isolate the precipitated isoniazid and dry.
5. Purify the isolated isoniazid by recrystallization with ethanol.
6. The melting range of pure isoniazid is 171-172 °C.

CHAPTER 12

WILLIAMSON'S ETHER SYNTHESIS

Williamson's ether synthesis involves the collision between nucleophile and unhindered sp3 carbon center with good leaving group. As the reaction proceeds, a new bond between carbon and nucleophile and cleavage of the bond between carbon and leaving group occurs. Several factors are responsible for the reaction progress

- The nucleophile must be small and should bear high electron density.
- The carbon (sp3) center must be unhindered.
- The leaving group must be weaker than the incoming nucleophile.
- Polar aprotic solvent is the choice (it should not over stabilize the charge).

Generally Williamson's ether synthesis utilizes phonoxide/alkoxide nucleophile. These S_N2 reactions are well promoted in polar aprotic solvents, whereas high nucleophilic oxygen allows the reaction to occur in protic solvents at high temperatures.

The reaction proceeds in the following way.

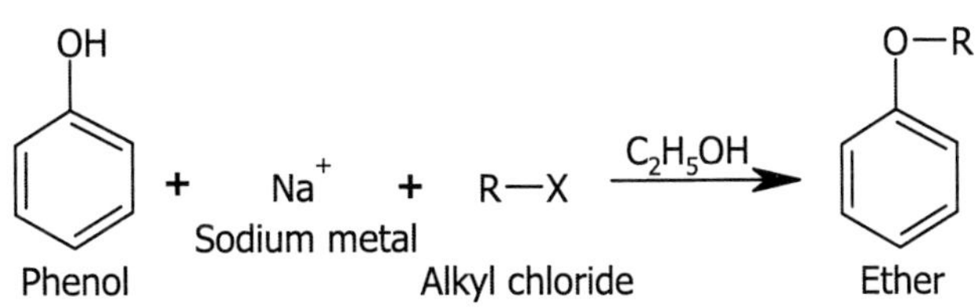

Formation of sodium ethoxide

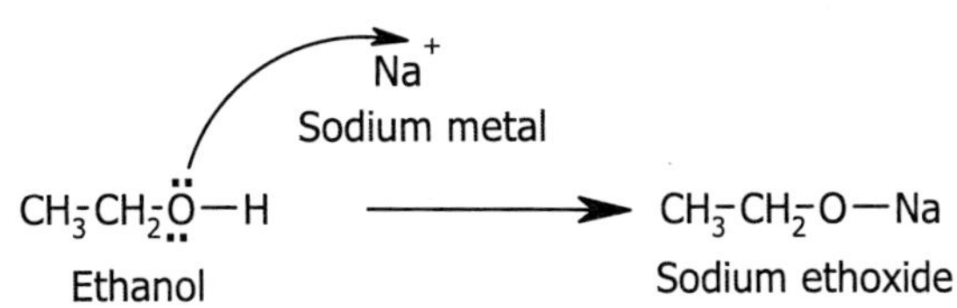

Formation of ether

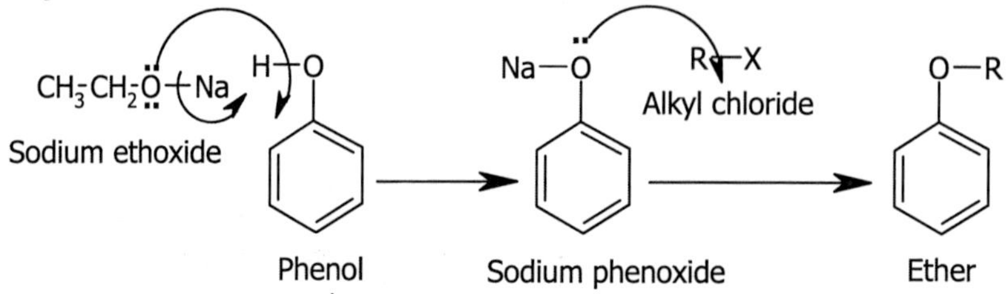

12.1 Preparation of Phenetole

***IUPAC Name*:** Ethyl phenyl ether

Chemicals: Phenol, Sodium hydroxide, Ethyl iodide, Sodium chloride and Diethyl ether.

Principle: Phenetole (phenolic ethyl ether) synthesis involves Williamson's ether synthesis. The phenol reacts with the sodium ethoxide to produce phenetole.

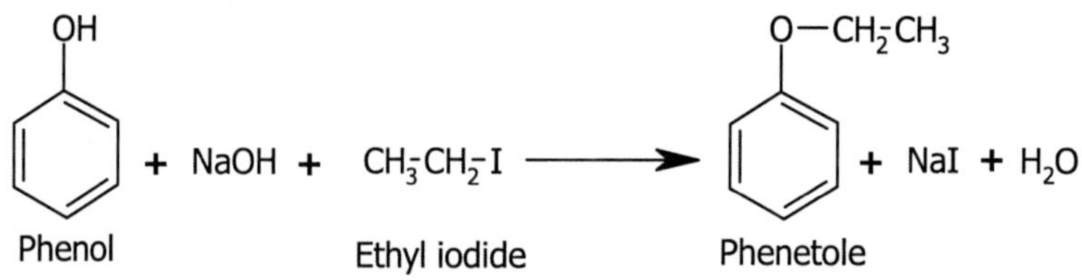

Williamson ether synthesis is suitable reaction for the preparation of unsymmetrical ether such as phenetole. Sodium ethoxide removes the hydrogen from the hydroxyl group of phenol and generates phenoxide ion. The negative charge on phenoxide oxygen is unstable, which favours the ethyl iodide to attach an ethyl group to the oxygen and forms unsymmetrical ether. The reaction requires the use of absolute ethanol, because sodium ethoxide is a very strong base.

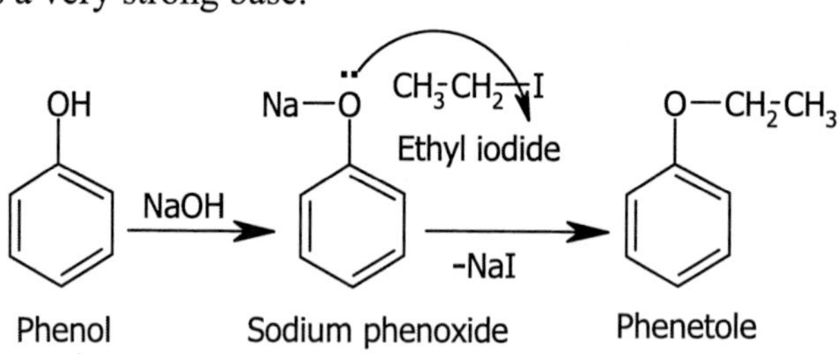

Procedure:

1. Place phenol (4.9 gm, 52 mmol) to the flask charged with solution of sodium hydroxide (2.4 gm in 12 ml water).
2. Stir the mixture to make a homogeneous solution.
3. Add ethyl iodide (4 ml, 0.5 mol) drop wise with vigorous stirring (maintain the temperature below 65 °C).

4. Attach reflux condenser and heat the contents for 3 hrs.
5. Transfer the contents of flask to a separatory funnel, add saturated sodium chloride (3 ml) solution and shake gently.
6. Drain the lower aqueous phase and extract the remaining mixture with ether (5 ml).
7. Distill the ether layer at 38-40 °C and collect phenetole.

12.2 Preparation of Phenacetin

IUPAC Name: N(4-Ethoxyphenyl) acetamide

Chemicals: *para*-Acetamido phenol (Paracetamol), Sodium metal, Ethanol and Ethyl iodide.

Principle: Phenacetin (acetamido phenolic ethyl ether) synthesis involves Williamson ether synthesis. The *p*-acetamido phenol (paracetamol) reacts with ethyl iodide in presence of sodium ethoxide and gives phenacetin.

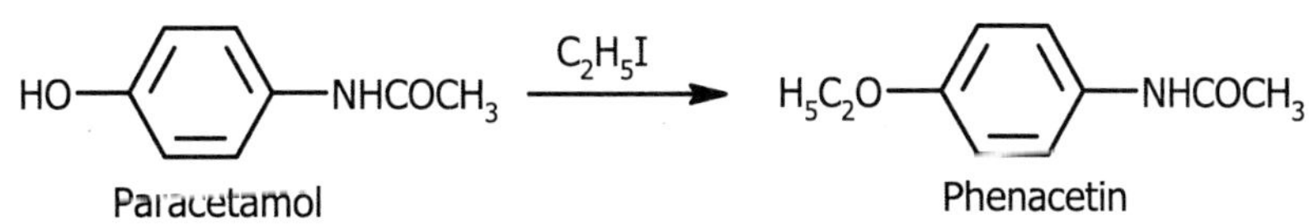

Williamson ether synthesis is suitable reaction for the preparation of unsymmetrical ether phenacetin. Sodium ethoxide removes the hydrogen from the hydroxyl group of *para*-acetamido phenol and generates phenoxide (alkoxide) ion. The negative charge on phenoxide oxygen is unstable, which favours the ethyl iodide to attach an ethyl group to the oxygen and forms Phenacetin. The reaction requires the use of absolute ethanol, because sodium ethoxide is a very strong base.

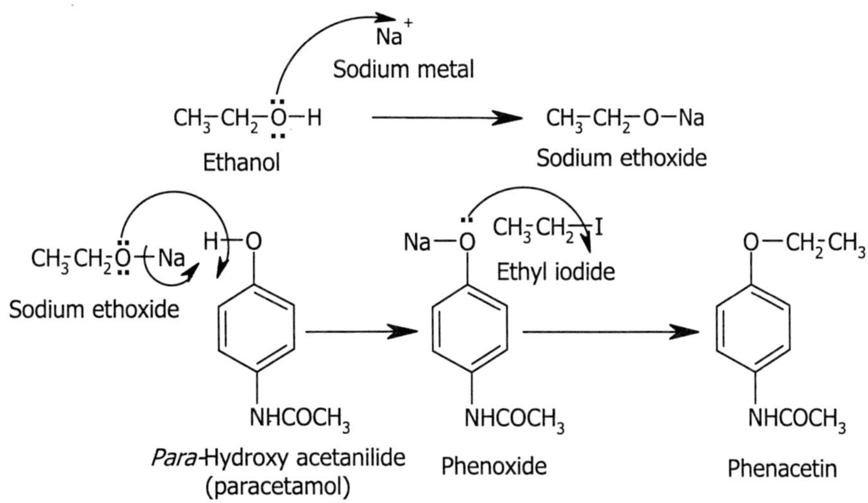

Procedure:

1. Place sodium metal pieces (155 mg, 6.8 mmol) to the flask containing absolute ethanol (4 ml).
2. Attach condenser and warm the flask (if necessary) to dissolve sodium completely.
3. Cool the mixture and add *para*-acetamido phenol (1 gm, 6.6 mmol) slowly followed by ethyl iodide (0.8 ml, 10 mmol), slowly through condenser.
4. Reflux the mixture for 1 hr.
5. Pour 10 ml of water through the condenser to prevent product crystallization (if crystals do separate, reflux the mixture until they dissolve).
6. Cool the flask by placing in an ice-bath, collect the phenacetin through vacuum filtration and wash with cold water.
7. Dissolve the crude phenacetin with 8 ml of ethanol (treat with charcoal boil and filter, if the solution appears colour).
8. Pour the clear solution obtained into beaker containing hot water (15 ml).
9. Allow it to cool and collect the solid product of pure phenacetin through filtration.
10. The melting range of pure phenacetin is 135-137 °C.

Note: Ref 3.2 (page 41) for the preparation of *para* acetamido phenol (paracetamol)

CHAPTER 13

CLAISEN-SCHMIDT CONDENSATION

Claisen-Schmidt condensation (Mixed aldol condensation)

The aldol reaction combines two carbonyl compounds with at least one α-hydrogen and generates carbon-carbon bonds. The product formed is β-hydroxyl carbonyl compound (aldol, aldehyde + alcohol). The aldol intermediate is used in the manufacture of drugs such as pentaerythritol, atorvastatin and amphotericin B.

The extension of simple aldol condensation reaction is the mixed aldol condensation reaction, which is known as Claisen-Schmidt condensation. The typical reaction occurs between an aldehyde with no α-hydrogen (which can't form enolate ion) and ketone compound. The aldehyde is more reactive towards nucleophiles of ketone (ketone enolate) and avoids the self condensation of ketone enolate. This can be achieved by the addition of ketone to mixture of aldehyde and base.

The condensation between aldehyde and ketone enolate generates β-hydroxyl carbonyl compound, which after eliminating water molecule forms conjucated double bond (C=C). Condensation of aldehyde and ketone by acid/base catalyzed reaction generates chalcones. Chalcones are 1,3-diphenyl-2-propene-1-one, in which two aromatic rings linked through three carbon α,β-unsaturated carbonyl system. The alternative names for the chalcones are phenyl styryl ketone, benzal acetophenone, β-phenylacrylophenonone, γ-oxo-α,γ-diphenyl-α-propylene and α-phenyl-β-benzoylethylene. Benzylidene acetophenone is the parent member of chalcone and can be prepared by the condensation of benzaldehyde and acetophenone.

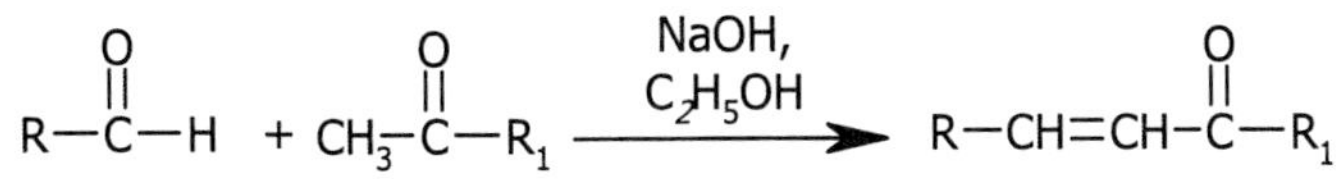

Aldehyde Methyl ketone Chalcone

Base-catalyzed preparation is the classical method for the preparation of chalcone. Piperidine mediated chalcone synthesis avoids multiple condensation, polymerization and rearrangement reactions. Base abstracts the proton of α-hydrogen atom of ketone and generates nucleophile (enolate ion). The nucleophile attacks the carbonyl carbon of aldehyde bearing no α-hydrogen and forms an intermediate. The intermediate β-hydroxyl carbonyl compound upon base mediated elimination of water molecule gives chalcone.

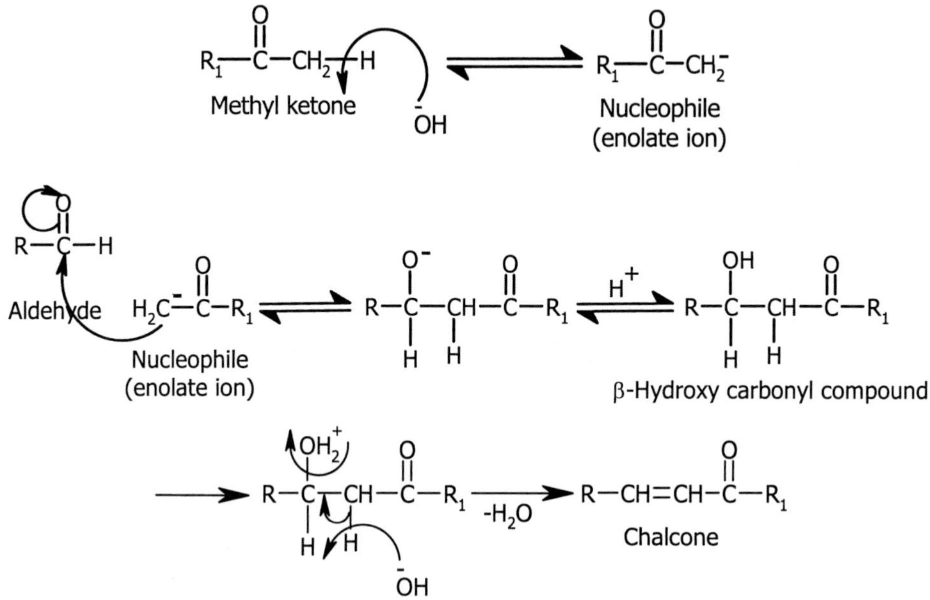

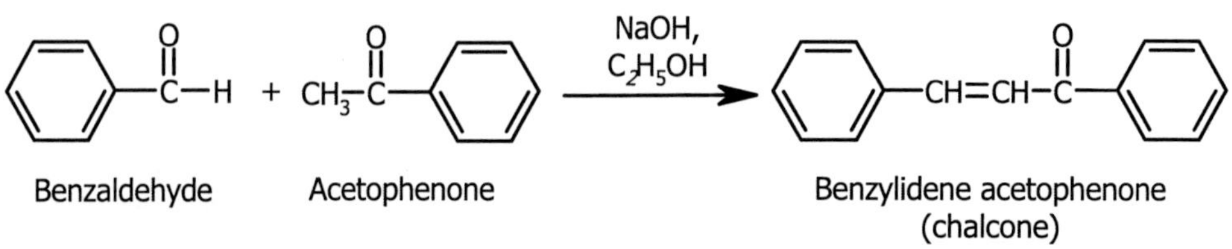

13.1 Preparation of Benzylidene Acetophenone

Traditional Name: Chalcone

Chemicals: Benzaldehyde, Acetophenone, Sodium hydroxide and Hydrochloric acid.

Principle: Benzylidene acetophenone (chalcone) can be prepared through the Claisen-Schimidt condensation. The condensation between benzaldehyde and acetophenone in sodium hydroxide and ethanol produces benzylidene acetophenone.

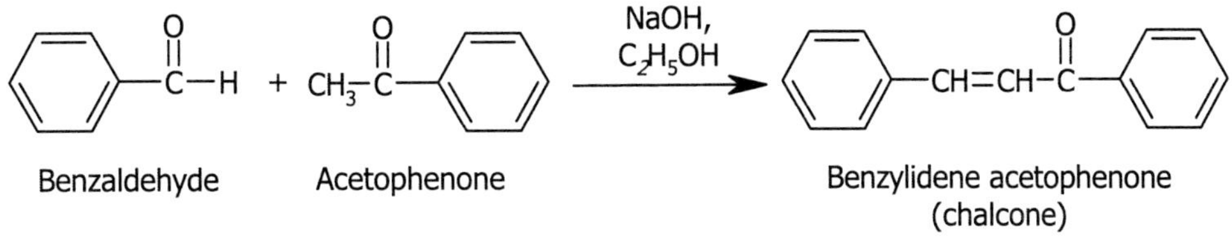

Reaction mechanism:

Generation of nucleophile

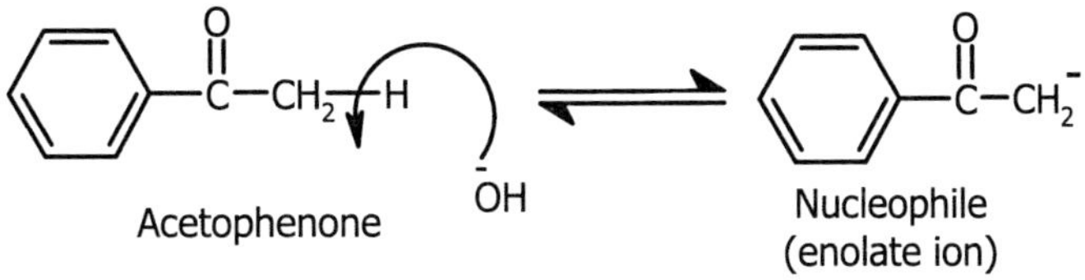

Condensation:

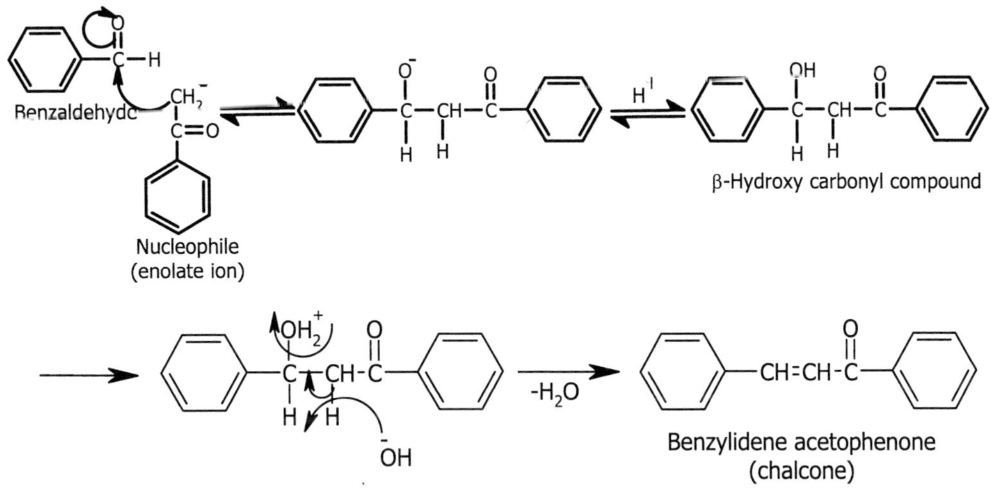

Procedure:

1. Place a solution of sodium hydroxide (2.2 gm, 55 mmol) in water (20 ml) and ethanol in a flask.
2. Immerse the flask on an ice-bath to maintain the temperature between 20-30 °C (optimum temperature is 25 °C),
3. Pour freshly distilled acetophenone (5 ml, 43 mmol) and add benzaldehyde (4.4 ml, 43 mmol).
4. Stir the mixture vigorously until it becomes thick (~ 3 hr).

5. Keep the reaction mixture in refrigerator for overnight.
6. Filter the solid separated through vacuum filtration, wash with cold water (until the washings are neutral to litmus) and with 20 ml of cold ethanol.
7. Recrystallize the product from ethanol.
8. The melting range of the pure benzylidene acetophenone is 56-57 °C.

Note:

- Benzylidene acetophenone is skin irritant, hence handle with gloves.
- If the tempeature is high (> 30 °C), then secondary reactions reduces the purity and yield.
- The product separates as an oil, if the temperature is low and high, and/srtirring is slow (which forms lumps).
- The suspension should be cooled gradually and then chilled.

CHAPTER 14

KNOEVENAGEL CONDENSATION

The condensation between carbonyl compounds (aldehydes and ketones) and active methylene compounds in presence of base catalyst is known as Knoevenagel condensation. Base catalysts include amines, ammonia, sodium ethoxides and piperidine in organic solvent. Mild basic media favours this reaction. This reaction is employed in the formation of C=C bond formation in fine chemicals. Normally active methylene groups in between two electron withdrawing groups and aldehydes/ketones containing no alpha hydrogen alone undergo this reaction.

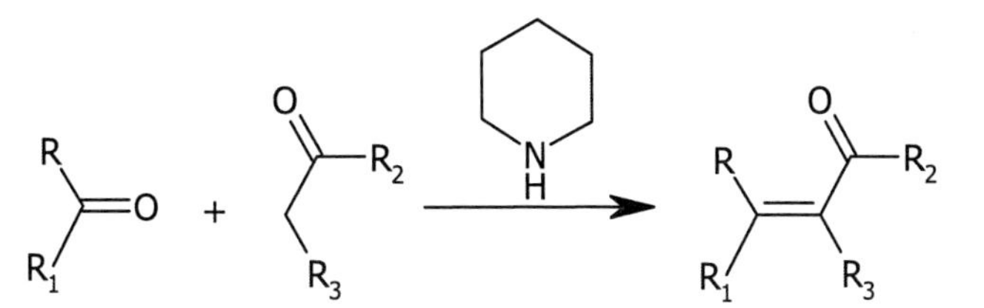

Carbonyl compound — Active methylene compound — Knoevenegel condensation product

Reaction mechanism

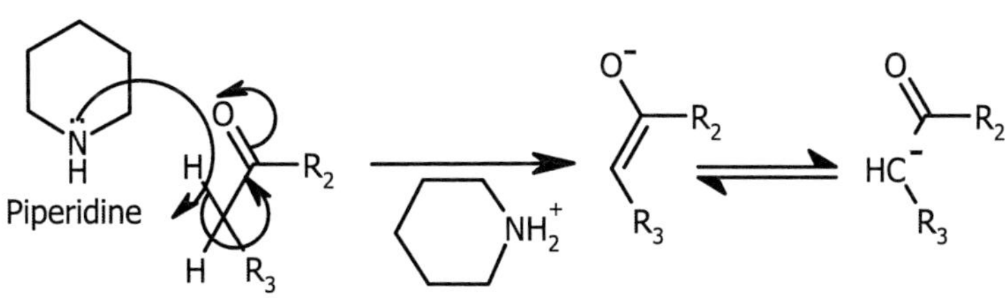

Active methylene compound

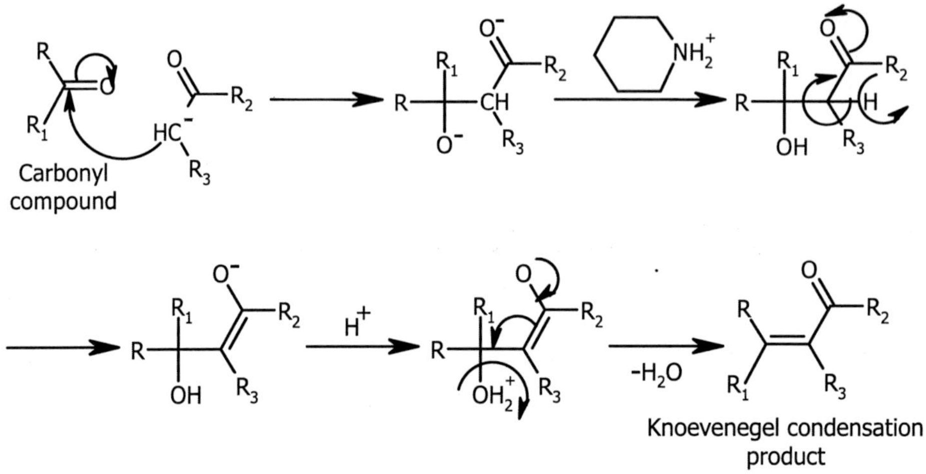

Pharmaceutical applications

- Preparation of pyrantal (anthelmintic).

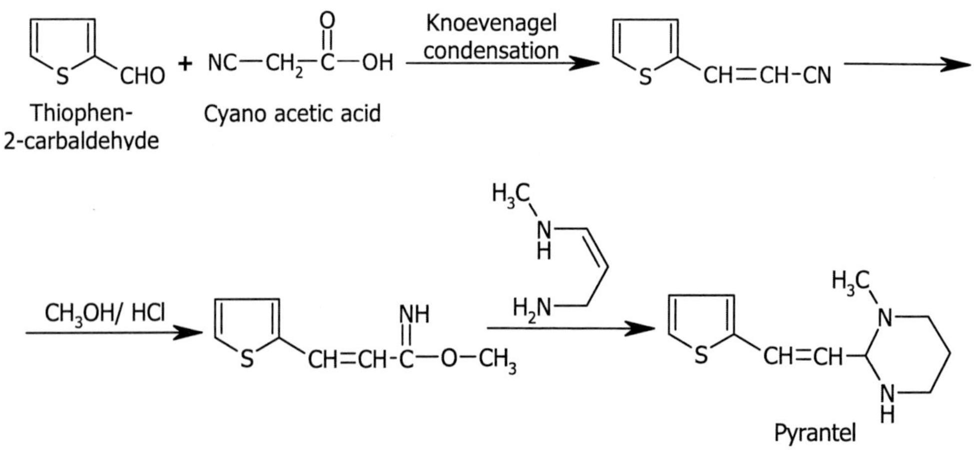

- Preparation of sulindac (anti-inflammatory drug).

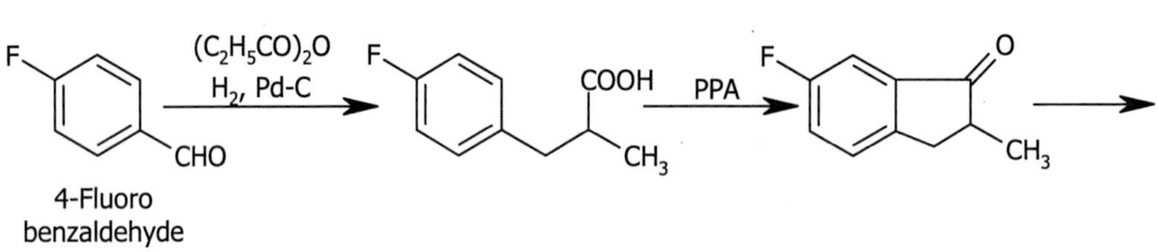

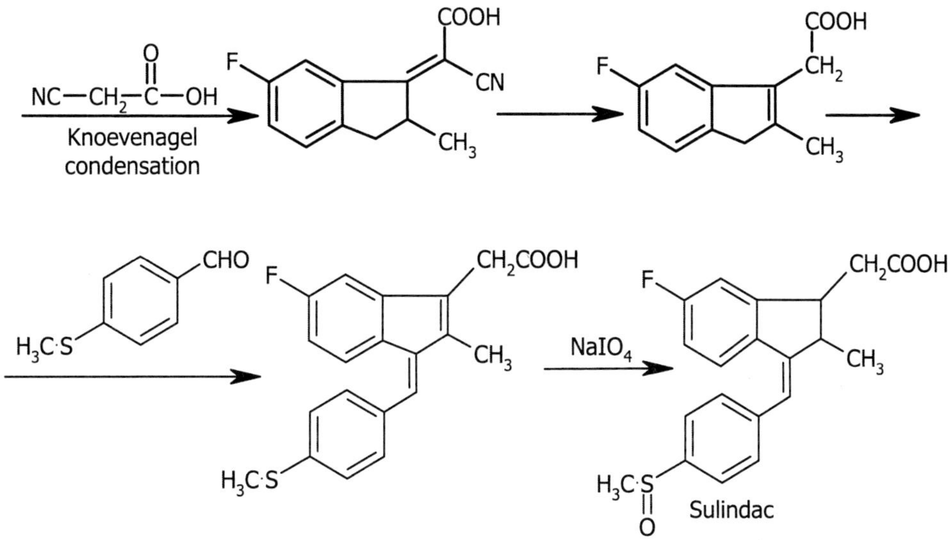

- Preparation of acenocoumarin (anti-coagulant).

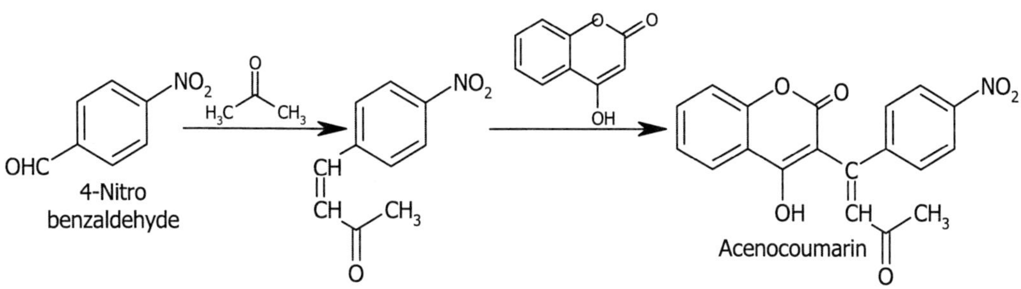

- Preparation of ethosuximide (anti-convulsant).

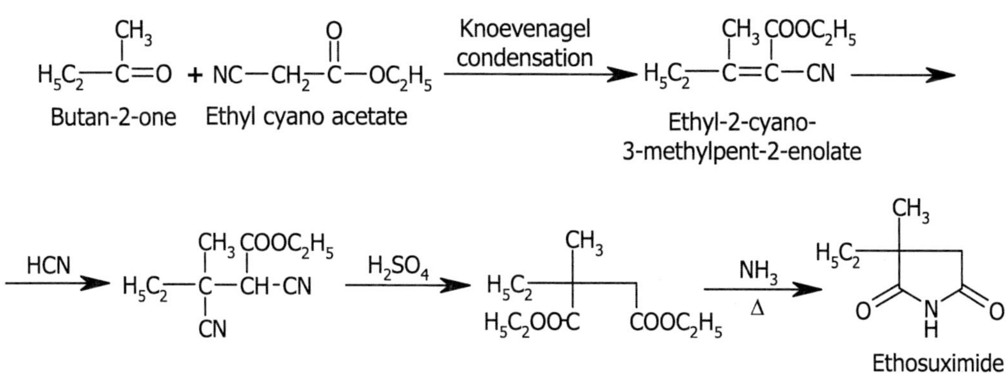

14.1 Preparation of 4-(benzylidene)-5-methyl--2,4-dihydro-3*H*-pyrazol-3-one

Chemicals: Ethyl acetoacetate, Hydrazine hydrate, Ethanol, Benzaldehyde, Glacial acetic acid and Sodium acetate (anhydrous).

Principle: 4-(Benzylidene)-5-methyl-2,4-dihydro-3*H*-pyrazol-3-one can be prepared from 5-methyl-2,4-dihydro-3*H*-pyrazol-3-one by Knoevenagel condensation. The condensation of ethyl acetoacetate and hydrazine hydrate, produce 5-methyl-2,4-dihydro-3*H*-pyrazol-3-one.

Base abstracts the active hydrogen of active methylene group of 5-methyl-2,4-dihydro-3*H*-pyrazol-3-one. The activated methylene group of 5-methyl-2,4-dihydro-3*H*-pyrazol-3-one attacks the carbonyl carbon of benzaldehyde and forms 4-(benzylidene)-5-methyl-2,4-dihydro-3*H*-pyrazol-3-one in presence of base catalyst.

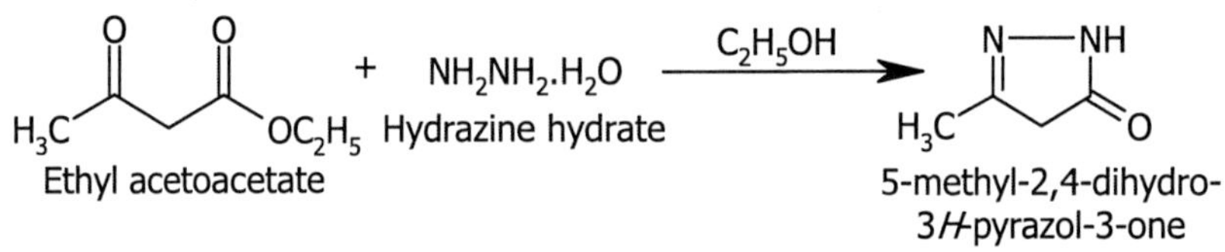

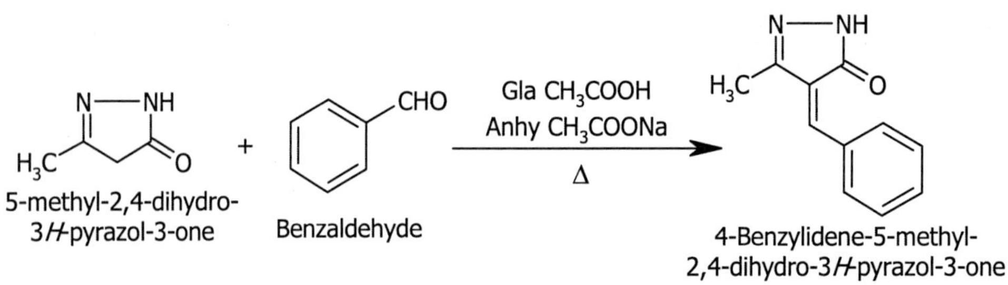

Reaction Mechanism

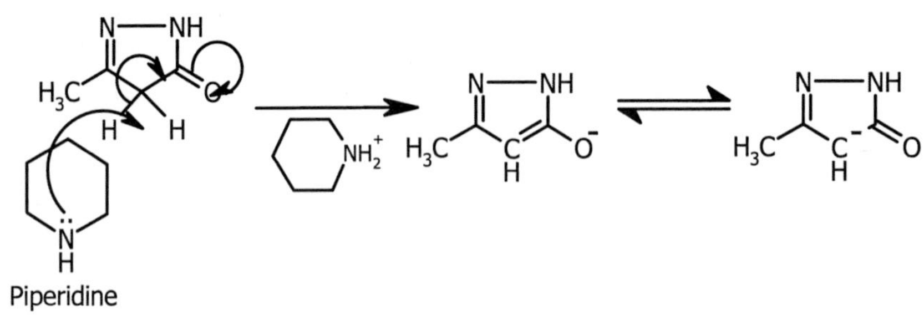

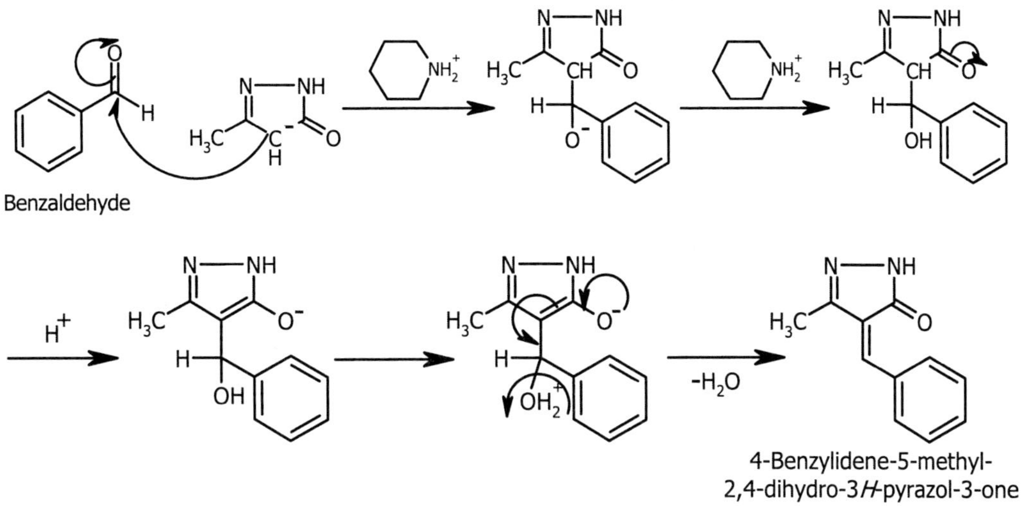

Preparation of 5-methyl-2,4-dihydro-3*H*-pyrazol-3-one

1. Add hydrazine hydrate (2 ml, 40 mmol) in ethanol (4 ml) drop wise with stirring to ethyl acetoacetate (2.6 gm, 20 mmol) loaded in a flask.
2. Stir the reaction mixture for 1 hr maintaining the temperature at 60 °C.
3. Cool the mixture to room temperature and place in an ice bath to effect the crystallization.
4. Filter the separated solid by filtration and wash with ice cold ethanol.
5. Recrystallize the product with ethanol.
6. The melting range of the pure 5-methyl-2,4-dihydro-3*H*-pyrazol-3-one is 202-205 °C.

Preparation of 4-(benzylidene)-5-methyl-2,4-dihydro-3*H*-pyrazol-3-one

1. Place 5-methyl-2,4-dihydro-3*H*-pyrazol-3-one (1.86 gm, 10 mmol), benzaldehyde (1 ml, 10 mmol), anhydrous sodium acetate (0.82 gm, 10 mmol) and glacial acetic acid (10 ml) in a round bottom flask.
2. Attach reflux condenser and reflux the reaction mixture for 10 hr.
3. Cool the mixture to room temperature and pour on to crushed ice.
4. Isolate the solid separated through vacuum filtration.
5. Purify the product by recrystallization with ethanol.
6. The melting range of the pure 4-(benzylidene)-5-methyl-2,4-dihydro-3*H*-pyrazol-3-one is 239-240 °C.

CHAPTER 15

PECHMANN REACTION

Coumarins are important synthons in the medicinal chemistry research. Coumarins can be synthesised by Pechmann, Perkin, Knoevenagel, Reformatsky and Wittig reactions. The most convenient method of synthesizing coumarins is Pechmann reaction. Condensation of phenols with a β-keto ester using acid condensing reagent produces coumarin and this reaction is known as Pechmann reaction. Acid condensing agents include sulphuric acid, phosphorous pentoxide, aluminium chloride, trichloro acetic acid and polyphosphoric acid.

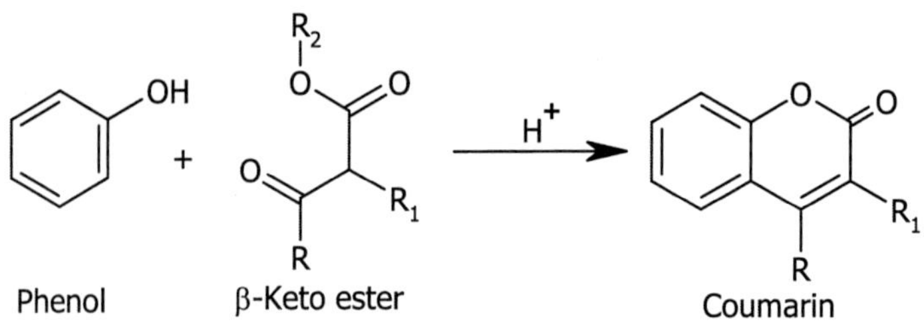

Reaction mechanism

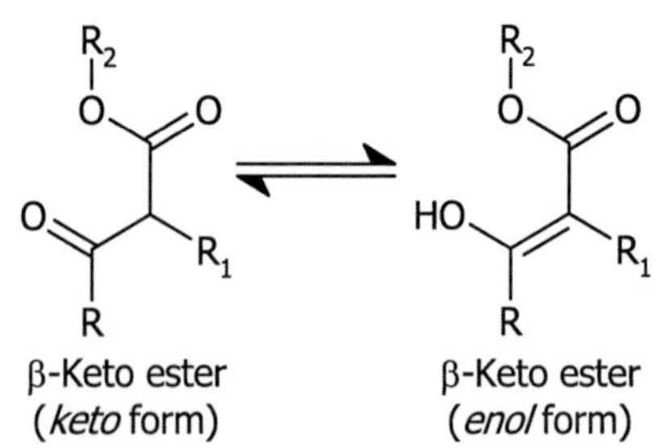

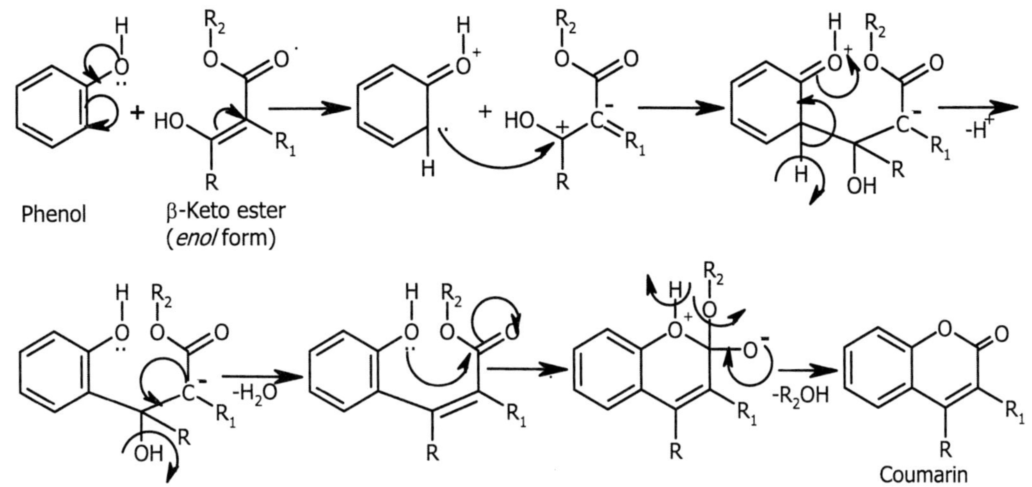

Pharmaceutical Applications

1. Preparation of umbelliferone (antioxidant)

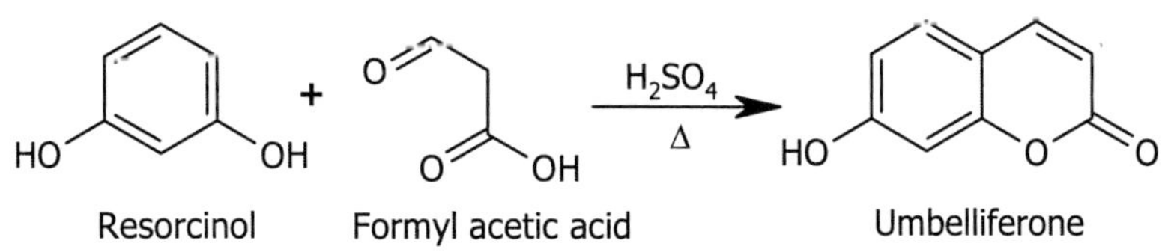

15.1 Preparation of 7-Hydroxy-4-Methyl Coumarin

Traditional Name: Hymecromone, 4-Methyl umbelliferone

Chemicals: Resorcinol, Ethyl acetoacetate, Sulphuric acid, Sodium hydroxide (10%) and Hydrochloric acid.

Principle: Condensation of resorcinol with ethyl acetoacetate using acidic condensing agent to produce 7-hydroxy-4-methyl coumarin is called as Pechmann reaction. Polyphosphoric acid is an effective condensing reagent in this reaction compared to sulphuric acid. It helps in the *trans*-esterfication and *keto-enol* tautomerism (β-keto ester).

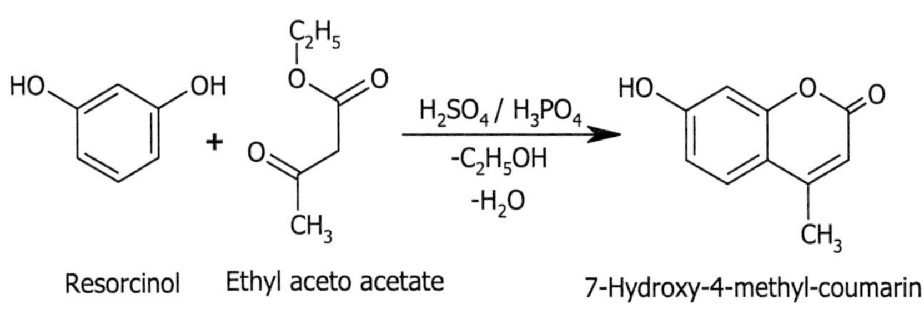

In ethyl acetoacetate, the inductive effect caused by two carbonyl oxygens weakens the C-H bonds and H-atom dissociates to give a stable anion. The acidity of the C-H bond is greatly enhanced by resonance on two carbonyl groups. Hence, ethyl acetoacetate exist in tautomeric mixture of *keto*- and *enol* form. The *enol* form of the ethyl acetoacetate (β-keto ester) readily condenses with resorcinol and renders 7-hydroxy-4-methyl coumarin.

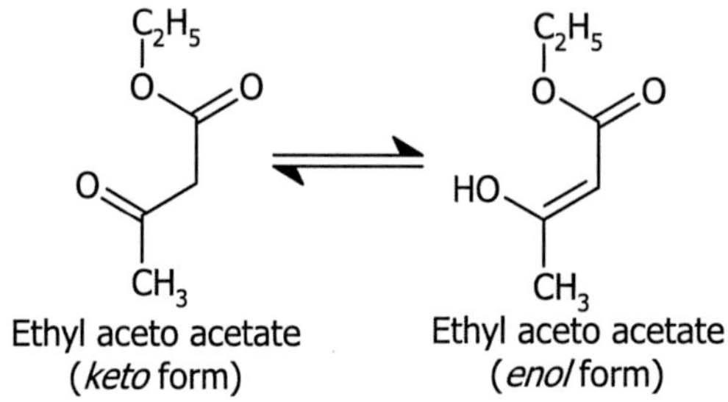

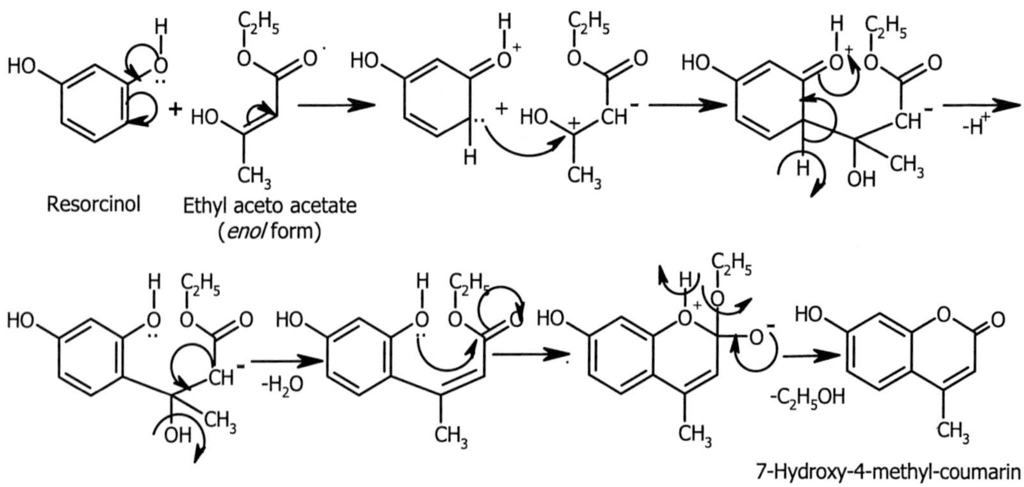

Procedure:

Method 1 (Using sulphuric acid as condensing agent)

1. Place sulphuric acid (10 ml) in a wide neck flask and stir mechanically by placing in an ice-salt bath, until the temperature of acid is below 10 °C.
2. Mix powdered resorcinol (1 gm, 9 mmol) and ethyl acetoacetete (1.3 ml, 10 mmol) by stirring to produce solution.
3. Add this mixture (resorcinol and ethyl acetoacetate) slowly to sulphuric acid containing beaker.
4. Maintain the temperature of mixture below 10 °C using ice-salt bath.

5. Continue stirring for 30 minutes and pour the reaction mixture onto crushed ice (20 gm).
6. The yellowish solid separates out and collect it by vacuum filtration.
7. Wash the solid with cold water (3 ml, three times)
8. Dissolve the solid in cold sodium hydroxide (15 ml, 5%)
9. Filter the solution and add dilute sulphuric acid (alternatively hydrochloric acid) the filtrate with vigorous stirring to precipitate the product.
10. Dry and recrystallize the product from ethanol.
11. The melting point of pure 7-hydroxy-4-methylcoumarin is 185 °C.

Method 2 (Using polyphosphoric acid as condensing agent)

1. Mix powdered resorcinol (1 gm, 9 mmol) and ethyl acetoacetete (1.3 ml, 10 mmol) in a flask by stirring to produce solution.
2. Add polyphosphoric acid (12.3 gm) to the flask containing ethyl acetoacetate and resorcinol mixture.
3. Stir the mixture mechanically and heat at 75 to 80 °C for 20 minutes.
4. Pour the reaction mixture into beaker containing ice-cold water.
5. Collect the precipitated yellow solid through vacuum filtration.
6. Wash the product with cold water and dry.
7. Purify the 7-hydroxy-4-methyl coumarin obtained by recrystallization with ethanol.
8. The melting point of pure 7-hydroxy-4-methyl coumarin is 185 °C.

CHAPTER 16

ALKYLATION

The replacement of active hydrogen of hydroxyl, amino (1° and 2°) and sulfhydryl group by an alkyl group with the help of alkylating agent is called as alkylation. Alkyl group can be transfered as a free radical, carbanion and carbene. Alkylation with one carbon alkylating agent is called as methylation and it founds major utility in synthetic organic chemistry. Alkylation can be named based on the hydrogen bearing group (which under goes replacement with alkylating agent). Alkylation with hydroxy group is named as *O*-alkylation, amino alkylation is known as *N*-alkylation and alkylation at sulfhydryl group is known as *S*-alkylation.

Alkylaing agents: Alkylating agents are grouped as:

1. ***Nucleophilic alkylating agent:*** These agents displace halide substitutents on carbon (alkyl and aryl halides). Alkylating agents of this category includes grignard reagent, organolithium reagent, organocopper reagent, organo sodium reagent.
2. ***Electrophilic alkylating agent:*** These agents alkylate aromatic compounds (Friedel-Crafts reaction), amines (C-N bonds), hydroxyl, carboxylic and thiol groups. Electrophilic alkylating agents are very toxic in nature.
3. ***Carbene alkylating agent:*** Carbenes are extremely reactive and attack even unactivated C-H bonds. Diazo group upon elimination produces carbenes.

For different reaction conditions, different alkylating agents are preferred.

- *Alkylating agents for acidic conditions*: Methyl iodide, methyl tosylate, dimethyl sulphate and methoxy acetate.
- *Alkylating agents for basic conditions*: Methanol acidified with hydrochloric acid and boron trifluride.
- *Alkylating agents for neutral conditions*: Diazo methane, triazene, dimethyl formamide acetal, isourea and orthoacetal.

16.1 Preparation of Methyl-2-Naphthyl Ether

Traditional Name: Nerolin ***IUPAC Name***: β-Napthol methyl ether

Chemicals: 2-Naphthol, Dimethyl sulphate and Sodium hydroxide and Ehanol.

Principle: Preparation of methyl-2-naphthyl ether involves O-alkylation reaction. Alkylation of 2-naphthol (β-naphthol) using dimethyl sulphate in presence of sodium hydroxide produces methyl-2-naphthyl ether.

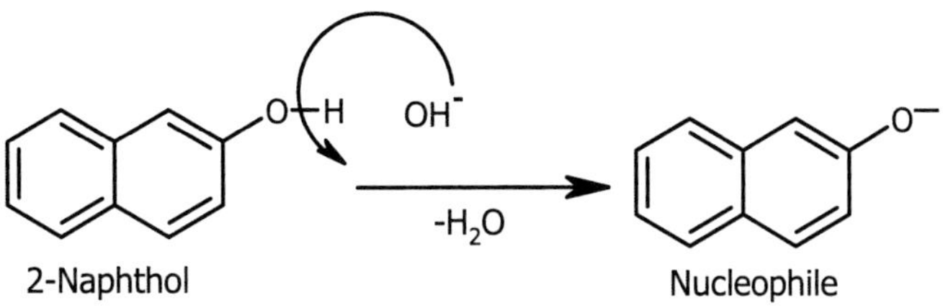

Sodium hydroxide abstracts hydrogen from 2-naphthol and generates nucleophile, 2-naphthoxide (sodium-2-naphthoxide). Sodium 2-naphthoxide undergoes alkylation with dimethyl sulphate and forms methyl-2-naphthyl ether.

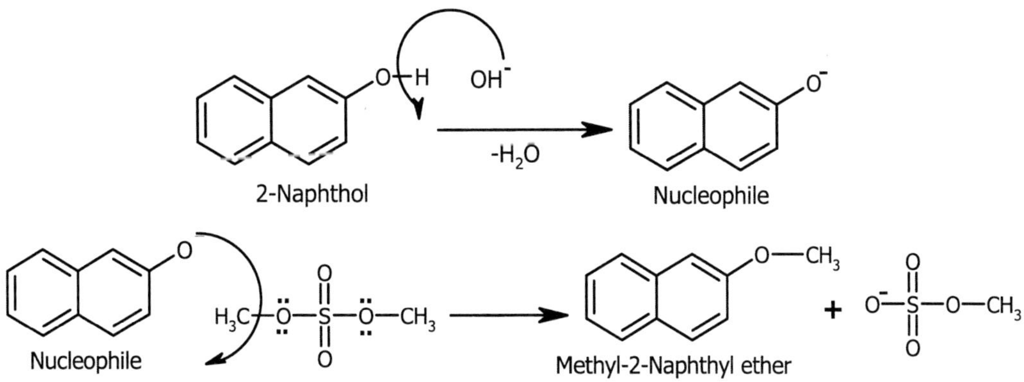

Procedure:

1. Dissolve 2-naphthol (1 gm, 6.9 mmol) in the solution of sodium hydroxide (0.3 gm) in water (5 ml) contained in the flask.
2. Stir, cool the warm mixture to about 10 °C by placing the flask in an ice-bath.
3. Add dimethyl sulphate (0.65 ml, 6.9 mmol) from dropping funnel cautiously, with vigorous stirring.
4. Warm the mixture for 1 hr at 70-80 °C with vigorous stirring.
5. Cool the reaction mixture and filter the methyl-2-naphthyl ether under suction.
6. Wash the residue with sodium hydroxide (10%) and then with water to remove unreacted reactants.
7. Dry the residue and recrystallize it from ethanol.
8. Melting point of the pure methyl-2-naphthyl ether is 72 °C.

Precautions:

1. Dimethyl sulphate should be used in fume cupboard and rubber gloves should be worn.
2. Both vapour and liquid dimethyl sulphate are highly poisonous and inhalation leads to giddiness.
3. If the liquid is accidently splashed on the skin, wash immediately with ammonium hydroxide and then rub gently with cotton wool soaked in ammonia.

16.2 Preparation of Antipyrine

Chemicals: Ethyl acetoacetate, Phenyl hydrazine, Diethyl ether, Sulphuric acid, Sodium hydroxide and Dimethyl sulphate.

Principle: The condensation of ethyl acetoacetate and phenyl hydrazine produce 3-methyl-1-phenyl-pyrazole-5-one, which upon methylation with dimethyl sulphate gives antipyrine. In ethyl acetoacetate, inductive effect caused by two carbonyl groups weakens the C-H bond of active methylene group. Thus the H-atom dissociates and generates stable anion. The acidity of the C-H bond is greatly enhanced by resonance, hence ethyl acetoacetate exists in *keto* and *enol* forms (tautomerism). The *enol* form of the ethyl acetoacetate readily condenses with phenyl hydrazine and forms 3-methyl-1-phenyl-pyrazole-5-one. The 3-methyl-1-phenyl-pyrazole-5-one upon methyl-tion with dimethyl sulphate gives antipyrine (2,3-dimethyl-1-phenyl pyrazole-5-one).

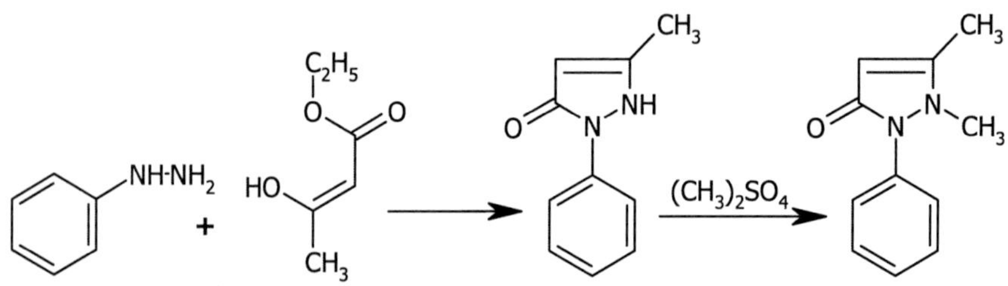

Phenyl hydrazine — Ethyl aceto acetate (*enol* form) — 3-Methyl-1-phenyl-pyrazole-5-one — Antipyrine (2,3-Dimethyl-1-phenyl-pyrazole-5-one)

Reaction mechanism

Formation of 3-methyl-1-phenyl-pyrazole-5-one

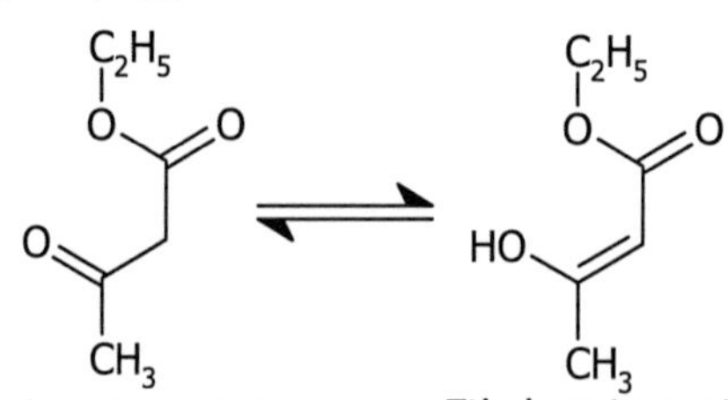

Ethyl aceto acetate (*keto* form) — Ethyl aceto acetate (*enol* form)

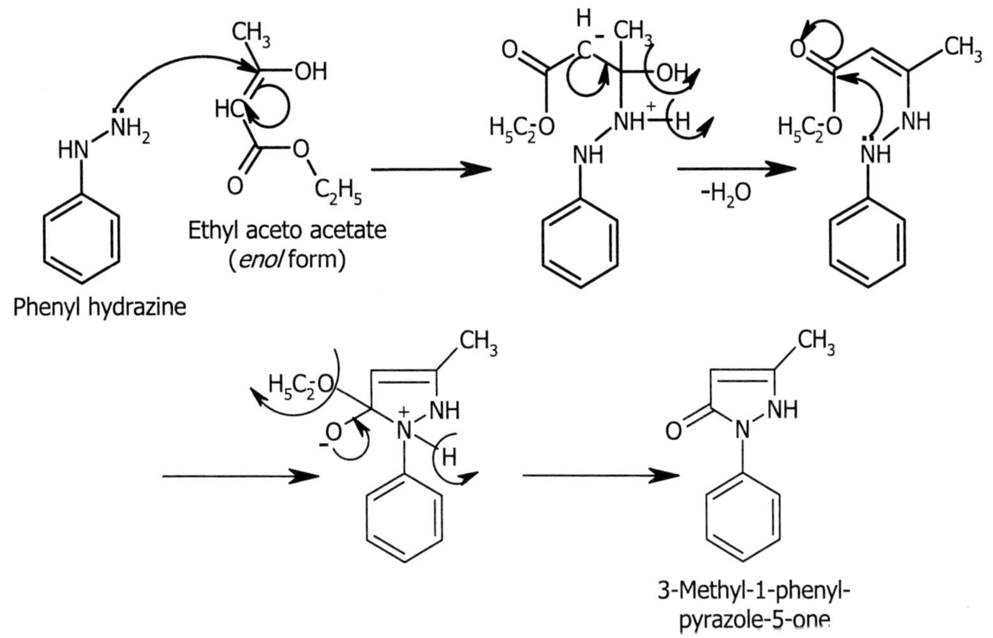

***N*-Methylation of 3-methyl-1-phenyl-pyrazole-5-one**

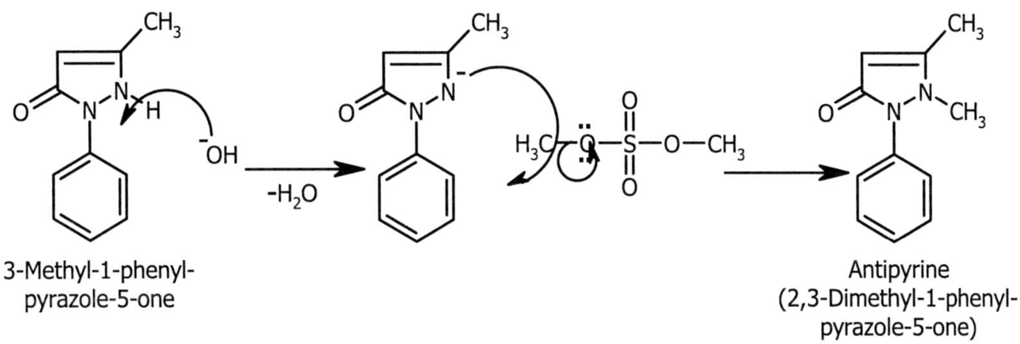

Procedure:

Step 1: Preparation of 3-methyl-1-phenyl pyrazol-5-one

1. Mix distilled ethyl acetoacetate (2 ml, 16 mmol) and phenyl hydrazine (1.5 ml, 15 mmol) in a china dish.
2. Heat the mixture by placing the dish on a water bath (set in a **fume** cupboard) for 2 hr with stirring using glass rod.
3. Allow the heavy red syrup to cool, add diethyl ether (4 ml) and stir the mixture vigorously.
4. Keep it aside for 15 minutes and filter the solid by vacuum filtration.
5. Wash the solid obtained with ether to remove coloured impurities.

6. Dry and recrystallize the product from hot water [or ethanol-water mixture (1:1)].
7. The melting point of 3-methyl-1-phenyl pyrazol-5-one is 127 °C.

Step 2: Preparation of 2,3-dimethyl -1-phenyl-pyrazol-5-one (antipyrine)

1. Place the solution of sodium hydroxide (230 mg) in water and the solution of 3-methyl-1-phenyl-pyrazol-5-one (1 gm, 6 mmol) in methanol (1 ml) into a round bottom flask, set in a fume hood.
2. Warm the mixture by placing the flask on a water bath and add dimethyl sulphate (0.66 ml, 7 mmol) from dropping funnel (alternatively burette can be used).
3. Reflux the mixture for 1 hr with stirring and allow it to cool to room temperature.
4. Evaporate the ethanol and add hot water to the residue.
5. Filter and extract the filtrate with benzene.
6. Evaporate the solvent and recrystallize the antipyrine with hot water or benzene.
7. The melting point of pure antipyrine is 113 °C.

16.3 Preparation of 7-Methoxy-4-Methyl-Coumarin

Chemicals: 7-Hydroxy-4-methyl coumarin, Dimethyl sulphate, Potassium carbonate, Acetone and Ethanol.

Principle:

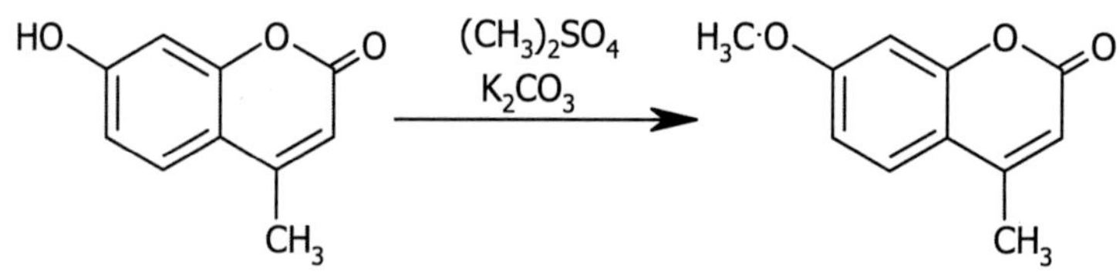

7-Hydroxy-4-methyl-coumarin 7-Methoxy-4-methyl-coumarin

Procedure:

1. Place 7-hydroxy-4-methyl coumarin (5.53 gm, 33 mmol) in round bottom flask containing acetone (75 ml).
2. Add dimethyl sulphate (3.55 ml, 0.34 mol), ignited potassium carbonate (12 gm) to the mixture and reflux for 4 hr under anhydrous condition.
3. Filter the acetone solution, remove the inorganic salts by washing with hot acetone (25 ml).
4. Distill the combined acetone solution and macerate the oily residue with ice.
5. Isolate the separated product, wash and crystallize with ethanol.
6. The melting range of pure 7-methoxy-4-methyl coumarin is 216-218 °C.

Note:

1. Refer 15.1 (page 127) for the preparation of 7-hydroxy-4-methyl coumarin.

CHAPTER 17

PERKIN CONDENSATION

Aromatic aldehydes upon heating with anhydrides of aliphatic acids containing two α hydrogen atoms in presence of its sodium salt (basic catalyst) condenses to give β-aryl-acrylic acid (α,β-unsaturated acid). This condensation reaction is named as Perkin condensation. Perkin condensation helps in the addition of two carbon atoms to the side chain of an aromatic ring. Base catalysts include sodium acetate, potassisum acetate, salts of carboxylic acid, carbonates of alkali metals and triethyl amines. Preparation of cinnamic acid by the condensation of acetic anhydride with benzaldehyde in presence of fused sodium/potassium acetate is a typical example for Perkin condensation. This method is suitable for the synthesis of β-acrylic acids (cinnamic acid and its derivatives).

Enolate anion of the acetic anhydride attack aromatic aldehyde and forms C=C bond. The reaction involves intermolecular aldol-type condensation followed by dehydration. Perkin condensation doesn't take place with aliphatic aldehydes. A special case of perkin reaction is the condensation of benzaldehyde with cyclic anhydrides such as succinic anhydrides and gives lactone.

Pharmaceutical Applications

1. Used in the preparation of x-ray contrast media - iopanoic acid (iodo aromatics).

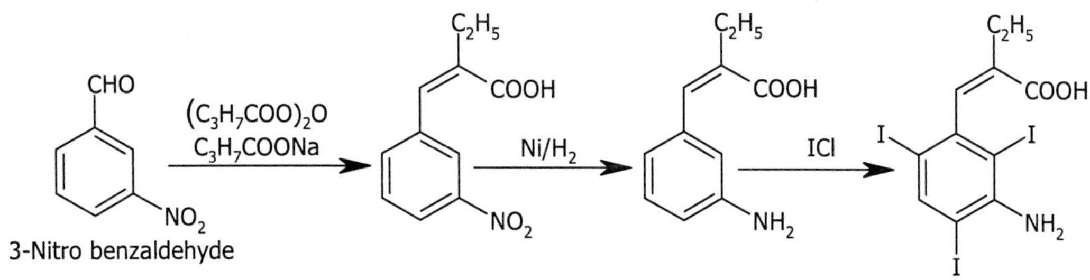

Iopanoic acid

17.1 Preparation of Cinnamic Acid

Traditional Name: Cinnamylic acid, Phenylacrylic acid. ***IUPAC Name*:** 3-Phenyl-prop-2-enoic-acid

Chemicals: Benzaldehyde, Acetic anhydride, Potassium acetate, Hydrochloric acid and Sodium carbonate.

Principle: Acetic anhydride condenses with benzaldehyde in presence of Potassium acetate and gives cinnamic acid.

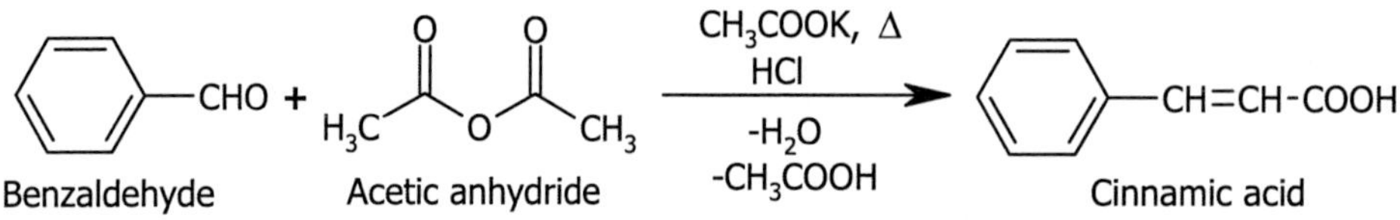

The potassium acetate (acyloxy ion) acts as a base and abstracts the proton from methyl group of acetic anhydride to generate nucleophile. Nucleophile of acetic anhydride attacks positively charged carbon (due to positive inductive effect) of carbonyl group in benzaldehyde and forms intermediate. This intermediate on heating eliminates water molecule and on hydrolysis removes acetate to yield cinnamic acid.

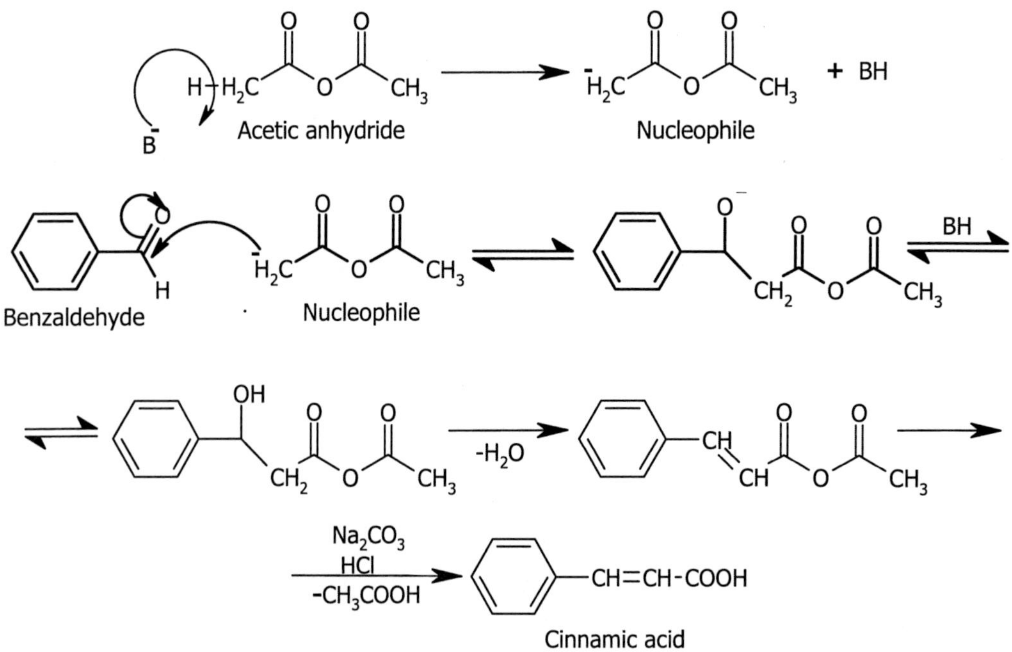

Procedure:

1. Place benzaldehyde (2 ml, 20 mmol), acetic anhydride (2.8 ml, 29 mmol) and finely powdered fused potassium acetate (1.2 gm, 12 mmol) in a dry round bottom flask.
2. Attach condenser which in turn fitted with calcium guard tube to the flask.

3. Mix well and heat the mixture in an oil bath/sand bath at 160 °C for 1 hr and at 170-180 °C for 3 hr.
4. Pour the hot mixture into round bottom flask containing water (10 ml).
5. Add saturated sodium carbonate solution with shaking until a drop of withdrawn solution turns red litmus into distinct blue.
6. Distill (steam distillation) the solution to remove unreacted benzaldehyde (distillate should be clear) and discard the distillate.
7. Cool the residual solution and filter to remove resinous materials.
8. Acidify the filtrate by adding hydrochloric acid slowly and with vigorous stirring (*until the evolution of carbon dioxide ceases*).
9. Cool, filter the precipitate and wash with cold water.
10. Dry and purify the residue by recrystallization with hot water or water-ethanol (3:1) mixture.
11. The melting point of pure cinnamic acid is 133 °C.

Note:

1. *Fused potassium acetate*: Heat potassium acetate in a porcelain dish to liquify, then allow it to solidify. Remove the residual water by heating in a large flame. Powder the solid and store in tightly stoppered bottle.
2. Sodium hydroxide should not be used in place of sodium carbonate to neutralize as it may generate benzoic acid.
3. Potassium acetate acts as a catalyst, sodium acetate, potassium carbonate and triethyl amine also can be used.

17.2 Preparation of *O*-Acetoxy Cinnamic Acid and Coumarin

IUPAC Name: 2-Acetoxy-3-phenyl propionic acid

Chemicals: Salicylaldehyde, Acetic anhydride, Triethyl amine and Hydrochloric acid.

Principle: Acetic anhydride condenses with salicylaldehyde in presence of triethyl amine and gives *ortho*-hydroxy cinnamic acid. The *ortho*-hydroxy cinnamic acid through *O*-acetylation by acetic anhydride generates *O*-acetyl cinnamic acid.

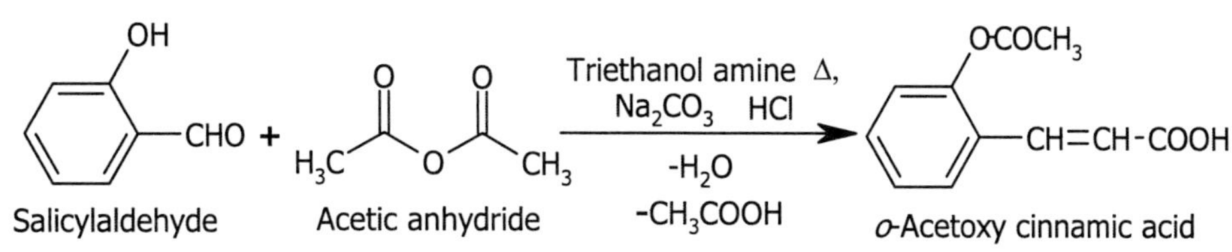

Triethyl amine acts as a base and abstracts the proton from methyl group of acetic anhydride and generate nucleophile. Nucleophile of acetic anhydride attacks positively charged carbon (due to positive inductive effect) of carbonyl group in salicylaldehyde and forms an intermediate. This intermediate on heating eliminates water molecule and on hydrolysis removes acetate to yield *ortho*-hydroxy cinnamic acid. The acetyl group formed from acetic anhydride acts as an electrophile and attacks hydroxyl group of *ortho*-hydroxy cinnamic acid. The *ortho*-hydroxy cinnamic acid through O-acetylation with acetic anhydride produces O-acetyl cinnamic acid.

Preparation of ortho-Hydroxy Cinnamic Acid

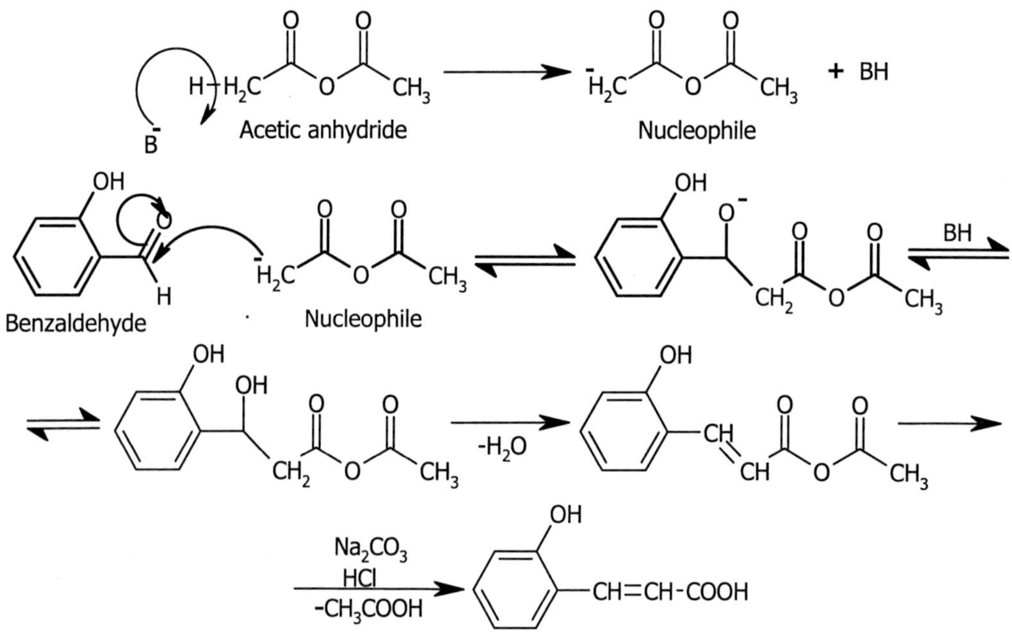

o-Hydroxy cinnamic acid

Preparation of O-Acetoxy Cinnamic Acid

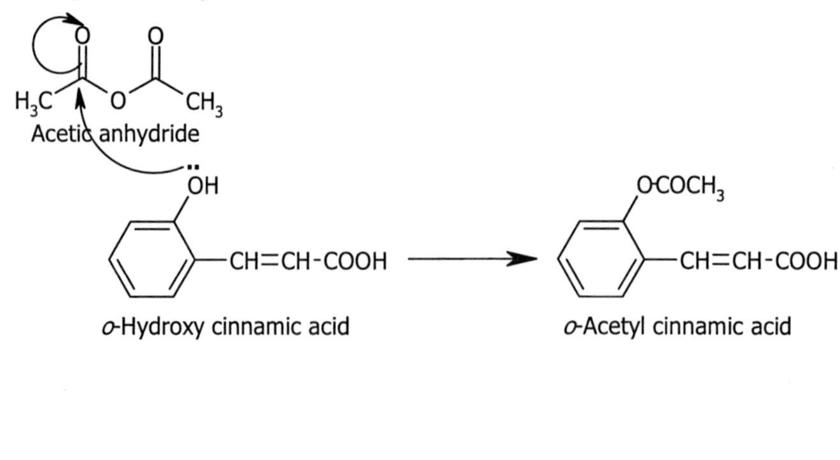

Procedure:

1. Place salicylaldehyde (2 ml, 19 mmol), anhydrous triethylamine (5.6 ml, 58 mmol) and acetic anhydride (2.5 ml, 26 mmol) in a round bottom flask.
2. Attach condenser which in turn fitted with calcium guard tube to falsk.
3. Mix well and reflux the mixture gently for 12 hr.
4. Cool the contents, steam distill and discard the distillate (to remove unreacted salicylaldehyde).
5. Add sodium bicarbonate slowly to the residue containing flask, until red litmus turns to blue.
6. Cool the contents by placing in an ice bath.
7. Filter the reaction mixture to isolate the solid product (coumarin) and preserve it separately.

Preparation of O-Acetoxy Cinnamic Acid

8. Filtrate on acidification with hydrochloric acid gives *o*-acetoxy cinnamic acid.
9. Purify the product by recrystallization with propanol (70%).
10. The melting range of pure *o*-acetoxy cinnamic acid is 153-154 °C.

Preparation of Coumarin

11. Boil the isolated solid product (Coumarin) with water (200 ml) and charcoal (0.2 g).
12. Filter the solution and concentrate the solution to 80 ml volume.
13. Cool the solution and collect the solid product crystallizing out.
14. Purify the coumarin by recrystallization with methanol.
15. The melting range of pure coumarin is 68 to 69 °C.

CHAPTER 18

WOHL-ZIEGLER REACTION

Wohl-Ziegler reaction involves the allylic or benzyllic bromination of hydorocarbons using an *N*-Bromo succinamide (NBS) in carbon tetrachloride. Bromination occurs in allylic position of olefinic bond. NBS is convenient source of cationic bromine, used in radical substitution and electrophilic addition reactions.

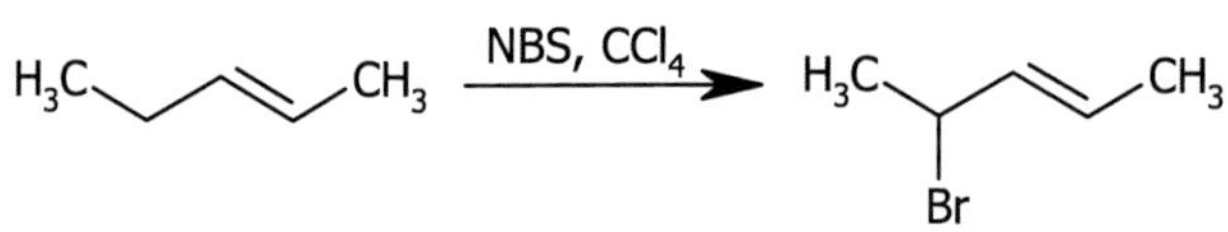

Pentene 4-Bromo pentene

Reaction mechanism

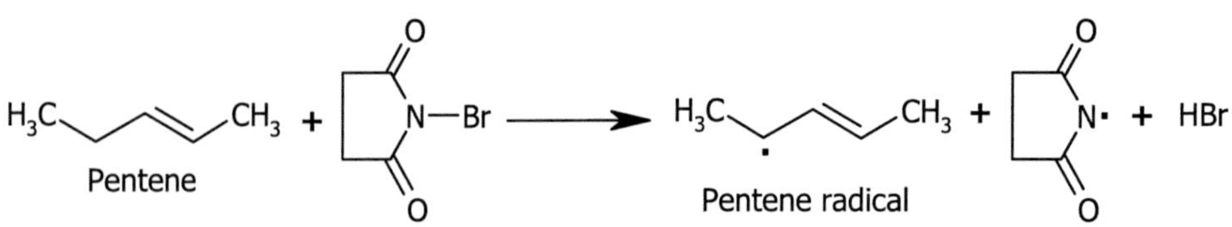

Pentene N-Bromo succinamide Pentene radical

N-Bromo succinamide

$$Br_2 \longrightarrow 2\,\dot{Br}$$

Bromine Bromine radical

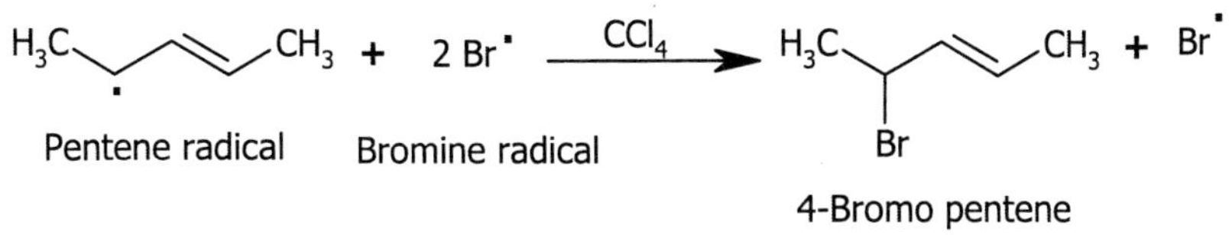

Precautions:

1. NBS decomposes over time giving off bromine, hence it should be stored in a refrigerator.
2. Reactions involving NBS are exothermic, extra precautions are required for large scale preparations.

Pharmaceutical Applications

1. Preparation of carbamazepine.

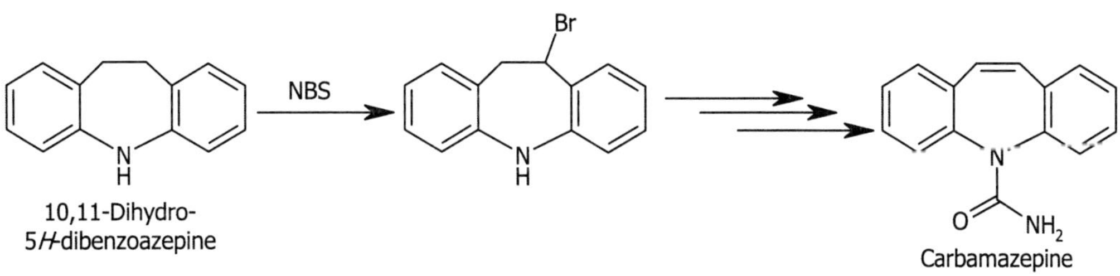

2. Preparation of Vitamin A_1 and A_2.

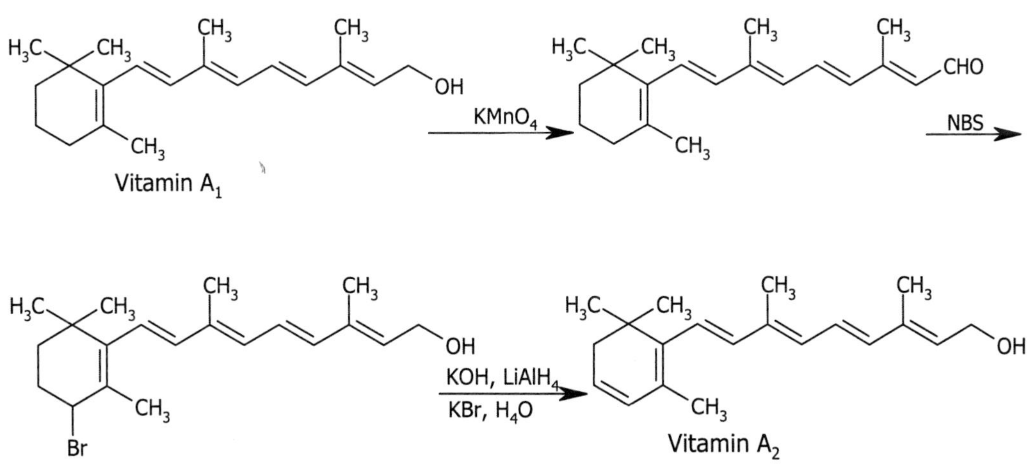

18.1 Preparation of 3-Bromo Cyclohexene

Chemicals: Cyclohexene, *N*-Bromo succinamide, Carbon tetrachloride and Benzoyl peroxide.

Principle: Bromination of cyclohexene with *N*-Bromo succinamide generates 3-bromo cyclohexene.

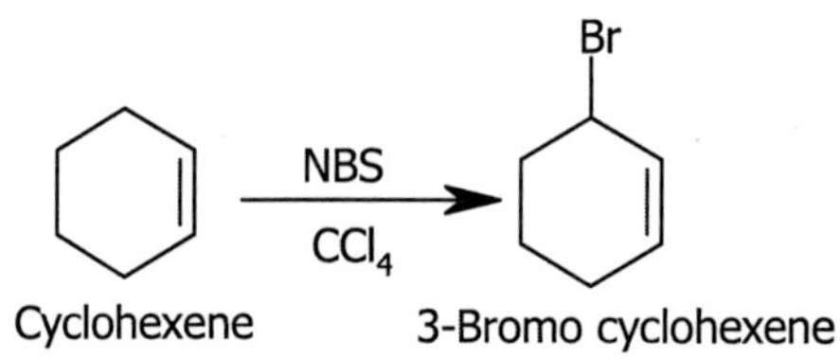

Abstraction of hydrogen by the *N*-Bromo succinamide forms hydrogen bromide and cyclohexene radical. Hydrogen bromide reacts with *N*-Bromo succinamide and gives bromine. Benzoyl peroxide converts bromine into bromine radical. The formed cyclohexene and bromine radicals associate to form 3-bromo cylcohexene. Bromination should be carried out in presence of anhydrous carbon tetrachloride or chloroform. Presence of water may hydrolyse *N*-bromo succinamide. Succinamide formed after the bromination is insoluble in carbon tetrachloride or chloroform, hence it can be removed by filtration.

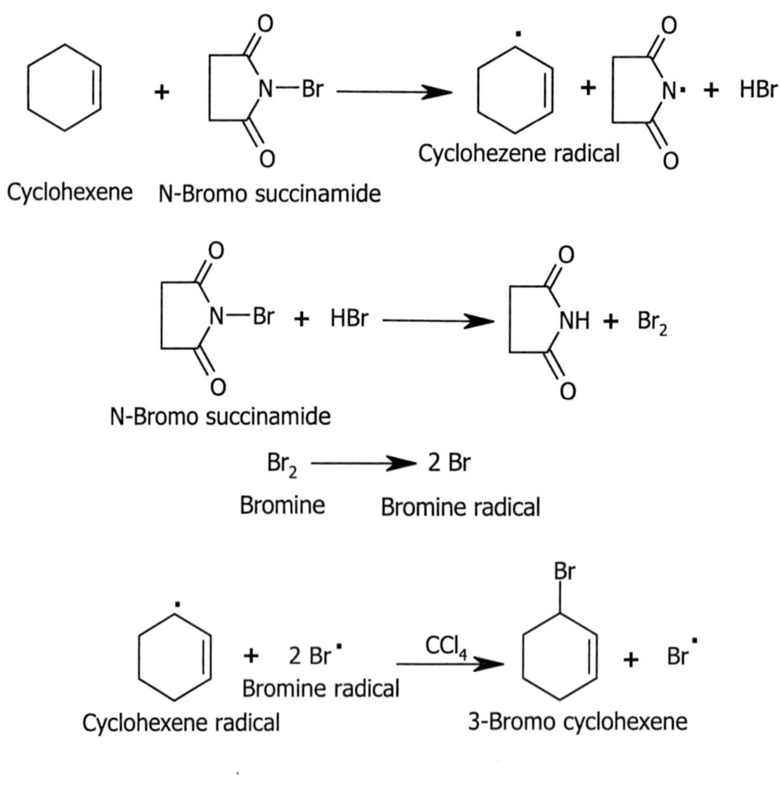

Procedure:

1. Place *N*-bromo succinamide (0.98 gm, 5 mmol), redistilled cyclohexene (1.23 gm, 15 mmol), carbon tetrachloride (4 ml) and benzoyl peroxide (12.5 mg) in a round bottom flask (*benzoyl peroxide must be previously dried by pressing between filter papers*).
2. Attach reflux condenser and allow it to stand at room temperature.
   (*The mixture becomes warm, yellowish N-bromosuccinamide becomes colourless and gets suspended in the flask*).
3. Transfer the flask to steam bath and heat under reflux for 2 hr (*until all the N- bromosuccinamide gets converted into succinamide*).
4. Cool, filter under suction and wash the residue with little carbon tetrachloride.
5. Distill the combined filtrates to remove carbon tetrachloride (b.p. 77 °C) and unreacted cyclohexane (b.p. 83 °C).
6. Distil the residue under reduced pressure to collect 3-bromo cyclohexene at 72-77 °C (*carbon tetrachloride and cyclohexane will come in first fraction*).
7. The following fraction goes brown on standing, subsequent distillation gives colourless 3-bromo cyclohexene.

Note: The completion of the reaction may be confirmed by treating starch-iodide paper with a drop of the reaction mixture. The appearance of no colour in starch iodide paper indicates the completion of the reaction.

Precautions:

1. Reactions involving N-bromo succinamide (NBS) are exothermic, therefore extra precautions should be taken in large scale preparations.
2. It should be stored in a refrigerator.
3. It will decompose over time giving off bromine.

CHAPTER 19

FISCHER INDOLE SYNTHESIS

Indole nucleus present in natural as well as synthetic molecules plays a vital role in the process of drug discovery and development. 2-Phenylindole present in the neurotransmitter serotonin (5-hydroxy tryptamine), anti-inflammatory drug indomethacin, dopamine D4 agonist cyanoindoles and lysergic acid. Psilocin [(4-hydroxy-dimethyltryptamine) a structural analogue of serotonin], sulphur analog of psilocin and N-methyl psilocin produces partial 5-HT_{2c} activation.

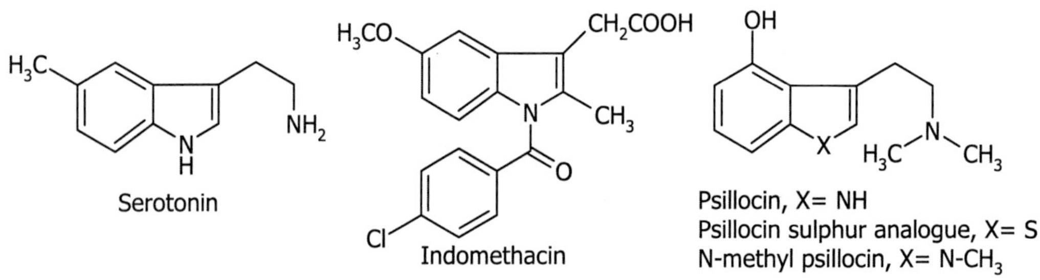

Formation of indoles on heating aryl hydrazones of aldehydes/ketones in the presence of acid catalysts (lewis/proton acids) is known as Fischer indole synthesis. Fischer indole synthesis involves the following two steps:

1. **Formation of aryl hydrazone:** Condensation of phenyl ketones, such as acetophenone with substituted hydrazines forms aryl hydrazones.
2. **Cyclization of aryl hydrazone:** Hydrazone undergoes cyclization in presence of acid catalysts and yields substituted 2-phenyl indole. One of the nitrogen present in the hydrazine will be eliminated as ammonia.

Acid catalysts such as zinc chloride, sulphuric acid, boron trifluride, methane sulphonic acid and polyphosphoric acid (PPA) are useful in cylcization of aryl hydrazones into phenyl indole.

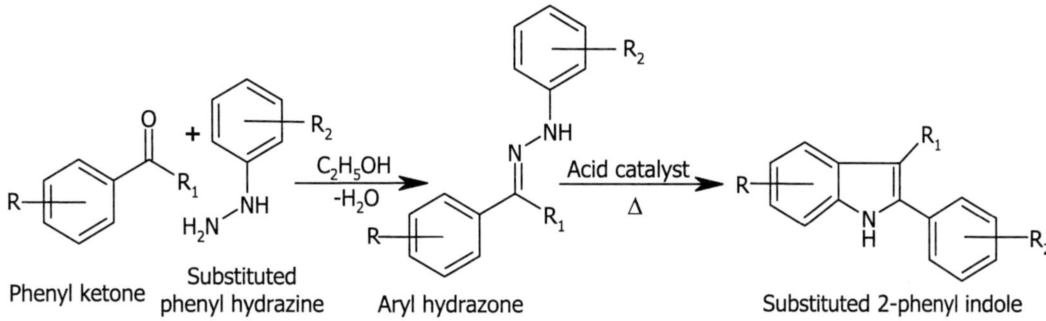

Reaction mechanism

M. Robinson and G.M. Robinson (1918) together explained the enolization (similar to that of carbonyl compounds) imine nitrogen. Tautomeric shift of hydrogen from the methyl group (α-carbon) to nitrogen makes it more basic. Protonation of imine nitrogen initiates the attack of α-carbon on the aromatic ring and cleaves N-N bond. Polyphosphoric acid is the source of proton and serves as a solvent and drying agent (water hydrolyses imine). Elimination of β-hydrogen restores the aromaticity and generates nucleophilic center on the nitrogen bound ring. Nucleophilic center attacks imine carbon and forms 5-membered ring. Loss of nitrogen as amine group generates 2-phenyl indole. Polyphosphoric acid forms ammonium phosphate with the liberated ammonia and drives the reaction to completion.

Formation of arylhydrazone

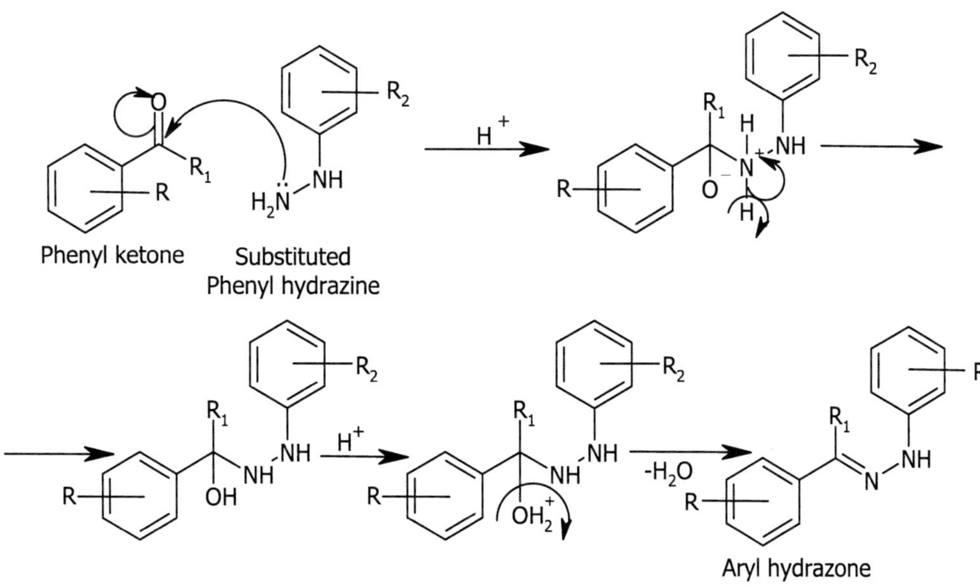

Cyclization into substituted 2-phenyl indole

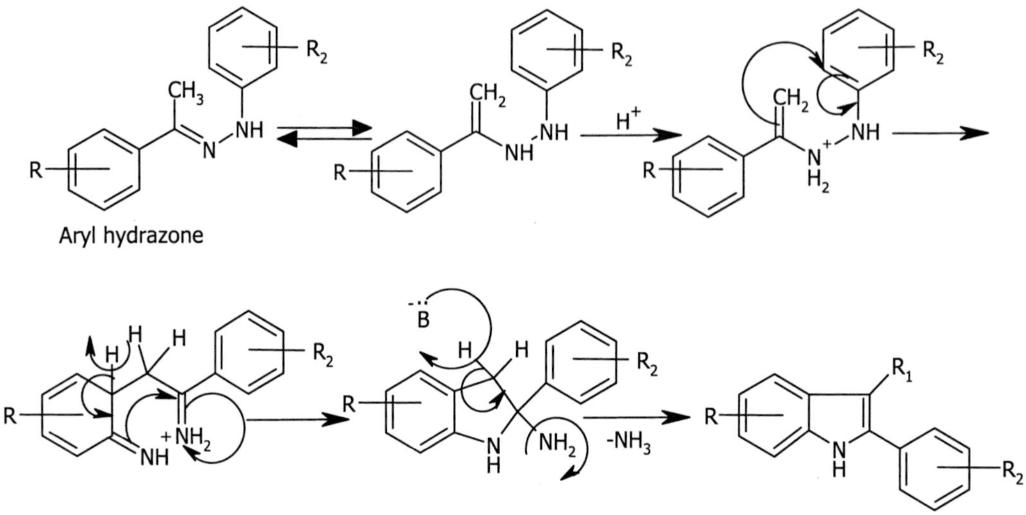

Pharmaceutical Applications

1. Indoles with substitution at 2nd position produces beneficial biological effects, where as substitution at 3rd position exhibit both positive and negative effects. Hence 3-substitued indoles are given importantance in drug discovery process.
2. An anti-inflammatory drug, indomethacin is a derivative of 2-phenyl indole.

19.1 Preparation of 2-Phenyl Indole

Chemicals: Acetophenone, Phenyl hydrazine, Ethanol, Glacial acetic acid, Hydrochloric acid and Polyphosphoric acid.

Principle: Polarization of acetophenone carbonyl group generates electron deficient carbonyl carbon. Lone pair of electrons present in nitrogen of phenyl hydrazine attacks (nucleophilic attack) electron deficient carbon and forms an intermediate. Protonation followed by dehydration produces aryl hydrazone. Polyphosphoric acid is the source of proton and serves as a solvent and drying agent. Imine nitrogen of aryl hydrazone undergoes enolization and tautomeric shift of the hydrogen of α-carbon to the imine nitrogen makes it more basic. Protonation of this nitrogen allows the attack of the α-carbon on the aromatic ring and breaks N-N bond of phenyl hydrazine.

Elimination of β-hydrogen restores the aromaticity and forms nucleophilic center on the nitrogen bound ring. Nucleophilic center attacks imine carbon and forms 5-membered ring. Loss of nitrogen as amine group generates 2-phenyl indole. Poly phosphoric acid forms ammonium phosphate with the liberated ammonia and drives the reaction to completion.

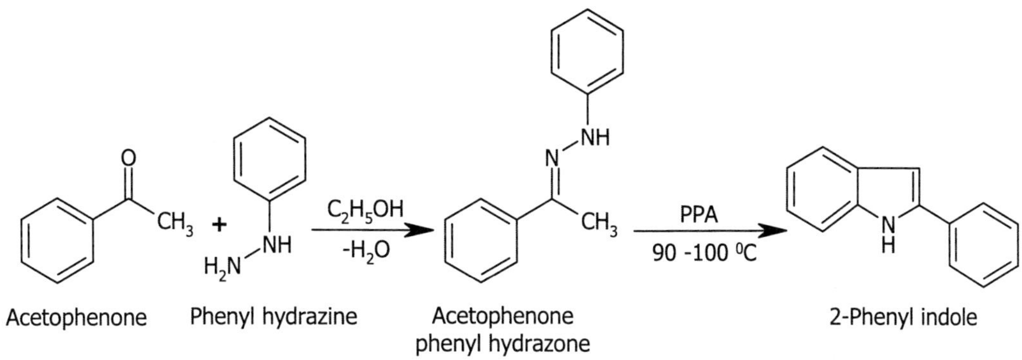

Formation of acetophenone phenylhydrazone

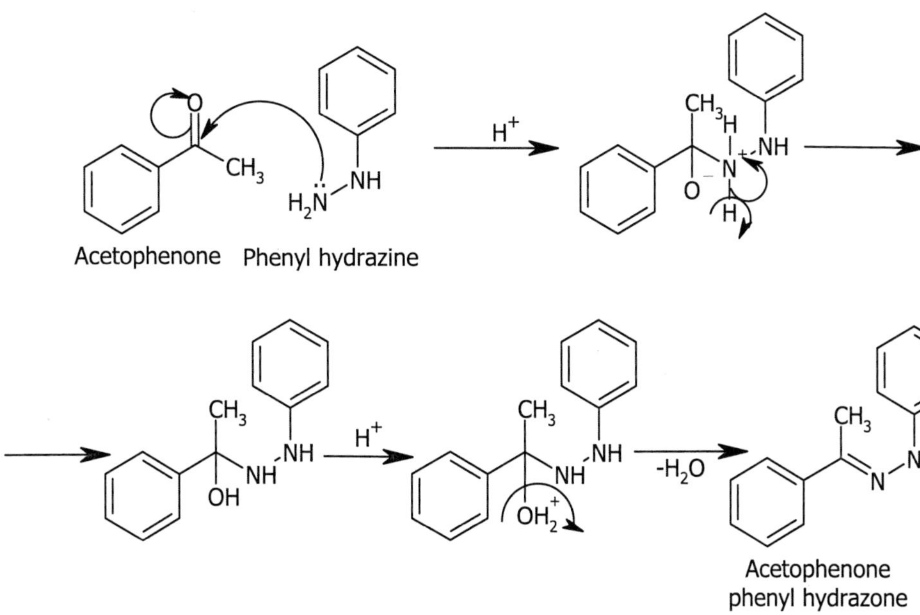

Cyclization into 2-phenyl indole

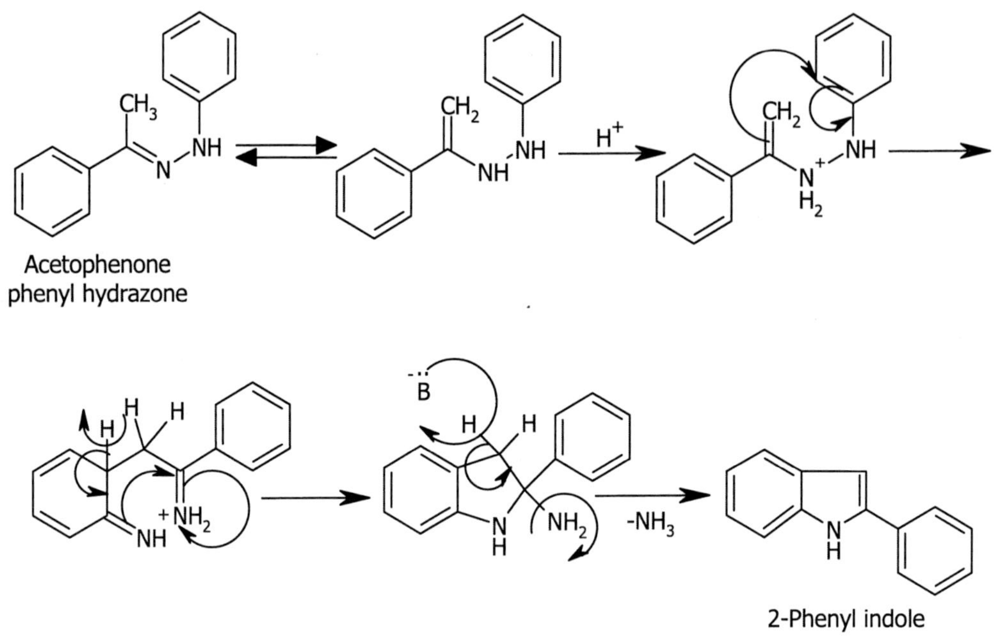

Procedure

Step 1: Preparation of acetophenone phenylhydrazone

1. Mix acetophenone (5 ml, 42 mmol) and phenyl hydrazine (4.1 ml, 42 mmol) with ethanol (15 ml) and few drops of glacial acetic acid.
2. Warm the mixture and cool to room temperature.
3. Filter the reaction mixture, wash the solid with dilute hydrochloric acid and cold ethanol (4 ml).
4. Purify the product by recrystallization with ethanol.
5. The melting point of pure acetophenone phenylhydrazone is 106 °C.

Step 2: Preparation of 2-phenyl indole

1. Place acetophenone phenylhydrazone (1 g) in a beaker containing polyphosphoric acid (6.4 gm).
2. Heat the mixture by placing on a water bath, stirr and maintain the temperature between 100-120 °C for about 10 minutes (exothermic reaction).
3. Add cold water (16 ml) to the reaction mixture, stir well, filter and wash the residue with water.
4. Add ethanol (11 ml) to the solid and reflux the mixture for 30 minutes.

5. Add charcoal to the reaction mixture and filter through preheated Buchner funnel.
6. Wash the residue with of hot ethanol (2 ml).
7. Cool the filtrate to room temperature and filter 2-phenyl indole separated.
8. Wash the isolated 2-phenyl indole with 1 ml portions of cold alcohol (three times) and dry in vacuum desiccator.
9. The melting range of pure 2-phenyl indole is 188 to 189 °C.

Precaution:

- Phenyl hydrazine is highly poisonous (carcinogen) and produces burns when it comes in contact with skin.
- If any liquid does come in contact with the skin, wash off immediately with 2% acetic acid then with soap and water.

CHAPTER 20

VILSMEIER-HAACK REACTION

Vilsmeier-Haack reaction is a typical aldehyde synthesis reaction using formylating agent derived from N,N-dimethylformamide (DMF) and phosphorous oxychloride ($POCl_3$). Aromatic compounds with activated nucleus can be formylated by chloro methylenium salt (Vilsmeier-Haack reagent), the product of DMF and $POCl_3$. *N*-methyl formanilide also can be used in place of DMF and this method works well with *ortho-* and *para-* phenolic ethers and dialkyl anilines [Ph-N-$(CH_3)_2$]. The method doesn't work with hydrocarbons, except with anthracene in which the aldehyde group enters 9th position. The formylation can be carried out by treating the anthracene compound with N-methyl formanilide and phosphorous oxychloride. Treatment with aqueous sodium acetate followed by the steam distillation gives the product.

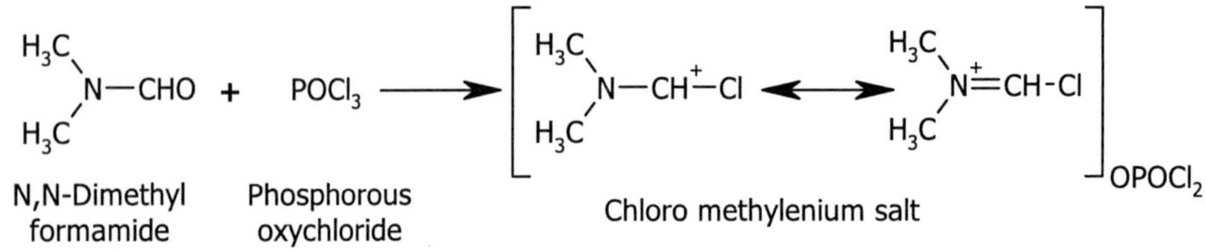

20.1 Preparation of 2-Phenyl Indole-3-carbaldehyde

Chemicals: 2-Phenyl indole, Phosphorous oxychloride, Dimethyl formamide and Sodium acetate.

Principle: Indole-3-carbaldehyde can be prepared by direct formylation of indole with dimethylformamide/*N*-methylformanilide using phosphorous oxychloride as a catalyst. Positively charged (electrophile) carbon in chloro methylenium salt (Vilsmeier-Haack reagent) will be attacked by the indole ring and forms schiff base. The schiff base upon hydrolysis by aqueous sodium acetate yields, indole-3-carbaldehyde and dimethyl amine.

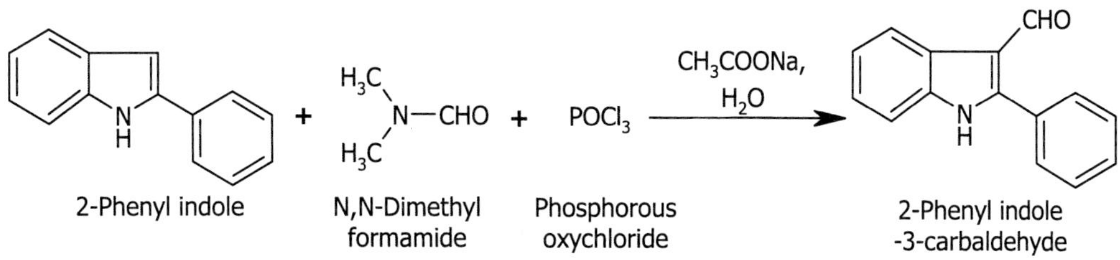

Reaction mechanism

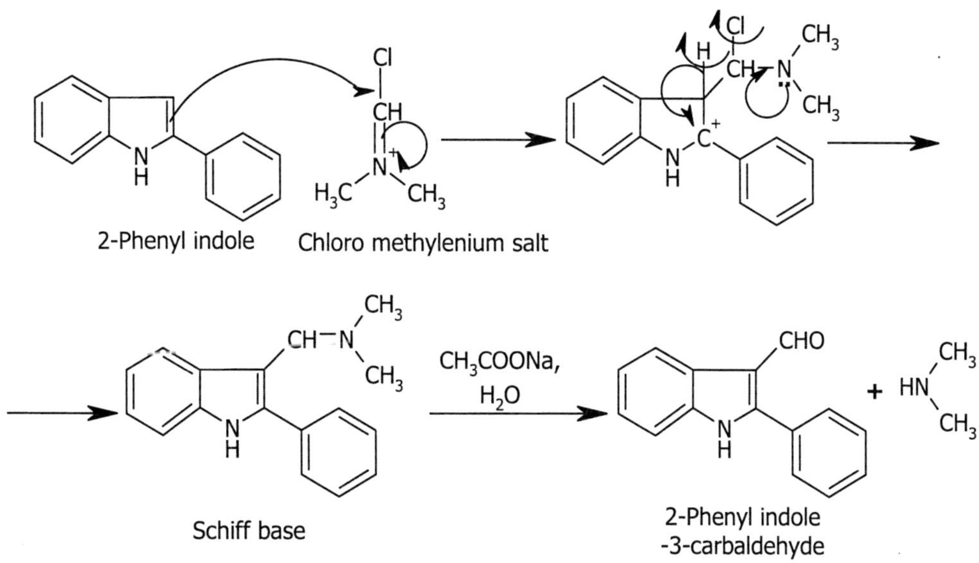

Procedure:

1. Place distilled dimethyl formamide (2.88 ml, 0.374 mol) in three necked flask.
2. Attach reflux condenser which in turn fitted with calcium guard tube to the flask.
3. Cool the contents by placing in an ice-salt bath for about 30 min and slowly add distilled phosphorous oxychloride (0.86 ml, 9.4 mmol) with stirring (*progress of formylation reaction develops pinkish colour*).
4. Slowly add the solution of 2-phenyl indole (1 gm, 8.5 mmol) in dimethyl formamide (1 ml, 13 mmol) to the reaction mixture and maintain the temperature below 10 °C.
5. Elevate the temperature to 35 °C, after the addition of 2-phenyl indole solution.
6. Stirr the syrup formed until yellow solution turns into opaque canary yellow paste (~1 hr).

7. Add crushed ice (3 gm) to the paste with stirring, which turns the paste into cherry-red solution.
   [*Cool the paste in an ice bath before adding the ice to it, because non-aqueous paste on treatment with ice generates heat (exothermic*)].
8. Transfer the solution into a three necked flask containing crushed ice (2 gm) provided with mechanical stirrer and dropping funnel containing a solution of sodium hydroxide (3.8 g in 100 ml of water).
9. Add one third of the sodium hydroxide drop wise with stirring and remaining rapidly with stirring (*after red colour changes to greenish blue or greenish yellow*).
10. Add water (10 ml) to break the cake formed and to ease the stirring.
11. Boil the suspension (develops yellow-orange solution), cool to room temperature and place it in the refrigerator overnight.
12. Collect the precipitate on a filter paper and resuspend in water (100 ml), filter, wash with three 30 ml portions of water and dry.
13. Purify the product by recrystallization with ethanol.
14. The melting range of pure 2-phenyl-indole-3-carbaldehyde is 196-197 °C.

CHAPTER 21

MANNICH REACTION

Condensation of compounds containing active hydrogen group with formaldehyde and amines (ammonia/primary amines/secondary amines) is known as Mannich reaction. Tertiary amines lacking –NH proton, will not undergo this reaction. The final product of Mannich reaction is β-amino-carbonyl compound [Mannich Base]. Mannich reaction involves in the synthesis of many pharmaceuticals, natural products and polymers. The product of Mannich reaction, 1,3-amino alcohols are called as Michael acceptor, which finds important role in the medicinal chemistry.

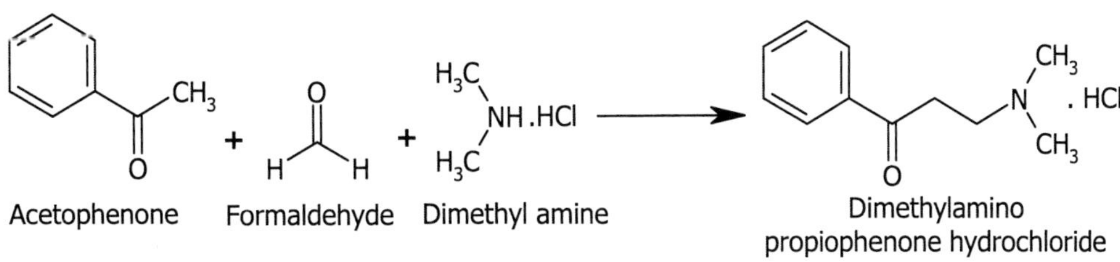

Amines activate the formaldehyde and generate intermediate enamine (methylene dimethylamine). Nucleophilic addition of an amine to a carbonyl compound, and dehydration produces Schiff base (methylene dimethylamine). This electrophilic schiff base reacts with compound containing acidic proton (acetophenone) through electrophilic addition reaction. Nitriles, acetylenes, aliphatic nitro compounds, α-alkyl pyridines and imines contain active hydrogen are the good substrates. Electron rich heterocycles such as furan, pyrrole, thiophen and indole also acts as a substrate. The π-electrons of ene group of substrate attack enamine and generates Mannich base.

Generation of methylene dimethylamine

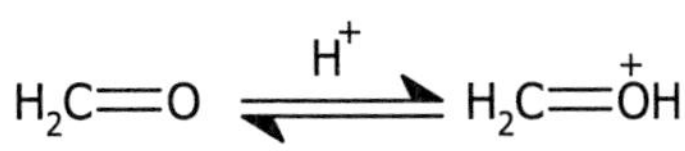

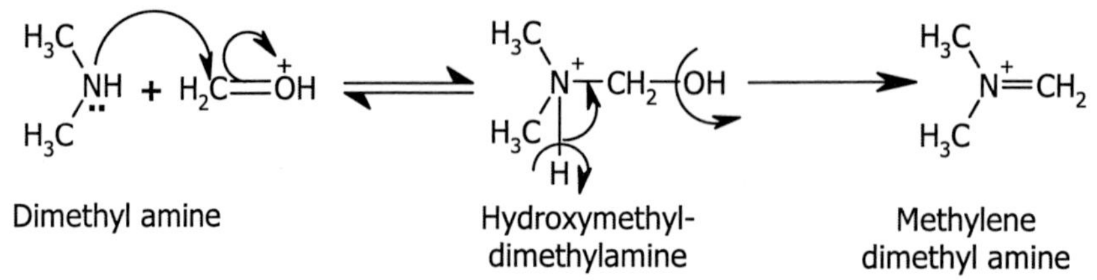

Formation of mannich base

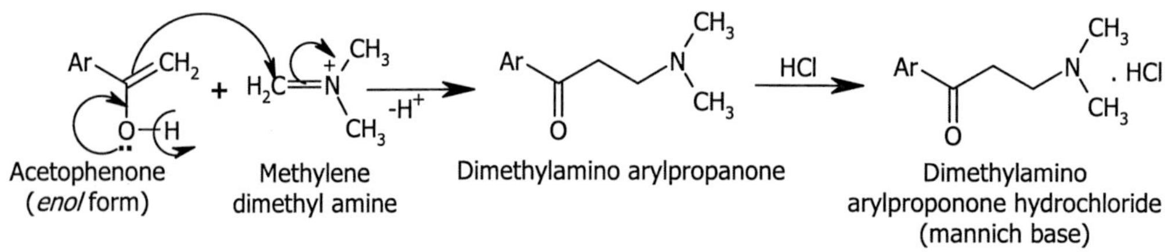

Pharmaceutical Applications

1. Preparation of Propoxyphen [Dextropropoxyphen (anti-inflammatory) and Levopropoxyphen (antitussive)].

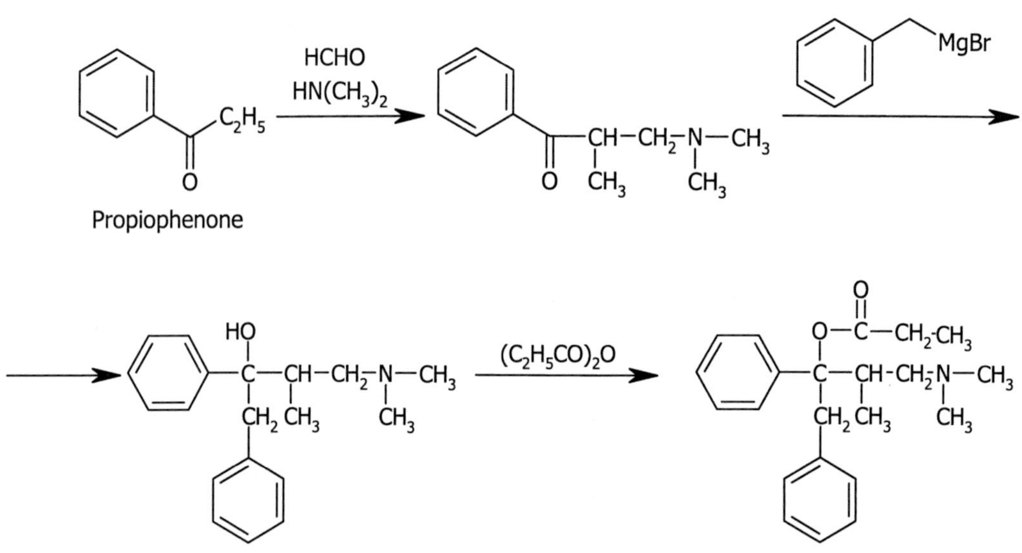

2. Preparation of Ranitidine (anti-ulcer drug).

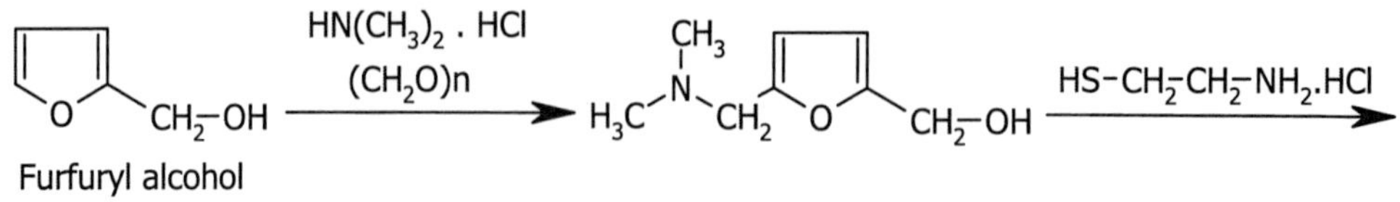

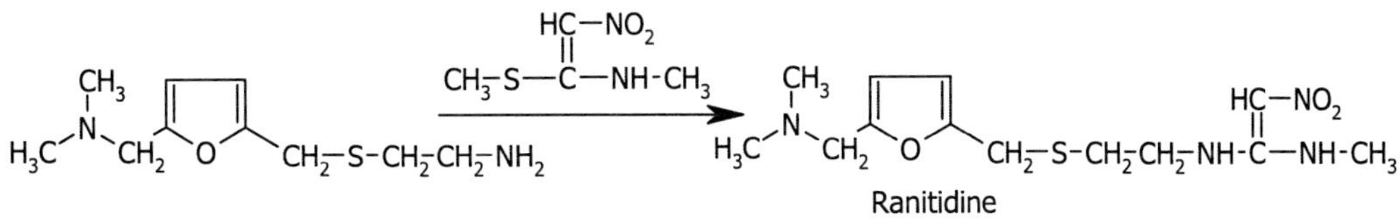

3. Preparation of Chlorothiazide (diuretic).

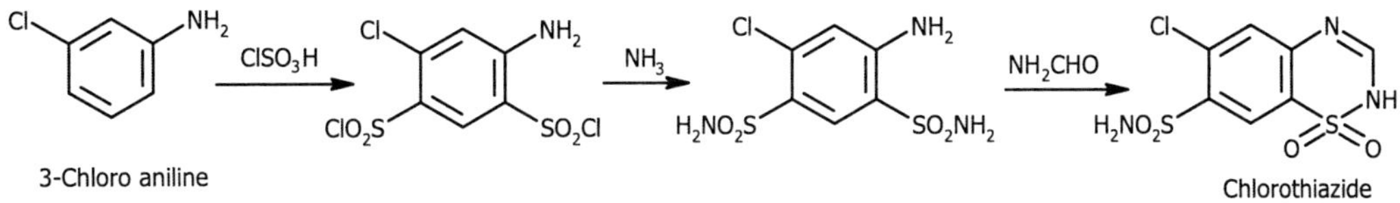

4. Preparation of Triprolidine (anti-histamine).

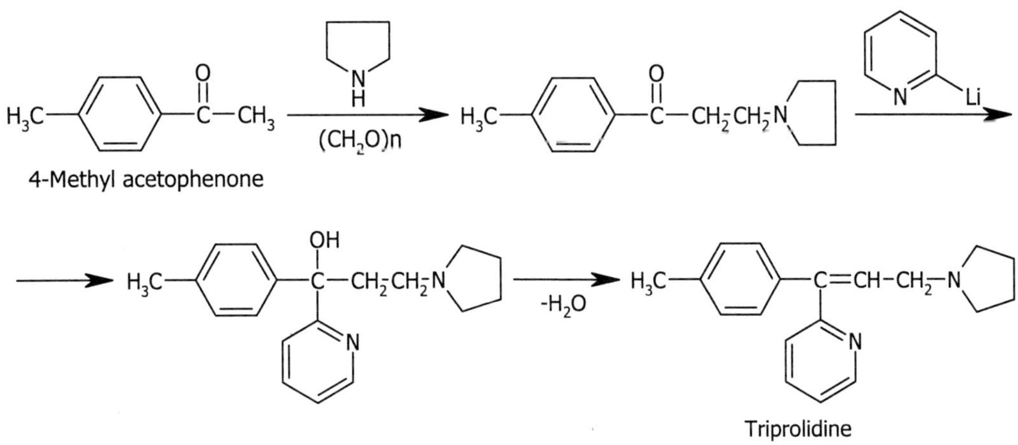

5. Preparation of Fluspirilene (neuroleptic).

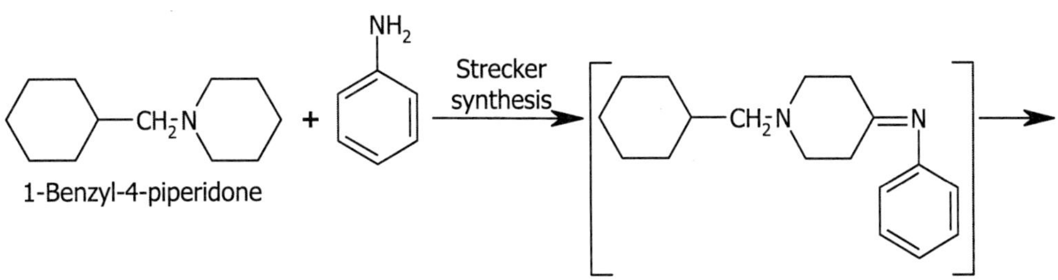

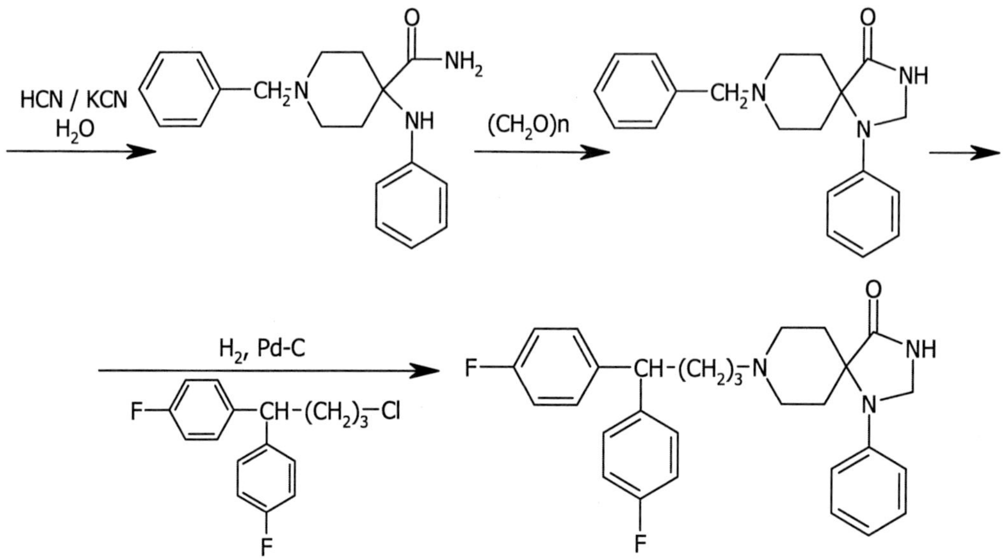

Fluspirilene

6. Preparation of Butizide (diuretic).

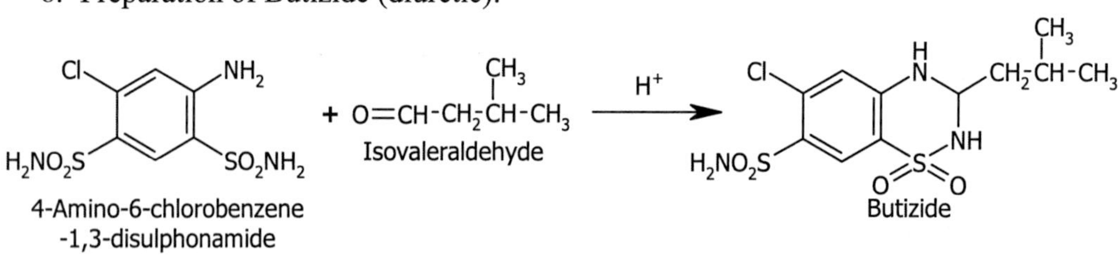

21.1 Preparation of Dimethyl Amino Propiophenone HCl

Chemicals: Acetophenone, Paraformaldehyde, Dimethyl amine hydrochloride, Hydrochloric acid and Acetone.

Principle: Acetophenone reacts with alcoholic solution of formaldehyde and dimethyl amine hydrochloride and forms β-dimethyl amino propiophenone hydrochloride, it is a typical example of the Mannich reaction. The active hydrogen may be derived from a methylene group, activated by neighbouring keto group. Dimethyl amine reacts with formaldehyde and gives an intermediate methyloamine (methylene dimethylamine). Methyloamine gets converted into methylene ammonium salt by the influence of acid. Enol form of the acetophenone (or with the carbanion derived from the ketone) condenses with methyloamine and yields β-dimethyl amino propiophenone hydrochloride.

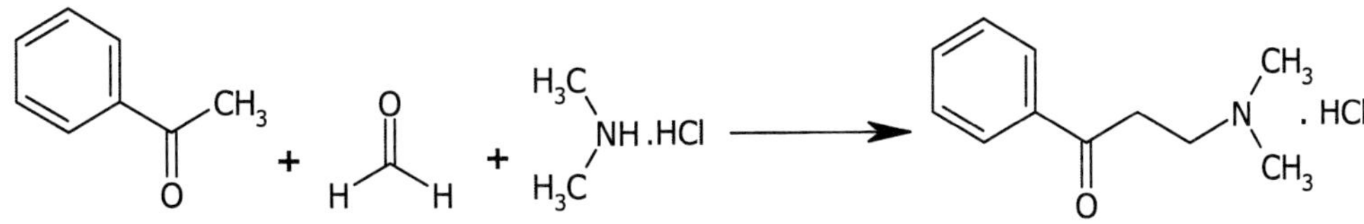

Acetophenone Formaldehyde Dimethyl amine Dimethylamino propiophenone hydrochloride

Reaction mechanism

Generation of methylene dimethylamine

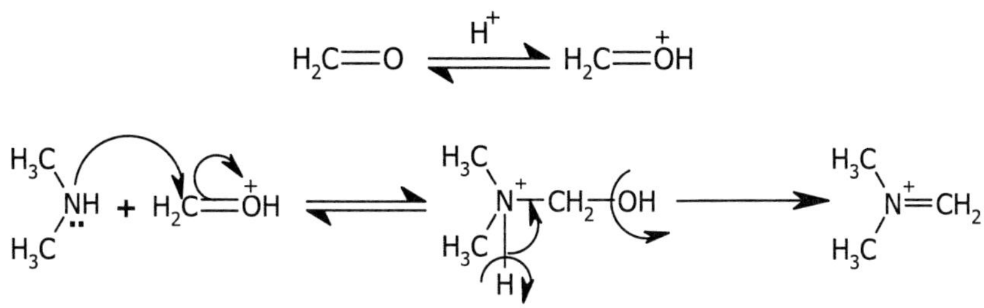

Dimethyl amine Hydroxymethyl-dimethylamine Methylene dimethyl amine

Formation of mannich base

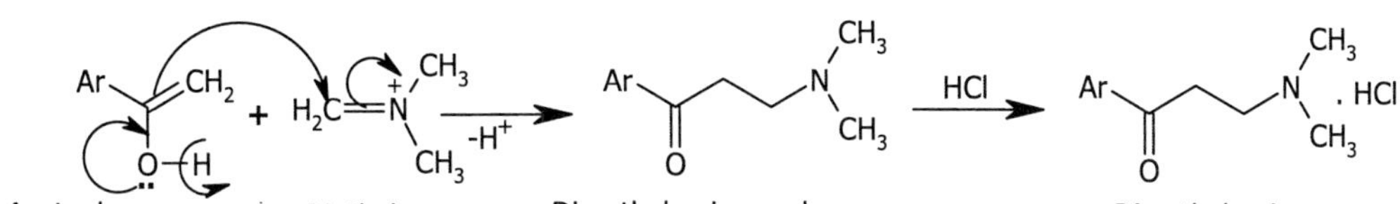

Acetophenone (*enol* form) Methylene dimethyl amine Dimethylamino arylpropanone Dimethylamino arylproponone hydrochloride (mannich base)

Procedure:

1. Place dry dimethylamine hydrochloride (2.7 gm, 33 mmol), powdered para formaldehyde (1 g, 33 mmol) and acetophenone (3 ml, 25 mmol) in a round bottom flask.
2. Add ethanol (4 ml, 95%) acidified with drop of hydrochloric acid to the flask.
3. Reflux the mixture by placing on a water bath for 2 hr (*until the reaction mixture becomes clear and homogeneous*).
4. Filter the yellowish solution (if necessary) through a hot water funnel.
5. Transfer the warm filtrate to wide mouthed flask and dilute with acetone (20 ml).
6. Cool to room temperature and leave in a refrigerator for overnight.

3. Reflux the mixture by placing on a water bath for 2 hr (*until the reaction mixture becomes clear and homogeneous*).
4. Filter the yellowish solution (if necessary) through a hot water funnel.
5. Transfer the warm filtrate to wide mouthed flask and dilute with acetone (20 ml).
6. Cool to room temperature and leave in a refrigerator for overnight.
7. Filter the crystals formed under vacuum, wash with acetone (1 ml and dry for 6 hr at 40-50 °C).
8. Purify the product by recrystalization with the mixture of hot ethanol and acetone (1:5).
   [*Add acetone slowly to the hot ethanol solution*]
9. Collect the solid which separates by vacuum filtration and dry at 70 °C.
10. The melting range of pure dimethylamino propiophenone HCl is 155 -156 °C.

CHAPTER 22

SCHIFF BASE SYNTHESIS

Condensation of primary amine with a carbonyl compound generates Schiff base. A Schiff base contains a carbon-nitrogen double bond (-CH=N-) functional group which is known as azomethane and aldimine.

The amine will be reactive in its un-ionized form and may require deprotonation, if it is protonated. The free amino group adds to the carbonyl group to form an intermediate known as a carbinolamine. Carbinolamine eliminates water molecule and forms a Schiff base.

$$R-NH_2 \xrightleftharpoons{H^+} R-NH_3^+$$

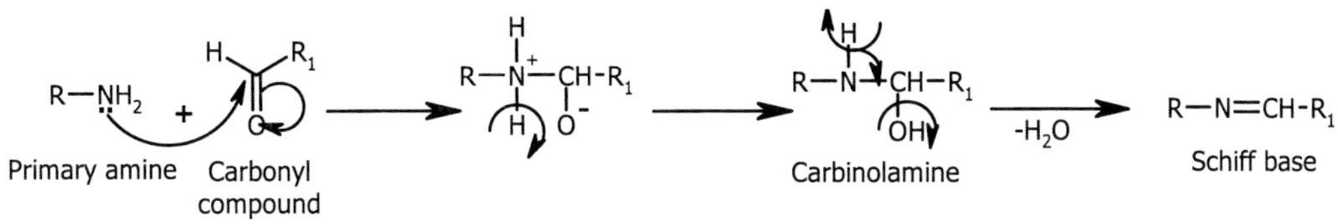

The nitrogen atom is connected to an aryl or alkyl group (not to hydrogen). The substitution on the nitrogen makes the schiff base stable imine. A Schiff base derived from aniline is called as anil. Schiff bases of aliphatic aldehydes are relatively unstable and are readily polymerizable. Whereas Schiff base of aromatic aldehydes having an effective conjugation system are more stable.

Pharmaceutical Applications

Schiff bases are useful in the identification and determination of aldehydes and ketones. Purification of carbonyl or amino compounds and protection of carbonyl or amino groups in sensitive reactions can be done through Schiff base formation. Schiff bases appear to be an important intermediate in a number of enzymatic reactions involving interaction of an enzyme with an amino or a carbonyl group of the substrate. Fructose 1,6-bipohosphate aldolase catalyzed reaction (glycolysis) and metabolism of amino acids generates Schiff base as an intermediate.

The Amadori rearrangement (carbohydrate chemistry) including non-enzymatic glycation and the Maillard browning reaction proceeds through Schiff base formation. The Mannich condensation useful in organic synthesis is an addition reaction of a quaternary Schiff base derived from formaldehyde with a carbonic center.

1. Preparation of Ambroxol [mucolytic].

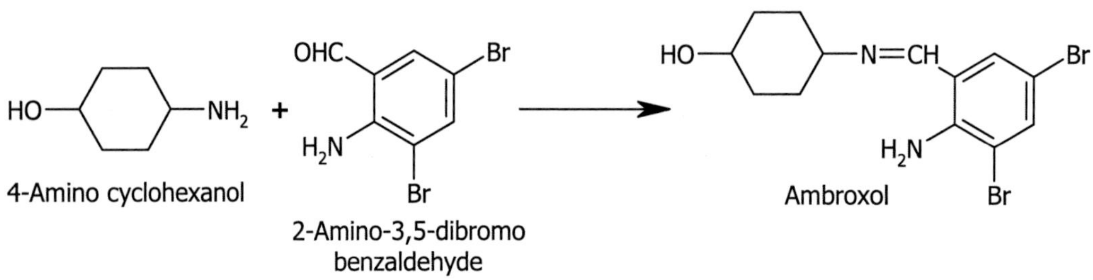

2. Preparation of Nifurtimox (antiprotozoal drug).

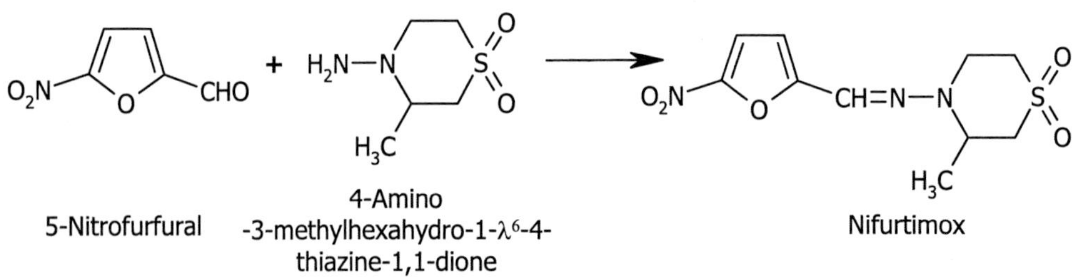

3. Preparation of Nitrofurantoin (urinary tract antiseptic).

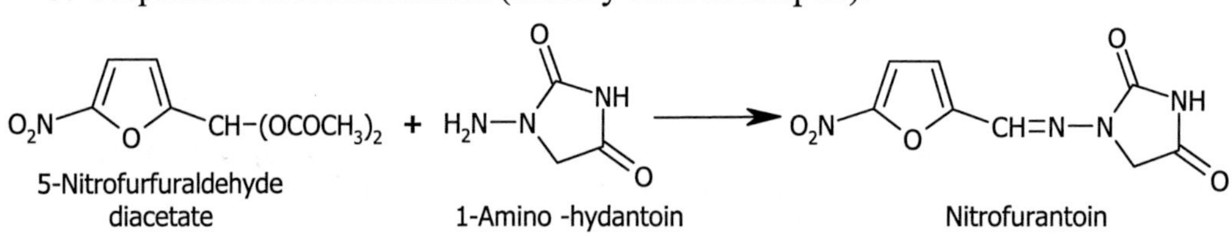

22.1 Preparation of N-(*m*-Nitrobenzylidene) Aniline

Chemicals: *m*-Nitro benzaldehyde, Aniline, Ethanol and Methanol.

Principle: N-(*m*-Nitrobenzylidene) aniline can be prepared by condensation of *m*-nitro benzaldehyde with aniline. The amino group of aniline in an ionized form attack carbonyl carbon of *m*-nitro benzaldehyde. The resulting carbinolamine intermediate eliminates water molecule produce N-(*m*-nitrobenzylidene) aniline (Schiff base).

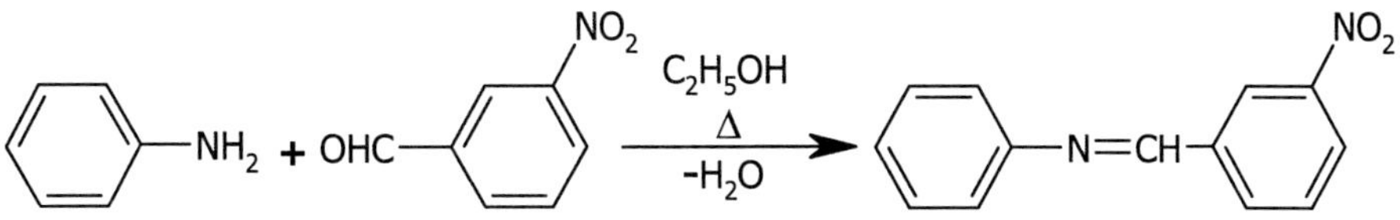

Aniline meta-Nitrobenzaldehyde N-(meta-Nitrobenzylidene) aniline

Schiff base formation

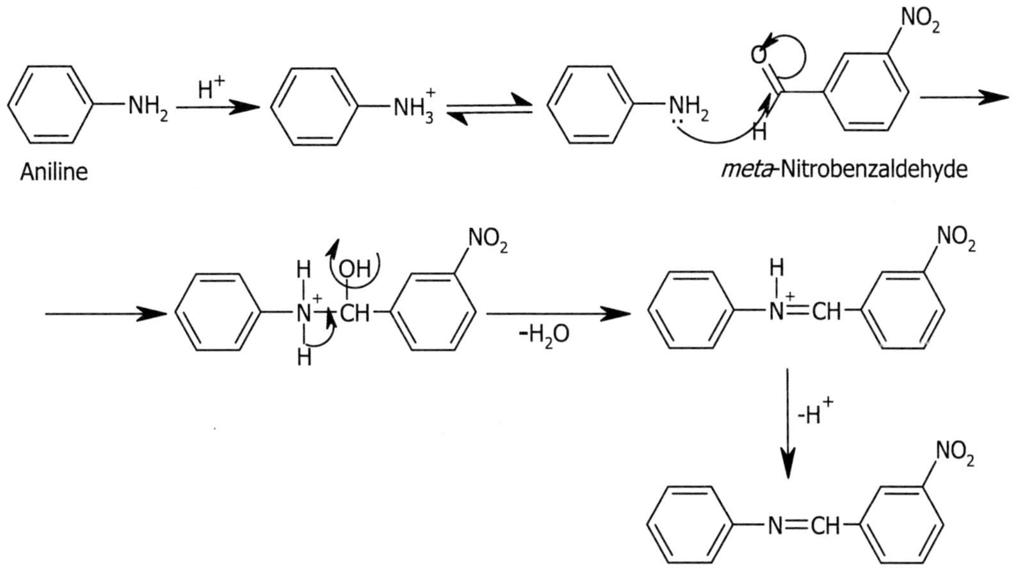

Aniline meta-Nitrobenzaldehyde

N-(meta-Nitrobenzylidene) aniline

Procedure:

1. Place *m*-nitro benzaldehyde (1 gm, 7 mmol), aniline (0.46 gm, 5 mmol) and ethanol (2 ml) in a round bottom flask.
2. Attach reflux condenser and reflux the mixture by placing on the water bath for 20 minutes.
3. Add water until a slight cloudiness persists and allow it to cool.
4. Rub the separated oil with glass rod to induce the crystallization.
5. Filter and wash the solid with cold aqueous ethanol.
6. Purify the product by recrystallization from aqueous methanol.
7. The melting range of pure N-(*m*-nitrobenzylidene) aniline is 65-66 °C.

(Schiff base can be prepared by heating the components in the absence of solvent at 100 °C for 15 min, cool and then stirring the product with methanol to induce crystallization).

CHAPTER 23

BIGINELLI REACTION

One pot condensation of an aldehyde, β-keto ester and urea/thiourea under acid catalysis gives 3, 4-dihydro pyrimidinones and is referred as Biginelli reaction. It is used to generate calcium channel blockers, mitotic kinesin Eg5 inhibitors, anti-viral and anti-bacterial compounds.

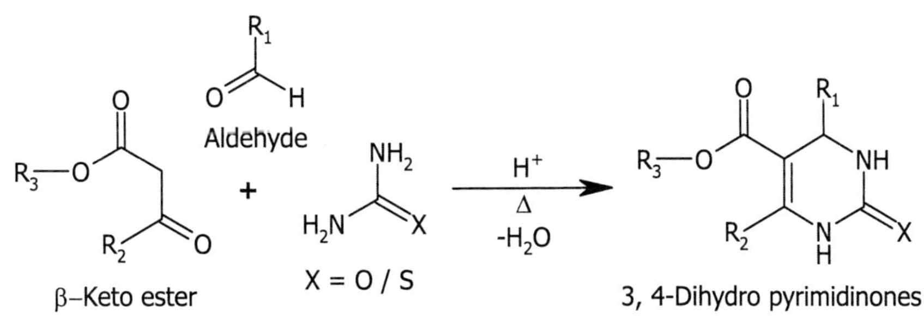

Mechanism

Three mechanistic proposals are explained for the Biginelli reaction.

1. Kappe mechanism.
2. Folkers and Johnson mechanism.
3. Sweet and Fissekis mechanism.

On various bases, Kappe mechanism is widely accepted and remaining are ruled out. Kappe explored the reaction mechanism using NMR spectroscopy, crystal X-ray analysis and trapping experiments. According to his hypothesis, aldehyde reacts with urea and forms an *N*-acyliminium intermediate *via* the unobservable hemiacetal. *N*-acyliminium reacts with enol tautomers of ethyl acetoacetate and gives the precursor to dihydro pyrimidine. Formation of hemiacetal is rate limiting step, which on cyclization gives 3, 4-dihydro pyrimidinone.

23.1 Preparation of 5-(Ethoxy Carbonyl)-4-Phenyl-6-Methyl-3,4-Dihydro Pyrimidine-2(1*H*)-One

Principle: Biginelli reaction is used in the synthesis of dihydropyrimidines (DHPMs). Benzaldehyde condenses with ethyl acetoacetate and urea in ethanol medium using calcium chloride ($CaCl_2$)/tetra-*n*-butylammonium bromide (TBAB) as catalyst and forms 5-(ethoxy carbonyl)-4-phenyl-6-methyl-3,4-dihydro pyrimidione-2(1*H*)-one.

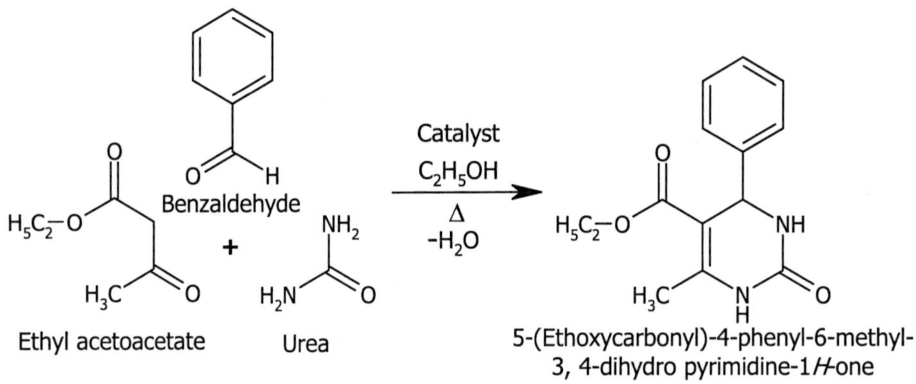

Benzaldehyde reacts with urea and forms a N-acyliminium intermediate *via* the unobservable hemiaminal. N-acyliminium reacts with enol tautomers of ethyl aceto acetate and gives the precursor to dihydropyrimidine. Formation of hemiacetal is rate limiting step, prevents the observation of intermediates by NMR. The formed precursor on cyclization gives 3,4-dihydro pyrimidinone.

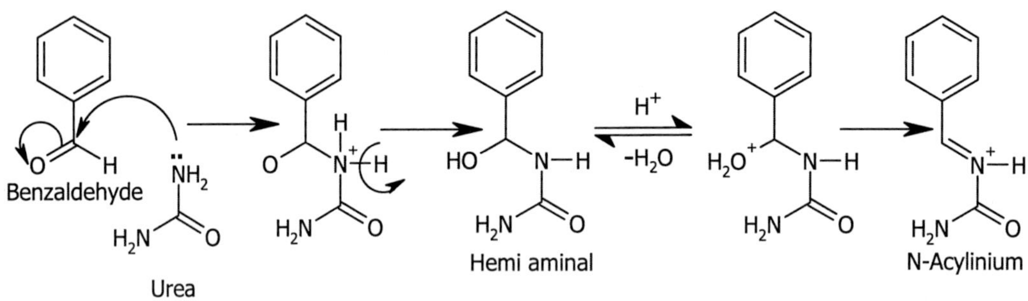

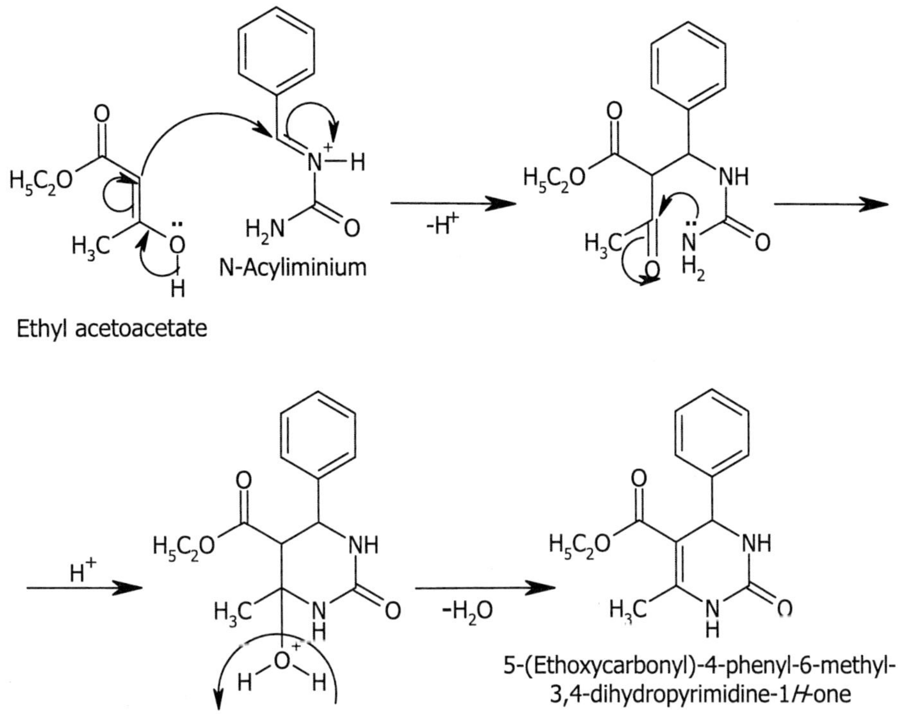

Procedure:

Method 1

Chemicals: Benzaldehyde, Ethyl acetoacetate, Ethanol and Calcium chloride.

1. Place ethyl acetoacetate (1.3 ml, 10 mmol), benzaldehyde (1 ml, 10 mmol) and urea (0.72 gm, 12 mmol) in a round bottom flask containing ethanol (10 ml).
2. Add calcium chloride (0.16 gm) to the flask as a catalyst.
3. Attach reflux condenser and reflux the mixture for 2 hr.
4. Pour the reaction mixture into beaker containing crushed ice, mix well to get a solid.
5. Filter the product under vacuum and purify the product by recrystallization with ethanol.
6. The melting range of pure 5-(ethoxy carbonyl)-4-phenyl-6-methyl-3,4-dihydro pyrimidine-2(*1H*)-one is 200-201 °C.

Method 2

Chemicals: Benzaldehyde, Ethyl acetoacetate, Glacial acetic acid, Tetra-n-butylammonium bromide and Ethanol.

1. Place ethyl acetoacetate (1.3 ml, 10 mmol), benzaldehyde (1 ml, 10 mmol) and urea (0.72 gm, 12 mmol) in a round bottom flask containing glacial acetic acid (10 ml).
2. Add tetra-*n*-butylammonium bromide (1.43 gm) to the flask as a catalyst.
3. Attach reflux condenser and reflux the mixture for 1 hr.
4. Pour the reaction mixture into beaker containing crushed ice and stir for 20 minutes.
5. Filter the product under vacuum and purify the product by recrystallization with ethanol.
6. The melting range of pure 5-(ethoxy carbonyl)-4-phenyl-6-methyl-3,4-dihydro pyrimidine-2(*1H*)-one is 200-201 °C.

CHAPTER 24

HANTZSCH 1,4-DIHYDRO PYRIDINE SYNTHESIS

Hantzsch 1,4-dihydropyridine (1,4-DHP) synthesis involves the condensation of aldehyde, β-keto ester and ammonia or ammonium salts to produce 1,4-dihydropyridine motif. Hantzsch 1,4-DHP reactions belong to the category of multi component condensation (MCC).

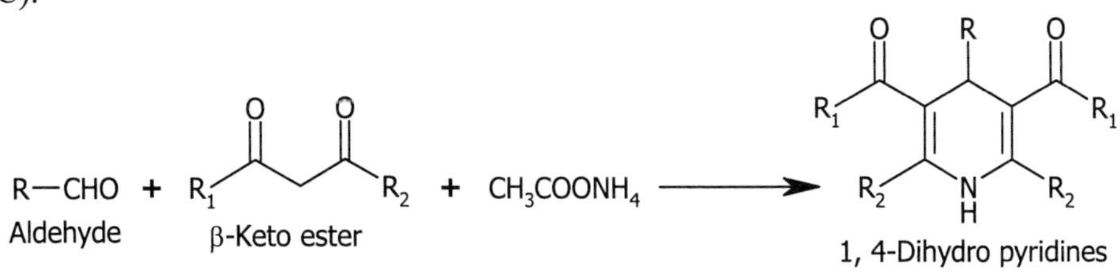

1,4-Dihydropyridines (1,4-DHPs) are well known as calcium channel modulators. Cardiovascular agents such as nifedipine, nicardipine and amlodipine contain 1,4-DHP scaffold. 1,4-Dihydropyridine derivatives possess a variety of biological activities such as vasodilator, bronchodilator, anti-atherosclerotic, antitumor, geroprotective, hepatoprotective, and anti-diabetic activity. Extensive studies have revealed the neuroprotective, anti-platelet, cerebral anti ischemic and chemosensitizer effect (tumor therapy) of 1,4-DHPs.

Hantzsch 1,4-dihydropyridine synthesis reaction mechanism

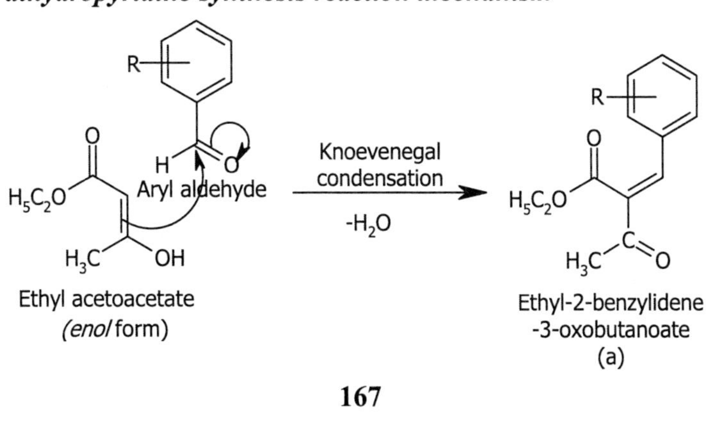

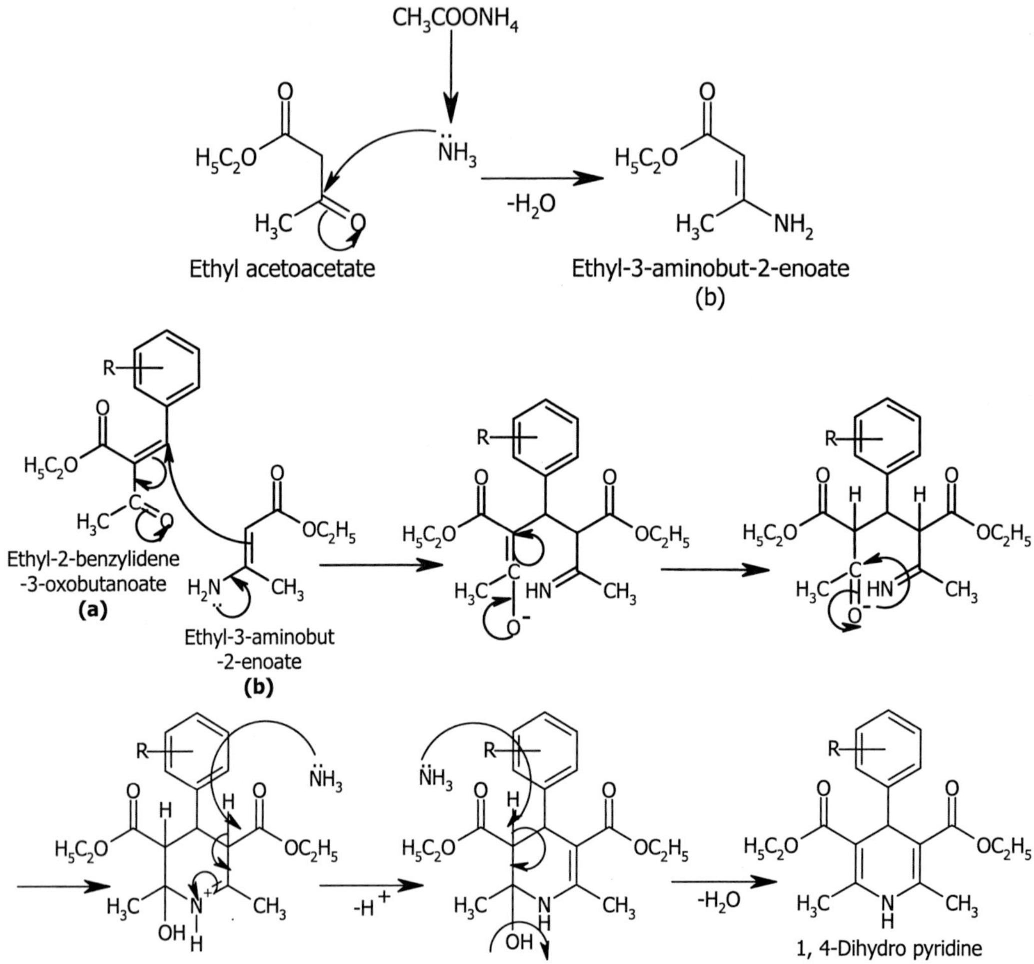

Pharmaceutical Applications

1. Preparation of Nifedipine (antihypertensive, anti-anginal drug).

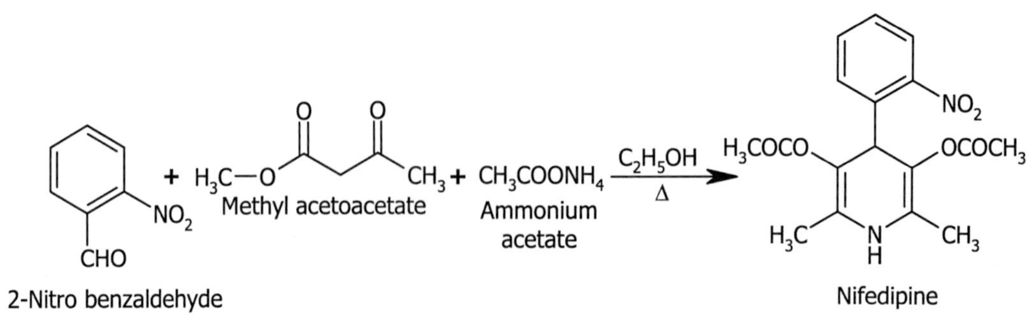

2. Preparation of Felodipine.

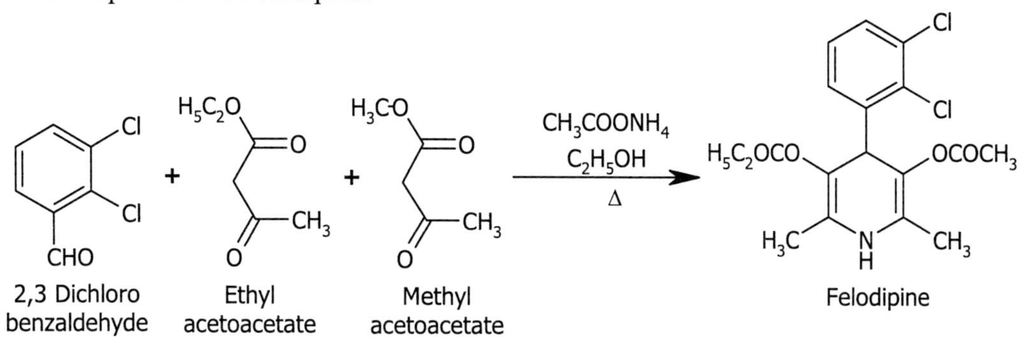

3. Preparation of Nitrendipine.

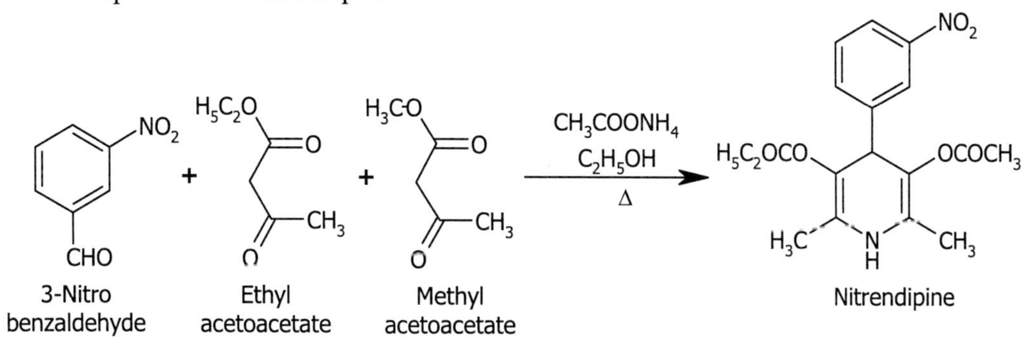

24.1 Preparation of Nifedipine

Chemicals: 2-Nitrobenzaldehyde, Methyl acetoacetate, Ammonium acetate, Ethanol and Tetra-*n*-butylammonium bromide.

Principle: Nifedipine, can be synthesized through the condensation of 2-nitro benzaldehyde with methyl acetoacetate and ammonium acetate in ethanol medium using tetra-*n*-butylammonium bromide (TBAB) as catalyst.

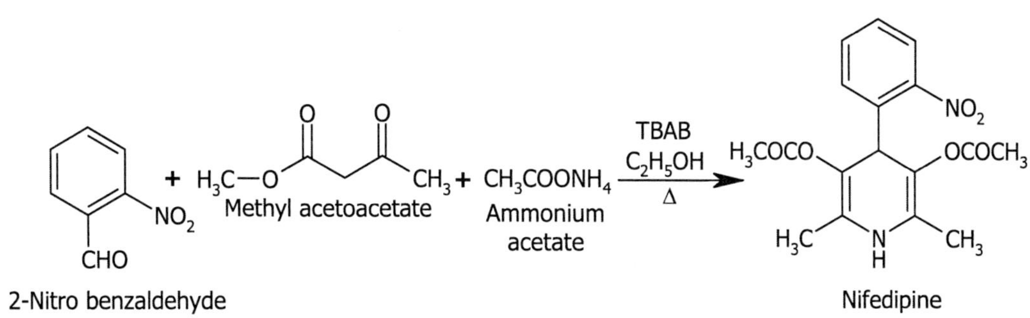

Procedure:

1. Place methyl acetoacetate (1.1 ml, 10 mmol), 2-nitro benzaldehyde (1.51 gm, 10 mmol), ammonium acetate (1.54 gm, 20 mmol), and ethanol (5 ml) in a round bottom flask.
2. Add tetra-*n*-butylammonium bromide (TBAB, 1.43 gm) to the flask and stir the mixture at 60 °C for 5 hr.
3. Pour the reaction mixture into beaker containing ethanol (2 ml) and crushed ice (10 gm) and keep aside for 1 hr.
4. Isolate the precipitated Nifedipine under vacuum and dry.
5. Purify the Nifedipine by recrystallization with ethanol.
6. The melting range of pure Nifedipine is 170-173 °C.

CHAPTER 25

RADZISZEWSKI REACTION

A five membered cyclic diene containing three carbon atoms and two nitrogen atoms are called as diazoles. Nitrogen present at 1st and 3rd positions are known as imidazole. Naturally occurring substituted imidazoles and synthetic derivatives have been found to have antihelmintic, analgesic, anti-bacterial, anti-fungal, anti-cancer, antiviral and antitubercular activities. The imidazole ring system is present in numerous compounds of therapeutic importance.

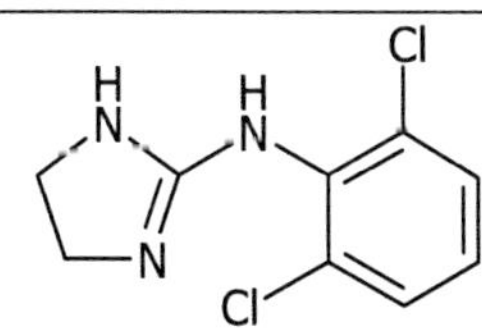 Clonidine (antihypertensive)	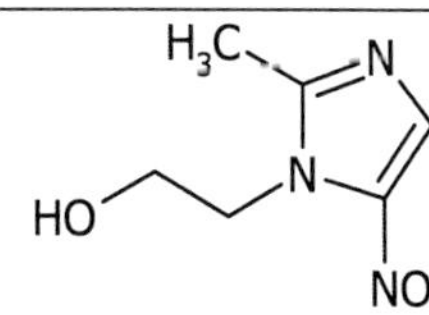 Metronidazole (antiamoebic)
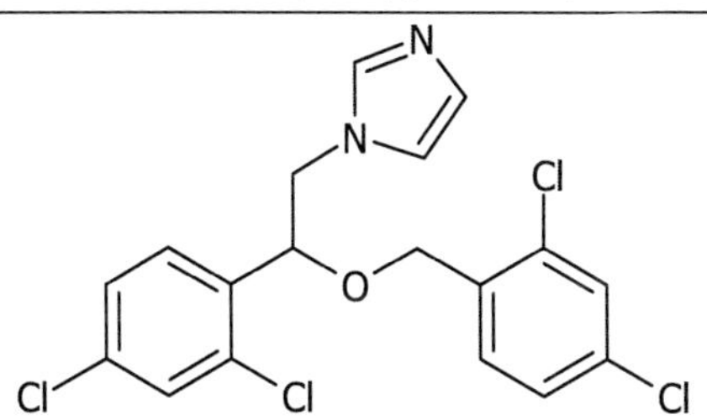 Miconazole (anti-fungal)	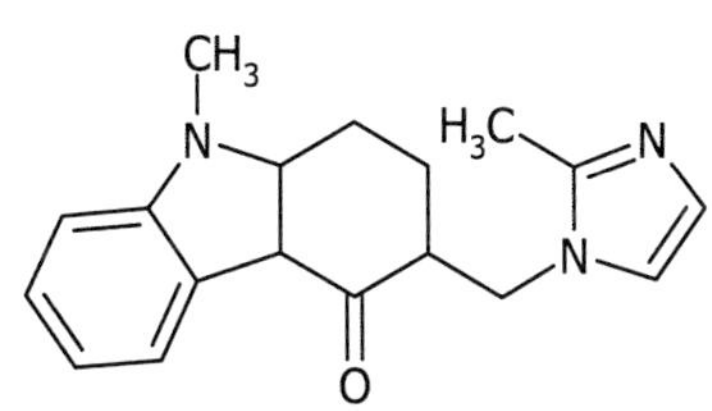 Ondansetron (antiemetic)
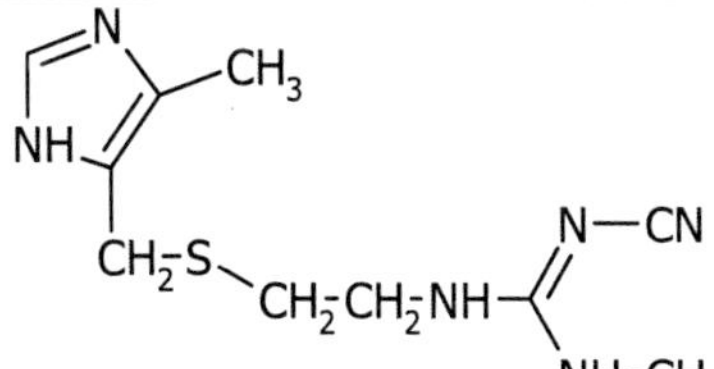 Cimetidine (antiulcer)	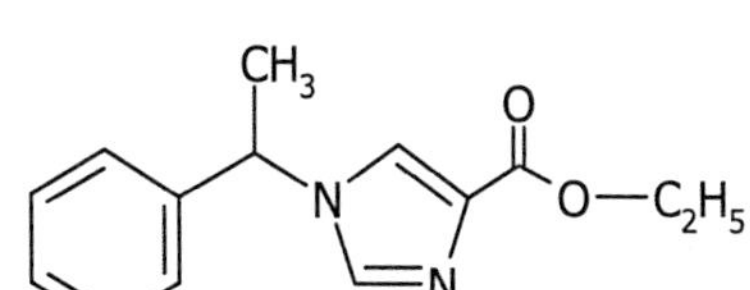 Etomidate (general anaesthetic)

Table *Contd...*

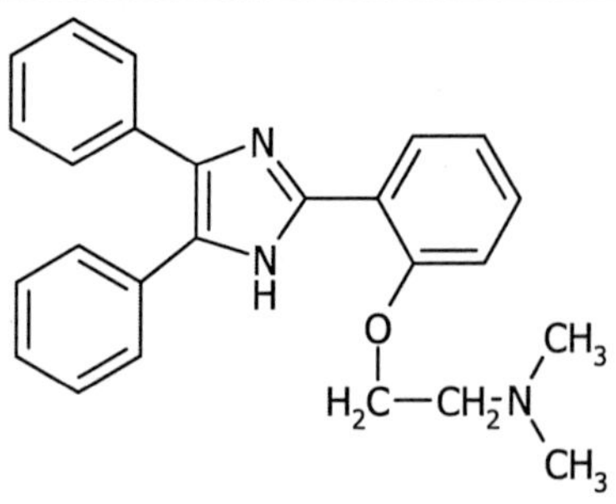

Trifenagrel
(platelet aggregation inhibitor)

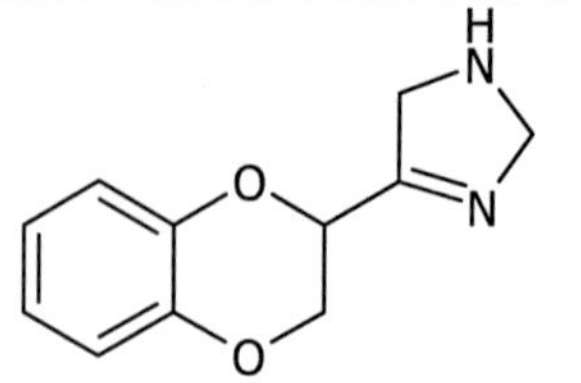

Idazoxan
(anti-schizophrenic)

Radziszewski reaction produces an imidazole scaffold by the condensation of α-dicarbonyl compound with aldehyde and two equivalents of dry ammonia/ammonium acetate in ethanol.

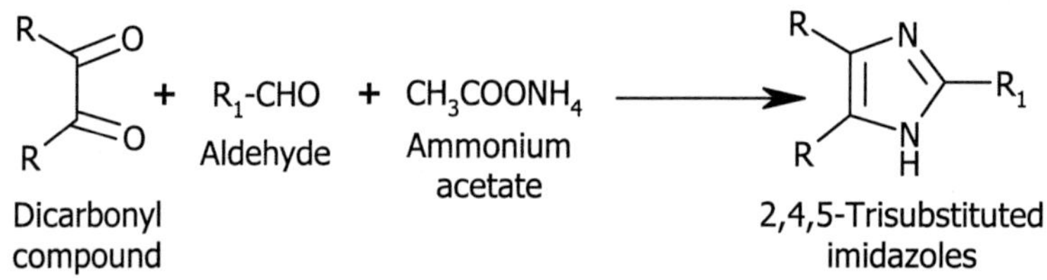

Reaction mechanism

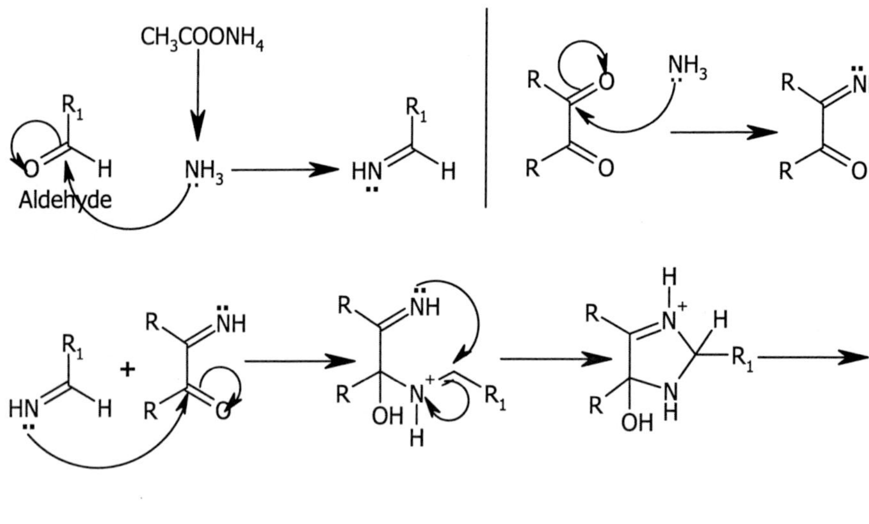

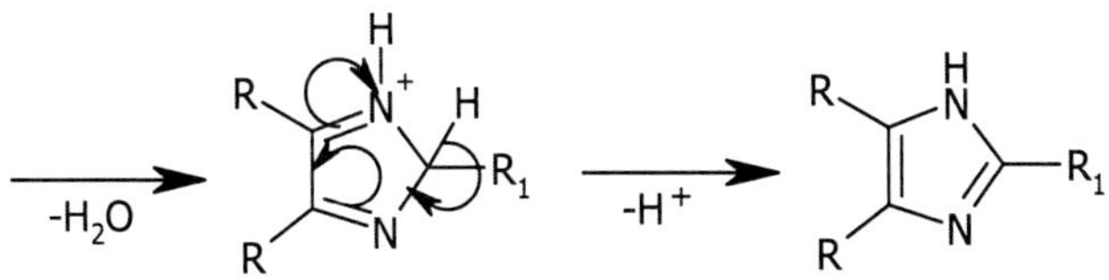

2,4,5-Trisubstituted imidazoles

25.1 Preparation of 2,4,5-Triphenyl Imidazole

Chemicals: Benzil, Benzaldehyde, Ammonium acetate, Glacial acetic acid, Polyethylene glycol 400 (PEG 400) and Ethanol.

Principle: Condensation of benzil with benzaldehyde and ammonium acetate in glacial acetic acid produces 2,4,5-triphenyl imidazole, and this reaction is typical example of Radziszewski reaction.

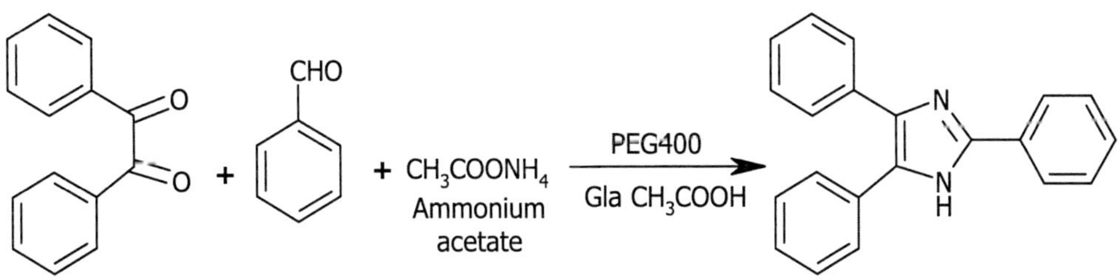

Benzil Benzaldehyde 2,4,5-Triphenyl imidazole

Reaction mechanism

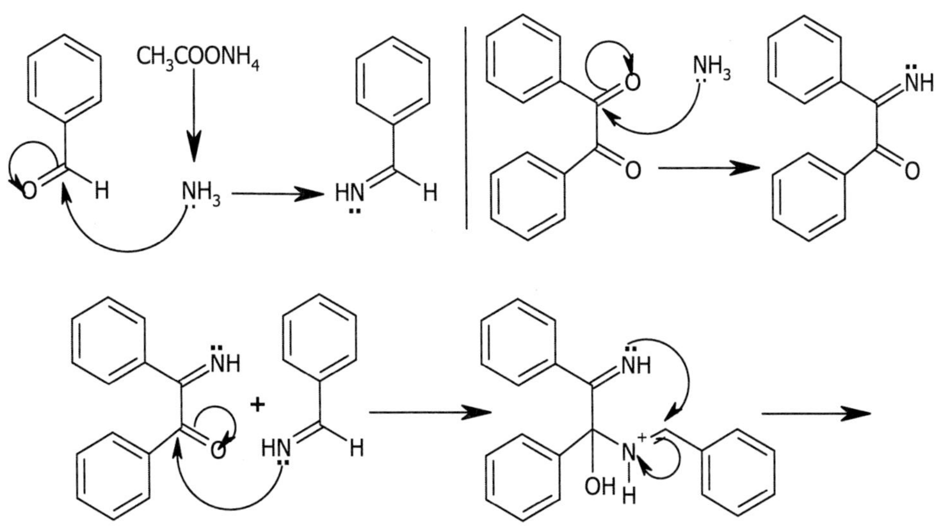

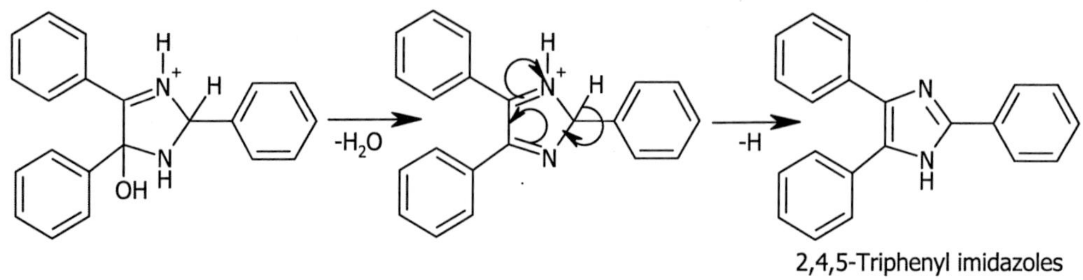

Procedure:

1. Place benzil (2.1 gm, 10 mmol), benzaldehyde (1.1 ml, 10 mmol), ammonium acetate (2 gm, 25 mmol) and glacial acetic acid (10 ml) in a round bottom flask.
2. Add PEG 400 (5 g) to the flask and reflux with stirring at 118 °C for 5 hr.
3. Pour the reaction mixture into beaker containing cold water (500 ml).
4. Isolate the precipitate formed through vacuum filtration, wash with cold water and then dry.
5. Purify the product by recrystallization with ethanol.
6. The melting range of pure 2,4,5-triphenyl imidazole is 267-269 °C.

25.2 Preparation of 4,5-Diphenyl Imidazole

Chemicals: Benzil, Hexamine, Ammonium acetate, Glacial acetic acid, Ammonium hydroxide, Charcoal and Ethanol.

Principle: The 4,5-diphenyl imidazole can be prepared by modified Radziszewski reaction. The condensation of benzil with ammonium acetate in presence of hexamine and acetic acid produces 4,5-diphenyl imidazole.

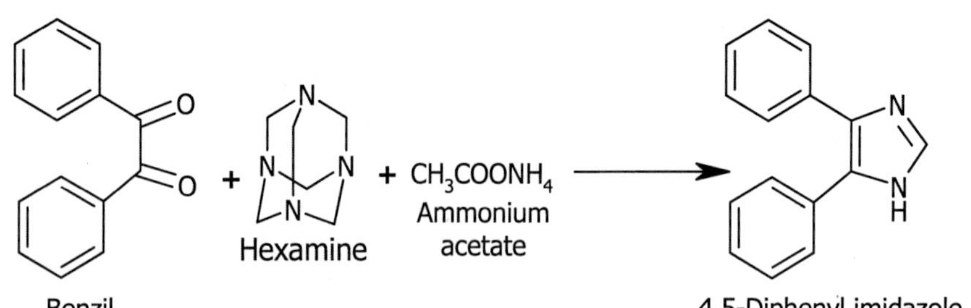

Hexamine, a special form of ammonia hydrolyses to formaldehyde and ammonia by thermal effect. This slows reaction compared with reaction using formaldehyde. The formaldehyde reacts with ammonium acetate and forms formamide, (methanimine) which offer carbon and nitrogen atoms for the ring formation.

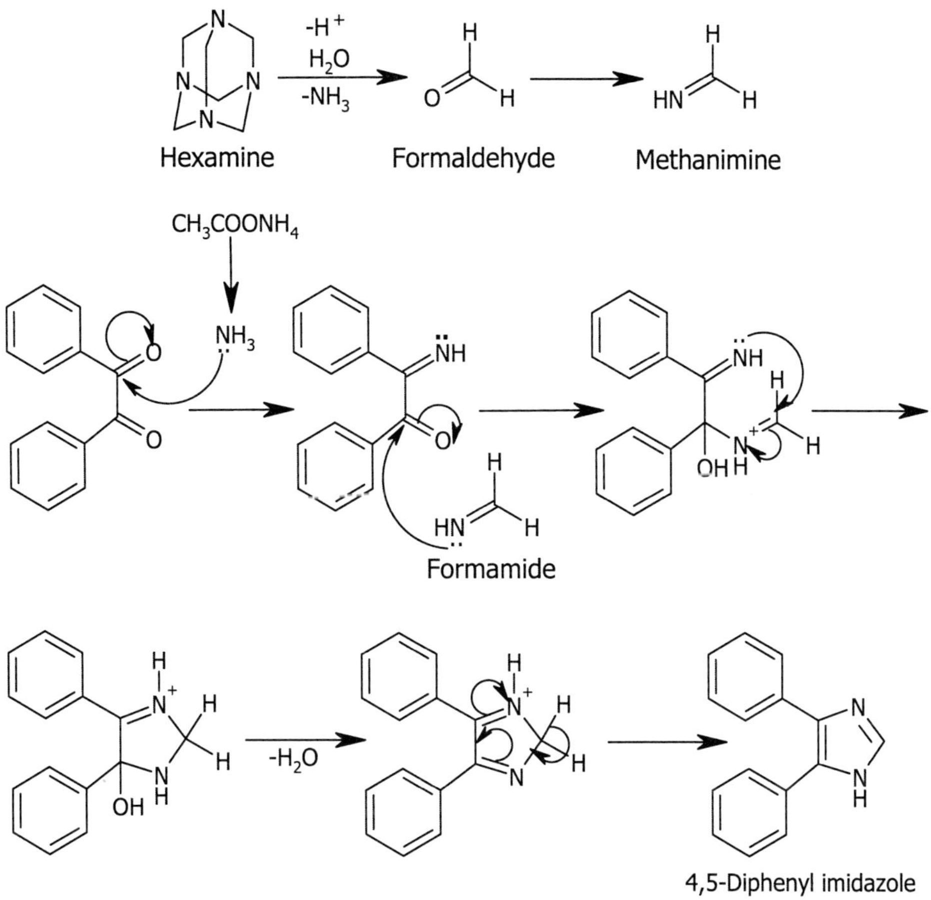

Procedure:

1. Place benzil (1.1 gm, 5 mmol), hexamine (13 gm, 0.928 mol), ammonium acetate (3 gm, 0.038 mol) and glacial acetic acid (25 ml) in a round bottom flask.
2. Attach reflux condenser to the flask and reflux the contents for 1 hr.
3. Pour the resulting mixture into water (20 ml).
4. Add charcoal and boil to remove the turbidity (if any formed).
5. Filter the mixture and basify the filtrate with ammonium hydroxide.

6. Cool to room temperature, filter the separated product and wash the residue with water.
7. Dry the product and recrystallize the product with ethanol.
8. The melting point of 4,5-diphenyl imidazole is 228-230 °C.

Precautions:

- Hexamine is sensitive to skin and cause allergic reactions (irritation, reddening and blistering).
- Use protective gloves and dust masks to protect against the allergic effects.

CHAPTER 26

PHILLIPS REACTION

Benzimidazole, a heterocyclic compound, consists of the fusion of benzene and imidazole rings. The most prominent benzimidazole compound in nature is N-ribosyl-dimethyl benzimidazole, which serves as an axial ligand for cobalt in vitamin B_{12}.

Phillips reaction involves the condensation of *o*-phenylene-diamines with organic acids in presence of dilute mineral acids to furnish benzimidazoles. This method produces good yields with aliphatic acids. Reaction in sealed tube increase yields for aromatic acids. Acetic acid can be condensed with *o*-phenylenediamine in presence of hydrochloric acid (4N). Benzoic acid can be condensed by carrying out the reaction in a sealed tube and maintaining the temperature 180 °C.

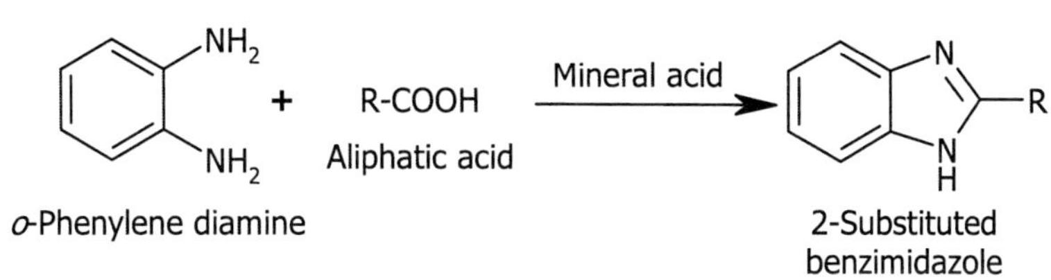

One of the amino group nitrogen undergoes acylation at first in presence of mineral acid and produces N-acylated compound. The second amino group nitrogen undergo acylation with carbonyl carbon acyl group of N-acylated compound and leads to ring closure. This ring closure generates 2-substituted benzimidazole.

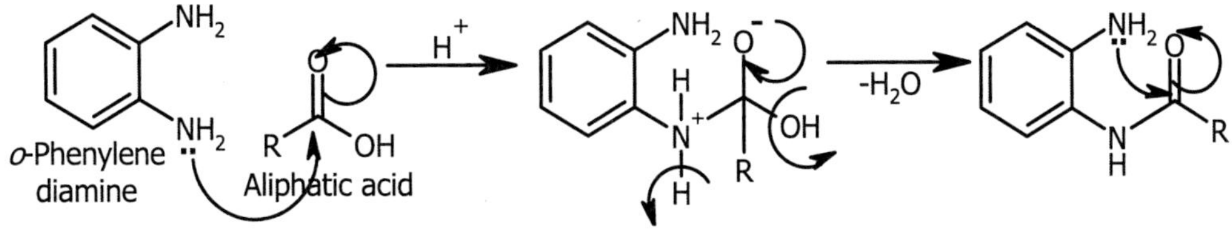

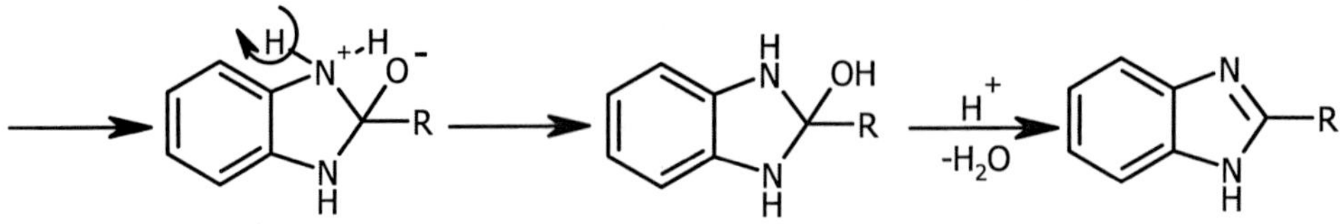

26.1 Preparation of Benzimidazole

Chemicals: *o*-Phenylene diamine, Formic acid, Charcoal and Sodium hydroxide.

Principle: The condensation of *o*-phenylene diamine with formic acid or the equivalent trimethyl orthoformate benzimidazole is typical example for Phillips reaction. 2-Substituted benzimidazoles can be prepared using different aliphatic and aromatic carboxylic acids.

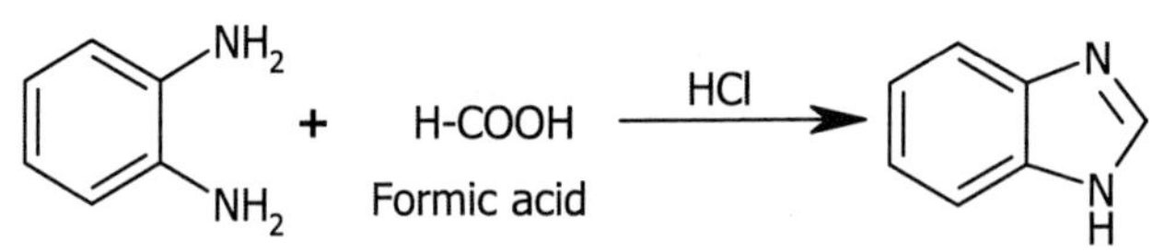

Reaction mechanism

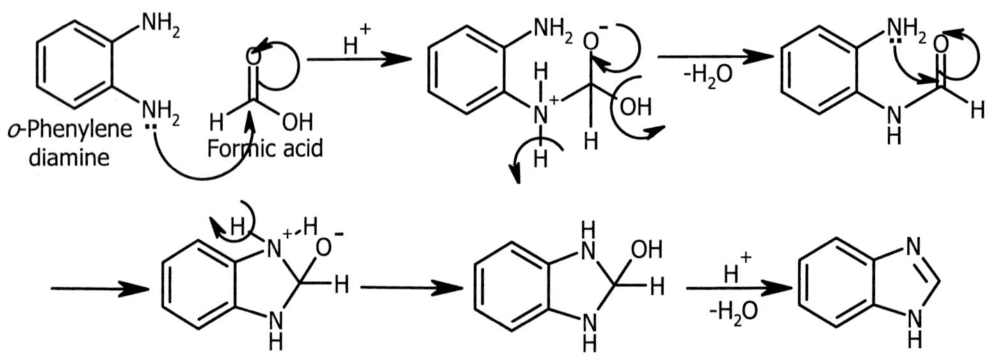

Procedure:

1. To the round bottom flask containing *o*-phenylene diamine (1.08 gm, 10 mmol) and catalytic amount of hydrochloric acid (4N) add formic acid (0.64 ml, 90%).
2. Heat the contents at 100 °C by placing on a water bath for 2 hr.
3. Cool and add sodium hydroxide solution (10%) slowly with constant stirring until the mixture becomes alkaline (litmus paper test).

4. Filter and wash the residue with ice cold water.
5. Dissolve the product in boiling water (16 ml), add charcoal (0.4 gm) and heat for 15 minutes.
6. Filter in hot condition, cool to 10 °C, filter and wash with cold water.
7. Dry and recrystallize the product.
8. The melting range of pure benzimidazole is 171-172 °C.

CHAPTER 27

BORSCHE-DRECHSEL CYCLIZATION

Condensation of phenyl hydrazine and cyclohexanone followed by the acid catalyzed rearrangement and ring closure leading to the formation of tetrahydro carbazole is known as Borsche-Drechsel cyclization.

Tetrahydro carbazole can be synthesized alternatively by Bucherer carbazole synthesis and Graebe-Ullmann reaction. Graebe-Ullmann reaction involves the formation of 1,2,3-triazole (unstable) from N-phenyl-1,2-diamino benzene. This unstable triazole eliminates nitrogen at higer temperature and forms carbazole.

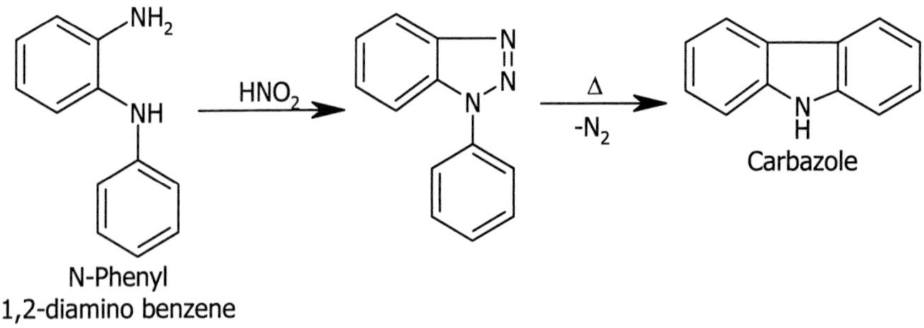

27.1 Preparation of 1,2,3,4-Tetra Hydrocarbazole

Chemicals: Cyclohexanone, Phenylhydrazine, Glacial acetic acid, Charcoal and Ethanol (75%).

Principle: The reaction involved in the preparation of 1,2,3,4-tetrahydro carbazole is known as Borsche-Drechsel cyclization. Phenyl hydrazine condenses with cyclohexanone and forms Schiff base [imine]. This Schiff base through sigmotropic rearrangement produce 1,2,3,4-tetrahydro carbazole. It involves the acid-catalyzed rearrangement of carbinolamine (cylclohexanone phenyl hydrazine complex) to tetrahydro carbazole.

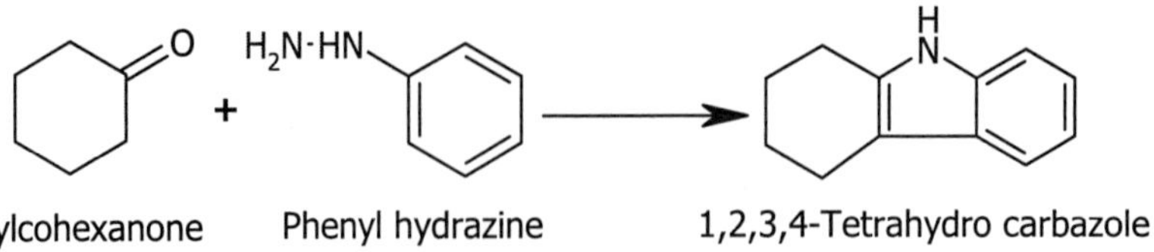

Nucleophilic addition of carbonyl group of cyclohexanone with amino group of phenyl hydrazine results schiff bases (imines). The addition of nucleophile amine to carbonyl group gives unstable carbinolamine. Carbinolamine undergoes acid-catalyzed (acetic acid) dehydration and forms imines by eliminating water molecule. This reaction proceeds through sigmotropic rearrangement.

Formation of carbinolamine

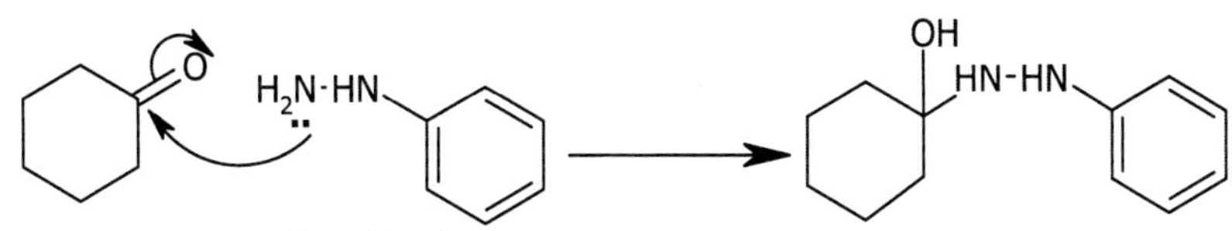

Cylcohexanone Phenyl hydrazine Carbinolamine

Formation of schiff base (imine)

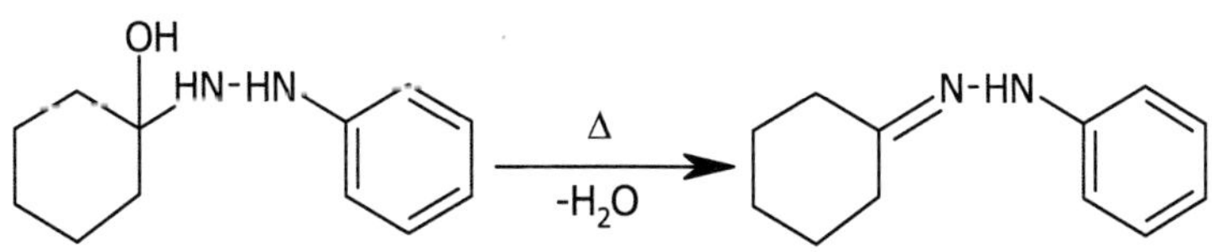

Carbinolamine Imine /Schiff base

Sigmotropic rearrangement

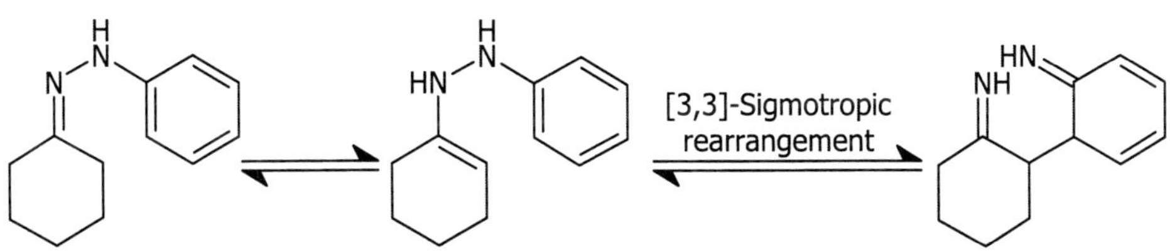

Imine /Schiff base

Formation of 1,2,3,4-tetrahydro carbazole

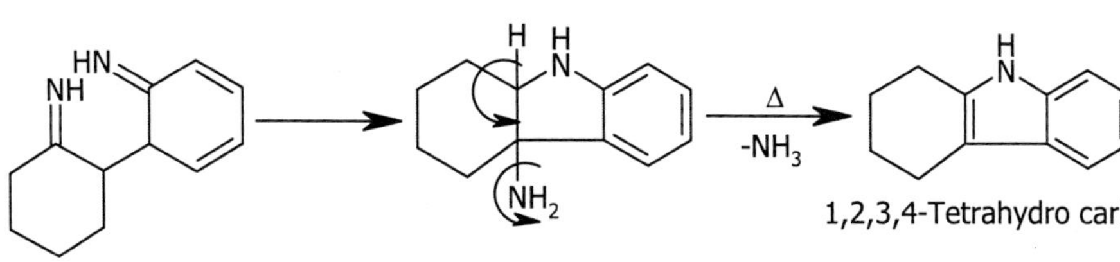

1,2,3,4-Tetrahydro carbazole

Procedure:

1. Place cyclohexanone (1 ml, 10 mmol), glacial acetic acid (3.5 ml) in a round bottom flask
2. Attach reflux condenser and reflux the contents for 1 hr with stirring.
3. Add phenylhydrazine (1 ml, 10 mmol) and continue stirring for further hour.
4. Pour the mixture into water (25 ml) and stir vigorously until the oily substance gets solidified.
5. Cool the mixture to 5 °C, filter the solid and wash it with water (5 ml) followed by ethanol (5 ml, 75%).

   [*Subject the solid for charcoal treatment if necessary*]
6. Dry the product and recrystallize the product with methanol.
7. The melting range of pure 1,2,3,4-tetrahydro carbazole is 116- 117 °C.

CHAPTER 28

PAAL-KNORR PYRROLE SYNTHESIS

The cyclization of acetonyl acetone (hexane-2,4-dione) and ammonia/primary amines produce 2,5-dimethyl pyrrole and N-substituted 2,5-dimethyl-pyrroles. This cyclization reaction is known as Paal-Knorr pyrrole synthesis.

Pyrrole nucleus is found in wide range of natural as well as synthetic drugs. It is the pyrolysis product of bone and present in endogenous pigments heme and chlorophyll. The marine derived pyrrole containing nakamuric acid and marinopyrroles are effective antibiotics against *Staphylococcus aureus*. Pentabromopseudilin, marine antibiotic found in soil-born actimycetes. Actinomycin and lydicamycin, platinsemycin, marinomycins (A, B, C and D), tirandamycin (A, B, C and D), lynamycin (A, B, C, D and E) and rabecamycin are some more important marine products. Tolmetin, atorvastatin and tallimustine are the synthetic therapeutics which contain pyrrole nucleus.

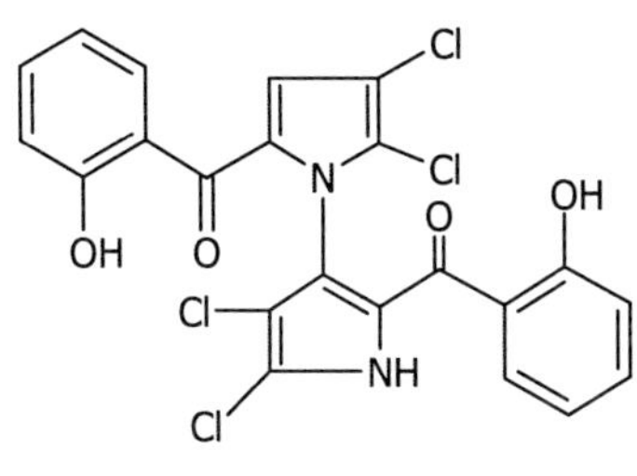

Marinopyrrole

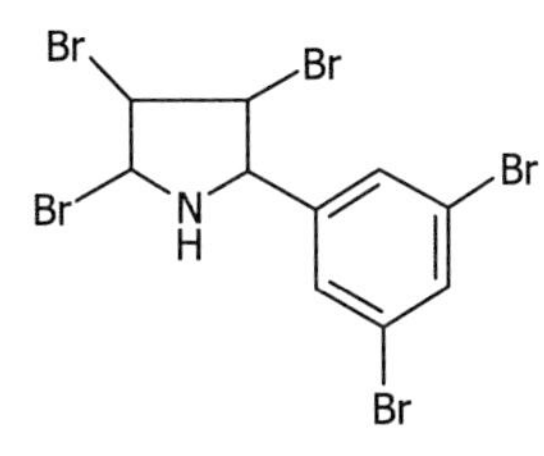

Penta bromopseudilin

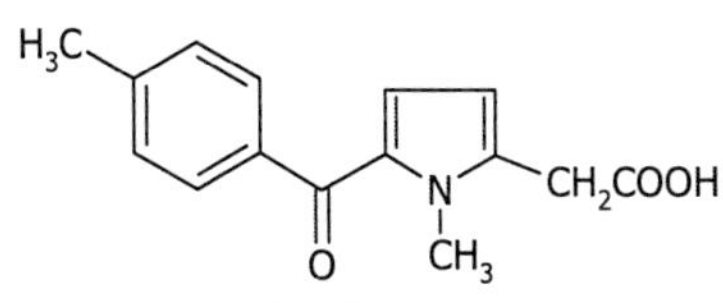

Tolmetin

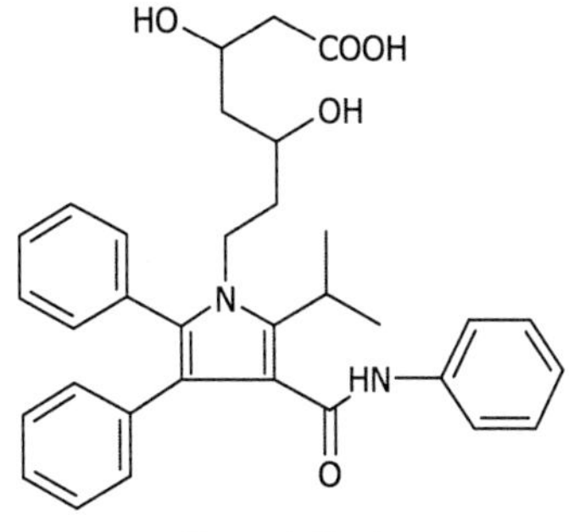

Atorvastatin

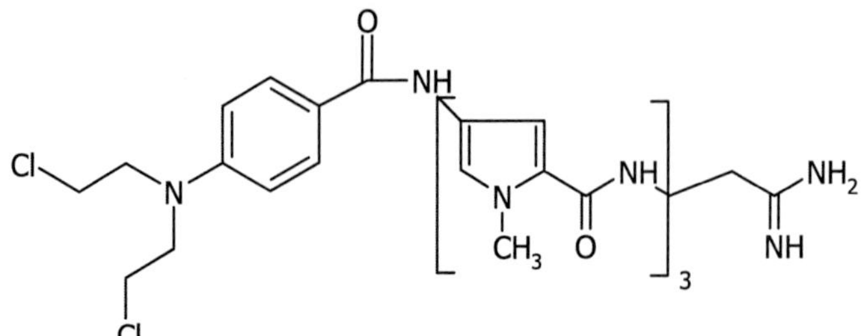

Tallimustine

28.1 Preparation of 2,5-Dimethyl Pyrrole

Chemicals: Acetonyl acetone (hexane-2,5-dione), Ammonia and Charcoal.

Principle: The cyclization reaction between acetonyl acetone and ammonia gives 2,5-dimethyl pyrrole. This reaction is typical example for Paal-Knorr Pyrrole synthesis. Acetonyl acetone exists in *keto-enol* tautomeric form. The lone pair of electrons present in the nitrogen of ammonia attacks enolic carbon of acetonyl acetone. This forms 5-amino-hexane-2-one, which on cyclization produces 2,5-dimethyl pyrrole.

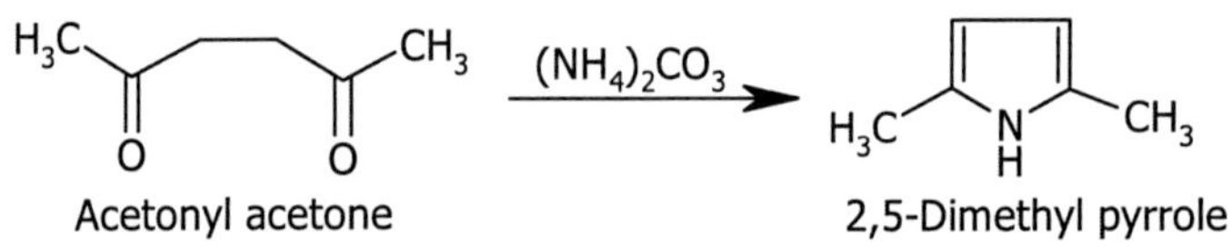

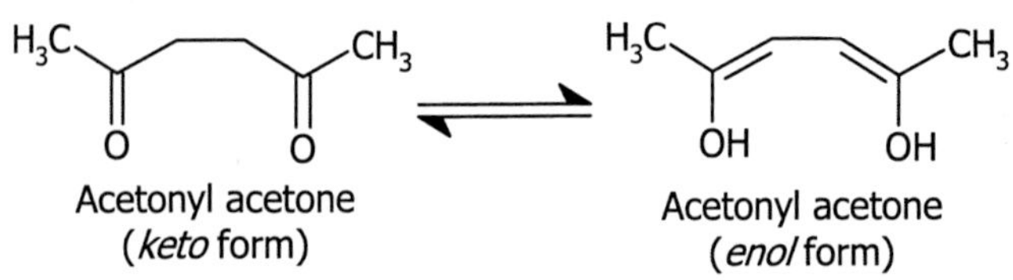

Reaction mechanism

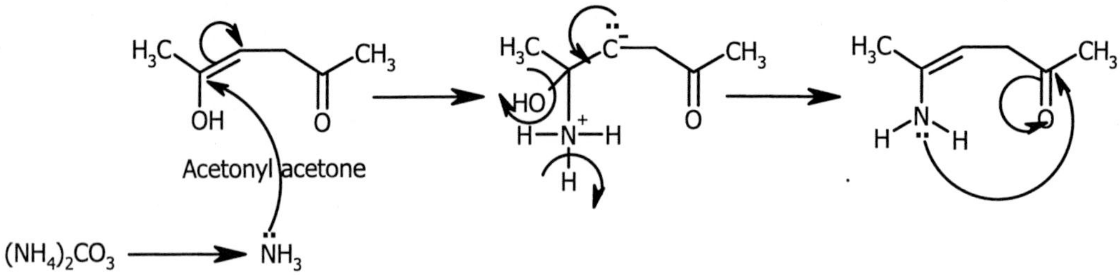

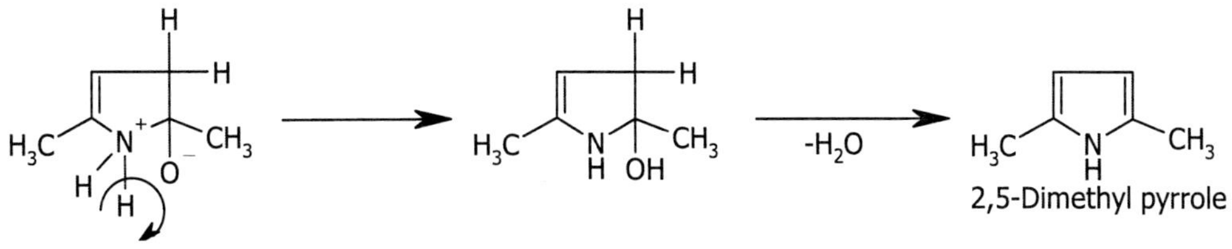

Procedure:

1. Place acetonyl acetone (1 ml, 8.5 mmol) and ammonium carbonate (19 gm) into the flask.
2. Attach air condenser and heat the mixture at 100°C [until effervescence stops, ~60–90 minutes]
3. The ammonium carbonate, which sublimes into the condenser, should be pushed back into the flask [use glass rod].
4. Replace air condenser with reflux condenser and reflux the mixture at 115°C for 30 minutes.
5. Dissolve the sublimed solid into the condenser with hot water (2 ml) and allow it to cool.
6. Separate the yellow layer (upper), and extract the lower layer with dichloromethane (2 ml).
7. Combine the two layers and dry over anhydrous sodium sulphate.
8. Distil the mixture and collect the 2,5-dimethyl pyrrole at 78 to 80°C.

CHAPTER 29

PAAL-KNORR THIOPHENE SYNTHESIS

Thiophen is an important building block in agrochemicals and pharmaceuticals. Many thiophen derivatives were identified as chemotherapeutic agents. Thiophen fused with various heterocyclic systems produces enhanced pharmacological activity. Thienopyrimidines, structural analogues of purines are potential nuleic acid antimetabolites. It can be synthesised by following methods:

1. *Gewald reaction*: Condensation of active methylene group containing ketone with β-ketonitrile in presence of base generates olefin, which upon cyclization with elemental sulphur produces amino thiophene.
2. *Volhard-Erdmann cyclization:* Cyclization of disodium succinate/1,4-difunctional compounds with phosphorous heptasulfide produces alkyl/aryl thiophenes.
3. *Paal-Knorr thiophene synthesis*: The generation of thiophenes by condensation of a 1,4-dicarbonyl compound in presence of an excess of phosphorous pentasulphide (P_2S_5) or Lawesson's reagent.
4. *Hinsberg synthesis*: Aldol condensation (two consecutive) of 1,2-dicarbonyl compound and diethyl thiodiacetates gives thiophen.
5. *Fiesselmann thiophen synthesis*: Condensation of thioglycollic acid and α, β-acetylenic esters produces 3-hydroxyl-2-thiophencarboxylic acid.

Paal-Knorr thiophene synthesis

Condensation of 1,4-dicarbonyl compound in the presence of an excess of phosphorous pentasulphide (P_2S_5) or Lawesson's reagent is known as Paal-Knorr thiophene synthesis. The preparation of 2,5-dimethyl thiophene from acetonyl acetone is the typical example of Paal-Knorr thiophene synthesis. Phosphorous pentasulphide and Lawesson's reagent act as sulphurizing and dehydrating agents. The reaction forms furan first and then gets converted into thiophene. Toxic hydrogen sulphide (H_2S) will form as a side product regardless of a sulphur source.

Precautions:

1. Phosphorous pentasulphide (P_2S_5) produce skin and eye irritation.
2. Phosphorous pentasulphide (P_2S_5) cause shortness of breath and pulmonary edema.
3. It also produce dizziness, headache and seizures.
4. It is an explosive chemical, hence should be handled in fume-cupboard.

29.1 Preparation of 2,5-Dimethyl Thiophen

Chemicals: Acetonyl acetone (hexane 2,5-dione), Phosphorus pentasulphide, Diethyl ether, Sodium carbonate and Magnesium sulphate.

Principle: The 2,5-dimethyl thiophene can be prepared by condensation of 1,4-dicarbonyl compound in the presence of an excess of a source of sulphur such as phosphorous pentasulphide (P_2S_5).

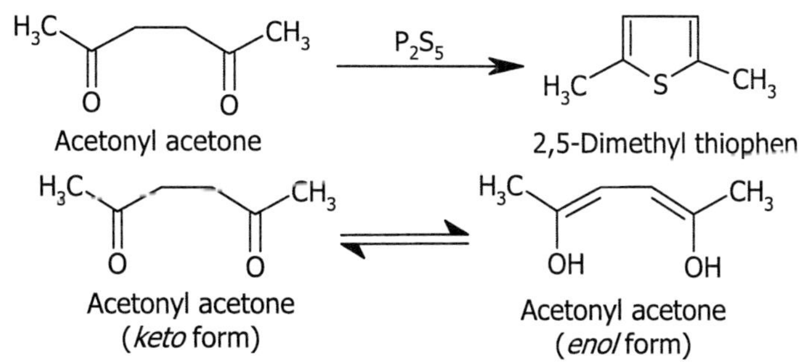

Acetonyl acetone exists in *keto-enol* form (tautomer), in presence of sulphurizing agent it forms thioketone (*keto-thiol tautomer*). Phosphorous pentasulphide acts as sulphurizing and dehydrating agents. Thioketone of acetonyl acetone on cyclization produces 2,5-dimethyl thiophene. The occurance of a bis-thioketone intermediate as assumed to be possible, but not necessary.

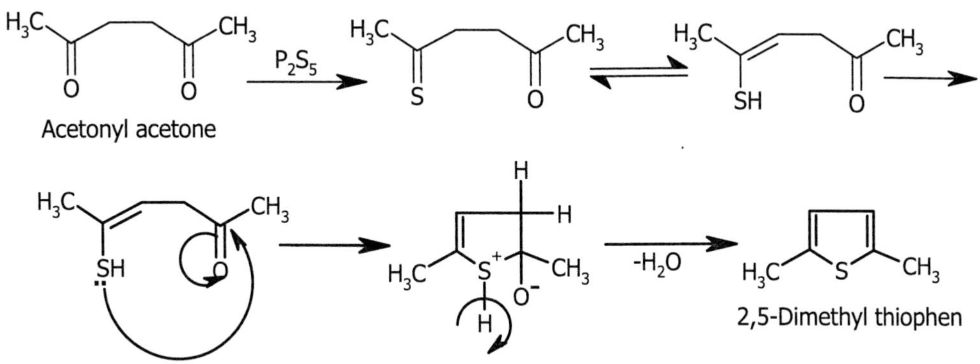

Procedure:

1. Place phosphorous pentasulphide (4.6 gm, 21 mmol) in a round bottom flask equipped with reflux condenser.
2. Add acetonyl acetone (2 ml, 17 mmol) dropwise and heat the reaction mixture with small luminous flame until the exothermic reaction commences.
3. Remove the flame and add acetonyl acetone (7 ml, 60 mmol) drop wise.
4. After the complete addition, reflux the reaction mixture for 1 hr.
5. Cool the reaction mixture, pour into ice cold water (20 ml) and extract the aqueous mixture with ether (5 ml) four times.
6. Combine the ether extracts with anhydrous sodium carbonate solution and water.
7. Dry the ether solution over anhydrous magnesium sulphate and distill to get 2,5-dimethyl thiophene.
8. The boiling range of pure 2,5-dimethyl thiophene is 135-136 °C.

CHAPTER 30

PAAL-KNORR PYRAZOLE SYNTHESIS

Paal-Knorr Pyrazole synthesis involves the condensation of 1,3-dicarbonyl compounds with hydrazines. The nitrogen of the hydrazine attacks carbonyl carbon and forms imine intermediate. The nitrogen of hydrazine also attacks other carbonyl carbon and forms second imine intermediate. The imine intermediate upon dehydrogenation forms pyrazole product. Reaction of semicarbazide or amino guanidines with acetyl acetone gives 1-carbamido or 1-carboxamide derivatives, which on hydrolysis and decarboxylation gives 3,5-dimethyl pyrazole. It can be also prepared from 1,2-pentanedien-4-one and hydrazine hydrate.

Celecoxib (anti-inflammatory), zaleplon (sedative and hypnotic), sildenafil (erectile dysfunction), zoniporide (cardioprotective) and acomplia (anti-obesity) are the important pharmaceuticals containing pyrazole nucleus.

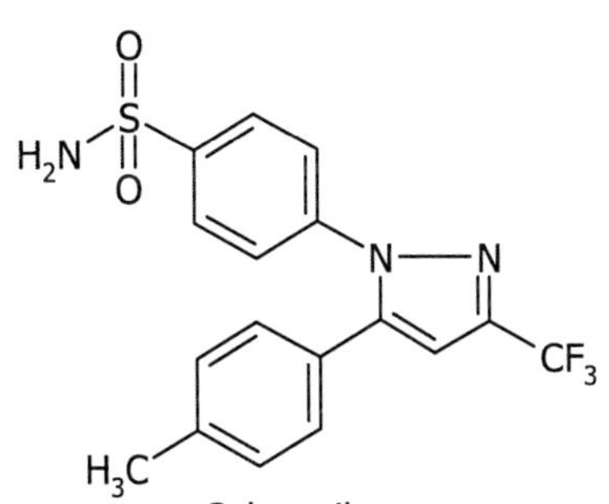

Celecoxib

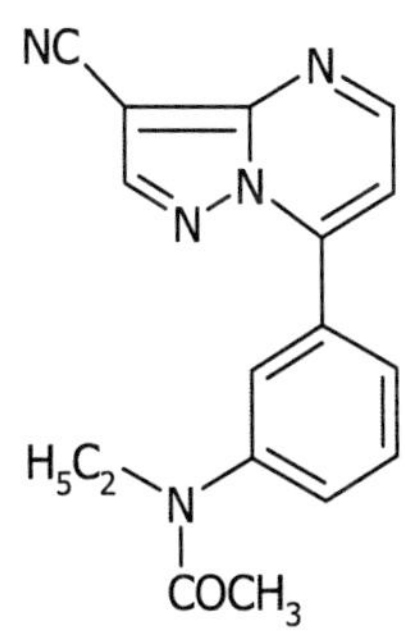

Zaleplon

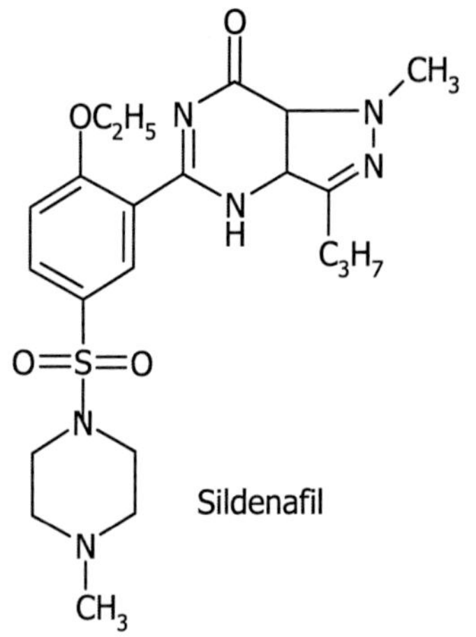
Sildenafil

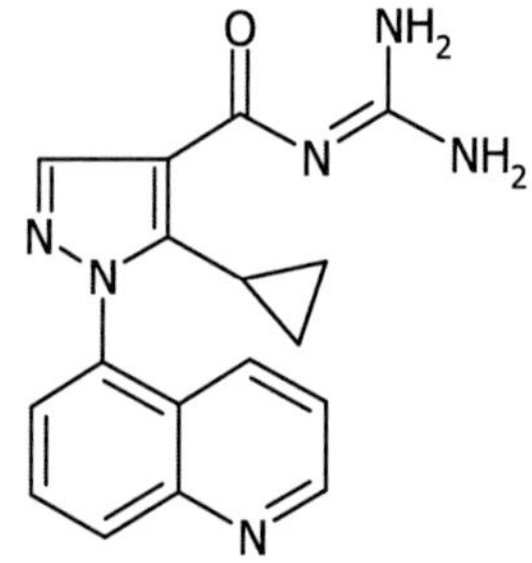
Zoniporide

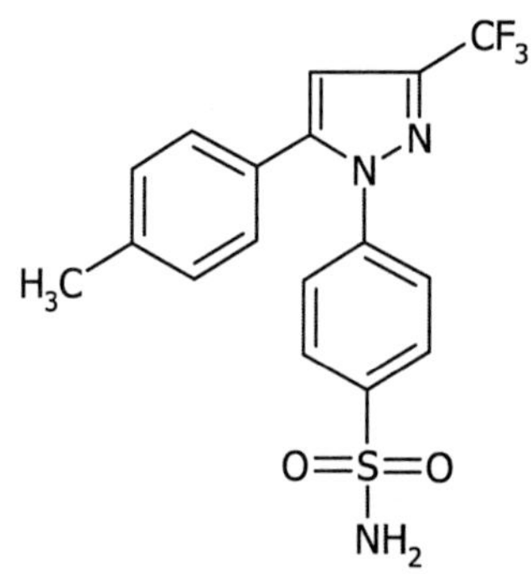
Acomplia

30.1 Preparation of 3,5-Dimethyl-1*H*-Pyrazole

Chemicals: Acetyl acetone, Hydrazine sulphate, Sodium hydroxide (2.5 M), Ether and Potassium carbonate.

Principle: The preparation of 3,5-dimethyl pyrazole involves the condensation of acetyl acetone (2,4-pentane dione) with hydrazine hydrate in ethanol or hydrazine sulphate in aqueous alkali. Hydrazine sulphate in alkali is preferred because the reaction with hydrazine hydrate is violent.

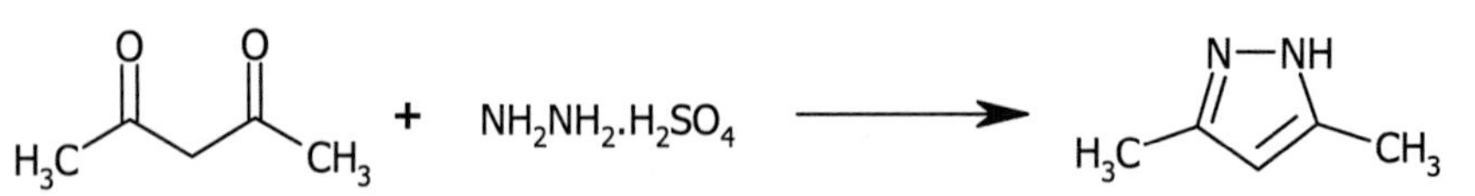
Acetyl acetone　　Hydrazine sulphate　　3,5-Dimethyl-1*H*-pyrazole

Acetyl acetone exists in *keto-enol* tautomeric form. The lone pair of electrons present in the nitrogen of hydrazine sulphate attacks enolic carbon of acetyl acetone. This forms

5-substituted pentane-2-one, and on cyclization eliminates two water molecules and produces 3,5-dimethyl pyrazole.

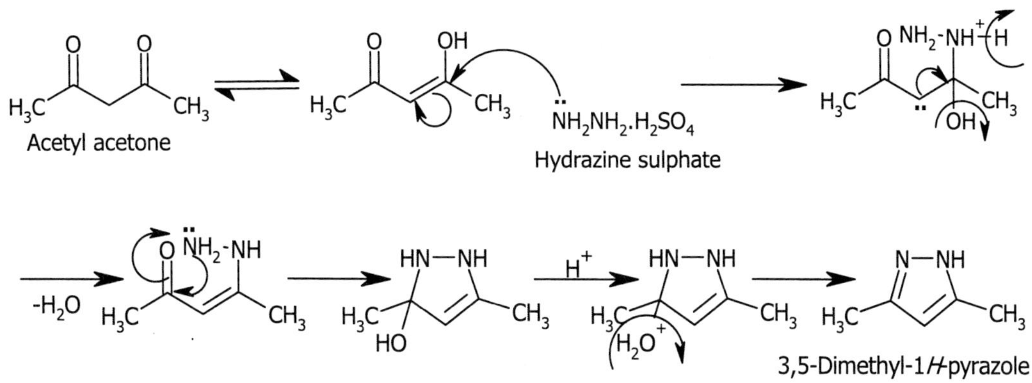

Procedure:

1. Place hydrazine sulphate (1.3 gm, 10 mmol) in sodium hydroxide solution (8 ml, 2.5 M) in a flask equipped with thermometer, magnetic stirrer and dropping funnel.
2. Immerse the flask in an ice-bath and when the temperature reaches 15 °C add acctyl acetone (pentane-2,4-dione) (1 ml, 10 mmol) dropwise with stirring while maintaining the same temperature.
3. Continue the stirring for further 1 hr at 15 °C (*dimethyl pyrazole separates during this period*).
4. Add 10 ml of water to dissolve the inorganic salts.
5. Transfer the contents to a separating funnel and extract the aqueous mixture with three 5 ml portions of ether.
6. Combine the ethereal layers and dry over anhydrous potassium carbonate.
7. Evaporate the ether to obtain pale yellow solid dimethyl pyrazole.
8. The melting range of pure dimethyl pyrrole is 107 to 108 °C.

CHAPTER 31

REIMER-TIEMANN REACTION

The conversion of phenols into salicylaldehyde by *o*-formylation reaction is typical example for Reimer-Tiemann reaction. Pyrroles will also undergo this reaction because of their relative electron richness. This reaction can be divided into normal and abnormal reactions.

Normal Reimer-Tiemann Reaction: It involves the chemical conversion of phenols into salicylaldehyde (2-hydroxy benzaldehyde, major product) and 4-hydroxy benzaldehyde (minor product). The mixture of chloroform and potassium hydroxide (or sodium hydroxide) mediates this formylation reaction.

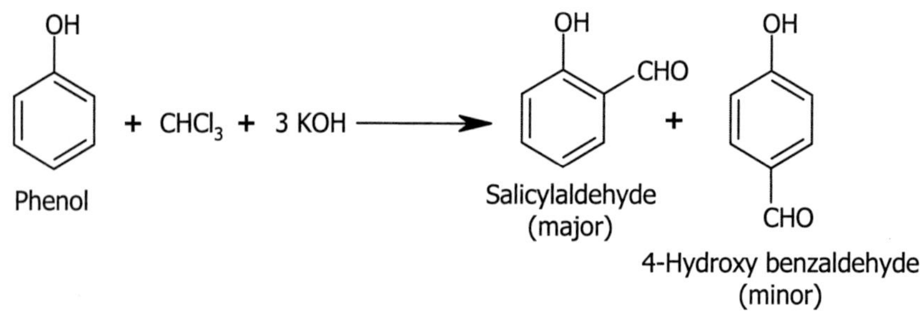

Reaction Mechanism

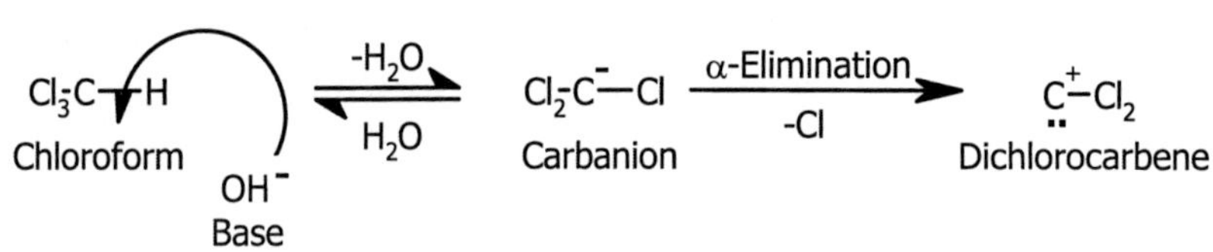

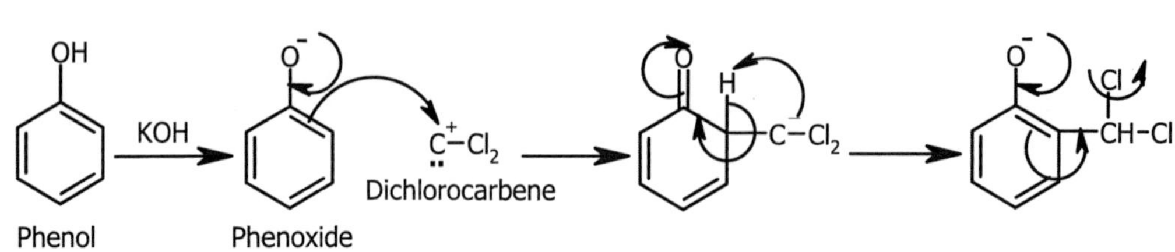

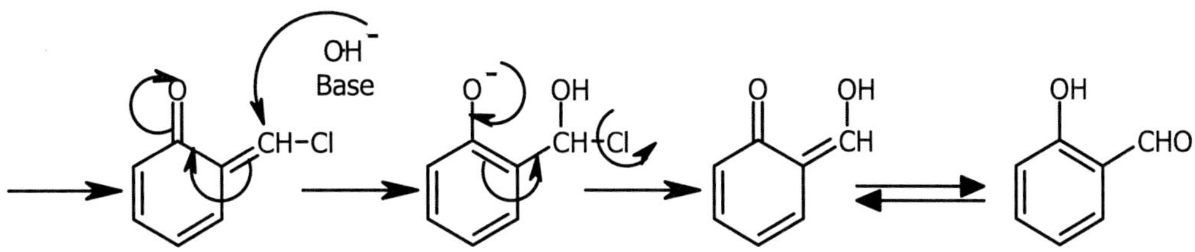

Salicylaldehyde

The base (–OH⁻) abstracts the proton of chloroform and forms carbanion. The carbanion gives off chlorine [α-elimination (slow)] and generates dichlorocarbene (reactive intermediate). The *o*-carbon of phenate attacks dichloro carbene and produces salicylaldehyde through base hydrolysis.

Abnormal Reimer-Tiemann Reaction: Phenols (2/4 substituted) produces disubstituted (2,2 or 4,4) cyclohexadienones along with the desired product.

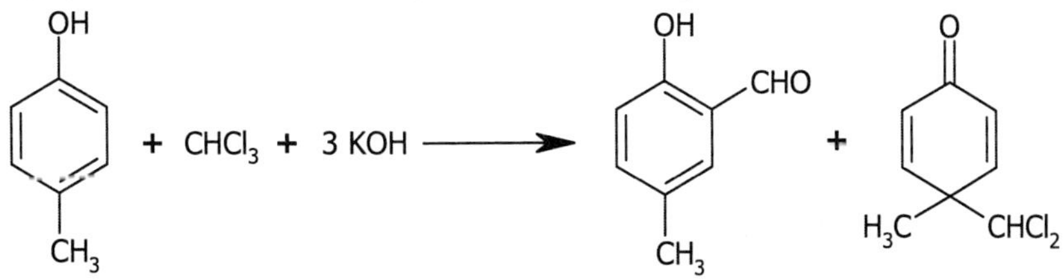

p-Cresol

5-Methyl salicylaldehyde (major)

4,4-Disubstituted cyclohexadienone

Pyrroles and similar five membered rings produces ring expansion products.

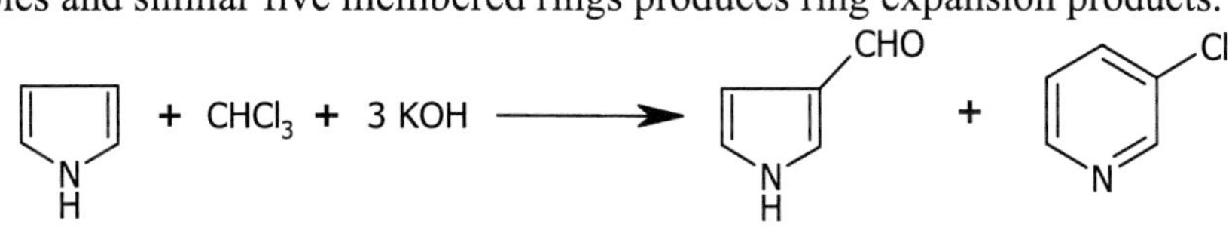

Pyrrole

Pyrrole-3-carbaldehyde

3-Chloro pyridine

31.1 Preparation of β-Hydroxy Naphthaldehyde

Chemicals: β-Naphthol, Chloroform, Sodium hydroxide, Ethanol, Hydrochloric acid, Magnesium Sulphate and Ether.

Principle: β-Naphthol undergoes formylation reaction with chloroform and sodium hydroxide to yield β-hydroxy naphthaldehyde.

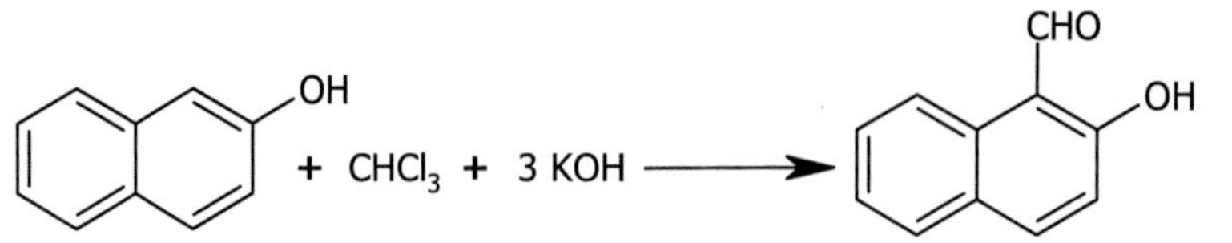

β-Naphthol β-Hydroxy naphthaldehyde

The sodium hydroxide (–OH⁻) abstracts the proton of chloroform and forms carbanion. The carbanion gives off chlorine [α-elimination (slow)] and generates dichlorocarbene (reactive intermediate). The *o*-carbon of β-naphthol attacks dichloro carbene and produces β-hydroxy naphthaldehyde after base hydrolysis.

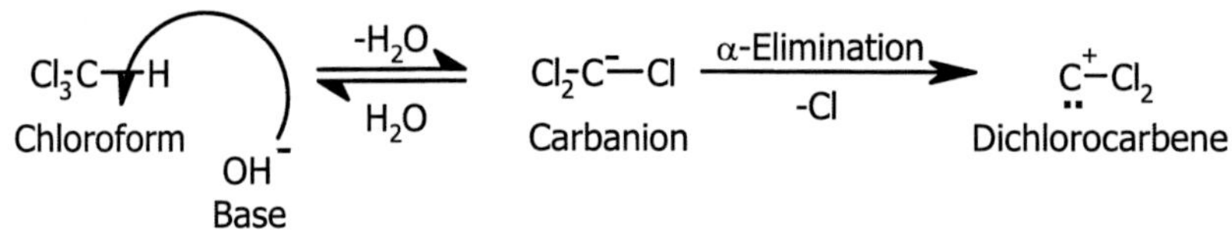

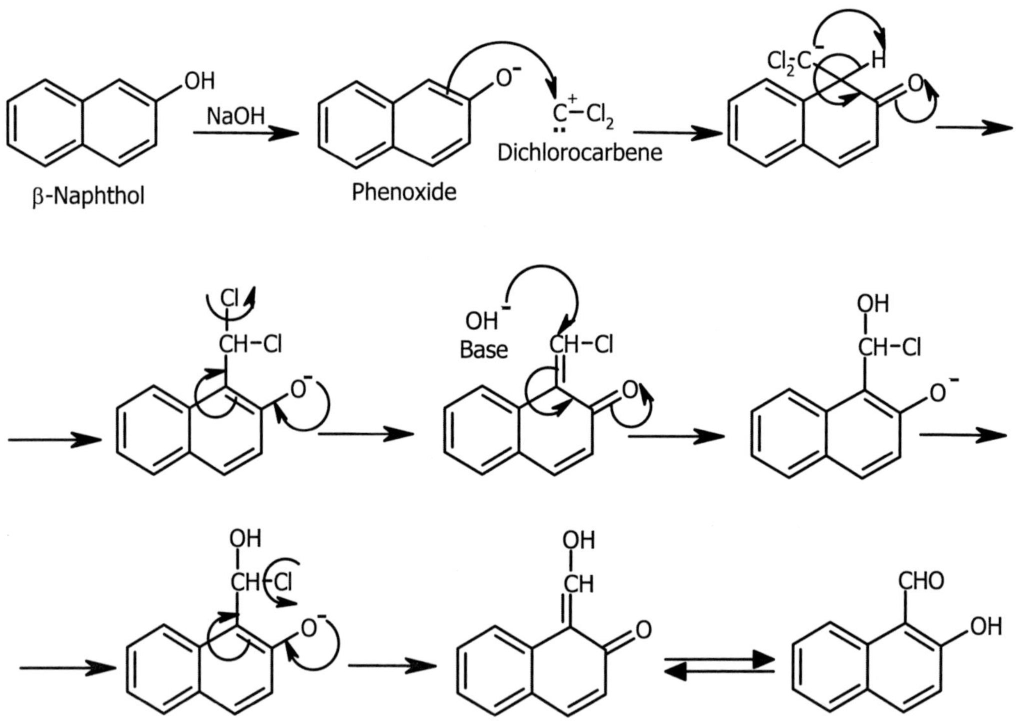

β-Hydroxy naphthaldehyde

Procedure:

1. Equip three necked flask with separating funnel, magnetic stirrer and double surface condenser.
2. Place 2-naphthol (1 gm, 7 mmol) and ethanol (3 ml) in the flask.

3. Add the solution of sodium hydroxide (2 gm in 4.2 ml of water) with rapid stirring.
4. Heat the mixture (60 to 70 °C) by placing the flask on a water bath.
5. Add chloroform (1 ml) drop wise.
   [*The development of deep blue colour indicates the commencement of reaction*].
6. Remove the flask from water bath and add chloroform (3 ml).
7. Reflux the mixture gently for 1.5 hrs.
   [*The sodium salt of phenolic aldehyde separates*].
8. Stirr the reaction mixture for 1 hr.
9. Acidify the residue/mixture with hydrochloric acid [use congo red paper].
10. Add water to dissolve the separated salt.
11. Extract the remaining oily layer with ether and wash the ether layer with water.
12. Dry the ether layer by passing through anhydrous magnesium sulphate.
13. Evaporate the ether layer and allow the residue to cool to obtain solid product.
14. Purify the β-hydroxy naphthaldehyde by recrystallization with ethanol.
15. The melting range of pure β-hydroxy naphthaldehyde is 80-81 °C.

CHAPTER 32

SKRAUP'S SYNTHESIS

Synthesis of quinoline by heating aniline with sulphuric acid, glycerol and an oxidizing agent is known as Skraup synthesis. Ferrous sulphate, iodine, arsenic and nitrobenzene can be used as an oxidising agent. Nitrobenzene also acts as solvent, but it is very violent in nature. Sulphuric acid dehydrates glycerol and forms acrolein. Acrolein reacting with aniline through 1,4-addition forms an intermediate, which eliminates water molecule and produces dihydro quinoline. Dihydroquinoline through oxidation converts into quinoline.

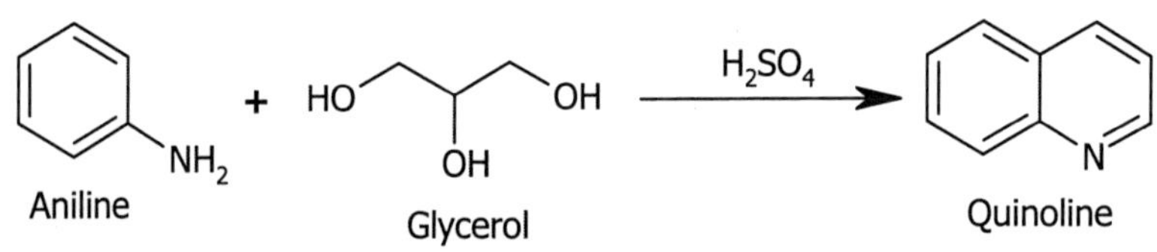

Reaction Mechanism
Formation of Acrolein

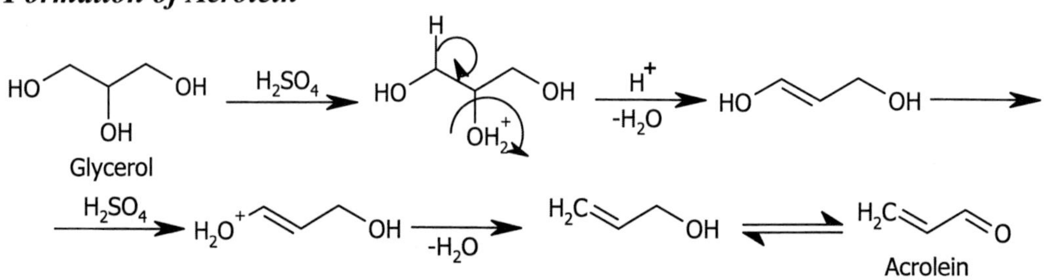

Formation of Quinoline

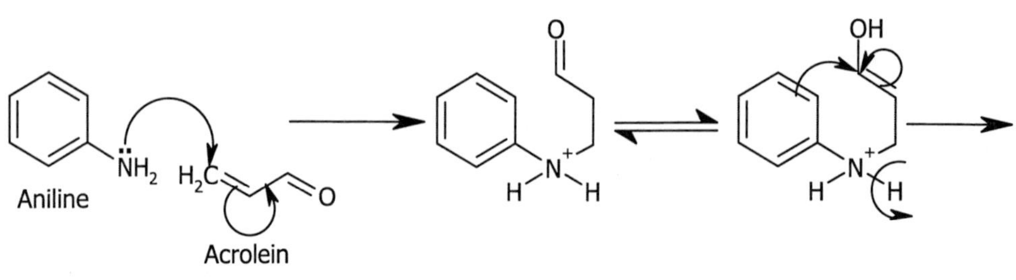

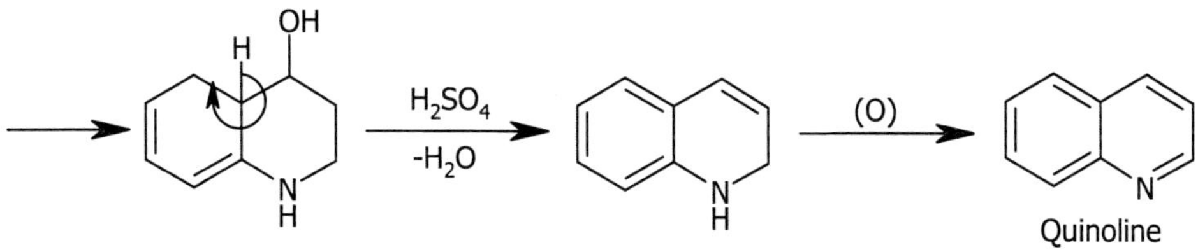

Pharmaceutical Applications

1. Preparation of anti-malarial drugs Primaquine and Pamaquine.
2. Preparation of Chlorquinaldol (anti-microbial).

32.1 Preparation of Quinoline

Traditional Name: 1-Azanaphthalene, Benzopyridine.

Chemicals: Aniline, Glycerol, Sulphuric acid, Iodine, Sodium hydroxide, Sodium nitrite, Starch iodide paper and Ether.

Principle: Aniline reacts with glycerol in presence of sulphuric acid and iodine to form quinoline.

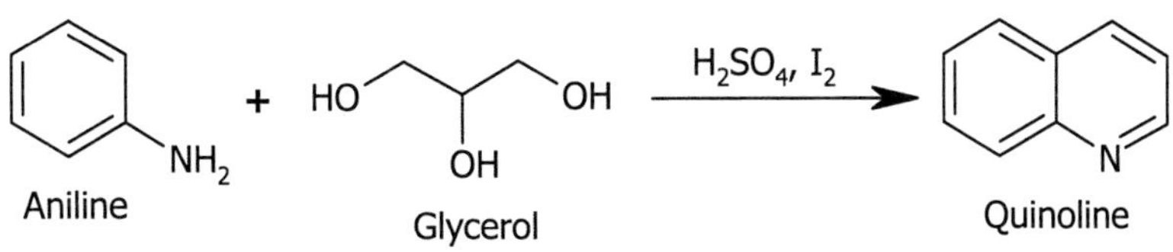

Sulphuric acid dehydrates glycerol and forms acrolein. Acrolein reacting with anilne through 1,4-addition forms an intermediate, which eliminates water molecule and produces dihydro quinoline. Iodine oxidizes (removal of hydrogen) dihydro quinoline into quinoline.

Procedure:

1. Transfer aniline (2 ml, 22 mmol), glycerol (1.3 gm, 14 mmol) and iodine (0.1 gm) into three necked round bottom flask.
2. Stir the reaction mixture and add sulphuric acid (3.3 ml) drop wise through the condenser. [*The temperature raises to 100 to 105 °C*]
3. Heat the reaction mixture by placing in an oil bath and maintain the temperature at 140 °C.
4. Sulphur dioxide vapour and iodine vapour will evolve and then continue the heating at170 °C for 1 hr.
5. Cool and make the solution alkaline by adding sodium hydroxide (17 ml, 5 M).

6. Distil (steam distillation) the quinoline until no oily drops passes over (it may contains small amount of aniline).
7. Purification of quinoline
   - Add sulphuric acid (2 ml) and cool (0-5 °C), then add solution of sodium nitirite (0.3 g) in water (2 ml) drop wise (until a drop of reaction mixture gives blue colour with potassium iodide starch paper).
   - Heat the diazotized solution for 1 hr and make it alkaline with solution of sodium hydroxide (1.5 g in 5 ml of water).
   - Distil (steam distillation), extract the distillate with ether and evaporate.
   - The melting point of pure quinoline is 236-238 °C.

32.2 Preparation of 8-Hydroxyquinoline

Traditional Name: 1-Azanaphthalene-8-ol, oxine.

IUPAC Name: Quinolin-8-ol, 8-quinolinol.

Chemicals: 2-Hydroxy aniline, Glycerol, Sulphuric acid and Ferrous sulphate.

Principle: 2-Hydroxy aniline reacts with glycerol in presence of sulphuric acid and ferrous sulphate to form 8-hydroxyquinoline.

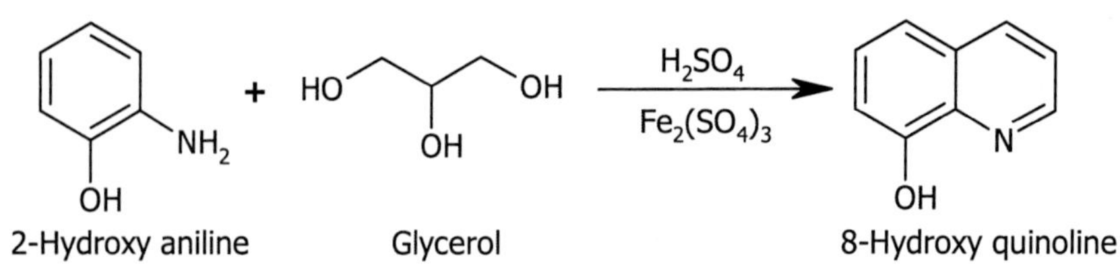

Sulphuric acid dehydrates glycerol and forms acrolein. Acrolein reacting with 2-hydroxy aniline through 1,4-addition forms an intermediate, which eliminates water molecule and produces 1,2-dihydro-8-hydroxy quinoline. Ferrous sulphate oxidizes (removal of hydrogen) 1,2-dihydro-8-hydroxyquinoline into 8-hydroxyquinoline.

Procedure:

1. Place sulphuric acid (2.3 ml), 2-hydroxy aniline (1 gm, 9 mmol) and anhydrous glycerol (2.5 gm, 27 mmol) in a flask [flask I].
2. Liquify the mixture present in flask I by placing on a steam bath.
3. Place 2-nitro phenol (0.7 g, 5 mmol) and ferrous sulphate (100 mg) in another flask [flask II] attached with reflux condenser.
4. Heat the contents of flask II to 100 - 120 °C.

5. Add the liquid mixture present in flask I to flask II in small portions.
6. Maintain the temperature between 135 to 150 °C during the addition process.
7. Reflux the mixture at 130 °C for 4 hrs after the addition process is complete.
8. Cool the reaction mixture and neutralize with sodium hydroxide solution with vigorous stirring to produce the pH 7.
9. Add crushed ice to the mixture to maintain the temperature below 40 °C.
10. Filter the precipitate and dry.
11. Distill the isolated product under reduced pressure [discard the first few ml of distillate (water)].
12. The distillate crystallizes on cooling.
13. The melting point of pure 8-hydroxyquinoline is 74 to 75 °C.

Anhydrous Glycerol: Heat glycerol (50 ml) in porcelain dish with stirring maintaining the temperature at 180 °C. Reduce the temperature to 100 °C, pour into beaker and keep in desiccator over sulphuric acid.

CHAPTER 33

VON BAEYER SYNTHESIS

German scientist Adolf Von Baeyer synthesized the first barbiturate from malonic acid. The condensation of urea and malonic acid in presence of absolute ethanol and sodium metal rendered barbiturate. It had no therapeutic effect, but served as lead for the development of many therapeutics.

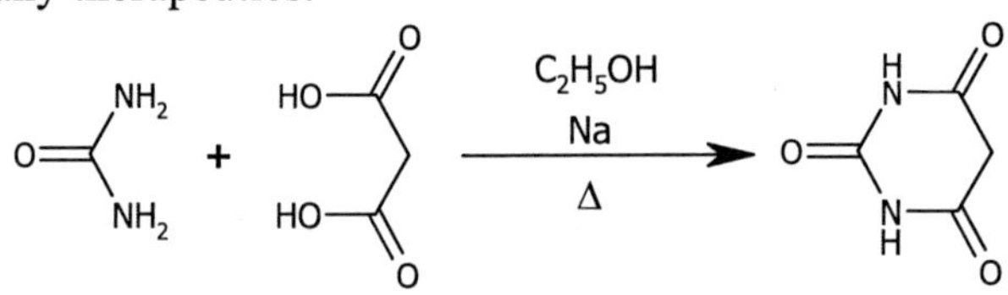

Urea   Malonic acid   Barbituric acid

The substitution of alkyl and cycloalkyl groups to the active hydrogen of methylene (C_5) portion of barbiturate produced therapeutically significant molecules.

***Barbiturate therapeutics*:**

1. *Sedative and hypnotics:* Pentobarbitone, Secobarbitone.
2. *Anti-convulsant:* Phenobarbitone
3. *General anaesthetic:* Thiopentone, Thiamylal.

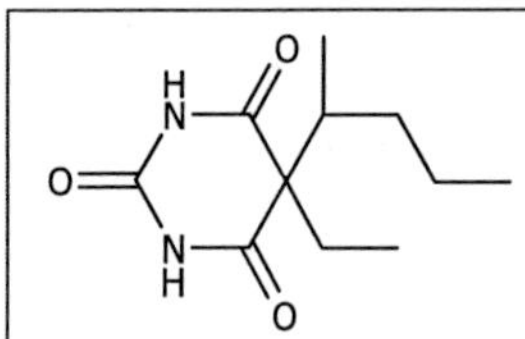 Pentobarbitone	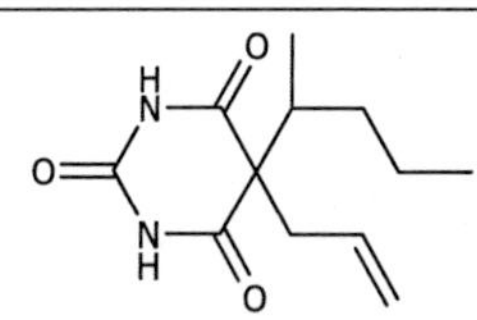 Secobarbitone	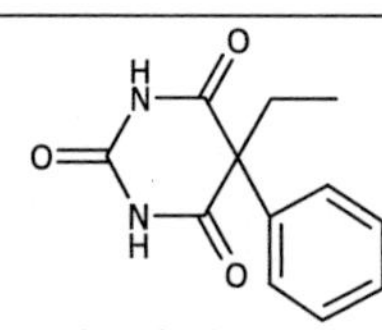 Phenobarbitone
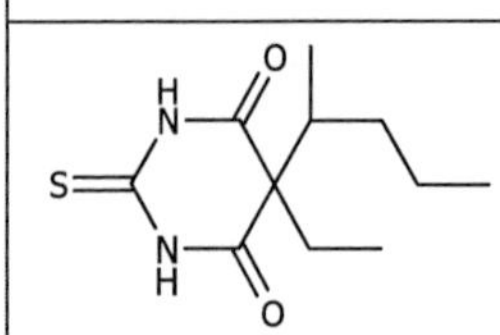 Thiopentone	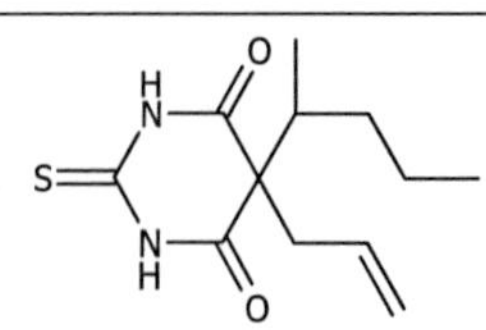 Thiamylal	

33.1 Preparation of Barbituric Acid

Traditional Name: Melonyl urea.

IUPAC Name: Pyrimidine-2,4,6-[1*H*, 3*H*, 5*H*]-triazine.

Chemicals: Urea, Sodium metal, Diethyl malonate, Ethanol and Hydrochloric acid.

Principle: Barbituric acid is chemically pyrimidine-2,4,6-trione. Knoevenagel condensation of barbituric acid generated large number of baribiturate derivatives. Barbiturates posses sedative, hypnotic, anti-psychotic and anti-depressant effects. Phenobarbital was the first marketed barbiturate. Barbituric acid also finds utility in the synthesis of riboflavin.

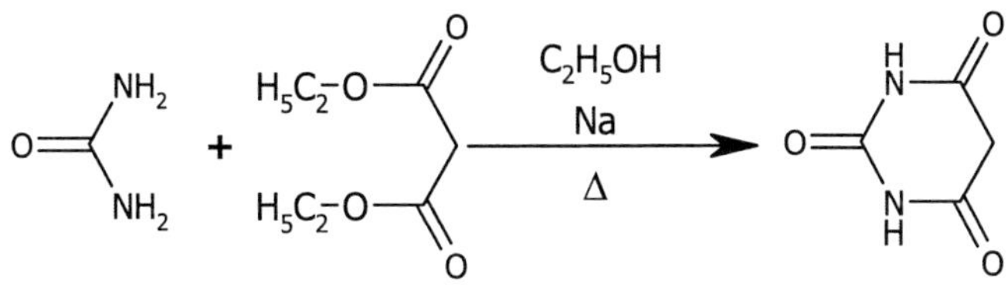

Urea Diethyl malonate Barbituric acid

Reaction Mechanism

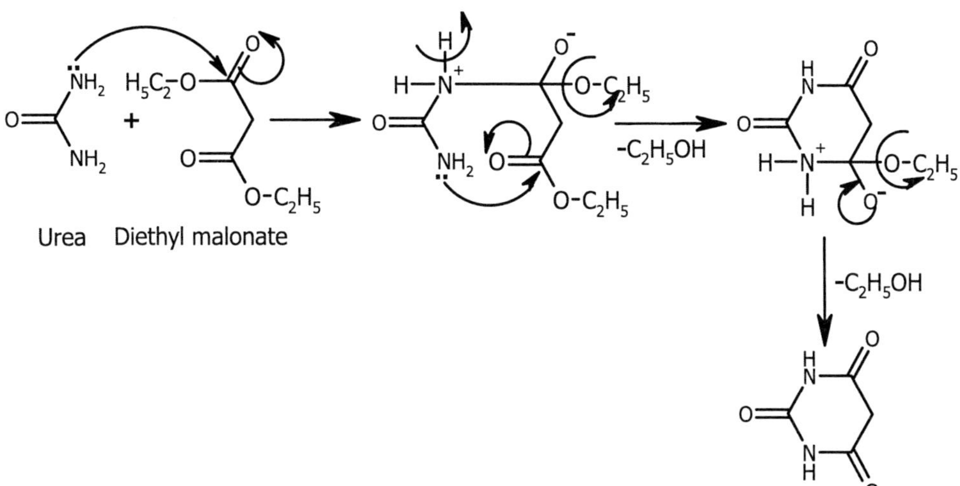

Urea Diethyl malonate

Barbituric acid

Procedure:

1. Add absolute ethanol (4 ml) and sodium metal (0.32 gm, 14 mmol) to the round bottom flask.
2. Cool the mixture by placing in an ice-bath (if mixture becomes vigorous).

3. Add diethyl malonate (2 ml, 14 mmol) and solution of urea (0.84 gm, 14 mmol) in hot absolute ethanol (4 ml).
4. A white solid separates out rapidly after being well shaken.
5. Swirl the contents to dissolve attach reflux condenser fitted with calcium guard tube and reflux the mixture for 7 hr at 110 °C.
6. Add hot water (8 ml, 50 °C) and hydrochloric acid to acidify the mixture.
7. Cool the resulting filter and keep the clear solution overnight in refrigerator.
8. Collect the separated barbituric acid by vacuum filtration and wash with cold water (5 ml).
9. The melting point of the pure barbituric acid is 245 °C.

CHAPTER 34

DIAZOTIZATION REACTION

Diazotization is an important chemical reaction in the synthesis of pharmaceuticals. It involves the reaction between aromatic primary amine ($-NH_2$) and nitrous acid in presence of excess of mineral or inorganic acids to form diazonium compounds. Sodium nitrite reacts with hydrochloric acid and gives nitrous acid (HONO) *in situ.*

Generation of nitrous acid: Sodium nitrite reacts with (mineral/inorganic acid) under ice cold condition and gives nitrous acid, *in situ.*

$$\underset{\text{Sodium nitrite}}{NaNO_2} + HCl \xrightarrow{0\text{-}5\ ^0C} \underset{\text{Nitrous acid}}{HONO} + NaCl$$

Formation of nitrosonium ion: Nitrous acid undergoes protonation with acid and forms nitrosonium ion.

$$\underset{\text{Nitrous acid}}{O{=}N{-}OH} + H_3O^+ \rightleftharpoons O{=}N{-}\overset{+}{O}H_2 + H_2O$$

Electrophilic attack on nitrosonium ion: The nitrosonium ion (electrophile) attacks electrons of amino group (aromatic primary amine) and forms an intermediate.

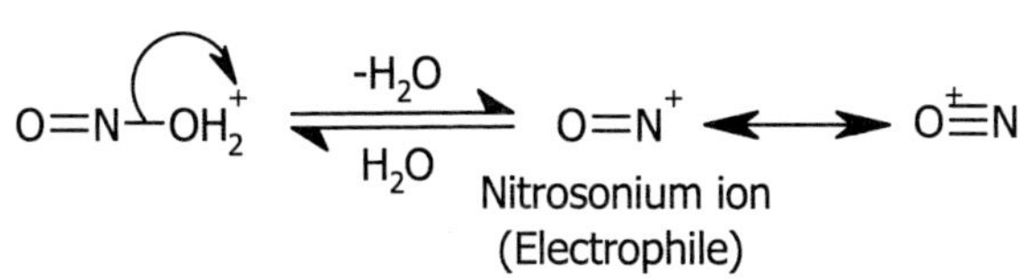

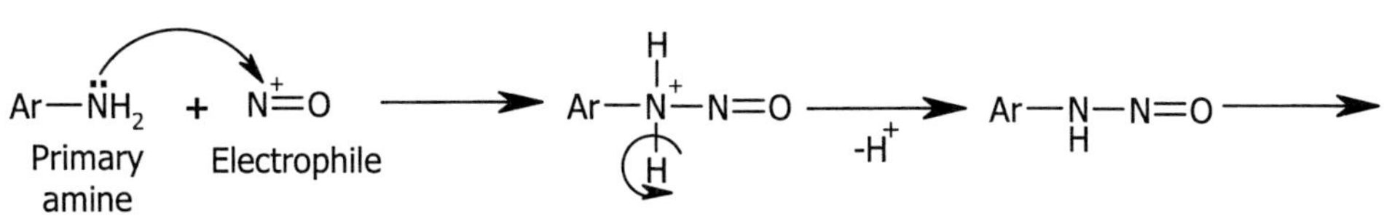

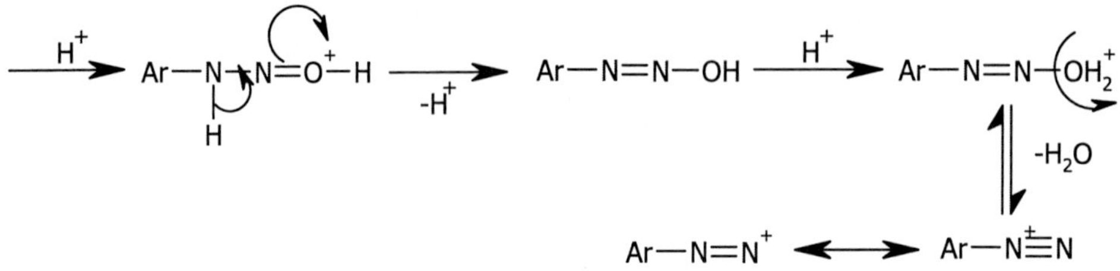

Pharmaceutical Applications

1. Preparation of Riboflavin (Vitamin B_2).

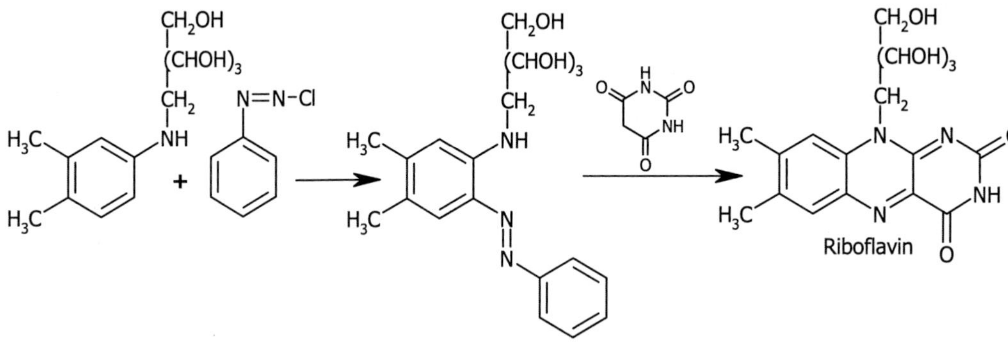

2. Preparation of Pentetrazol (respiratory stimulant).

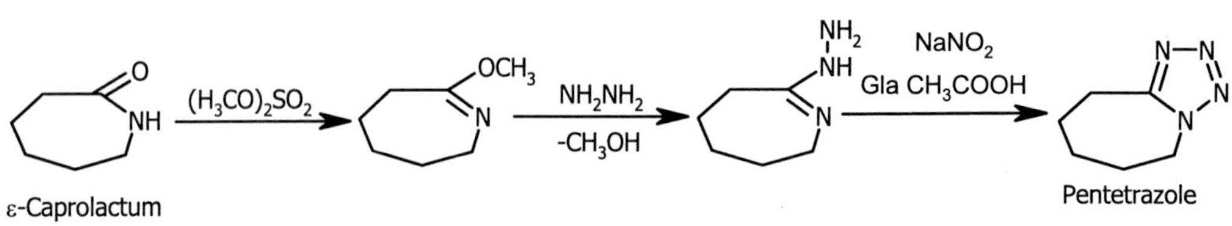

3. In the preparation of diazepam - 'Japp-Klingemann reaction'.

34.1 Preparation of Benzotriazole

Chemicals: *o*-Phenylene diamine, Sodium nitrite, Glacial acetic acid and Charcoal.

Principle: The preparation of benzotriazole involves diazotization followed by cyclization. Diazotization of *o*-phenylene diamine with nitrous acid formed by the reaction between sodium nitrite and glacial acetic acid gives diazonium salt. The formed diazonium group couples with adjacent amino group and gives benzotriazole.

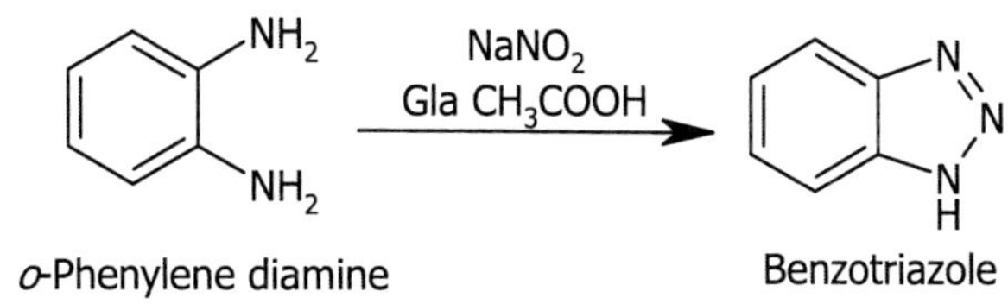

Formation of diazonium compound: Intermediate formed in previous stage undergoes rearrangement and forms diazonium compound by reacting with acid.

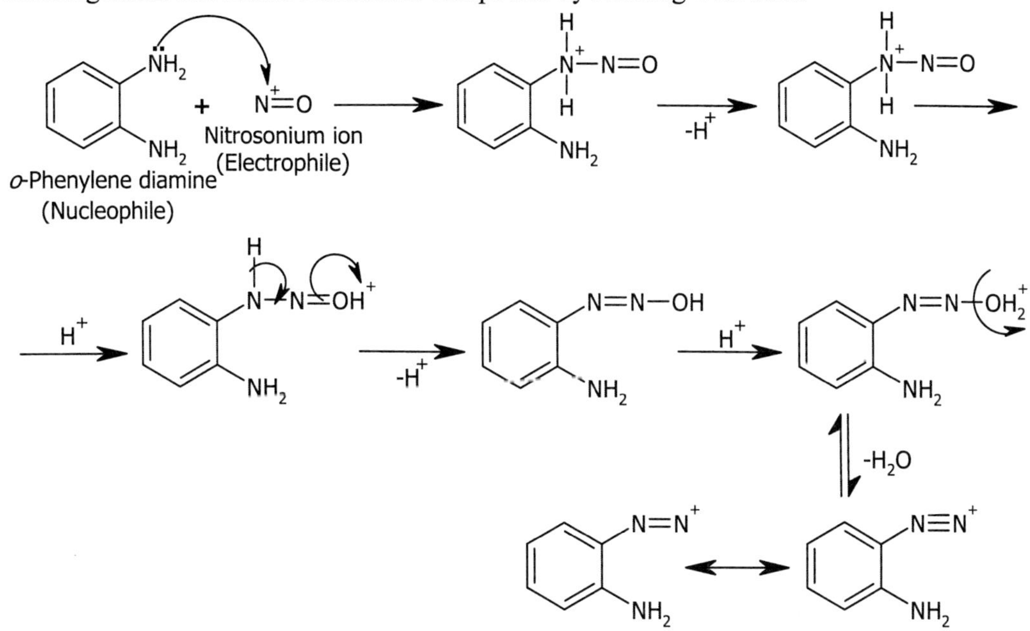

Cyclization into benzotriazole: Diazonium compound upon further reaction in acid medium undergoes cyclization and forms benzotriazole.

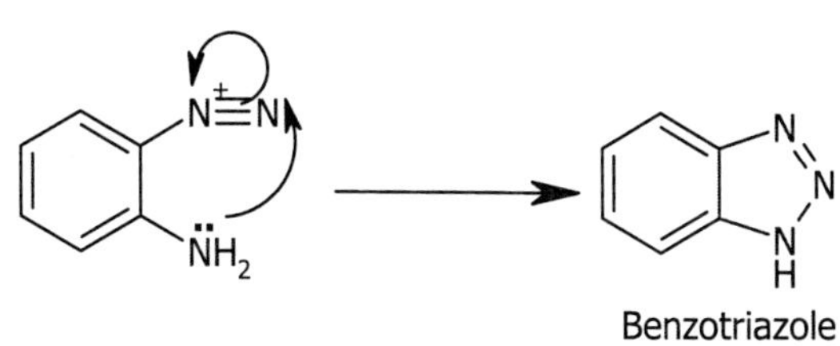

Procedure:

1. Dissolve *o*-phenylene diamine (1 gm, 9 mmol) in a mixture of glacial acetic acid (1 ml) and water (3 ml) contained in a beaker.
   [*Warm the solution if necessary to effect the solution*].
2. Cool the clear solution to 15 °C, stir magnetically and then add a solution of sodium nitrite (0.70 gm, 10 mmol) in water (2 ml) to the beaker.

3. Continue stirring for 15 minutes and place it in an ice bath for 30 minutes.
4. Collect the pale brown solid by vacuum filtration and wash thoroughly with ice-cold water.
5. Dissolve the solid in boiling water (15 ml), add charcoal for decolourising.
6. Filter and allow the filtrate to cool by placing in an ice bath.
7. Collect the pale straw-coloured needles of benzotriazole.
8. Dry the product and recrystallize the product from benzene.
9. The melting range of pure benzotriazole is 90-95 °C.

CHAPTER 35

QUINOXALINE SYNTHESIS

A quinoxaline, (also called as benzopyrazine) is a ring complex made up of benzene ring and pyrazine ring. Quinoxaline is isomeric with quinazolines and are used as dyes and pharmaceuticals. The antibiotics such as echinomycin, levomycin and actinoleutin contains quinoxaline nucleus. In general, quinoxaline can be prepared by condensing *o*-diamines with 1,2-diketones. Substituted quinoxaline can be prepared by replacing diketones with α-ketonic acids, α-chloroketones, α-aldehyde alcohols and α-ketone alcohols. The condensation reaction between *o*-phenylene diamine and benzil in the presence of rectified spirit gives 2,3-diphenyl quinoxaline.

35.1 Preparation of 2,3-Diphenyl Quinoxaline

Chemicals: Benzil, *o*-phenylene diamine and Ethanol.

Principle: The condensation of *o*-phenylene diamine and benzil in the presence of ethanol/rectified spirit produce 2,3-Diphenyl quinoxaline.

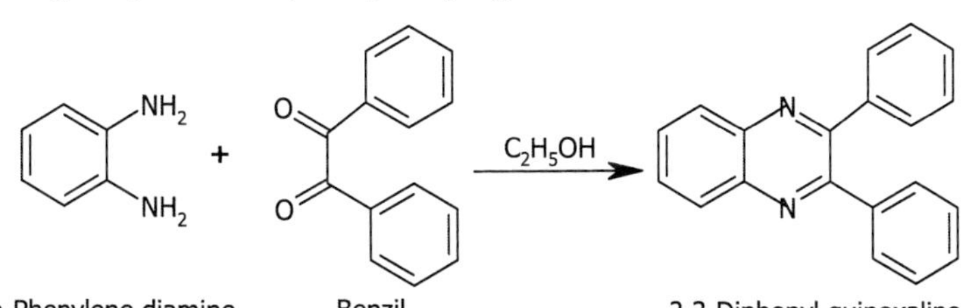

o-Phenylene diamine   Benzil   2,3-Diphenyl quinoxaline

Reaction Mechanism

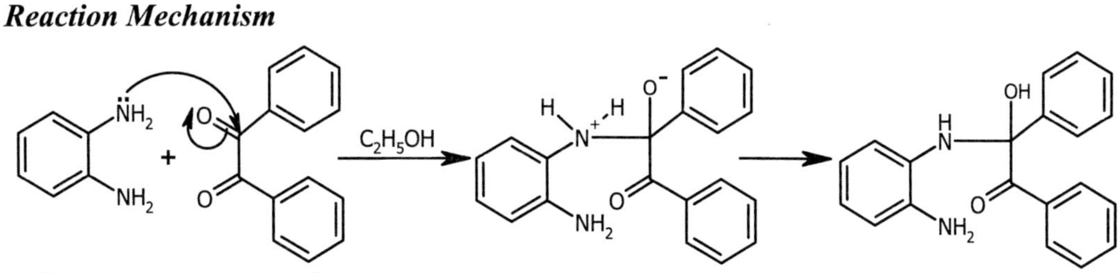

o-Phenylene diamine   Benzil

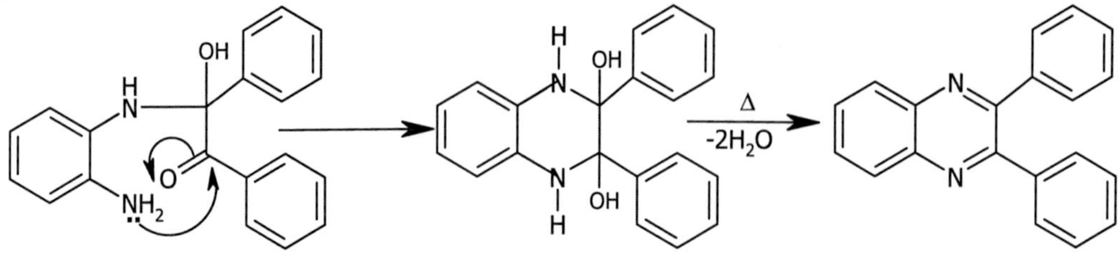

2,3-Diphenyl quinoxaline

Procedure:

1. To the warm solution of benzil (1 gm, 5 mmol) in ethanol (4 ml) in a round bottom flask, add solution of *o*-phenylene diamine (0.55 gm, 5 mmol) in ethanol (4 ml).
2. Warm the mixture by placing in water bath for 30 minutes.
3. To the warm mixture add water until a slight cloudiness persists and allow it to cool.
4. Filter the resulting mixture and recrystallize the residue from aqueous ethanol.
5. The melting range of 2,3-diphenyl quinoxaline is 125-126 °C.

CHAPTER 36

BERNTHSEN SYNTHESIS

Phenothiazine is the important pharmacophore present in many centrally (CNS) acting drugs. It includes antiemetics (promethazine), antihistamines (trimeprazine), antipsychotics (chlorpromazine), and antimigraines (prochlorperazine). The basic nucleus can be synthesised by the ring closure of diphenyl amine with elemental sulphur at 250 °C. This reaction is known as Bernthsen synthesis. It can be synthesized through other methods such as Ullman-type cyclization and Smile rearrangement reaction. Metal halides, aluminium chloride, zinc chloride and elemental iodine are the catalysts used in the preparation.

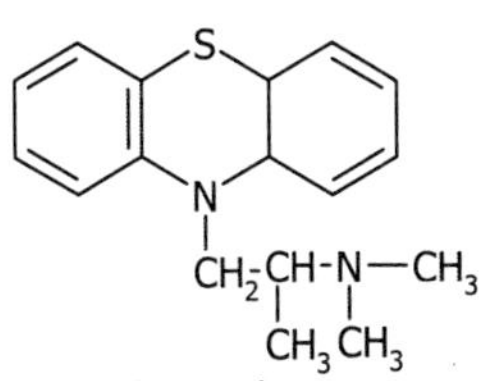
Promethazine

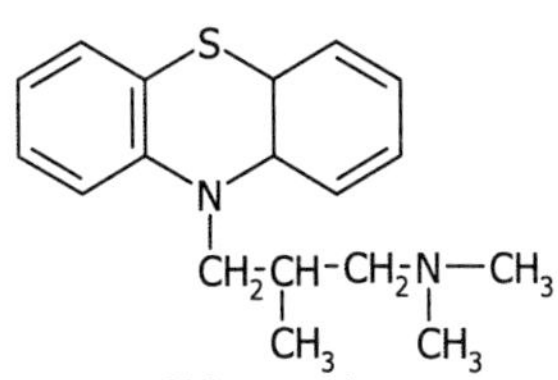
Trimeprazine

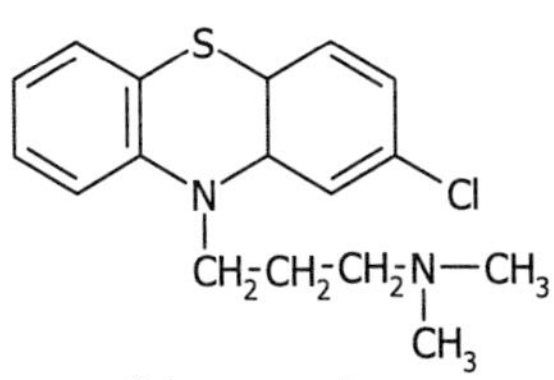
Chlorpromazine

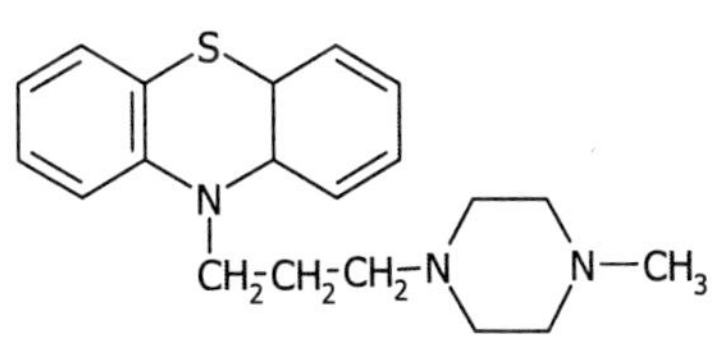
Prochlorperazine

36.1 Preparation of Phenothiazine

Traditional Name: Thiodiphenylamine.

***IUPAC Name*:** Dibenzo-[b,e]-1,4-thiazine.

Chemicals: Diphenylamine, Sulphur and Resublimed iodine.

Principle: Phenothiazine is the basic structure for large number of antihistamines, antipsychotics, sedatives and antiemetics. It is prepared by melting diphenylamine with elemental sulphur with catalysts such as metal halides (e.g.: $AlCl_3$ and $ZnCl_2$) or by using elemental iodine. Elemental iodine gives better yields.

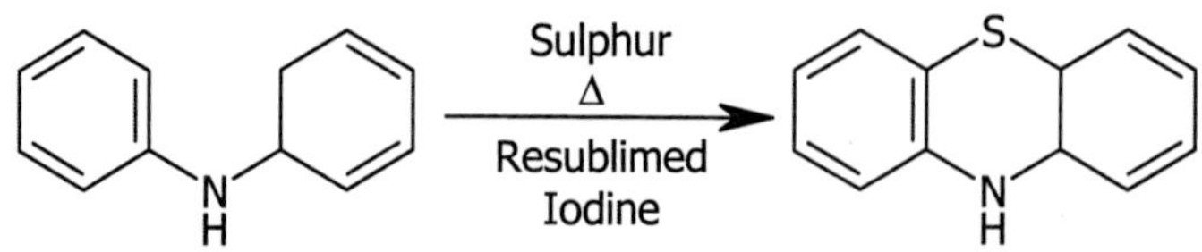

Diphenyl amine — Phenothiazine

Procedure:

1. Place diphenylamine (8.4 gm) and elemental sulphur (3.2 gm) in a round bottom flask.
2. Add resublimed iodine (0.15 gm) to the flask as a catalyst.
3. Melt the mixture by keeping in an oil bath at 190-200 °C for 30 minutes.
4. Cool the molten solid and powder it using mortor and pestle.
5. Recrystallize the product from ethanol.
6. The melting range of pure phenothiazine is 184-185 °C.

CHAPTER 37

URACIL SYNTHESIS

Uracil a nucleobase present in ribo nucleic acid (RNA). The methylated uracil (thymine) is component of deoxy ribonucleic acid (DNA). It was isolated from yeast nuclein and also present in bovine thymus, spleen and wheat germ.

The most common way of synthesizing uracil is condensing maleic acid with urea in fuming sulphuric acid. Hydrolysis of cytosine produces uracil and ammonia, double decomposition of thiouracil with chloro acetic acid also generates uracil. Photodehydrogenation of 5,6-diuracil (condensation product of β-alanine and urea) also produces uracil.

Pharmaceutical Applications

- 5-Fluoro uracil, cidofovir and trifluridine are uracil containing anti-cancer drugs.

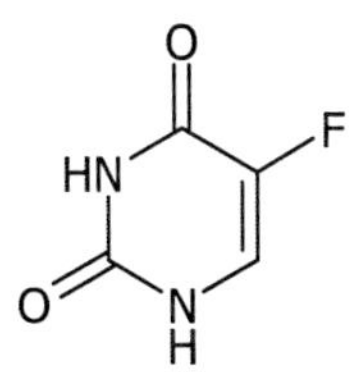

5-Fluoro uracil

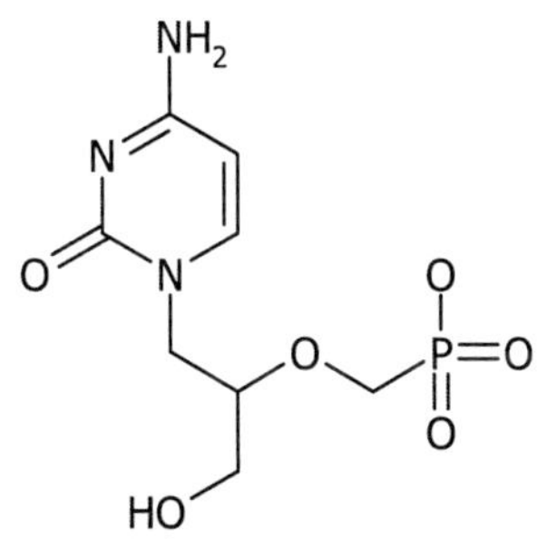

Cidofovir

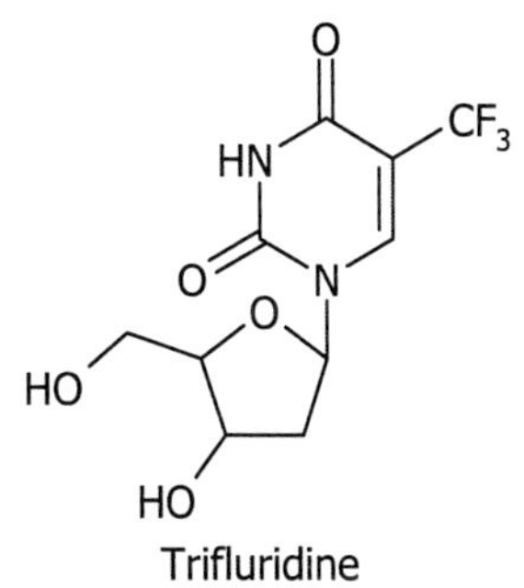

Trifluridine

- Caffeine (CNS stimulant) can be synthesized from uracil.
- Uracil is an allosteric regulator of and coenzyme for many biochemical reactions.
- Uracil plays major role in the biosynthesis of polysaccharides and transportation of aldoses (carbohydrate metabolism).
- Propylthiouracil and methylthiouracil are effective anti-thyroid drugs.
  (*Most serious side effect of thiouracil therapy is agranulcytosis*).

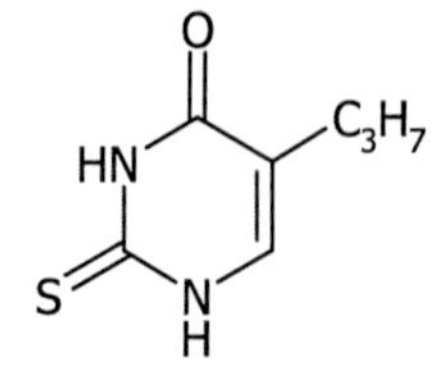

Propylthiouracil

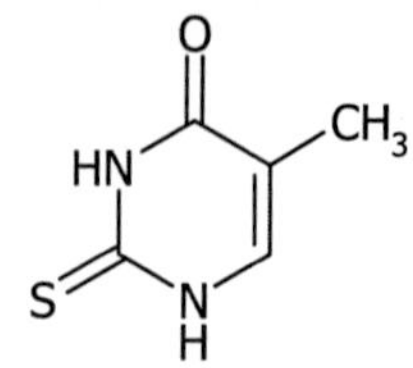

Methylthiouracil

37.1 Preparation of 6-Methyl Uracil

Chemicals: Ethyl acetoacetate, Thiourea, Ethanol, Hydrochloric acid, Potassium hydroxide and Chloroacetic acid.

Principle: 6-Methyl uracil can be prepared by condensing thiourea with ethyl acetoacetate followed by desulphuration reaction with chloroacetic acid in alcoholic potassium hydroxide. The *enol* form of the ethyl acetoacetate readily condenses with thiourea and forms 1,2,3,4-tetrahydro thiopyrimidine, which upon reaction with chloroacetic acid forms 6-methyl uracil.

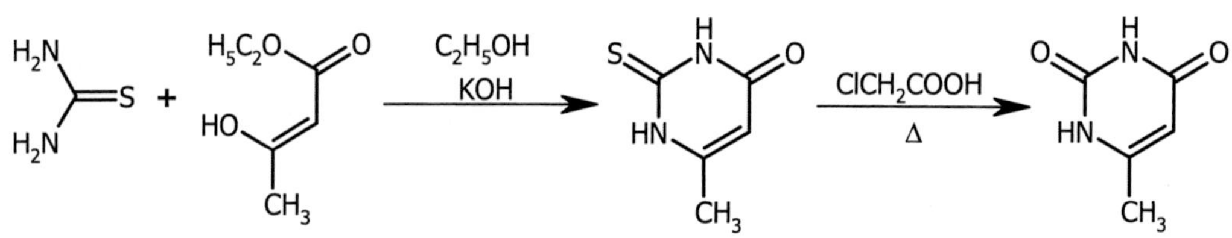

Thiourea Ethyl acetoacetate (*enol form*) 6-Methyl-4-oxo-1,2,3,4-tetrahydrothiopyrimidine 6-Methyl uracil

In ethyl acetoacetate, inductive effect of two carbonyl groups weakens the C-H bonds of active methylene group. Thus the H-atom dissociate and generate stable anion. The acidity of the C-H bond is greatly enhanced by resonance, hence ethyl acetoacetate exist in *keto* and *enol* forms (tautomeric mixture). Abstraction of proton (H^+) from thiourea by base ($-OH^-$) generates nucleophile, which attacks electrophilic center of ethyl acetoacetate. This upon cyclization produces 6-methyl-4-oxo-1,2,3,4-tetrahydro thiopyrimidine. In 6-methyl-4-oxo-1,2,3,4-tetrahydro thiopyrimidine, sulfhydryl group reacts with chloroacetic acid and forms thio ether complex, which on hydrolysis (desulfuration) forms the 6-methyl uracil.

Generation of nucleophile

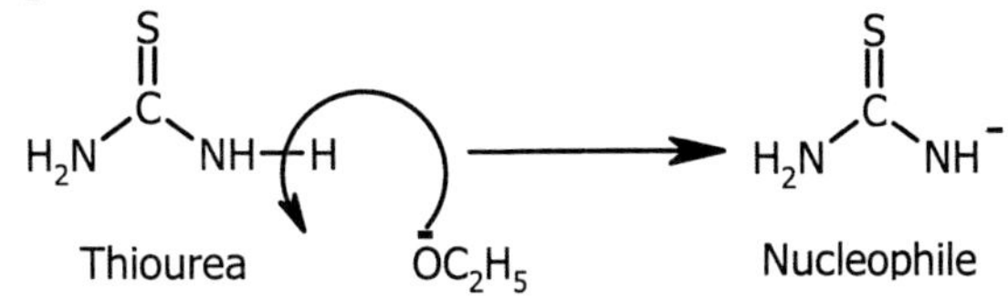

Nucleophilic attack on ethly acetoacetate

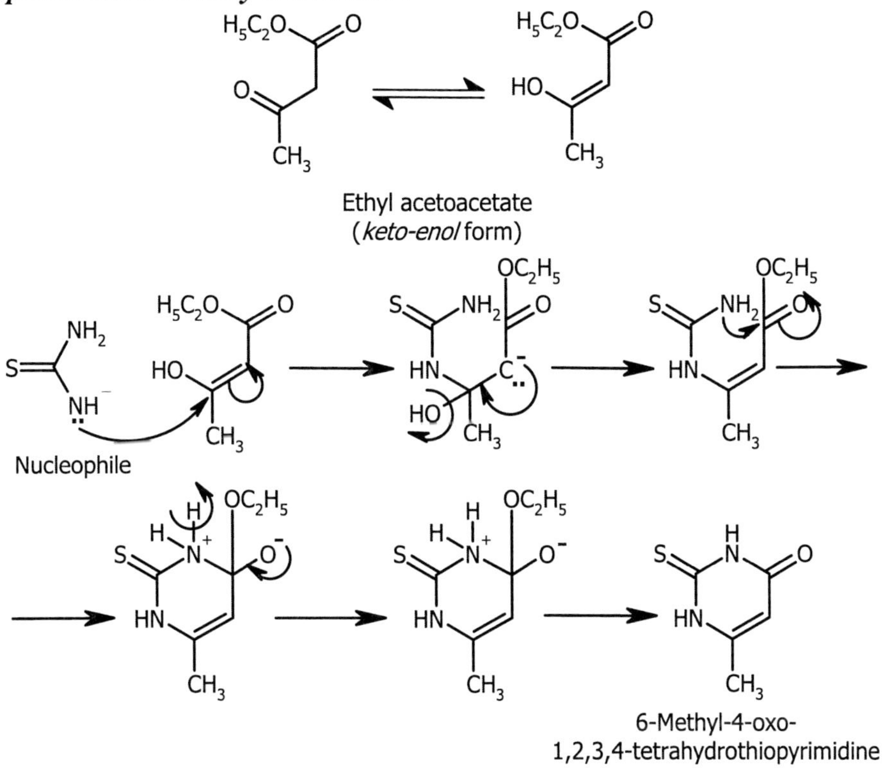

Preparation of 6-methyl uracil (desulfuration)

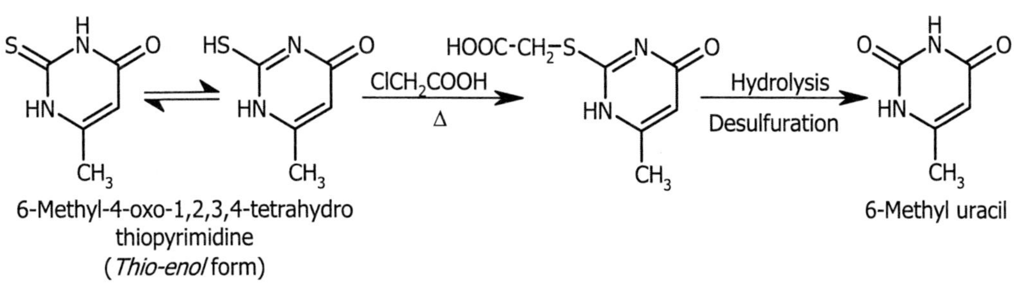

Procedure:

Step* 1*: Preparation of 6-methyl-4-oxo-1,2,3,4-tetra-hydro thio-pyrimidine

1. Mix thiourea (1 gm, 13 mmol) and ethyl acetoacetate (2 ml, 16 mmol) in ethanol (7 ml).
2. Add a solution of potassium hydroxide (0.8 gm, 13 mmol) in water (3 ml) and reflux it for 2 hr.
3. Acidify the mixture with hydrochloric acid to neutralize the potassium hydroxide.
4. Cool to room temperature, filter and dry the product.

Step* 2: *Preparation of 6-methyl uracil

1. Place 6-methyl-4-oxo-1,2,3,4-tetrahydro thiopyrimidine (1 gm, 7 mmol) and solution of chloroacetic acid (1.4 gm, 15 mmol) in water (35 ml) in a round bottom flask.
2. Heat the mixture for 2 hr in fume cupboard, filter the hot solution and allow the product to crystallize.
3. Dry and recrystallize the product from hot ethanol, the melting point of 6-methyl uracil is 318 °C.

CHAPTER 38

SULPHONAMIDE SYNTHESIS

Sulphonamides, the first antimicrobial drugs produced revolution in antibiotic medicines. The term sulphonamide is a generic name employed for the *para* amino sulphonamide derivatives. Prontosil is the first sulphonamide, had strong protection against *Streptococci* infections. But the beneficial effect is observed only in *in vivo* not in *in vitro* model. The research team of Pasteur Institute discovered that the metabolite of prontosil, sulphanilamide is responsible for the effects observed in *in vivo* model.

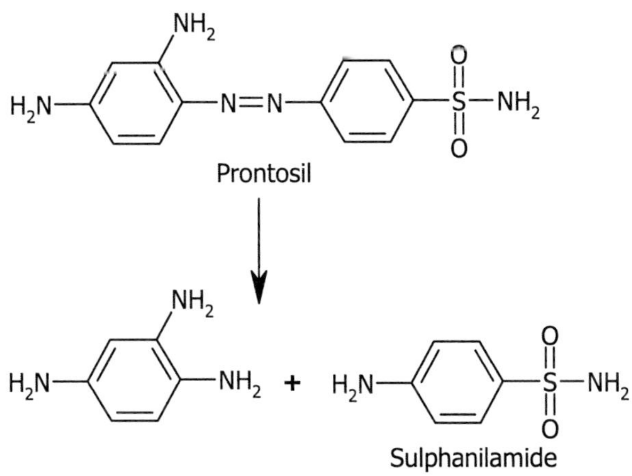

This discovery led to the development of thousands of clinically useful sulphonamides. These sulphonamides are useful in the treatment of urinary tract infection, tonsilitis, burn therapy, conjunctivitis, peptic ulcer and bacillary dysentry. The sulphanilamide portion of sulphonamides competes with *para*-amino benzoic acid, an essential component in the synthesis of folic acid. Folic acid is essential for the DNA and RNA synthesis. Thus sulphonamides inhibits the folic acid synthesis and arrests the bacterial growth.

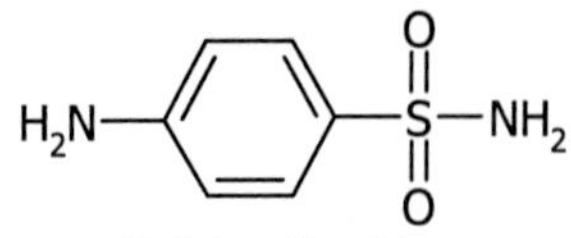

Sulphanilamide

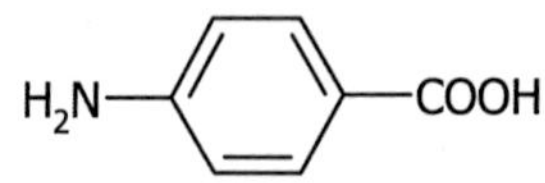

Para-Aminobenzoic acid

Clinically useful sulphonamides

1. Sulphisoxazole: Used in the treatment of otis media in combination with erythromycin (antibiotic).

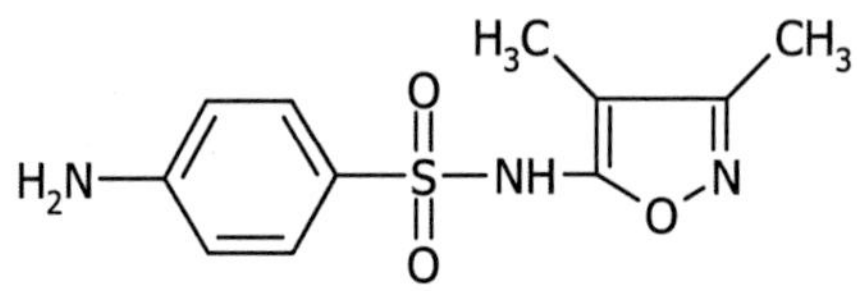

Sulphisoxazole

2. Sulphadiazine: Used in the treatment of urinary tract infection.

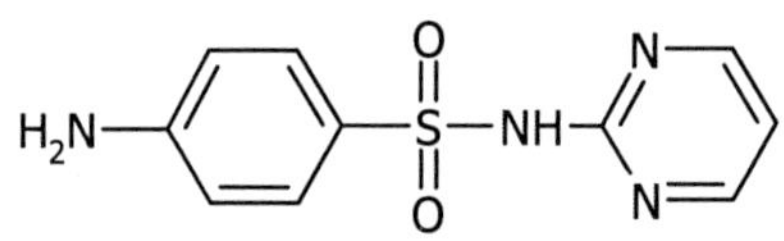

Sulphadiazine

3. Mafenide (diuretic)

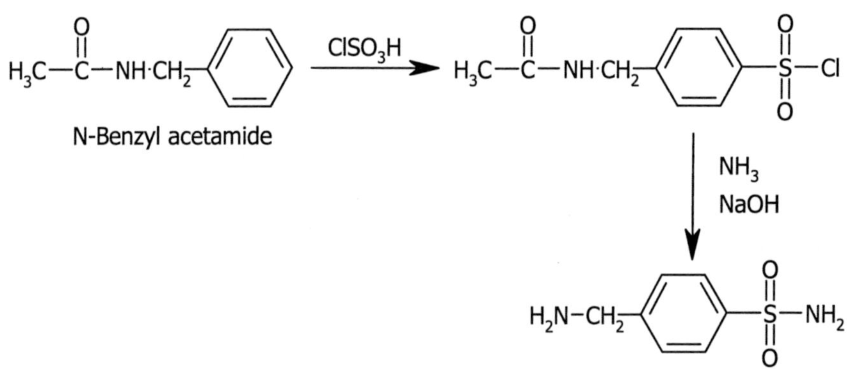

Mafenide

4. Sulphamethoxazole: Used in the treatment of urinary tract infection in combination with trimethoprim and also used in the treatment of sinusitis.

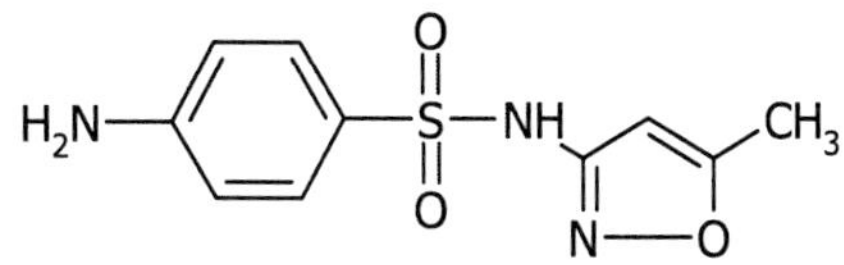

Sulphamethoxazole

5. Silver sulphadiazine: Used in the treatment of burns.

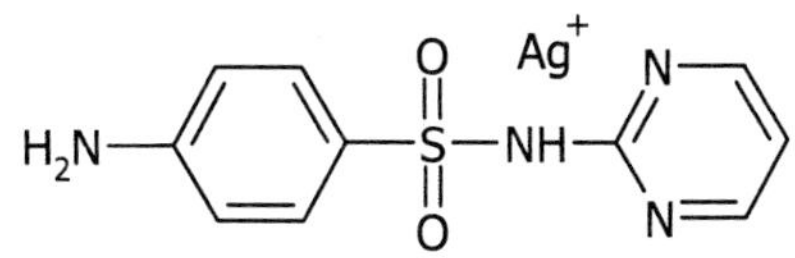

Silver sulphadiazine

6. Bumetanide (diuretic)

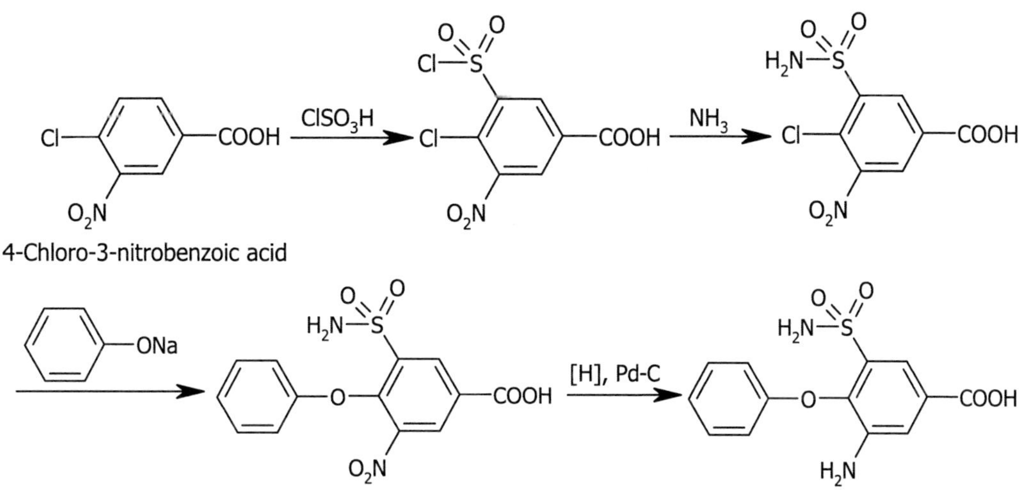

4-Chloro-3-nitrobenzoic acid

Bumetanide

38.1 Preparation of Sulphanilamide

***IUPAC Name*:** 4-Amino benzene sulphonamide.

Chemicals: Acetanilide, Chlorosulphonic acid, Hydrochloric acid, Sulphuric acid, Ammonia, Sodium bicarbonate and Charcoal.

Principle: Acetanilide reacts with chlorosulphonic acid and renders *p*-acetamido benzenesulphonyl chloride, which in turn reacts with ammonia and forms sulphanilamide. Protection of amino group of aniline by acetylation directs the chloro sulphonation at *para*-position (otherwise amino group gets chloro sulphonation). Weaker basic nature of the acetamido group favours the attack of chloro sulphonyl chloride only at *para* position. Reaction between *para*-acetamido benzene sulphonylchloride with excess of ammonia produces *para*-acetamido benzene sulphanilamide. Deacetylation of the former with hydrochloric acid forms sulphanilamide.

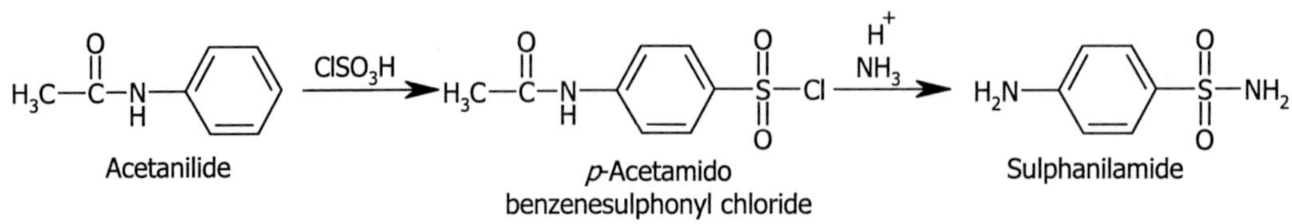

Step* 1: *Formation of para-acetamido benzenesulphonyl chloride (chloro sulphonation)

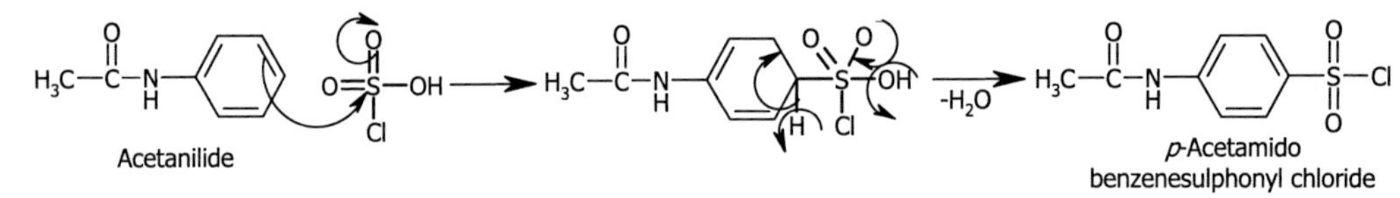

Step* 2: *Foramtion of para-acetamido benzene sulphonamide (amination)

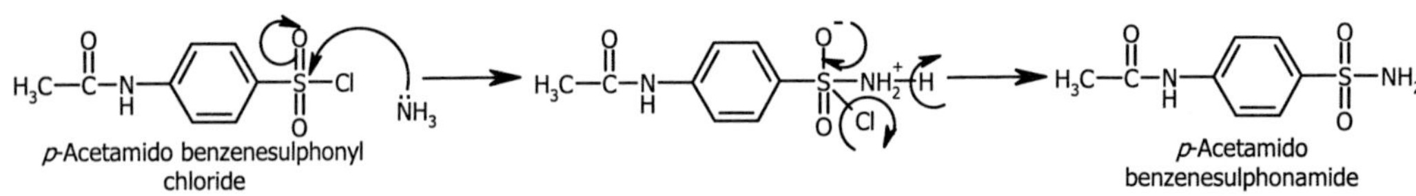

Step 3: *Foramtion of sulphanilamide (deacetylation)*

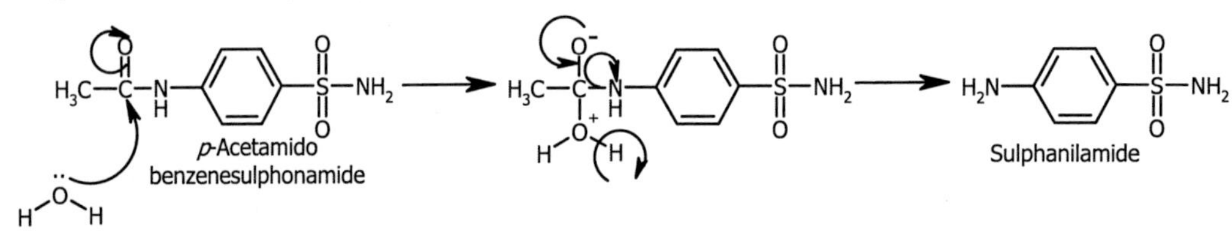

Procedure:

Step* 1: *Preparation of para-acetamido benzenesulphonyl chloride

1. Place dry acetanilide (1.35 gm, 10 mmol) in a two necked round bottom flask [fitted with dropping funnel and reflux condenser].
2. Add chlorosulphonic acid (3.4 ml, 5.2 mol) in small portions with the help of dropping funnel fitted with calcium guard tube.
3. Shake the flask occasionally to ensure the mixing.

4. Heat the reaction mixture by placing on a water bath for 1 hr.
5. Cool the contents and pour the oily mixture in a thin stream with stirring into a beaker containing crushed ice.
6. Rinse the flask with cold water and transfer into the beaker.
7. Stir vigorously to dissolve the lumps formed and obtain white solid suspension.
8. Filter the suspension and dry the residue.

Precaution:

- Excess of chlorosulphonic acid reacts with water and produces harmful vapours.
- Chlorosulphonic acid is highly corrosive and may cause serious burns, hence wear gloves when handling and exercise extreme care.

Step 2: Preparation of para-acetamido benzenesulphonamide

1. Mix the crude product with ammonia (3 ml, 23.8 mmol) and water (3 ml).
2. Heat the mixture just below boiling point for 15 minutes with occasional swirling.
3. Cool the suspension in ice-bath, acidify with dilute sulphuric acid.
4. Filter the precipitate and wash with cold water.

Step 3: Preparation of sulphanilamide (deacetylation)

1. Transfer the *para*-acetamido benzenesulphanilamide to round bottom flask and add hydrochloric acid (0.7 ml) and water (2.1 ml).
2. Attach reflux condenser, and reflux for 45 minutes.
3. Cool the solution and boil with charcoal.
4. Filter and add sodium bicarbonate (1.08 gm, 19 mmol) to the filtrate to neutralize the hydrochloric acid.
5. Cool to room temperature and filter the precipitate formed.
6. Dry and recrystallize from water.
7. The melting range of pure sulphanilamide is 161-163 °C.

38.2 Preparation of Sulphaguanidine

IUPAC Name: 4-Amino-N-[amino(imino)] methylbenzene sulphonamide

Chemicals: Sulphanilamide, Guanidine nitrate, Sodium hydroxide, Methanol, Cyclohexanol, Hydrochloric acid (dilute).

Principle: Sulphaguanidine can be prepared from sulphanilamide by the reaction with guanidine nitrate in basic medium.

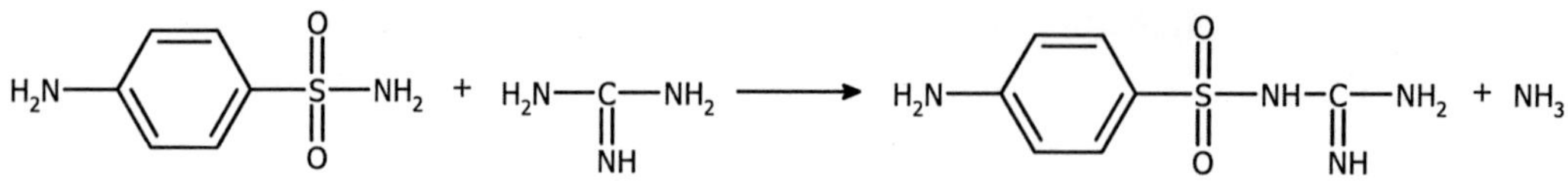

Sulphanilamide Guanidine Sulphaguanidine

Procedure:

1. Dissolve guanidine nitrate in the solution of sodium hydroxide (2.4 gm) in methanol (12.5 ml).
2. Stir the reaction mixture, maintaining the temperature at 50-60 °C for 15 minutes.
3. Cool the reaction mixture to 20 °C.
4. Filter to remove nitrite and wash the residue with two portions of 3 ml of methanol.
5. Transfer the combined methanolic extract into 3-necked flask and add sulphanilamide (500 mg, 60 mmol).
6. Stir the reaction mixture for 15 minutes at 50-55 °C into distil the methanol.
7. Add cyclohexanol (2.1 ml) and elevate the temperature to 180-190 °C and continue the distillation.
8. Maintain the temperature of reaction mixture between 150-160 °C for 2 hr and pour into dilute hydrochloric acid (50 ml).
9. Stir, alkalinify by adding sodium hydroxide slowly with stirring.
10. Cool the mixture, filter the separated sulphaguanidine and wash with cold water.
11. Purify the sulphaguanidine by recrystallization with hot water.
12. The melting range of the pure compound is 189-190 °C.

CHAPTER 39

ISOXAZOLE SYNTHESIS

A five membered cyclic diene containing three carbon atoms, one nitrogen and one oxygen is called an oxazole. If the two hetero atoms (O, N) are adjacent to each other then it is known as isoxazole. Isoxazoles are generally prepared by:

1. The reaction of 1,3-diketones with hydroxyl amines
2. 1,3-Dipolar cycloaddition of nitrile oxides with alkynes
3. The reaction of α,β-unsaturated aldehydes and ketones with hydroxyl amines

Isoxazole based heterocycles are potential bioactive molecules and are associated with wide spectrum of biological activities such as antibacterial, antifungal, antitubercular, antiviral, analgesic, anti-inflammatory, antitumor, antidiabetic, mosquito larvicidal antioxidant, antihypertensive and antidepressant. The isoxazole ring system is present in numerous compounds of therapeutic importance such as sitaxentan (anti-hypertensive), valdecoxib (anti-inflammatory), leflunomide (anti-arthritic), glisoxepide (anti-diabetic) and isocarboxazid (anti-depressant).

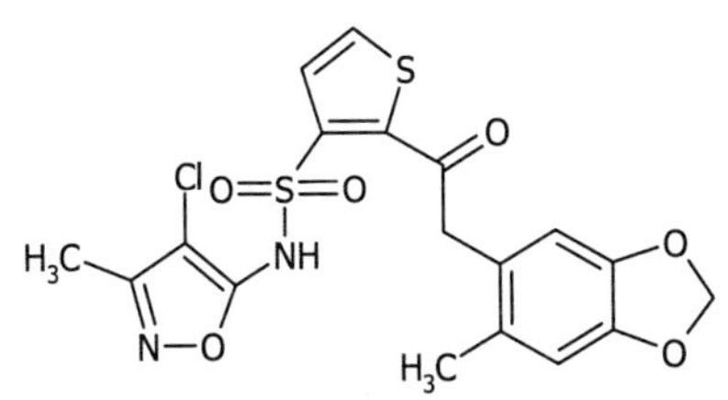

Sitaxentan

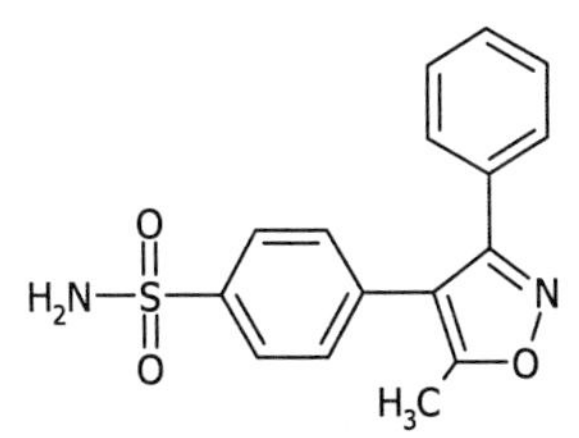

Valdecoxib

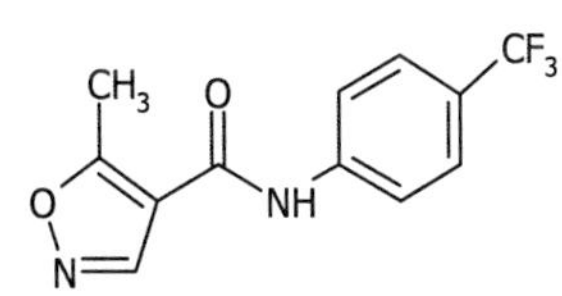

Leflunomide

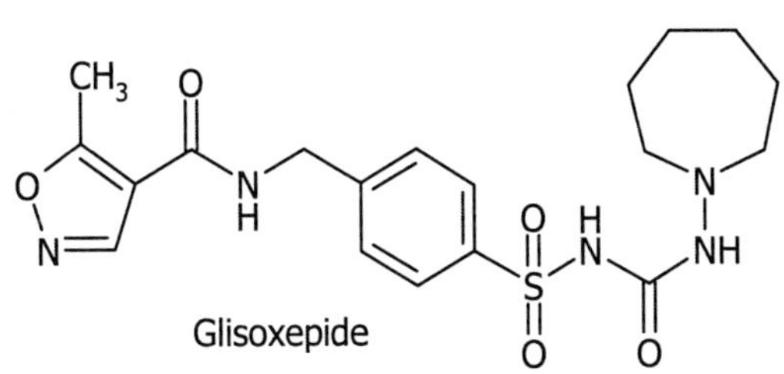

Glisoxepide

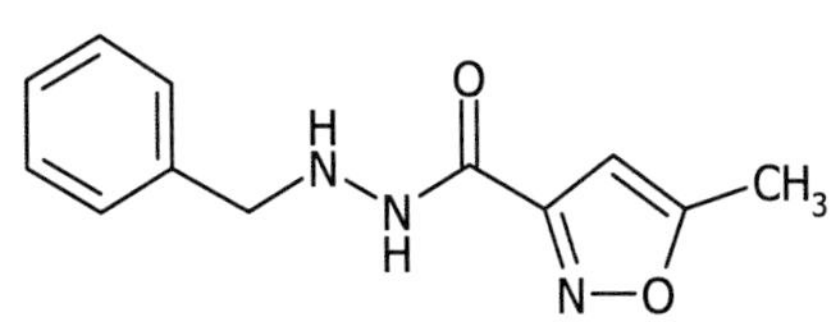

Isocarboxazid

39.1 Preparation of 3,5-Dimethyl Isoxazole

***Chemicals*:** Acetyl acetone, Hydroxylamine, Diethyl ether, Magnesium sulphate and Ethanol.

***Principle*:** The condensation of active methylene group containing β-keto esters/β-diketones with amino compounds yields cycloaddition products. Acetyl acetone reacts with hydroxyl amine and gives 3,5-dimethyl isoxazole. Reaction between acetyl acetone and hydroxylamine gives monoxime, β-isonitroso butyric acid, which undergoes cyclization and generates 3, 5-dimethyl isoxazole.

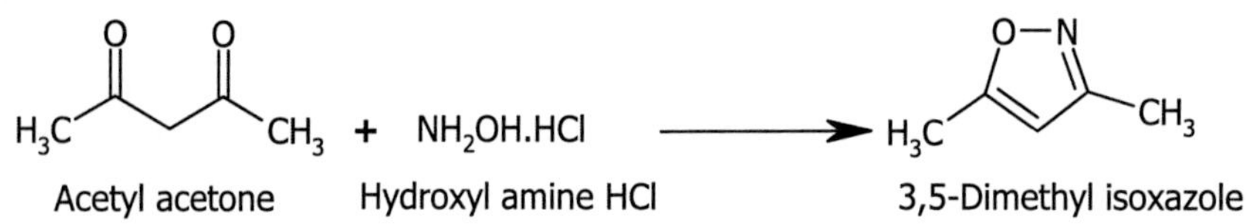

Reaction mechanism

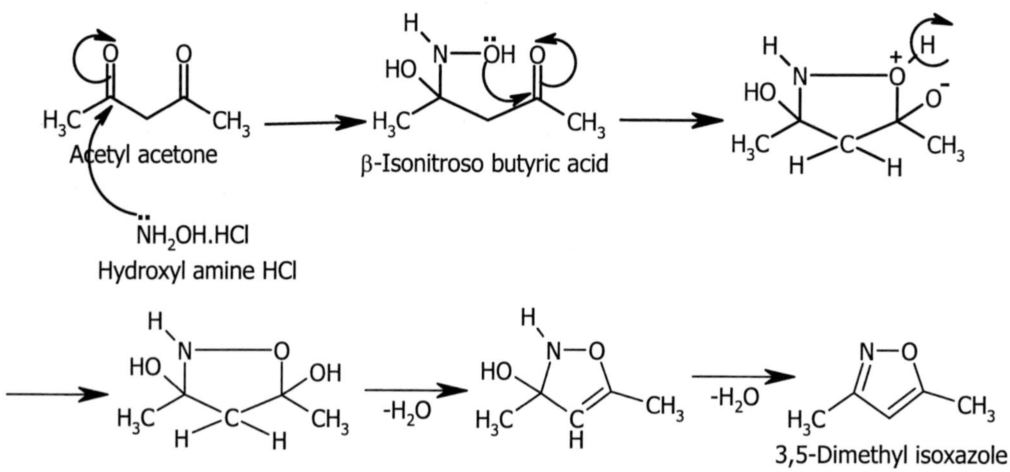

***Procedure*:**

1. Place hydroxylamine (7 gm) in a round bottom flask containing water (15 ml) and add ethanol (10 ml) to the flask.
2. Reflux the contents until the negative ferric chloride test is obtained
3. Pour the mixture into cold water (60 ml) contained in beaker.
4. Extract the aqueous mixture with three portions of 5 ml ether and combine the etheral layers.
5. Evaporate the ether layer to obtain brown oil residue.
6. Distill the oil to get colourless oil, the boiling point of 3,5-dimethyl isoxazloe is 140-142 °C.

CHAPTER 40

SULPHONIC ESTERS SYNTHESIS

Sulphonic esters can be conveniently prepared by the reaction between sulphonyl halides with alcohols (alkyl or aryl alcohols) in the presence of base. This chemical conversion produces tosylates, brosylates and similar sulphonic esters from alcohols.

$$\underset{\text{Alcohol}}{R-OH} + \underset{\text{Sulphonyl chloride}}{R_1-SO_2-Cl} \xrightarrow{\text{Base}} \underset{\text{Sulphonic ester}}{R-O-SO_2-R_1}$$

Pyridine is the most preffered base and it functions as nucleophilic catalyst. Alcohols react most rapidly and the reactivity of primary alcohols is higher. Primary alcohols can be selectively sulphonated in the molecules containing secondary and tertiary alcoholic groups.

Reaction mechanism: Pyridine abstracts the hydrogen from alcohol and generates nucleophile (RO^-). This nucleophile attacks the sulphur of sulphonyl chloride and forms sulphonic esters. The pyridine base combines with liberated chlorine and forms pyridinium chloride.

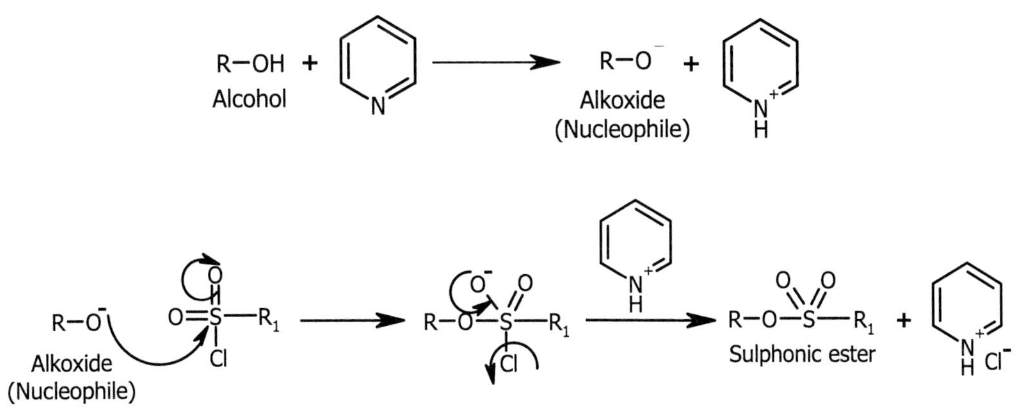

40.1 Preparation of Busulphan

Chemicals: Butane-1,4-diol, Methanesulphonyl chloride and Pyridine.

Principle: Butane-1,4-diol reacts with methanesulphonyl chloride in presence of pyridine base (nucleophilic catalyst) and generates busulphan (anti-cancer drug).

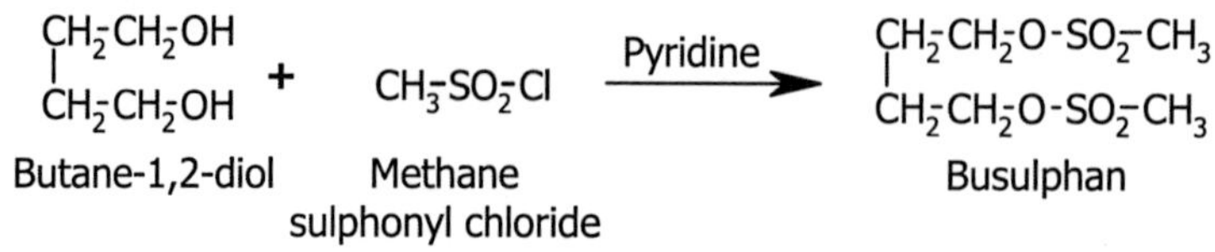

Pyridine abstracts the hydrogen from butane1,4-diol and generates alkoxide nucleophile. The nucleophile attacks the sulphur of methanesulphonyl chloride and forms busulphan (sulphonic ester). The pyridine base combines with liberated chlorine and forms pyridinium chloride.

Reaction mechanism:

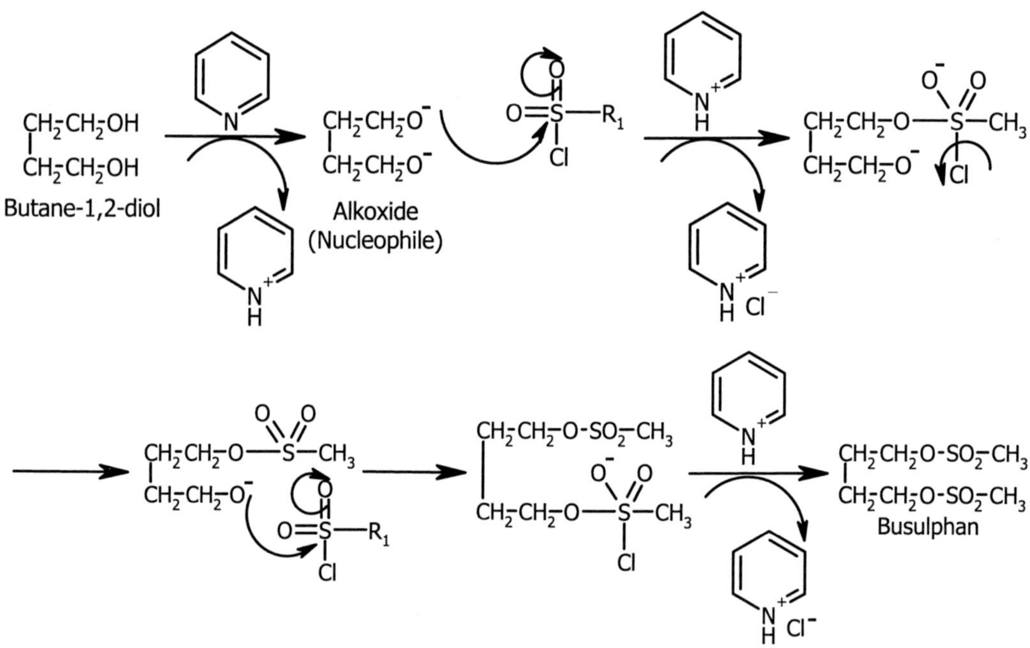

Procedure:

1. Dissolve butane-1,4-diol (3.6 ml, 40 mmol) in flask containing pyridine (10 ml).
2. Place the mixture in an ice-bath to reduce the temperature below 20 °C.

3. Add methanesulphonyl chloride (6 ml, 80 mmol) dropwise and maintain the temperature 18 ± 2 °C.
4. Allow the reaction mixture to stand at room temperature for 30 minutes (exothermic).
5. Cool the mixture by placing in ice-bath and add ice cold water (30 ml).
6. Stir the reaction mixture with glass rod to obtain white crystalline solid.
7. Filter the solid under vacuum and wash with ice-cold water.
8. The melting point of pure busulphan is 110 °C.

Precautions:

1. Methanesulphonyl chloride is highly toxic, corrosive and lachrymator.
2. Hence, conduct the reaction in fume cupboard.

GREEN SYNTHESIS

CHAPTER 41

GREEN SYNTHESIS

Green chemistry, a newer chemical approach prevents pollution and sustains earth. It uses environment friendly reagents in place of hazardous traditional chemicals. Green chemistry can be defined as the invention, design and application of chemical products and processes to reduce or to eliminate the use and generation of hazardous substances.

Green chemistry principles

- Atom economy and energy efficiency.
- Less hazardous chemical synthesis.
- Design and use of safer chemicals including solvents.
- Use of protection and deprotection strategy to reduce derivatives.
- Catalytic reaction to speed up the reaction and increase the yield.
- Design for degradation of chemical products and prevention of waste.
- Design of pollution prevention and accident prevention strategies.

Green chemical technologies

1. Solvent free reactions (neat reaction).
2. Water reactions (replacing organic solvents with water).
3. Organic reactions in micro-reactors.
4. Microwave assisted synthesis.
    - Solid phase synthesis.
    - Catalysis.

Advantages of green chemistry

1. Improved yield.
2. Reduced purification need.
3. Reduced energy need.
4. Shortened reaction time.

41.1 Preparation of Benzoic Acid

Chemicals: Benzaldehyde, Acetic acid, Sodium hypochlorite

Principle: Aldehydes can be oxidized into corresponding acids by variety of oxidizing agents such as dichromate, perchlorate and osmium tetroxide. But many of them are hazardous and corrosive in nature this demands benign oxidizing agent. The sodium hypochlorite (NaOCl) is a mild oxidizing agent and is the best alternative. Sodium hypochlorite should be never mixed with ammonia as it releases deadly chlorine gas.

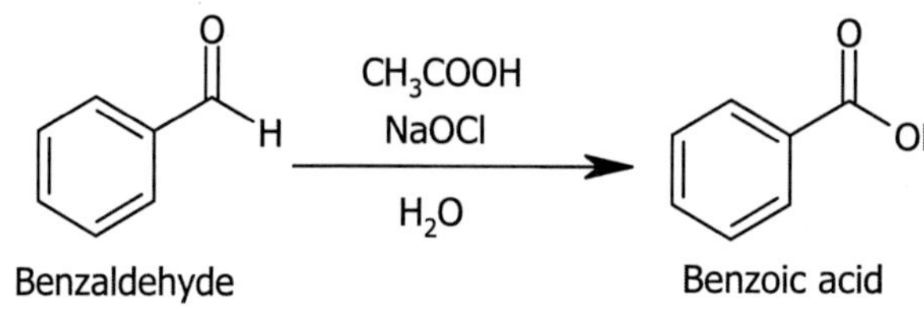

Procedure:

1. Place benzaldehyde (2 ml) and acetic acid (1 ml) into a flask.
2. Heat the mixture with stirring using mechanical stirrer.
3. Add sodium hypochlorite (30 ml) in 5 ml portions over 5 minutes.
4. Heat the reaction mixture with stirring at 80-90 °C for 45 minutes.
5. Cool to room temperature, filter the product under vacuum and wash it with ice-cold water.
6. Dry and recrystallize the product from boiling water.
7. The melting point of pure benzoic acid is 121 °C.

Water reactions

Organic reactions in water medium have attracted much attention. Water induces reactivity and selectivity and reduces the use of harmful organic solvents. The most important advantages of water reactions are listed below:

- ***Cost:*** Water reactions are economical.
- ***Safety:*** Non-flammable, benign and recyclable.
- ***Synthetic efficiency:*** Useful in carbohydrate and protein chemistry with water soluble substrates. Avoids the protection and deprotection of functional groups.
- ***Operation simplicity:*** Isolation of products can be achieved by simple phase separation.

41.2 Preparation of Nifedipine

Chemicals: 2-Nitro benzaldehyde, Methyl acetoacetate, Ammonium acetate, Water and Tetra-n-butylammonium bromide.

Principle: Nifedipine, an antihypertensive drug can be synthesized through Hantzsch reaction. 2-Nitro benzaldehyde condenses with methyl acetoacetate and ammonium acetate in water using tetra-n-butylammonium bromide (TBAB) as catalyst and forms Nifedipine.

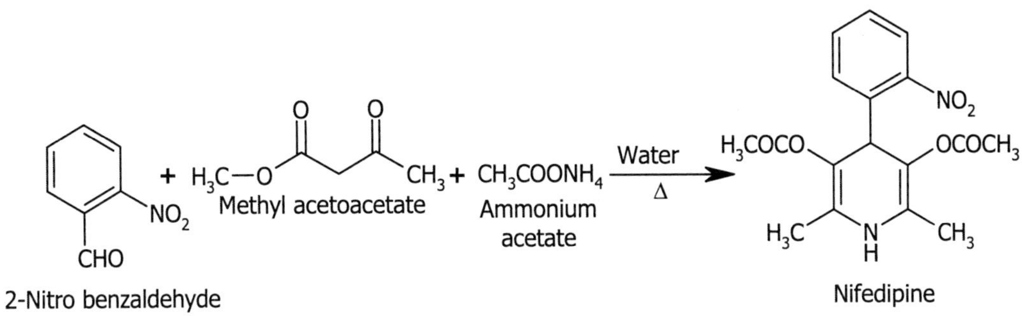

Procedure:

1. Place methyl acetoacctate (1.1 ml, 10 mmol), 2-nitro benzaldehyde (1.51 gm, 10 mmol), ammonium acetate (1.54 gm, 20 mmol), and water (2 ml) in a round bottom flask.
2. Add tetra-n-butyl ammonium bromide (1.43 gm) to the flask and stir the mixture at 60 °C for 5 hr.
3. Pour the reaction mixture into beaker containing ethanol (2 ml) and crushed ice (10 gm) and keep aside for 1 hr.
4. Isolate the precipitated Nifedipine under vacuum and dry.
5. Purify the Nifedipine by recrystallization with ethanol.
6. The melting range of pure Nifedipine is 170-173 °C.

Microwave reactions

Microwave-induced organic reaction enhancement (MORE) has emerged as efficient and ecofriendly method for the synthesis of large number of organic molecules. Most organic reactions uses traditional heat transfer systems such as oil baths, sand bath and heating mantles. These heating sources create temperature gradient and produce localized overheating of the product, substrate and reagent leading to their degradation. Microwave dielectric heating overcomes these problems. Microwave radiation is non-ionizing and does not alter molecular structure of the reagent and reactants.

Microwave heating finds variety of applications in pharmaceutical research. It includes synthesis of drugs, synthesis of drug intermediates, activation of chromate-graphic adsorbents, drying of granules, drying and sterilization of glass wares, drying of medicinal plants and extraction of essential oils. Microwave heating is effectively utilized for the combinatorial libraries development, known as microwave assisted combinatorial library synthesis (MICROCOS). The major advantages of MORE are higher yields and shorter reaction times.

Mode of action

Microwave radiation pass into materials in beam lines of radiation. Molecules gets polarized by electrically or magnetically in microwave frequency radiation. Principles involved in microwave heating are dipolar polarization, conduction mechanism and interfacial polarization. Rotational changes in the molecules by microwave radiation occur by dipolar polarization and dielectric conduction.

Absorption of energy occurs when dipolar molecules rotate and align themselves with fluctuating electric field.

Solvents

Polar solvents [dimethyl formamide (DMF), dimethyl sulphoxide (DMSO), ethanol and water] are good microwave absorbers and will heat efficiently. But non-polar or less polar solvents [toluene and tetra hydrofuron (THF)] will not heat.

Microwave techniques

Techniques used in microwave chemical reactions are:

1. Solvent mediated reactions.
2. Dry media reactions.
   (a) Neat reactions.
   (b) Solid supported reactions.

41.3 Preparation of 7-Hydroxy-4-Methyl Coumarin

Chemicals: Resorcinol, Ethyl acetoacetate, Sodium hydroxide (10%), *para*-toluene sulphonic acid, Sulphuric acid and Hydrochloric acid (dilute).

Principle: Condensation of phenols with a β-keto ester using condensing reagents is called as Pechmann reaction. Sulphuric acid (H_2SO_4), phosphorous penta oxide (P_2O_5), polyphosphoric acid (H_3PO_4) and *para* toluene sulphonic acid (*p*-TSA) are most common condensing reagents. *Para*-Toluene sulphonic acid (*p*-TSA) is effective condensing reagents. In ethyl acetoacetate, inductive effect caused by two carbonyl oxygens weakens the C-H bonds. Thus ethyl acetoacetate exist in tautomeric mixture of keto and enol forms. The enol form of the ethyl acetoacetate readily condenses resorcinol and leads to the formation of 7-hydroxy-4-methylcoumarin.

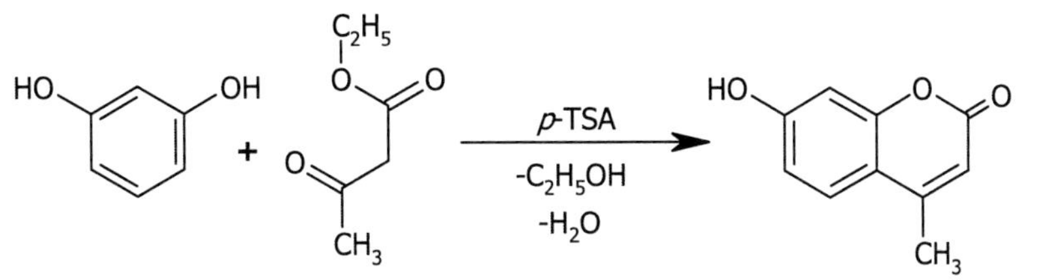

***Microwave synthesis* of *7-hydroxy-4-methylcoumarin*:**

1. Mix resorcinol (1 gm), ethyl acetoacetete (1.3 mmol), and *para*-toluene sulphonic acid (*p*-TSA) (10 mmol) in a conical flask.
2. Cover the flask with glass funnel.
3. Place the flask in microwave oven and irradiate at 300 W for 1 min.
4. Wash the solid obtained with water and purify by recrystallization with ethanol.
5. The melting point of 7-hydroxy-4-methylcoumarin is 185 °C.

41.4 Preparation of Phenytoin

Chemicals: Benzil, Urea, Sodium hydroxide, Ethanol and Hydrochloric acid

Principle: *Refer section 9.1*

Microwave synthesis of phenytoin

1. Mix benzil (0.53 gm), urea (0.3 gm), aqueous sodium hydroxide (30%, 1.5 ml) and ethanol (7.5 ml) in a 25 ml conical flask.
2. Cap the flask with glass funnel to avoid the excess of solvent evaporation.
3. Place the flask in microwave oven and irradiate at 450 W for 1 minute.
4. Cool the reaction mixture to room temperature and pour into water (25 ml).
5. Mix thoroughly and allow it to stand for 15 minutes.
6. Filter under suction to remove any insoluble by-product.
7. Acidify the filtrate with hydrochloric acid, cool in an ice bath and immediately filter the precipitate under vacuum pump.
8. Dry and recrystallize the product from ethanol.
9. The melting range of phenytoin is 297-298 °C.

Note: The procedure described is for use with domestic microwave ovens.

ESTIMATION OF DRUGS PRESENT IN FORMULATIONS

CHAPTER 42

ESTIMATION OF DRUGS BY ACID-BASE TITRATION

In acid-base titrations, either solutions of free bases are titrated against standard acid solutions (acidimetry), or solutions of free acids are titrated against standard alkali solutions (alkalimetry). Titration of base with excess of acid and the residual acid is back titrated using a standard base.

Acidimetric titration

- ***Direct:*** Titration of base with a known excess of standard acid.
- ***Residual:*** Titration of excess of acid with standard base.

Alkalimetric titration

- ***Direct:*** Titration of acid with a known excess of standard base.
- ***Residual:*** Titration of excess of base with standard base.

Indicators used in acid base titrations: Compounds, which shows well-defined colour change in certain pH intervals are known as indicators. Indicators change colour according to the H^+ ion concentration (pH) of the solution. They are weak acids and bases, which exhibit different colour in their conjugate bases and acid forms (two colour). Some indicators produce no colour in either of these forms.

In acid solution, the excess of H^+ ions depress the ionization of weakly acidic indicator. The concentration of ionized molecule of indicator will be very small and produces colour of unionized form of indicator. But alkali promotes the ionization and gives colour of ionized form of indicator.

42.1 Estimation of Ibuprofen Present in Tablets

Chemicals: Ibuprofen tablets, Chloroform, Ethanol (95%), Sodium hydroxide (0.1M) and Phenolphthalein.

Principle: The amount of ibuprofen (free acid) present in the tablets can be determined by alkalimetry method. This can be done by extracting ibuprofen present in the tablets with chloroform. The ethanolic solution of ibuprofen can be titrated against 0.1M sodium hydroxide using phenolphthalein as an indicator.

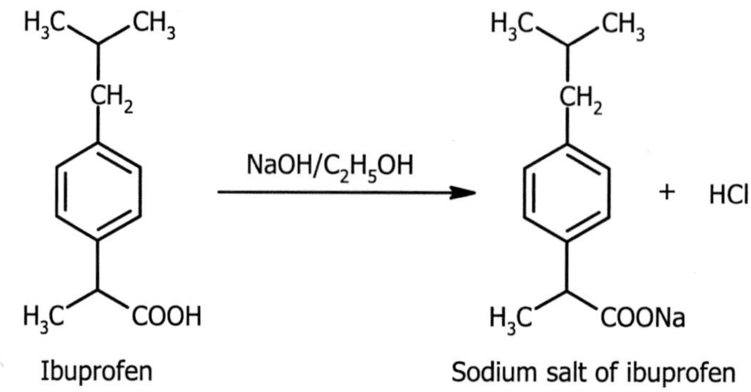

Phenolphthalein (weakly acidic indicator) exhibits no colour in acidic medium, where as in basic medium readily undergoes ionization hence produces pink colour in the solution. Sodium hydroxide (0.1 M) converts Ibuprofen into its sodium salt. The slight excess of sodium hydroxide introduces basic medium and produces pink colour (end point).

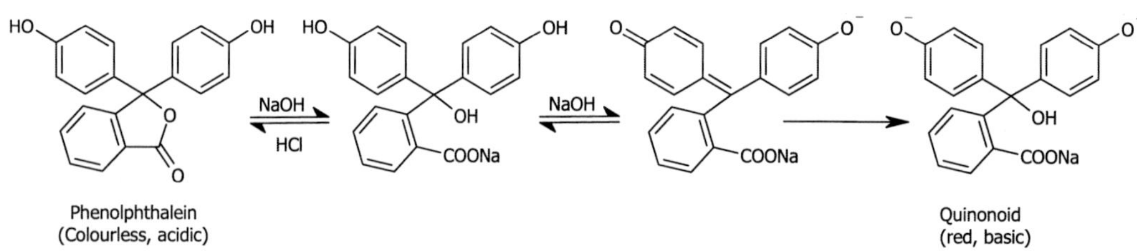

Procedure:

1. Weigh and powder 20 ibuprofen tablets.
2. Weigh accurately a quantity of powder equivalent to 0.5 gm of ibuprofen.
3. Extract with two 30 ml volumes of chloroform and collect the chloroform layer.
4. Wash the residue with 3 quantities of each of 10 ml of chloroform.
5. Mix the chloroform extracts and evaporate the solvent to dryness.
6. Dissolve the residue in ethanol (95%, 10 ml).
7. Titrate with sodium hydroxide (0.1M) using phenolphthalein as an indicator **(a)**.
8. Appearance of pale pink colour indicates the end point.
9. Repeat the operation without the substance being examined (**b**).

   *[(**a** – **b**) is the volume of sodium hydroxide consumed to neutralize the ibuprofen present in the given formulation]*

$$\text{Amount of ibuprofen present in the tablets powder} = \frac{\text{Equivalent Weight Factor} \times \text{Molarity of NaOH} \times \text{Volume of NaOH}}{\text{Standard molarity of NaOH}}$$

Each ml of 0.1M sodium hydroxide is equivalent to 0.02063 gm of ibuprofen.

42.2 Estimation of Salicylic Acid

Requirements: Salicylic acid, Sodium hydroxide (0.1M), Phenol red, Oxalic acid and Phenolphthalein.

Principle: The amount of salicylic acid (free acid) can be determined by alkalimetry method. The ethanolic solution of salicylic acid can be titrated against 0.1M sodium hydroxide using phenol red as an indicator. Colour change from yellow to red indicates the endpoint of the titration. Sodium hydroxide converts salicylic acid into its sodium salt. The slight excess of sodium hydroxide introduces basic medium and converts yellow colour into red (end point).

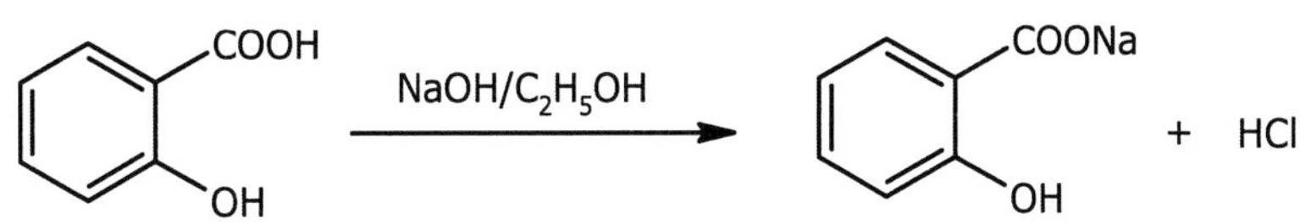

Salicylic acid   Sodium salt of salicylic acid

Procedure:

1. Weigh 0.1 gm of salicylic acid and transfer it into a clean conical flask.
2. Add 15 ml of ethanol to the flask and dilute it with 7 ml of water.
3. Titrate the solution against sodium hydroxide (0.1 M) using phenol red as an indicator.
4. Change in the colour from yellow to red is the end point.

The % purity of salicylic acid present in the sample

$$= \frac{\text{Equivalent weight factor} \times \text{Molarity of sodium hydroxide} \times \text{Volume of sodium hydroxide} \times 100}{\text{Weight taken} \times \text{Actual molarity of sodium hydroxide}}$$

Each ml of 0.1M sodium hydroxide is equivalent to 0.01381 gm of salicylic acid.

42.3 Estimation of Acetyl Salicylic Acid Present in Tablets

Chemicals: Acetyl salicylic acid tablets, Sodium hydroxide (0.5M), Hydrochloric acid (0.5 M) and Phenol red.

Principle: The amount of acetyl salicylic acid (aspirin) present in tablets can be determined by acidimetric method (residual titration method). Aspirin upon boiling with sodium hydroxide undergoes hydrolysis and generates sodium salt of salicylic acid and acetic acid. Excess of base (sodium hydroxide) can be titrated against hydrochloric acid by residual titration method (back titration). The volume of sodium hydroxide consumed by the aspirin for the hydrolysis can be calculated from the volume of hydrochloric acid required to neutralize the excess sodium hydroxide.

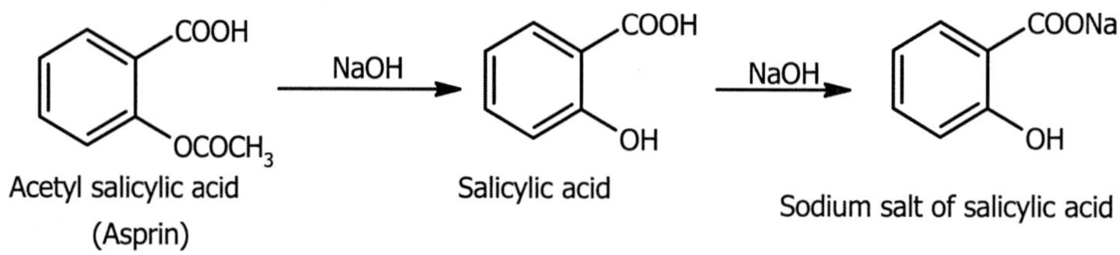

Procedure:

1. Weigh and powder 20 tablets.
2. Weigh the powder equivalent to 0.5 gm of acetyl salicylic acid and transfer it into a clean conical flask.
3. Add 30 ml of sodium hydroxide (0.5M) to the flask, boil the mixture for 10 minutes and cool to room temperature.
4. Titrate the excess of sodium hydroxide (0.5M) with hydrochloric acid (0.5M) using phenol red as an indicator.
5. Change in the colour from yellow to red is the end point.

The amount of acetyl salicylic acid present in the tablet powder =

$$\frac{\text{Equivalent weight factor} \times \text{Molarity of sodium hydroxide} \times \text{Volume of sodium hydroxide}}{\text{Actual molarity of sodium hydroxide}}$$

Each ml of 0.5M sodium hydroxide is equivalent to 0.04504 gm of acetyl salicylic acid.

42.4 Estimation of Busulphan

Chemicals: Busulphan, Sodium hydroxide (0.1N) and Phenolphthalein.

Principle: The amount of busulphan can be estimated through acid-base titration. Busulphan upon hydrolysis produces butane-1,4-diol and methane sulphonic acid. The liberated methane sulphonic acid can be titrated against sodium hydroxide (0.1N) using phenolphthalein as an indicator. After the complete conversion of methane sulphonic acid into its sodium salt, in the slight excess of sodium hydroxide gives pale pink colour.

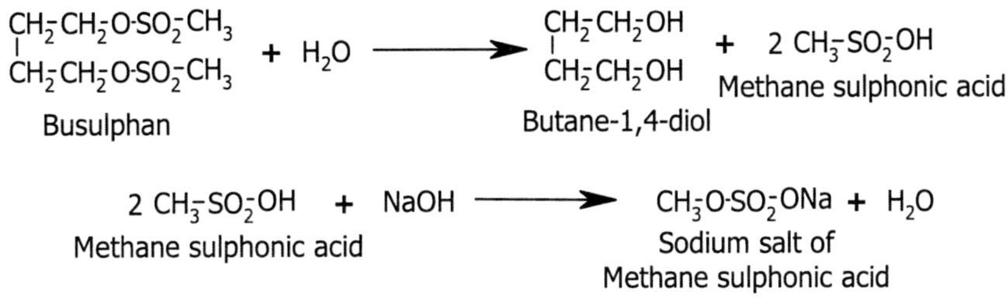

Procedure:

1. Dissolve the accurately weighed busulphan (0.25 gm) in 25 ml of distilled water.
2. Boil the mixture under reflux condenser for 30 minutes.
3. Wash the condenser with distilled water and cool the mixture.
4. Titrate the mixture with 0.1N sodium hydroxide using phenolphthalein as an indicator.
5. The appearance of pale pink colour indicates the end point of the titration.

The % purity of busulphan present in the sample =

$$\frac{\text{Equivalent weight factor} \times \text{Molarity of sodium hydroxide} \times \text{Volume of sodium hydroxide}}{\text{Weight of busulphan} \times \text{Actual normality of sodium hydroxide}} \times 100$$

Each ml of sodium hydroxide (0.1N) is equivalent to 0.01232 gm of busulphan.

42.5 Estimation of Codeine Phosphate Present in Syrup

Chemicals: Codeine phosphate syrup, Dilute ammonia, Chloroform, Ethanol (95%), Hydrochloric acid (0.05M), Sodium hydroxide (0.05M) and Methyl red.

Principle: The amount of codeine phosphate present in syrup can be determined by alkalimetry method (residual titration method). Excess of acid is titrated against sodium hydroxide by residual titration method (back titration). Codeine phosphate upon treatment with ammonium hydroxide liberates the free base of codeine. The liberated free base reacts with hydrochloric acid and forms chloride salt. The unreacted hydrochloric acid can be titrated against sodium hydroxide. The volume of hydrochloric acid consumed by the codeine base can be calculated from the volume of sodium hydroxide required to neutralize the acid.

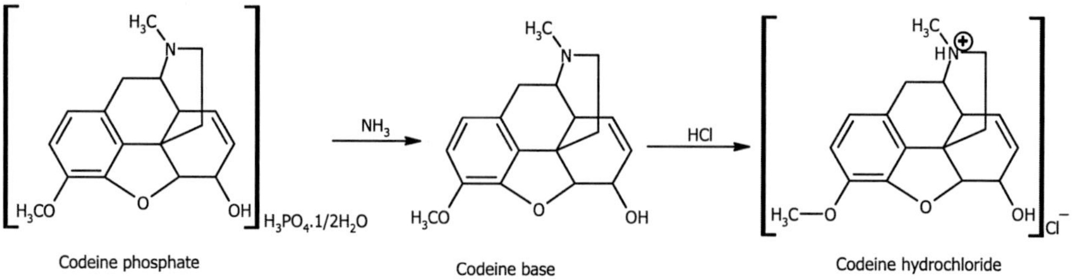

***Procedure*:**

1. Neutralize 10 gm of codeine phosphate syrup with dilute ammonia until the solution is alkaline to litmus paper.
2. Extract the alkaline solution with 4 quantities of each 25 ml of chloroform and collect the chloroform layer.
3. Wash each extract with 10 ml of water and combine all the extracts.
4. Gently evaporate the solvent to dryness in water bath.
5. To the residue add 5 ml of ethanol (95%) and evaporate the solvent to dryness in water bath.
6. Dissolve the residue in 5 ml of 0.05M hydrochloric acid.
7. Titrate the excess of acid with 0.05M sodium hydroxide using methyl red as an indicator **(a)**.
8. At the end point (basic medium) colour changes from orange red to golden yellow.
9. Repeat the operation without the substance being examined **(b)**.
10. **(b – a)** is the volume of sodium hydroxide consumed for the codeine phosphate.

Amount of codeine phosphate present in the given sample =

$$\frac{\text{Equivalent weight factor} \times \text{Molarity of hydrochloric acid} \times \text{Volume of hydrochloric acid}}{\text{Standard molarity of hydrochloric acid}}$$

Each ml of 0.05M hydrochloric acid is equivalent to 0.02 gm of codeine phosphate ($C_{18}H_{21}NO_3 . H_3PO_4 . \frac{1}{2} H_2O$).

CHAPTER 43

ESTIMATION OF DRUGS BY NON-AQUEOUS TITRATION

Acetous perchloric acid titrations are used in the estimation of inorganic and organic salts in non-aqueous acidic or inert solvent medium. Even salts of strong acids can be estimated by this method. The end point of acetous perchloric acid titrations can be found potentiometrically. A visible end point determination using crystal violet in glacial acetic acid is also possible.

In general, the halide ions (chloride, bromide and iodide) present in the salt form of drugs are very weakly basic in character, so that they cannot react quantitatively with acetous perchloric acid. In order to overcome this problem, mercuric acetate is usually added (it remains undissociated in acetic acid solution) to a halide salt. The replacement of halide ion by an equivalent amount of acetate ion occurs.

Crystal violet gives violet colour with substances which behave basic in acetic acid. Addition of acetous perchloric acid changes it into blue and then into green to yellow. The colour change from blue to bluish green is the commonly observed in majority of the drugs.

$$\underset{\text{Mercuric acetate}}{(CH_3COO)_2Hg} + \underset{\text{Halide ion of drug}}{2\,X} \longrightarrow HgX_2 + \underset{\text{Acetate of drug}}{2\,CH_3COO^-}$$

$$HClO_4 + CH_3COOH \rightleftharpoons ClO_4^- + CH_3COOH_2^+$$

$$2\,CH_3COOH_2^+ + 2\,CH_3COO^- \rightleftharpoons 4\,CH_3COOH$$

Advantages

1. Suitable for the estimation of very weak acids and bases
2. Enhances the (weak) reactivity of the substances

Drugs assayed

- Ephedrine
- Methyldopa
- Cetrizine
- Chlorpromazine
- Amitryptyline

43.1 Estimation of Amitryptyline

Chemicals: Amitryptyline hydrochloride, Mercuric acetate, Crystal violet, Perchloric acid (0.1N), Glacial acetic acid.

Principle: In this estimation, the amitryptyline is first converted into base form.

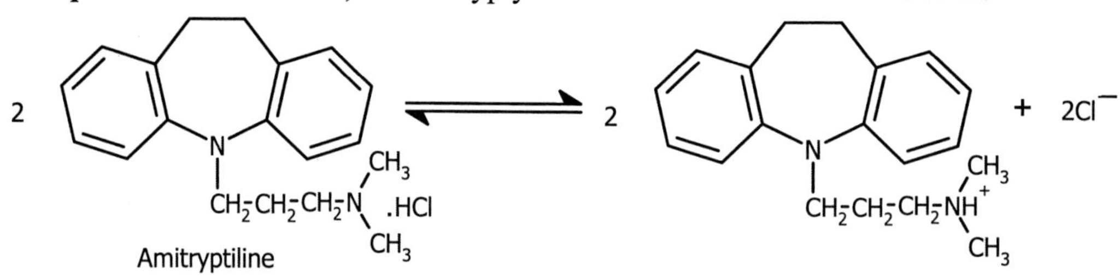

In general, the halide ions (chloride, bromide and iodide) are very weakly basic in character, so that they cannot react quantitatively with acetous perchloric acid. In order to overcome this problem, mercuric acetate is usually added (it remains undissociated in acetic acid solution) to a halide salt thereby causing the replacement of halide ion by an equivalent amount of acetate ion, which serves as a strong base in acetic acid.

$$\underset{\text{Mercuric acetate}}{(CH_3COO)_2Hg} + \underset{\text{Chloride ion of drug}}{2\ Cl^-} \longrightarrow HgCl_2 + \underset{\text{Acetate of drug}}{2\ CH_3COO^-}$$

$$HClO_4 + CH_3COOH \rightleftharpoons ClO_4^- + CH_3COOH_2^+$$

$$2\ CH_3COOH_2^+ + 2\ CH_3COO^- \rightleftharpoons 4\ CH_3COOH$$

Procedure:

1. Dissolve accurately weighed amitryptyline (1 gm) in 50 ml of glacial acetic acid (warm if necessary).
2. Cool the mixture and add mercuric acetate solution (10 ml).
3. Titrate the mixture against 0.1N perchloric acid using crystal violet as an indicator (a).

4. The colour change from violet to green indicates the end point.
5. Perform blank titration to make necessary corrections (b).

The % purity of amitryptyline present in the given sample

$$\frac{\text{Equivalent weight factor} \times \text{Normality of } HClO_4 \times \text{Volume of } HClO_4 (a-b) \times 100}{\text{Weight taken} \times \text{Standard normality of perchloric acid}}$$

Each ml of 0.1N perchloric acid is equivalent to 0.03139 gm of $C_{20}H_{23}N.HCl$.

43.2 Estimation of Phenytoin Present in Tablets

Chemicals: Phenytoin tablets, Acetous perchloric acid (0.1 M), Glacial acetic acid, Acetic anhydride, Acetous mercuric acetate, Acetous crystal violet.

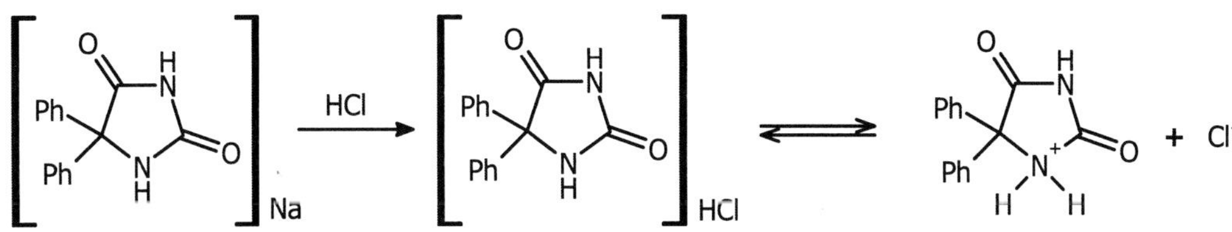

Phenytoin sodium   Phenytoin hydrochloride

Principle: Phenytoin is estimated by non-aqueous titration. Perchloric acid is the most widely used acid for the titration of weak bases. In this estimation phenytoin is converted into its hydrochloride salt.

In general, the halide ions (chloride, bromide and iodide) are very weakly basic in character, so that they cannot react quantitatively with acetous perchloric acid. In order to overcome this problem, mercuric acetate is usually added (it remains undissociated in acetic acid solution) to a halide salt thereby causing the replacement of halide ion by an equivalent amount of acetate ion, which serves as a strong base in acetic acid as shown in the chemical reactions.

$$\underset{\text{Mercuric acetate}}{(CH_3COO)_2Hg} + \underset{\text{Chloride ion of drug}}{2\,Cl^-} \longrightarrow HgCl_2 + \underset{\text{Acetate of drug}}{2\,CH_3COO^-}$$

$$HClO_4 + CH_3COOH \rightleftharpoons ClO_4^- + CH_3COOH_2^+$$

$$2\,CH_3COOH_2^+ + 2\,CH_3COO^- \rightleftharpoons 4\,CH_3COOH$$

Procedure:

1. Weigh and powder 20 tablets.

2. Weigh accurately a quantity of the powder equivalent to 0.5 gm of phenytoin sodium.
3. Shake with 40 ml sodium hydroxide (0.1N) for 5 minutes and add sufficient sodium hydroxide (0.1N) to produce 50 ml.
4. Centrifuge, acidify 25 ml of the clear solution with 10 ml of hydrochloric acid (0.1M) and extract successively with 50, 40, 25 ml of ether.
5. Wash the combined extracts with 10 ml of water, evaporate to dryness and dry the residue at 105 °C.
6. Dissolve in 50 ml of glacial acetic acid warming slightly, if necessary, to effect the solution.
7. Cool, add mercuric acetate solution (10 ml) and titrate with perchloric acid (0.1N) using acetous crystal violet (1-naphtholbenzein) as indicator.
8. At the end point color changes from violet to bluish-green.

   Amount of phenytoin present in the tablet powder

$$= \frac{\text{Equivalent weight factor} \times \text{Normality of } HClO_4 \times \text{Volume of } HClO_4}{\text{Standard normality of perchloric acid}}$$

Each ml of 0.1N perchloric acid is equivalent to 0.02743 gm of phenytoin.

43.3 Estimation of Methyl Dopa

Chemicals: Methyl dopa, acetous perchloric acid (0.1M), Glacial acetic acid, Acetic anhydride, Acetous mercuric acetate, Acetous crystal violet.

Principle: Methyl dopa can be estimated by non-aqueous titration. Perchloric acid is the most widely used acid for the titration of weak bases. In this estimation methyl dopa is converted into its hydrochloride salt.

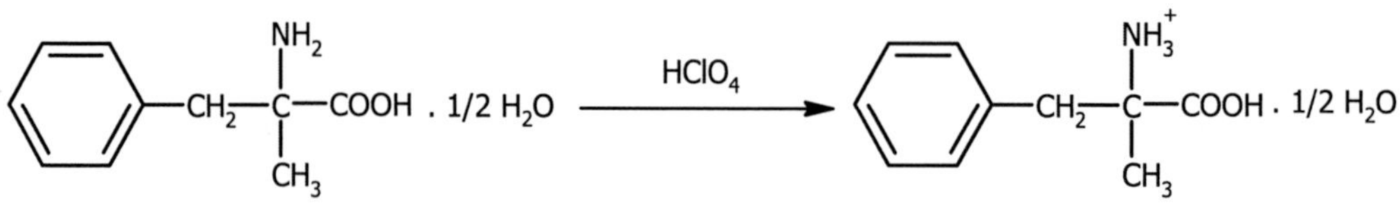

Methyl dopa

Procedure:

1. Weigh accurately 0.2 gm of methyl dopa and dissolve in the mixture of anhydrous formic acid (15 ml), glacial acetic acid (30 ml) and dioxane (30 ml).
2. Add 0.1 ml of crystal violet as an indicator and titrate against 0.1N perchloric acid.
3. At the end point color changes from violet to bluish-green.

4. Perform blank determination to make necessary corrections (b).

Each ml of the 0.1N perchloric acid is equivalent to 0.02112 gm of methyl dopa

$$\text{\% purity of methyl dopa present in the given sample} = \frac{\text{Equivalent weight factor} \times \text{Normality of } HClO_4 \times \text{Volume of } HClO_4 (a-b)}{\text{Weight of the sample} \times \text{Standard normality of perchloric acid}} \times 100$$

Note: Titrations should be carried out at the same temperature, otherwise erroneous results may come.

43.4 Estimation of Ethambutol Present in Tablets

Chemicals: Ethambutol, acetous perchloric acid (0.1M), Glacial acetic acid, Acetous mercuric acetate, Acetous α-naptholbenzein.

Principle: Ethambutol can be estimated by non-aqueous titration. Perchloric acid is the most widely used acid for the titration of weak bases reaction.

$$\text{\% purity of ethambutol present in the given sample} = \frac{\text{Equivalent weight factor} \times \text{Normality of } HClO_4 \times \text{Volume of } HClO_4 \ (a-b)}{\text{Weight of the sample} \times \text{Standard normality of perchloric acid}} \times 100$$

***Procedure*:**

1. To the portion of powdered tablet in a 250 ml of Erlenmeyer flask, add 50 ml of glacial acetic acid and dissolve.
2. Stir the contents for 30 minutes on a rotary shaker.
3. Add 10 ml of acetous mercuric acetate (5%) and 10 drops of acetous α-naptholbenzein (0.1%).
4. Titrate the solution against 0.1N acetous perchloric acid (a).
5. At the end point color changes from violet to bluish-green.
6. Perform blank determination to make necessary corrections (b).

Each ml of the 0.1N perchloric acid is equivalent to 0.01386 gm of ethambutol.

CHAPTER 44

ESTIMATION OF DRUGS BY DIAZOTIZATION TITRATION

An important pharmaceutical application of sodium nitrite titration is the analysis of sulphonamides by diazotization of primary aromatic amino group present in this class of drugs. Benzocaine, procaine hydrochloride, dapsone and solapsone also can be estimated by this method. This analysis is based on the reaction between aromatic primary amine ($-NH_2$) and nitrous acid (HONO) in presence of excess of mineral or inorganic acid. Sodium nitrite reacts with hydrochloric acid and gives nitrous acid (HONO). Aromatic primary amino group ($-NH_2$) can be analyzed quantitatively by measuring the volume of sodium nitrite solution required to convert them into diazonium salts. Unreacted nitrous acid remains in the solution and can be detected by using starch iodide paper or starch paste as an external indicator. Nitrous acid reacts with indicator and liberates iodine this in turn gives blue colour with starch.

$$\underset{\text{Sodium nitrite}}{NaNO_2} + \underset{\text{Hydrochloric acid}}{HCl} \xrightarrow{0\text{-}5\ ^{\circ}C} \underset{\text{Nitrous acid}}{HONO} + NaCl$$

$$\underset{\text{Primary amine}}{R{-}NH_2} + \underset{\text{Nitrous acid}}{HONO} \xrightarrow[0\text{-}5\ ^{0}C]{NaCl} \underset{\text{Diazonium compound}}{R{-}N{=}N{-}Cl} + H_2O$$

$$KI + HCl \longrightarrow HI + KCl$$

$$2\ HI + 2\ HONO \longrightarrow I_2 + 2\ NO + 2\ H_2O$$

$$I_2 + \text{Starch} \longrightarrow \text{Blue colour (end point)}$$

44.1 Estimation of Sulphanilamide

Chemicals: Sulphanilamide, Sodium nitrite (0.1M), Hydrochloric acid (0.2 M), Starch iodide paper.

Principle: Sulphanilamide can be estimated by diazotizing amino group (4-amino) present in it. This analysis is based on the reaction between aromatic primary amine (-NH_2) and nitrous acid in presence of mineral acid. Sodium nitrite reacts with hydrochloric acid and gives nitrous acid (HONO). Aromatic primary amino group (-NH_2) can be analysed quantitatively by measuring the volume of sodium nitrite solution required to convert them into diazonium salts. Unreacted nitrous acid remains in the solution and can be detected by using starch iodide paper or paste as external indicator. Nitrous acid reacts with indicator and liberates iodine this in turn gives blue colour with starch.

$$\underset{\text{Sodium nitrite}}{NaNO_2} + \underset{\text{Hydrochloric acid}}{HCl} \xrightarrow{0\text{-}5\ ^{o}C} \underset{\text{Nitrous acid}}{HONO} + NaCl$$

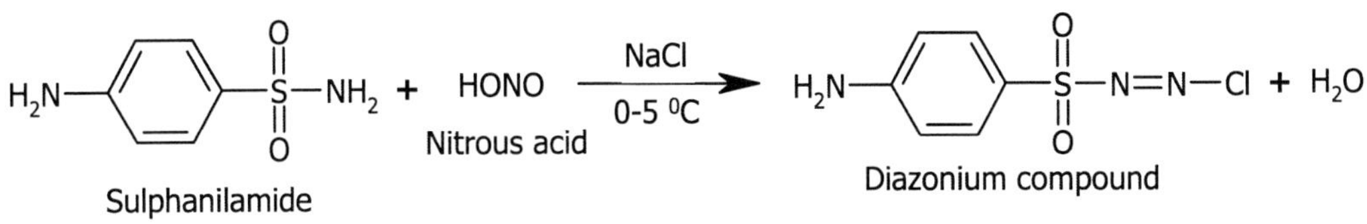

$$KI + HCl \longrightarrow HI + KCl$$

$$2\,HI + 2\,HONO \longrightarrow I_2 + 2\,NO + 2\,H_2O$$

$$I_2 + \text{Starch} \longrightarrow \text{Blue colour (end point)}$$

Estimation of sulphanilamide

1. Weigh accurately about 1 gm of sulphanilamide.
2. Dissolve it in the mixture of hydrochloric acid (40 ml) and water (100 ml).
3. Cool the solution to 0 – 5 °C and titrate against 0.1M sodium nitrite.
4. Continue the addition of sodium nitrite until the drop of solution immediately gives blue colour with starch iodide paper (starch iodide paste).
5. Each ml of 0.1M sodium nitrite is equivalent to 0.01722 gm of sulphanilamide

   The % purity of sulphanilamide present in the given sample.

$$= \frac{\text{Equivalent weight factor} \times \text{Normality of sodium nitrite} \times \text{Volume of sodium nitrite}}{\text{Weight taken} \times \text{Standard normality of sodium nitrite}} \times 100$$

Preparation of starch iodide paper: Starch iodide paper or paste is the starch solution with equal volume of 5% potassium iodide solution in water.

44.2 Estimation of Dapsone Present in Tablets

Chemicals: Dapsone tablets, Sodium nitrite (0.1 M), Hydrochloric acid (0.2 M), Starch iodide paper.

Principle: Dapsone is estimated by diazotizing amino group of dapsone by diazotization or nitrite titration. This analysis is based on the reaction between aromatic primary amine ($-NH_2$) and nitrous acid in presence of mineral acid. Sodium nitrite reacts with hydrochloric acid and gives nitrous acid (HONO). Aromatic primary amino group ($-NH_2$) can be analysed quantitatively by measuring the volume of sodium nitrite solution required to convert them into diazonium salts. Unreacted nitrous acid remains in the solution and can be detected by the starch iodide paper or paste as external indicator. Nitrous acid reacts with indicator and liberates iodine.

$$\underset{\text{Sodium nitrite}}{NaNO_2} + \underset{\text{Hydrochloric acid}}{HCl} \xrightarrow{0\text{-}5\ ^{o}C} \underset{\text{Nitrous acid}}{HONO} + NaCl$$

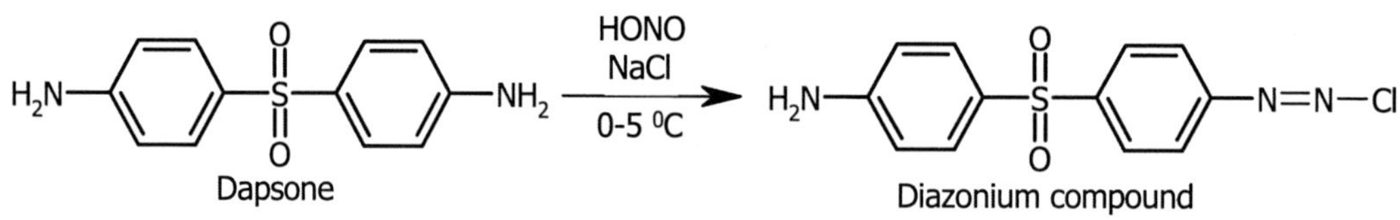

$$KI + HCl \longrightarrow HI + KCl$$

$$2\ HI + 2\ HONO \longrightarrow I_2 + 2\ NO + 2\ H_2O$$

$$I_2 + \text{Starch} \longrightarrow \text{Blue colour (end point)}$$

***Procedure*:**

1. Weigh and powder 20 tablets.
2. Weigh accurately a quantity of the powder equivalent to 0.25 gm of dapsone.
3. Dissolve it in a mixture of 15 ml of water and 15 ml of hydrochloric acid (2M).
4. Cool the solution to below 15 °C (0-15 °C) and titrate with 0.1M sodium nitrite, shaking continuously using starch iodide paper as external indicator.
5. Appearance of blue colour instantaneously in starch iodide paper, when the drop of titrated solution is placed is the end point.
6. Perform a blank determination.
7. Each ml of 0.1M sodium nitrite is equivalent to 0.01242 gm of dapsone.

   Amount of dapsone present in the given sample

$$= \frac{\text{Equivalent weight factor} \times \text{Normality of sodium nitrite} \times \text{Volume of sodium nitrite}}{\text{Standard normality of sodium nitrite}}$$

Preparation of starch iodide paper: Starch iodide paper or paste is the starch solution with equal volume of 5% potassium iodide solution in water.

CHAPTER 45

ESTIMATION OF DRUGS BY REDOX TITRATION

Redox titration involves simultaneous oxidation and reduction processes. It transfers electron from one species (reducing agent) to another species (oxidizing agent). The electron donating species gets oxidized during the course of reducing other species. The species (oxidizing agent) which accepts electron undergoes reduction.

Redox indicators

1. Self
   (i) Potassium permanganate: Imparts pink colour at the end point
   (ii) Cerium (IV) sulphate: Imparts yellow colour at the end point
   (iii) Iodine: Imparts brown colour at the end point
2. External
   (i) Potassium ferricyanide
   (ii) Potassium ferrocyanide
   (iii) Potassium hexacyanoferrate: Used in the titration of ferrous ions with potassium chromate. The presence of ferrous ion can be detected by the appearance of deep prussian blue colour with indicator. Towards the end point the colour disappears.
   (iv) Starch
3. Internal
   (i) Diphenyl amine
   (ii) Sodium diphenyl amino sulphonate
   (iii) Starch solution
   (iv) N,N-Diphenyl benzidine
   (v) 1,10-Phenanthroline iron sulphate
   (vi) 5,6-Dimethyl ferroin
   (vii) 4,7-Dimethyl ferroin

(viii) Methylene blue

(ix) Nitro ferroin

Important redox titrations

- Permanganometric titrations [uses potassium permanganate as titrant]
- Iodine titrations
  - Iodimetric: Titration against iodine (direct)
  - Iodometric: Titration against liberated iodine (indirect)
- Cerimetric titrations [uses cerium salts as titrant]
- Bromatometric titrations [uses bromine as titrant]
- Chromatometric titrations [uses potassium dichromate as titrant]

IODINE TITRATIONS

Iodine (I_2), a mild oxidizing agent oxidizes other species, which donate electrons and gets reduced into iodide (I^-) by accepting electrons. The acidic, neutral and dilute acetic acid medium provides better results.

$$I_2 + 2e^- \rightleftharpoons 2I^-$$

***Iodimetric titrations*:** It uses standard iodine solution as oxidizing agent. The substances with strong reducing property easily estimated through iodimetric titrations.

***Iodometric titrations*:** Iodine liberated from potassium iodide by the action of oxidizing agents is used in the titrations. Copper sulphate ($CuSO_4.5H_2O$), potassium dichromate ($K_2Cr_2O_7$), cerium sulphate ($Ce(SO_4)_2$), potassium iodate (KIO_3) and potassium bromate ($KBrO_3$) are the important oxidizing agents for liberating the iodine from potassium iodide. Weakly reducing agents can be estimated through iodometric titrations.

45.1 Estimation of Ascorbic Acid

Method applicable for both ascorbic acid powder and tablets

Chemicals: Ascorbic acid, Sulphuric acid (1M), Iodine (0.05M), Starch solution.

Principle: Ascorbic acid (vitamin C) is an essential antioxidant plays essential role in the protection of body from infections and diseases. It is not synthesized in the body, hence should be supplied through diet. The amount of ascorbic acid can be determined by redox-titration using iodine (iodimetric titration). Ascorbic acid reduces iodine into iodide and gets oxidized to dehydroascorbic acid. Iodine reduction to iodide will continue as long as ascorbic acid is present. The excess of iodine can be detected by the reaction with starch indicator, which gives blue colour with iodine. This method is applicable for ascorbic acid present in fruits, food and pharmaceutical formulations.

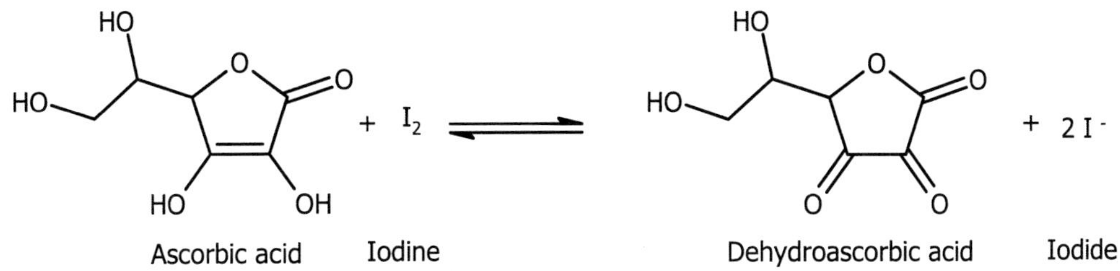

Estimation of ascorbic acid

1. Dissolve the accurately weighed ascorbic acid powder (0.1 gm) in 100 ml of distilled water (freshly boiled and cooled).
2. Add 25 ml of sulphuric acid (1M) and immediately titrate the mixture with iodine (0.05 M) using starch solution as indicator.
3. The appearance of persistent blue-colour is the end point.
4. Each ml of iodine (0.05M) is equivalent to 0.008806 gm of ascorbic acid.

   Amount of ascorbic acid present in the given sample

$$= \frac{\text{Equivalent weight factor} \times \text{Molarity of Iodine} \times \text{Volume of Iodine}}{\text{Standard Molarity of Iodine}}$$

Note: *Ascorbic acid is susceptible to oxidation by atmospheric oxygen; hence solution should be prepared immediately. Alternatively, oxidations can be minimized by adding small amount of oxalic acid.*

Preparation of iodine solution: Add 1.3 gm of iodine into beaker containing 2 gm of potassium iodide. Dissolve the mixture with small amount of distilled water with swirling. Transfer the iodine solution in to 1000 ml volumetric flask and make up the volume to 1000 ml with distilled water. Beaker can be rinsed with water and transfer the rinsing in to volumetric flask before volume make up.

Preparation of starch indicator (0.5%): Boil 0.25 gm of starch with 50 ml of distilled water with stirring. Cool and use for the experiments.

CERIMETRIC TITRATION

Ceric ammonium sulphate [$Ce(SO_4)_2$, $2(NH_4)_2SO_4$, $2H_2O$] is very strong oxidizing agent. The solutions of ceric ammonium sulphate are highly stable and have high oxidation potential, hence it is advantageous over potassium permanganate ($KMnO_4$). Cerimetric titrations are used in the estimation of tablets of ferrous gluconate, ascorbic acid, chlorpromazine and paracetamol.

General principle: Reducing compounds can be estimated by cerimetric tirtarion. The known excess of ceric ammonium sulphate oxidizes reducing agents (e.g.: ascorbic acid into dehydro ascorbic acid) and undergoes reduction. Upon oxidizing (loss of electrons) the reducing agents, the ceric ion (Ce^{4+}, oxidized state) of ceric ammonium sulphate gains electrons (reduction) and converts into cerous state (Ce^{3+}, reduced state).

$$\underset{}{\text{Reducing agent}} + \underset{\text{Ceric ion}}{Ce^{4+}} \longrightarrow \underset{\text{(Oxidized form)}}{\text{Reducing agent}} + \underset{\text{Cerous ion}}{Ce^{3+}} + \underset{\text{Ceric ion (unreacted)}}{Ce^{4+}}$$

The unreacted ceric ion (Ce^{4+}) oxidizes ferrous of ferroin indicator (1,10-phenanthroline-ferrous complex) into ferric and itself reduces into cerous ion (Ce^{3+}). Upon oxidation the colour of ferroin sulphate indicator changes from orange-yellow into pale blue colour. The apperanace of pale blue colour is the end point of the titration.

***Preparation of ferroin indicator*:** Dissolve 1.485 gm of 1,10-phenanthroline in 100 ml of 0.025M ferrous sulphate solution.

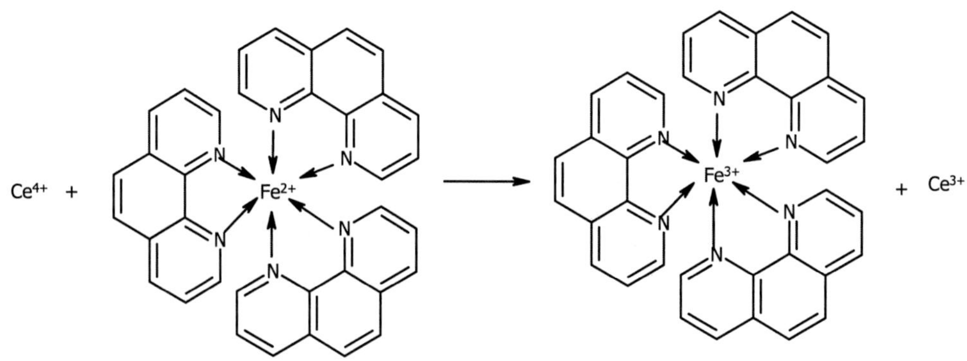

Ceric ion (Unreacted) — 1,10-Phenanthroline-Ferrous complex (Orange-yellow color) — 1,10-Phenanthroline-Ferrous complex (Pale blue color) — Cerous ion

45.2 Estimation of Ascorbic Acid Present in Tablets

Chemicals: Ascorbic acid tablets, Sulphuric acid (1M), Ceric ammonium sulphate (0.1M), Ferroin sulphate.

Principle: The amount of ascorbic acid present in the tablets can be estimated by cerimetric titration. The known excess of ceric ammonium sulphate oxidizes ascorbic acid into dehydroascorbic acid. Upon oxidizing (loss of electrons) the ascorbic acid, the ceric ion (Ce^{4+}, oxidized state) of ceric ammonium sulphate gains electrons (reduction) and converts into cerous state (Ce^{3+}, reduced state).

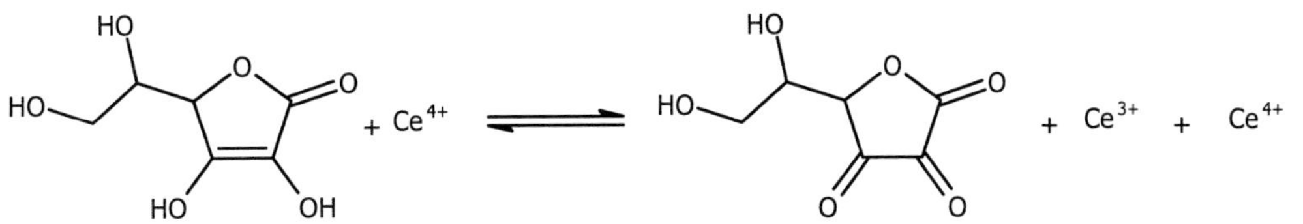

Ascorbic acid Ceric ion Dehydroascorbic acid Cerous ion Ceric ion (unreacted)

The unreacted ceric ion (Ce^{4+}) oxidizes ferrous of ferroin indicator (1,10-phenanthroline-ferrous complex) into ferric and itself reduces into cerous ion (Ce^{3+}). Upon oxidation the colour of ferroin sulphate indicator changes from orange-yellow into pale blue colour. The apperanace of pale blue colour is the end point of the titration.

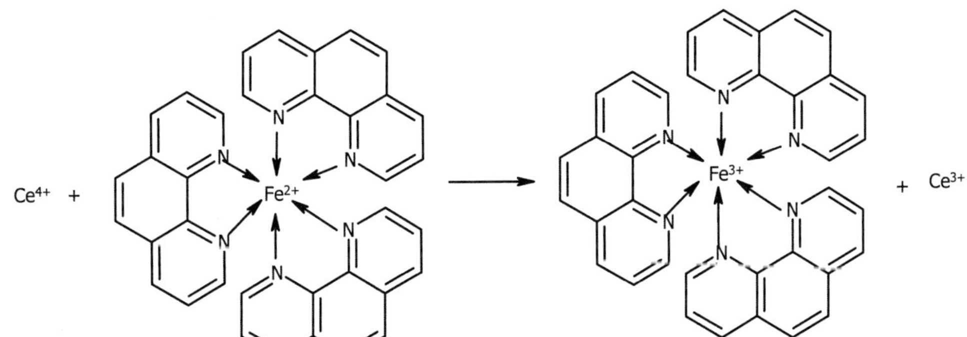

Ceric ion (Unreacted) 1,10-Phenanthroline-Ferrous complex (Orange-yellow color) 1,10-Phenanthroline-Ferrous complex (Pale blue color) Cerous ion

Estimation of amount of ascorbic acid

1. Weigh and powder 20 ascorbic acid tablets.
2. Dissolve accurately weighed ascorbic acid powder (150 mg) in the mixture of distilled water (30 ml) and sulphuric acid (1M, 20 ml).
3. Titrate the mixture with ceric ammonium sulphate (0.1M) using ferroin sulphate (0.1 ml) as indicator.
4. The colour change of orange-yellow into persistent pale blue-colour indicates the end point of titration.
5. Perform the blank titration for necessary correction.
6. Each ml of ceric ammonium sulphate (0.1M) is equivalent to 0.008806 gm of ascorbic acid.

Amount of ascorbic acid present in the given sample

$$= \frac{\text{Equivalent weight Factor} \times \text{Molarity of ceric ammonium sulphate} \times \text{Volume of ceric ammonium sulphate}}{\text{Standard molarity of iodine}}$$

45.3 Estimation of Paracetamol

Chemicals: Paracetamol, Distilled water, Sulphuric acid (1M), Hydrochloric acid (2M), Ceric ammonium sulphate (0.1M), Ferroin sulphate.

Principle: The amount of paracetamol present in the tablets can be estimated by cerimetric titration. The sulphuric acid hydrolyses paracetamol into *para*-amino phenol and acetic acid.

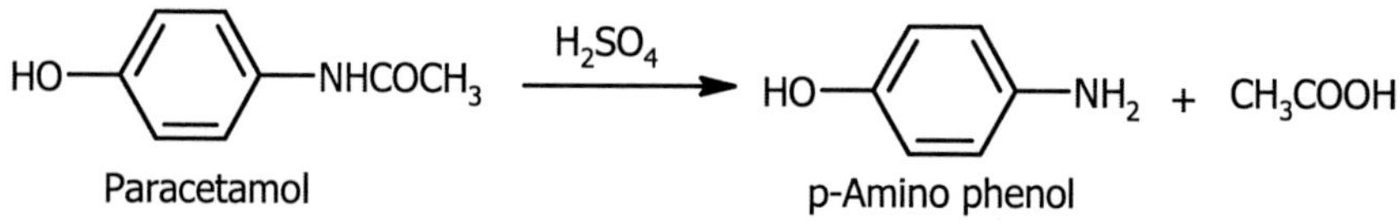

The known excess of ceric ammonium sulphate oxidizes *para*-amino phenol into *para*-imino quinone. Upon oxidizing (loss of electrons) the *para*-amino phenol, the ceric ion (Ce^{4+}, oxidized state) of ceric ammonium sulphate gains electrons (reduction) and converts into cerous state (Ce^{3+}, reduced state).

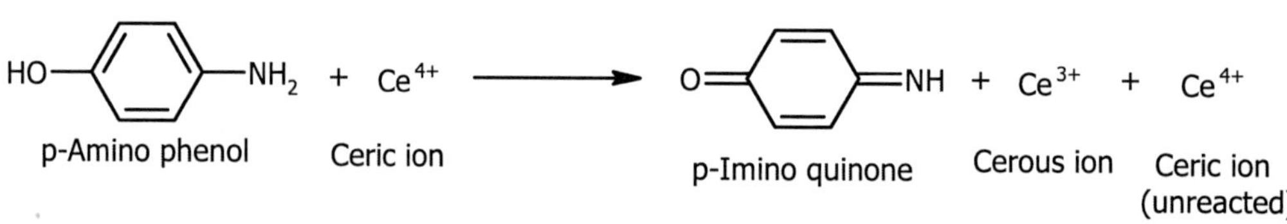

The unreacted ceric ion (Ce^{4+}) oxidizes ferrous of ferroin indicator (1,10-phenathroline-ferrous complex) into ferric and itself reduces into cerous ion (Ce^{3+}). Upon oxidation the colour of ferroin sulphate indicator changes from orange-yellow into pale blue colour. The apperanace of pale blue colour is the end point of the titration.

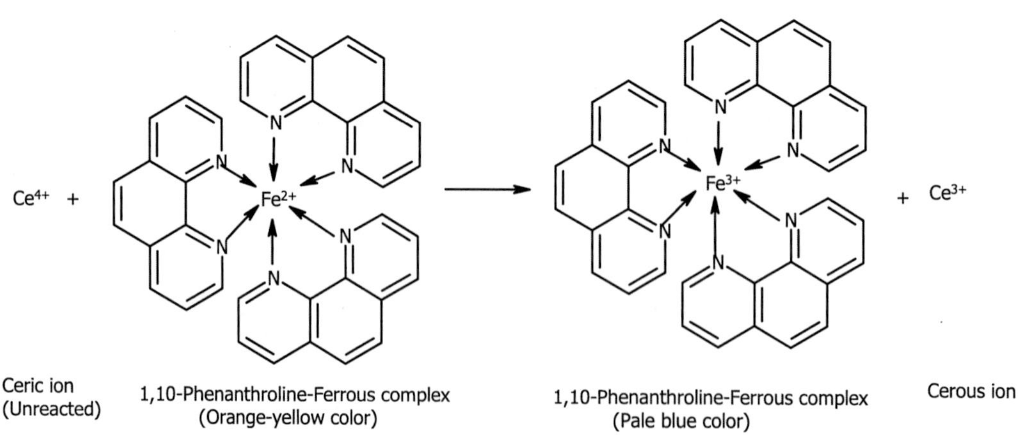

Procedure:

1. Dissolve accurately weighed paracetamol powder (0.5 gm) in the mixture of water (10 ml) and sulphuric acid (1M, 50 ml).
2. Boil the mixture under reflux for 1 hr, cool and dilute to 100 ml with distilled water.
3. To the 20 ml of above solution add water (40 ml), ice (40 gm) and hydrochloric acid (2M, 15 ml).
4. Titrate the mixture immediately with ceric ammonium sulphate (0.1M) using ferroin sulphate (0.1 ml) as indicator.
5. The colour change of orange-yellow into persistant pale blue-colour indicates the end point of titration.
6. Perform the blank titration for necessary correction.
7. Each ml of ceric ammonium sulphate (0.1M) is equivalent to 0.00756 gm of paracetamol.

   Amount of paracetamol present in the given sample

$$= \frac{\text{Equivalent weight Factor} \times \text{Molarity of ceric ammonium sulphate} \times \text{Volume of ceric ammonium sulphate}}{\text{Standard molarity of ceric ammonium sulphate}}$$

QUALITATIVE ANALYSIS OF INORGANIC AND ORGANIC COMPOUNDS

CHAPTER 46

QUALITY CONTROL OF PHARMACEUTICALS

Any organic material present in medicinal preparations, besides the active ingredient /excipient is considered as an impurity. Impurities include starting materials, by-products, interaction products, intermediates, penultimate intermediates, degradation products, reagents, ligands, catalysts and residual solvents. All the medicinal substances should be free from dangerous and/or objectionable impurities. The presence of sodium bromide in potassium bromide (hypnotic and sedative) will not produce any harm, but in large proportion reduces the medicinal quality of potassium bromide. The impurities remain in active pharmaceutical ingredients (APIs) are inevitable, and even in trace amount they may influence the efficacy and safety of APIs. They may produce teratogenic, carcinogenic and mutagenic effects. The pharmaceutical industry is heavily regulated and led to the mandatory examination of all substances from starting materials to finished products. Regulation generally includes:

- Proof of product stability.
- Drug-release profiles.
- Active Pharmaceutical Ingredient (API) authentification
- Product safety including sterility.
- Foreign materials including heavy-metals.

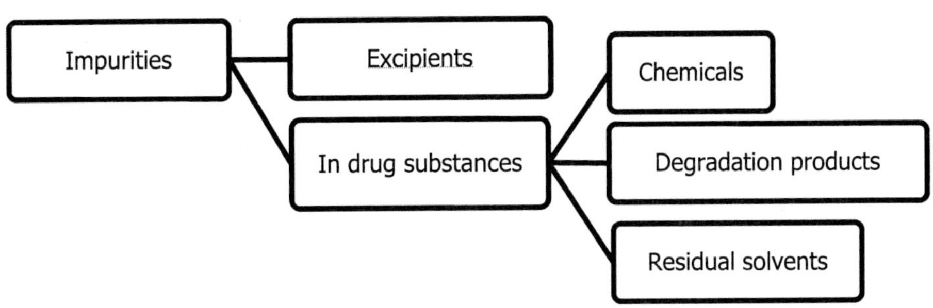

Different types of impurities known are:

1. ***Specified impurities*:** An impurity that is individually listed and limited with a specific acceptance criterion in the specification.
2. ***Unspecified impurities*:** An impurity that is limited by a general acceptance criterion, but not individually listed with its own specific acceptance criterion in the specification.
3. ***Identified impurities*:** An impurity for which a structural characterization has been achieved.
4. ***Unidentified impurities*:** An impurity for which a structural characterization has not been achieved and that is defined solely by qualitative analytical properties.

Limit test

One of the most common product safety-related analytical test is Limit Test. It is the quantification of inorganics and/or heavy metals present in pharmaceuticals. Most of the heavy metals have significant health risks from organ failure to death. Lead, mercury, bismuth, arsenic, antimony, tin, cadmium, silver, copper, and molybdenum are the important heavy metals. Their presence in pharmaceuticals should be limited. Limit test involves the simple comparison of opalescence, turbidity and colour intensity of the samples with fixed standards.

In general, all the limit tests are carried out with an aqueous solution of the substance. But special methods are applied for selective cases:

- An alkaline substance has to be neutralized with nitric acid or hydrochloric acid.
- Water insoluble substance has to be dissolved in acid (alternatively aqueous extract can be used).

46.1 The Chloride Limit Test

***Requirements*:** Nessler's cylinders, pipettes, measuring cylinders, glass rods, test sample, standard solution, dilute nitric acid, silver nitrate, distilled water.

***Principle*:** The chloride limit test determines the allowable limit of chloride contained in pharmaceutical products.

- ***Chemical interaction*:** Limit test for chloride depends upon the interaction of chloride impurities present in the sample with silver nitrate in presence of dilute nitric acid.
- ***Formation of opalescence*:** Chloride impurity reacts with silver nitrate and forms white opalescence, due to the formation of silver chloride.

$$NaCl + AgNO_3 \xrightarrow{HNO_3} AgCl + NaNO_3$$

- ***Comparison of test sample with standard*:** The opalescence produced by sample under test is compared with opalescence produced by standard.

***Procedure*:**

S. No.	Standard Nessler Cylinder	Test Nessler Cylinder
1.	Dilute 1 ml of 0.05845% w/v solution of sodium chloride with 10 ml of water	Dissolve 1 gm of sample in 10 ml of water
2.	Add 10 ml of dilute nitric acid	Add 10 ml of dilute nitric acid
3.	Add 1 ml of silver nitrate reagent	Add 1 ml of silver nitrate reagent
4.	Dilute to 50 ml with distilled water	Dilute to 50 ml with distilled water

***Comparison of test sample with standard*:** The opalescence produced by sample under test is compared with opalescence produced by standard.

1. Greater the opalescence in test nessler cylinder – sample fails the limit test for chlorides.
2. Greater the opalescence in standard nessler cylinder – sample passes the limit test for chlorides.

Comparison of opalescence can be carried out by placing two cylinders on a white tile and viewing vertically downwards through the colours of liquid in the tube.

46.2 The Sulphate Limit Test

Requirements: Nessler's cylinders, pipettes, measuring cylinders, glass rods, barium sulphate, ethanolic sulphated standard solution (10 ppm SO_4), acetic acid (5M), barium chloride, hydrochloric acid (dilute).

Principle: The sulphate limit test determines the allowable limit of sulphate contained in pharmaceutical products.

- ***Chemical interaction:*** Limit test for sulphate depends upon the interaction of soluble sulphate with barium chloride in presence of alcohol and potassium sulphate.
- ***Formation of turbidity:*** Deposition of solid particles of sulphate as barium sulphate produces turbidity.
- The small amount of barium sulphate formed in the reagent acts as seeding agent.
    1. Barium chloride, alcohol and a small amount of potassium sulphate is used for this purpose.
    2. Alcohol prevents super saturation.
    3. Potassium sulphate increases sensitivity of the test by giving ionic concentration in the reagent.

$$SO_4^{2-} + BaCl_2 \xrightarrow{HNO_3} BaSO_4 + 2Cl^-$$

- ***Comparison of test sample with standard*:** The turbidity produced by test is compared with standard sample.

***Procedure*:**

S. No.	Standard Nessler Cylinder	Test Nessler Cylinder
1.	Dilute 1 ml of 0.1089% w/v solution of potassium sulphate with 2 ml of dilute hydrochloric acid	Dissolve 1 gm of sample in 2 ml of dilute hydrochloric acid
2.	Add 40 ml of distilled water	Add 40 ml of distilled water
3.	Add 5 ml of barium sulphate reagent	Add 5 ml of barium sulphate reagent
4.	Dilute to 50 ml with distilled water	Dilute to 50 ml with distilled water

***Comparison of test sample with standard*:** The turbidity produced by sample under test is compared with turbidity produced by standard.

1. Greater the turbidity in test nessler cylinder – fails the limit test.
2. Greater the turbidity in standard nessler cylinder – passes the limit test.

46.3 The Iron Limit Test

Requirements: Nessler's cylinders, pipettes, measuring cylinders, glass rods, standard solution of iron (20 ppm Fe), sample 20% w/v iron free citric acid, thioglycolic acid, iron free ammonia, distilled water.

Principle: The iron limit test is to determine the allowable limit of iron contained in pharmaceuticals.

- ***Chemical interaction:*** Limit test for iron is based upon the chemical reaction of iron in ammonical solution in presence of iron free citric acid with thioglycolic acid. Thioglycolic acid reduces ferric (Fe^{3+}) to ferrous (Fe^{2+}) and forms ferrous thioglycolate complex.

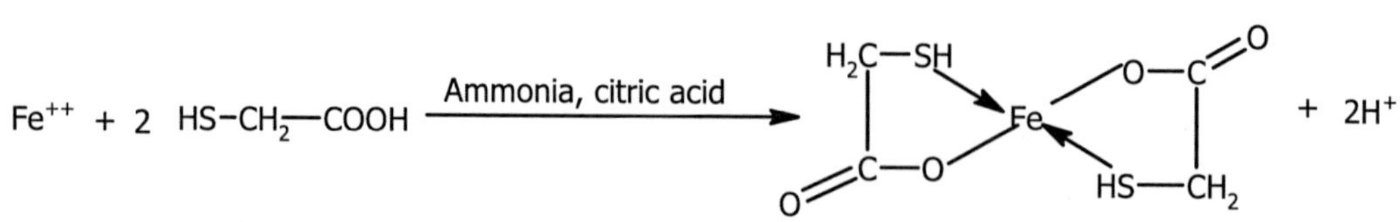

Thioglycolic acid

Ferrous thioglycolate complex

- ***Formation of purple colour:*** Ferrous thioglycolate complex forms pale pink to deep reddish purple colour in alkaline medium. Citric acid (20%) forms complex with other metal ion and eliminates their interference. Ammonia solution forms precipitate with iron, therefore citric acid is added to form ammonium citrate buffer which stabilizes ferrous thioglycolate complex.

- ***Comparison of test with standard***: The purple colour produced by test is compared with the colour produced by the standard.

Procedure:

S. No.	Standard Nessler Cylinder	Test Nessler Cylinder
1.	Dilute 2 ml of standard iron solution with 40 ml of distilled water	Dissolve 1 gm of sample in 40 ml of distilled water
2.	Add 2 ml of 20% iron free citric acid	Add 2 ml of 20% iron free citric acid
3.	Add 2 drops of thioglycolic acid	Add 2 drops of thioglycolic acid
4.	Make alkaline with iron free ammonia	Make alkaline with iron free ammonia
5.	Dilute to 50 ml with distilled water	Dilute to 50 ml with distilled water

Comparison of test sample with standard: The purple colour produced by sample under test is compared with colour produced by standard.

1. Greater the purple colour in test nessler cylinder – sample fails the limit test for iron.
2. Greater the purple colour in standard nessler cylinder – sample passes the limit test for iron.

Comparison of colour intensity can be carried out by placing two cylinders on a white tile and viewing vertically downwards through the colours of liquid in the tube.

Note: The purple colour disappears upon oxidation by atmospheric oxygen.

46.4 The Arsenic Limit Test

Requirements: Arsenic limit test apparatus, cotton wool, sample, standard arsenic, zinc, hydrochloric acid, stannous chloride, potassium iodide lead acetate solution, mercuric chloride/bromide paper.

Theory

The Arsenic Limit Test is a limit test for arsenic contained in pharmaceuticals. The limit is expressed in terms of arsenic trioxide (As_2O_3). In each monograph, the permissible limit for arsenic (As_2O_3) is described in terms of ppm in parentheses. Arsenic is not usually found as a pure metal, but in combination with other materials. Eyeliner, eye-shadow, lipstick and mascara contain cadmium, beryllium and arsenic. Toxins absorbed through the skin, leads to hormone disruption, cancer, kidney problems, lung damage, dermatitis and hair loss. There are two types of arsenic; it includes organic (harmless) and inorganic (harmful) arsenic.

Apparatus

The arsenic limit test apparatus consists of the following parts:

1. *Generator bottle*: Used to keep all the reactants (sample, reducing agents).
2. *Rubber bung*: To insert glass tube and to seal the container.
3. *Glass tube*: Tube with 6.5 mm internal diameter, 18 mm external diameter and small aperture at the lower end.
4. *Lead acetate cotton wool*: Cotton wool moistened with lead acetate solution to trap hydrogen sulphide gas.
5. *Mercuric chloride/bromide paper*: To trap and identify the arsine gas formed.
6. *Spring clip*: To fix the mercuric chloride/bromide paper in place.

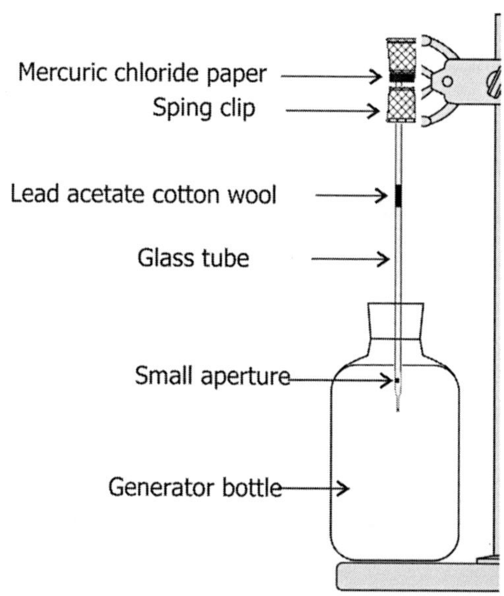

Principle:

- ***Reduction of arsenic:*** The hydrochloric acid reduces arsenic present in the sample into arsenious acid (in case arsenic is trivalent) or arsenic acid (in case arsenic is pentavalent). The combined effect of hydrochloric acid and stannous chloride reduces arsenic acid (if any) into arsenious acid. The nascent hydrogen produced by the reaction of zinc and hydrochloric acid reduces arsenious acid into arsine gas (AsH_3, arsenious hydride).

$$\underset{\text{Arsenic (Trivalent)}}{As^{3+}} \xrightarrow{H^+} \underset{\text{Arsenious acid}}{H_3AsO_3} \qquad \underset{\text{Arsenic (Pentavalent)}}{As^{5+}} \xrightarrow{H^+} \underset{\text{Arsenic acid}}{H_3AsO_4}$$

$$\underset{\text{Stannous chloride}}{SnCl_2} + HCl \longrightarrow H_2SnO_2$$

$$\underset{\text{Arsenic acid}}{H_3AsO_4} + H_2SnO_2 \longrightarrow \underset{\text{Arsenious acid}}{H_3AsO_3} + H_2SnO_3$$

$$\underset{\text{Arsenious acid}}{H_3AsO_3} + H_2SnO_3 \longrightarrow \underset{\text{Arsine}}{AsH_3} + 3H_2O$$

- ***Stain formation*:** The nascent hydrogen gas carries the arsine into mercuric chloride/bromide paper present in the drying tube and produces yellow colour stain on the paper. The intensity of the colour depends upon the quantity of arsenic present. This chemical reaction is known as 'Gutzeit test'. The cotton wool moistened with lead acetate present in the drying tube traps hydrogen sulphide formed. Otherwise hydrogen sulphide produces yellow stain with mercuric chloride paper (instead of AsH_3) and leads to erroneous conclusions.

$$\underset{\text{Arsine}}{AsH_3} \xrightarrow{\text{Mercuric chloride / bromide paper}} \text{Yellow stain}$$

- ***Comparison of test sample with standard*:** The stain produced by sample under test is compared with stain produced by the standard sample.

Preparation of Reagents

- **Hydrochloric acid:** Dilute 10 ml of hydrochloric acid with 50 ml of distilled water and add 5 ml of ammonium thiocyante to the above and stir immediately.
- **Mercuric chloride paper:** Soak 25 mm width white paper in a saturated mercuric chloride solution and dry at 60 °C.
- **Potassium iodide:** Dissolve 10 gm of potassium iodide in 25 ml of hydrochloric acid and 25 ml of water and add 5 drops of stannous chloride solution to the above.

Procedure:

S. No.	Standard	Test
1.	Dilute 1 ml of As_2O_3 solution (0.132 gm of As_2O_3 in 100 ml of water) to 100 ml	Place solution to be examined (5 ml of 1 gm solution)
2.	Add 5 ml of 1M of potassium iodide/5 ml of stannous chloride acid and 10 gm of zinc.	Add 5 ml of 1M of potassium iodide/5 ml of stannous chloride acid and 10 gm of zinc.
3.	Allow it to remain for 40 minutes	Allow it to remain for 40 minutes

***Comparison of test sample with standard*:** The yellow stain produced by sample under test is compared with standard stain in day light immediately.

1. Greater the yellow stain in test bottle – sample fails the limit test for arsenic.
2. Greater the yellow stain in standard bottle – sample passes the limit test for arsenic.

Comparison of colour intensity can be carried out by placing two cylinders on a white tile and viewing vertically downwards through the colours of liquid in the tube.

Marsh's test: The yellow stain produced by the Gutzeit method, is inferior in terms of accuracy and results. The stain is light sensitive and gets affected by foreign gases and moisture. Hence the modified method called 'Marsh's test' is developed. Extremely small quantities of arsenic can be detected by this method.

The apparatus consists of drying tube loaded with lead acetate wool followed by calcium chloride to dry the evolved gas. The hard glass exit tube is connected to the tube, the exit tube is heated at one point using micro burner. The evolved arsine forms the mirror of arsenic beyond the heated part of the tube.

Fleitmann's test: The arsine gas can be detected by using filter paper moistened with silver nitrate solution. The formation of greyish-black stain of metallic silver on the paper indicates the presence of arsenic. The stain disappears upon treatment with dilute acid.

46.5 The Heavy Metals Limit Test

Requirements: Nessler's cylinders, pipette, measuring cylinder, glass rods, standard solution of lead and sample to be tested.

***Principle*:** The lead limit test is designed to determine the allowable limit of lead contained in pharmaceuticals. The more obvious source of lead in chemicals are sulphuric acid used in the manufacture and the use of lead apparatus. Limit test of heavy metals is based on the reaction of metallic impurities with hydrogen sulfide in acidic medium to form brownish colour solution. Metals that response to this test are lead, mercury, bismuth, arsenic, antimony, tin, cadmium, silver, copper, and molybdenum. The metallic impurities in substances are expressed as parts of lead per million parts of the substance. The usual limit as per Indian Pharmacopoeia is 20 ppm.

Note:

1. Nessler cylinder made of lead free glass is preferred for the limit test for lead.
2. Noticeable colour difference between two cylinders can be rectified by the addition of solution of burnt sugar.

***Procedure*:**

Method 1:

This method is followed for the substances which give clear and

S. No.	Standard Nessler Cylinder	Test Nessler Cylinder
1.	Dilute 1 ml of standard lead solution (20 ppm) with 25 ml of distilled water	Dissolve specified quantity of sample as per monograph in 25 ml of distilled water
2.	Adjust the pH to 3 to 4 using dilute acetic acid/ammonia	Adjust the pH to 3 to 4 using dilute acetic acid/ammonia
3.	Dilute with distilled water (35 ml) and mix	Dilute with distilled water (35 ml) and mix
4.	Add 10 ml of freshly prepared hydrogen sulphide solution.	Add 10 ml of freshly prepared hydrogen sulphide solution.
5	Make up the volume upto 50 ml with distilled water and set aside for 5 minutes	Make up the volume upto 50 ml with distilled water and set aside for 5 minutes
6	View downwards over a white surface	View downwards over a white surface

Method 2:

S. No.	Standard Nessler Cylinder	Test Nessler Cylinder
1.	Dilute 1 ml of standard lead solution (20 ppm) with 20 ml of distilled water	Dissolve specified quantity of sample as per monograph in 20 ml of distilled water.
2.	Add 5 ml of dilute sodium hydroxide	Add 5 ml of sodium hydroxide
3.	Dilute to 50 ml with distilled water	Dilute to 50 ml with distilled water
4.	Add 5 drops of sodium sulphide solution and set aside for 5 minutes	Add 5 drops of sodium sulphide solution and set aside for 5 minutes

Method 3:

S. No.	Standard Nessler Cylinder	Test Nessler Cylinder
1.	Dilute 10 ml of standard lead solution (1 ppm/ 2 ppm) with 25 ml of distilled water	Pippete 12 ml of test solution (prepared as directed by monograph)
2	Add 2 ml of test solution and mix	
3	Add 2 ml of acetate buffer (pH 3.5) and mix	Add 2 ml of acetate buffer (pH 3.5) and mix
4	Add 1.2 ml of thioacetamide reagent	Add 1.2 ml of thioacetamide reagent
5	Allow to stand for 2 minutes and view downwards over a white surface	Allow to stand for 2 minutes and view downwards over a white surface

Ammonium citrate, potassium cyanide, hydroxylamine hydrochloride is used to make pH optimum so interference and influence of other impurities have been eliminated. Phenol red is used as indicator to develop the color at the end of process. Lead present as an impurities in the substance, gets separated by extracting an alkaline solution with a dithizone extraction solution.

CHAPTER 47

QUALITATIVE ANALYSIS OF INORGANIC COMPOUNDS

Qualitative analysis is the process of identifying a chemical substance. In inorganic chemical analysis, it includes the identification of acidic radicals and basic radicals.

These tests are of two types:

1. Dry tests
2. Wet tests

Dry tests: Heating the sample in non-luminous flame on a platinum wire or on charcoal or on borax or on fusing with sodium carbonate.

Wet tests: Results obtained are more reliable than dry tests. Solution of the sample treated with different reagents or chemicals.

Qualitative Analysis of Inorganic Compounds

1	**Colour**	
	• Blue or green • Light green • Dark green • Dark brown • Yellow • Light pink • Pink violet • Colourless	• Copper or Nickel salt • Ferrous salts • Chromium salts • Ferric salts • Ferric salts, ferrous oxalate, ferrous carbonate or chromates • Manganese salts • salts • Al, Zn, Ca, Cr, Ba, Mg, Na, Ze, NH_4 salts etc absent
2	**Heat test:** Heat the given mixture in dry test tube	
	Violet vapour turned starch iodide paper blue	May be iodine
	Yellow colour changes to brown on cooling	May be copper

Table Contd...

	Colourless and/or odourless gas turns lime water milky • Decrepitation (cracking noise) • Fusion accompanied by condensation of water drops on the sides of the test tube • Smell of ammonia	May be carbonates and bicarbonates • Crystalline salts of Na, K, Pb • Water due to hygroscopic salts, water of crystallization of hydroxide decomposition • Ammonium salts
3	**Flame test:** Introduce the glass rod dipped in the paste of sample (made with hydrochloric acid) into the non-luminous flame.	
	• Persistent golden yellow • Pale violet • Brick red/orange • Yellowish-grassy green • Bluish green/blue	• Sodium • Potassium • Calcium • Barium • Copper
4	**Dilute sulphuric acid test:** Add dilute sulphuric acid to the sample placed in a dry test tube and heat gently.	
	• White fumes evolve and turns blue litmus red • Colourless gas with smell of vinegar	• May be chlorides • May be acetates
	Evolution of carbon dioxide with brisk effervescence, which turns lime water milky	May be carbonates
	Evolution of hydrogen sulphide (smell of rotten eggs), which turns lead acetate paper black	May be sulphides
	Brown gas turns ferrous sulphate solution slowly	May be nitrites
5	**Concentrated sulphuric acid test:** Heat the sample with concentrated sulphuric acid in a dry test tube.	
	Brown gas turns ferrous sulphate solution in to black	May be nitrates
	Violet vapour turns moist iodide paper dark blue, sublimes solid on the sides of the test tube	May be iodides
	Colourless gas with pungent odour	May be chlorides
	Reddish fumes with pungent odour	May be bromides
6	**Manganese dioxide test:** To small amount of sample, add manganese dioxide and 2 ml of concentrated sulphuric acid and heat gently.	
	Greenish yellow fumes evolve, which turns blue litmus into red and bleaches	May be chlorides
	Brown fumes evolve, which turns starch iodide paper into blue and starch paper into yellow.	May be bromides
	Violet vapours evolve, which turns starch paper into bluish black	May be iodides

47.1 Detection of Acid Radicals

Preparation of extract of acid radicals

- Take 1 gm of substance in small beaker, add 3 gm of sodium carbonate and 15 ml of water. Boil the contents for five minutes, cool and filter. The filtrate is known as sodium carbonate extract.

  Neutralize the extract by adding dilute nitric acid (till it becomes slightly acidic), then add ammonia to turn slightly alkaline to litmus. Boil the solution to expel off ammonia and cool. Then perform the following tests

1	**Silver nitrate test:** To the neutralized extract add 1 ml of silver nitrate solution	
	White precipitate soluble in ammonia	Presence of chlorides
	$Cl^- + AgNO_3 \longrightarrow AgCl\downarrow + NO_3^-$	
	White or brown precipitate soluble in nitric acid	Presence of carbonates
	$CO_3^{2-} + 2\,AgNO_3 \longrightarrow Ag_2CO_3\downarrow + 2NO_3^-$	
	White precipitate insoluble in nitric acid	Presence of sulphates
	$SO_4^{2-} + 2\,AgNO_3 \longrightarrow Ag_2SO_4\downarrow + 2NO_3^-$	
	Pale yellow precipitate soluble in ammonia with difficulty	Presence of bromides
	$Br^- + AgNO_3 \longrightarrow AgBr\downarrow + NO_3^-$	
	Yellow precipitate soluble in excess of ammonia	Presence of iodides
	$I^- + AgNO_3 \longrightarrow AgI\downarrow + NO_3^-$	
	Black precipitate	Presence of sulphides
	$S^{2-} + 2\,AgNO_3 \longrightarrow Ag_2S\downarrow + 2NO_3^-$	
2	**Barium nitrate test:** To the neutralized extract add 1 ml of barium nitrate reagent.	
	White precipitate - insoluble in nitric acid	Presence of sulphates
	$SO_4^{2-} + Ba(NO_3)_2 \longrightarrow BaSO_4\downarrow + 2NO_3^-$	
	Soluble in dilute acids with effervescence	Presence of carbonates
	$CO_3^{2-} + Ba(NO_3)_2 \longrightarrow BaCO_3\downarrow + 2NO_3^-$	
3	**Ferrous sulphate test:** Neutralize 1 ml of extract with acetic acid and add freshly prepared ferrous sulphate solution.	
	A brown solution is formed	Presence of nitrates

Table Contd...

	$3\ NO_3^- + 2\ FeSO_4 \longrightarrow Fe_2(NO_2)_3\downarrow + SO_4^{2-}$	
4	To 1 ml of extract add equal amount of sulphuric acid, cool and add ferrous sulphate solution from the sides of the test tube.	
	A brown ring at the junction of the two layers	Presence of nitrates
5	**Lead acetate test:** To 1 ml of neutralized extract add lead acetate solution.	
	White precipitate- Soluble in hot water	Presence of chlorides and/bromides
	$Pb(CH_3COO)_2 + 2\ Cl^- /Br^- \longrightarrow PbCl_2/\ PbBr_2\downarrow + 2\ CH_3COO^-$	
	White precipitate- Insoluble in dilute acids, but soluble in ammonium acetate	Presence of sulphates
	$SO_4^{2-} + Pb(CH_3COO)_2 \longrightarrow Pb_2SO_4\downarrow + 2\ CH_3COO^-$	
	Yellow precipitate- Soluble in boiling water, on cooling separates golden yellow crystals	Presence of iodides
	$2I^- + Pb(CH_3COO)_2 \longrightarrow PbI_2\downarrow + 2\ CH_3COO^-$	
	Black precipitate	Presence of sulphides
	$Pb(CH_3COO)_2 + S^{2-} \longrightarrow PbS\downarrow + 2\ CH_3COO^-$	

47.2 Confirmatory Tests for Acid Radicals

Confirmatory Tests for Carbonates

1	To an aqueous solution of substance add dilute sulphuric acid.	
	Brisk effervescence turns lime water milky	Confirms the presence of carbonates
	$CO_3^{2-} + H_2SO_4 \longrightarrow CO_2\uparrow + SO_4^{2-} + H_2O$	
2	To an aqueous solution of the substance add barium nitrate solution.	
	White precipitate soluble in dilute acids with effervescence	Confirms the presence of carbonates
	$CO_3^{2-} + Ba(NO_3)_2 \longrightarrow BaCO_3\downarrow + 2NO_3^-$	
3	**Distinct test for carbonates and bicarbonates:** To an aqueous solution of the substance add magnesium sulphate solution.	
	White precipitate	Confirms the presence of carbonates
	$CO_3^{2-} + MgSO_4 \longrightarrow MgCO_3\downarrow + SO_4^{2-}$	
	Clear solution	Confirms the presence of bicarbonates
	$2HCO_3^- + MgSO_4 \longrightarrow Mg(HCO_3)_2\downarrow + SO_4^{-2}$	

Confirmatory Tests for Chlorides

1	**Chromyl chloride test:** Heat the sample with potassium dichromate and 5 ml sulphuric acid in a test tube. Pass the vapours to the lead acetate solution.	
	Yellow precipitate of lead chromate is formed	Confirms the presence of chloride
	$K_2Cr_2O_7 + 3\,H_2SO_4 + 4Cl^- \longrightarrow 2CrO_2Cl_2 + 2SO_4^{2-} + K_2SO_4 + 3\,H_2O$ $CrO_2Cl_2 + 2\,H_2O \longrightarrow H_2CrO_4 + 2\,HCl$ $H_2CrO_4 + Pb(CH_3COO)_2 \longrightarrow PbCrO_4\downarrow + 2\,CH_3COOH$	
2	Heat the solution of the substance with manganese dioxide and sulphuric acid.	
	A greenish yellow vapour turns blue litmus red and then bleaches	Confirms the presence of chloride
	$MnO_2 + 3\,H_2SO_4 + 2\,Cl^- \longrightarrow MnSO_4 + 2HSO_4^- + Cl_2\uparrow + 2\,H_2O$	

Confirmatory Tests for Bromides

1	Heat the solution of the substance with manganese dioxide and sulphuric acid.	
	Brownish vapour turns starch iodide paper blue and starch paper yellow	Confirms the presence of bromides
	$MnO_2 + 3H_2SO_4 + 2\,Br^- \longrightarrow MnSO_4 + 2HSO_4^- + Br_2\uparrow + 2H_2O$	

Confirmatory Tests for Iodides

1	Heat the solution of the substance with manganese dioxide and sulphuric acid.	
	Violet vapour turns starch paper blue	Confirms the presence of iodide
	$MnO_2 + 3H_2SO_4 + 2I^- \longrightarrow MnSO_4 + 2HSO_4^- + I_2\uparrow + 2H_2O$	
2	To the solution of the substance add mercuric chloride solution.	
	Red precipitate of mercuric iodide formed	Confirms the presence of iodides
	$HgCl_2 + 2\,I^- \longrightarrow HgI_2\downarrow + 2\,Cl^-$	

Confirmatory Tests for Sulphides

1	**Sodium nitroprusside test:** To the solution of the substance add sodium nitroprusside solution.	
	Appearance of violet colour	Confirms the presence of sulphides
	$Na_2[Fe(CN)_5NO] + Na_2S \longrightarrow Na_4[Fe(CN)_5NOS]$	

Confirmatory Tests for Sulphates

1	**Barium chloride test**: To the solution of the substance add barium chloride solution.	
	White precipitate, which is insoluble in hydrochloric acid and nitric acid	Confirms the presence of sulphates
	$BaCl_2 + SO_4^{2-} \longrightarrow BaSO_4\downarrow + 2Cl^-$	

47.3 Detection of Basic Radicals

Preparation of the sample solution

Solution in water: Take a small quantity of sample in the test tube and add about 1/3rd of water and then boil. If the sample dissolves completely, the components are water soluble. In such a case the aqueous solution of mixture is used as such for analysis.

Solution in dilute hydrochloric acid: (If the sample is not completely soluble in water): Take small quantity of sample in a test tube and add 10 ml of dilute hydrochloric acid, boil it. If the sample dissolves completely, use this solution for analysis from group II onwards. If gas like CO_2 and H_2S evolved, boil to remove the gas completely.

Solution in hydrochloric acid: Take a small quantity of sample in a test tube and add 5 ml of concentrated hydrochloric acid. Boil, if any gas evolves boil to remove. Dilute after cooling, with water and use for further analysis.

Solution in aqua regia (3 parts of Con. HCl and 1 part of Con. HNO_3): If the sample is insoluble in all mineral acids, then make the solution with aqua regia. Take a small quantity of sample in a test tube and add 2-3 ml of aqua regia. Boil the contents, cool and then dilute with water. This solution is used for analysis.

If any insoluble compounds exits, filter it and observe the colour of that residue.

Observation	Inference
1. White	AgCl, $BaSO_4$, Al_2O_3
2. Yellow	CaF_2, $PbSO_4$, AgBr
3. Green	Chromium
4. Black	Mercuric
5. Brown	Fe_2O_3

Reagent	Observation	Characteristics
Group I: Mercurous, Lead, Gold, Silver		
Dilute hydrochloric acid	White precipitate of respective chlorides	Insoluble in water
$Ag^{+} + 2HCl \longrightarrow AgCl_2 \downarrow$	$Hg^{2+} + 2HCl \longrightarrow HgCl_2 \downarrow$	$Pb^{2+} + 2HCl \longrightarrow PbCl_2 \downarrow$
Group II: Mercuric, Copper, Lead, Bismuth, Cadmium, Arsenic, Strontium, Stannous		
Dilute hydrochloric acid and Hydrogen sulphide	Black precipitation of sulphides (HgS, CuS)	Insoluble in dilute acid
$Hg^{2+} + HCl + H_2S \longrightarrow HgS \downarrow$	$Cu^{2+} + HCl + H_2S \longrightarrow CuS \downarrow$	
Group IIIa: Aluminium, Ferric, Ferrous, Chromium		
Excess of ammonium chloride and ammonium hydroxide	• White precipitate [$Al(OH)_3$] • Reddish brown precipitate [$Fe(OH)_3$] • Dirty green precipitate [$Fe(OH)_3$]	Hydroxides insoluble in presence of ammonium chloride
$Fe^{3+} + NH_4Cl + 3NH_4OH \longrightarrow Fe(OH)_3 \downarrow$	$Al^{3+} + NH_4Cl + 3NH_4OH \longrightarrow Al(OH)_3 \downarrow$	
Group IIIb: Zinc, Manganese, Cobalt, Nickel		
Excess of ammonium chloride, ammonium hydroxide and ammonium/hydrogen sulphide	• Whiter precipitate - ZnS • Pinkish precipitate - MnS • Black precipitate - CoS • Black precipitate - NiS	Sulphides insoluble in presence of ammonium chloride and ammonium hydroxide
$Zn^{2+} + NH_4Cl + NH_4OH + (NH_4)_2S / H_2S \longrightarrow ZnS \downarrow$		
Group IV: Calcium, Barium, Strontium		
Excess of ammonium chloride, ammonium hydroxide and ammonium carbonate	White precipitate [$CaCO_3$, $BaCO_3$, $SrCO_3$]	Carbonates insoluble in presence of ammonium chloride and ammonium hydroxide
$Ca^{2+} + NH_4Cl + NH_4OH + (NH_4)_2CO_3 \longrightarrow CaCO_3 \downarrow$ $Ba^{2+} + NH_4Cl + NH_4OH + (NH_4)_2CO_3 \longrightarrow BaCO_3 \downarrow$		

Table Contd...

$Sr^{2+} + NH_4Cl + NH_4OH + (NH_4)_2CO_3 \longrightarrow SrCO_3\downarrow$		
Group V: Magnesium		
Excess of ammonium chloride, ammonium hydroxide and disodium hydrogenphoshate	White precipitate [$MgNH_4PO_4$]	Phosphates insoluble in presence of ammonium chloride and ammonium hydroxide
$Mg^{2+} + NH_4Cl + NH_4OH + Na_2HPO_4 \longrightarrow MgNH_4PO_4\downarrow$		
Group VI: Ammonium, Sodium, Potassium, hydrogen		
If no characteristic reaction in above cases, indicates that sample may be either of these four.		

47.4 Confirmatory Tests for Basic Radicals

Confirmatory Tests for Group I

Confirmatory tests for mercurous (Group I)		
1	Dissolve the sample in dilute hydrochloric acid and wash (several times) the precipitate with hot water. Pour 10-15 ml ammonium hydroxide on the precipitate and collect filtrate.	
	Black precipitate on filter paper	Confirms the presence of mercurous
2	Dissolve the black precipitate by boiling with 5 ml of hydrochloric acid and add crystals of potassium chlorate. Boil off all gases and dilute the solution.	
	Stannous chloride test: To the part of above solution add equal volume of stannous chloride solution.	
	White precipitate turns grey on standing	Confirms the presence of mercurous
	$Hg^+ + SnCl_2 \longrightarrow HgCl_2\downarrow$	
	Copper turning test: To the part of above solution add a few copper turnings and allow it to stand	
	Grey film of mercury becoming silvery on rubbing	Confirms the presence of mercurous
3	To the solution of the substance add potassium iodate solution.	
	Greenish yellow precipitate	Confirms the presence of mercurous
	$Hg^+ + KIO_3 \longrightarrow HgI\downarrow$	

Table Contd...

4	To the solution of substance add sodium hydroxide solution.	
	Black precipitate	Confirms the presence of mercurous
	$Hg^+ + 2NaOH \longrightarrow Hg(OH)_2\downarrow + Hg_2O + H_2O + 2Na^+$	
If there is no black precipitate on the filter paper, (confirms the absence of mercurous) carry out the following tests		
Confirmatory tests for silver (Group I)		
1	Acidify the part of filtrate with dilute nitric acid (until it shows red colour with blue litmus).	
	White precipitate turning grey on sun light	Confirms the presence of silver
	$Ag^+ + HNO_3 \longrightarrow AgNO_3\downarrow$	
2	**Potassium iodate test**: Add potassium iodate solution to the part of filtrate.	
	Pale yellow precipitate	Confirms the presence of silver
	$Ag^+ + KIO_3 \longrightarrow AgI\downarrow$	
Confirmatory tests for lead (Group I)		
1	Cool the filtrate (obtained previously) under tap water	
	White crystals settles down	Confirms the presence of lead
2	**Potassium iodide test**: To the hot filtrate add potassium iodide solution.	
	Yellow precipitate (dissolves on boiling)	Confirms the presence of lead
	$Pb^{2+} + 2KI \longrightarrow PbI_2\downarrow$	
3	**Potassium chromate test**: To the hot filtrate add potassium chromate solution.	
	Brilliant yellow precipitate (dissolves in sodium hydroxide)	Confirms the presence of lead
	$Pb^{2+} + K_2Cr_2O_7 \longrightarrow PbCr_2O_7\downarrow$	
4	To the solution of the substance add sodium hydroxide solution.	
	White precipitate (dissolves on excess of sodium hydroxide)	Confirms the presence of lead
	$Pb^{2+} + 2NaOH \longrightarrow Pb(OH)_2\downarrow$	
5	To the solution of the substance add dilute sulphuric acid.	
	White precipitate	Confirms the presence of lead
	$Pb^{2+} + H_2SO_4 \longrightarrow PbSO_4\downarrow$	

Confirmatory Tests for Group II

Heat the filtrate (group I) and pass hydrogen sulphide (H_2S) gas. Filter and pass the H_2S gas until no precipitation. Use this precipitate for the further analysis.

Confirmatory tests for mercury (Group II)		
1	Wash the precipitate with hot distilled water and pour it into 10 ml of ammonium sulphate. Transfer the contents in to boiling tube and heat at 50-60 °C for 4 min with shaking. Filter and preserve the filtrate for group II-B analysis and use the precipitate for the following tests.	
2	Transfer the part of precipitate to a beaker and add 15 ml dilute nitric acid. Boil for 3 min, cool and filter.	
	Formation of black precipitate	Confirms the presence of mercury
3	Dissolve the part of precipitate, in hydrochloric acid with a crystal of potassium chlorate. Boil off gases, dilute and divide into two parts	
4	To one part of above solution add stannous chloride solution	
	White precipitate turns grey on cooling	Confirms the presence of mercury
	$2HgCl_2 + 2SnCl_2 \longrightarrow Hg_2Cl_2\downarrow + SnCl_4$ $Hg_2Cl_2 + SnCl_2 \longrightarrow 2Hg\downarrow + SnCl_4$	
5	**Copper turning test:** To one part of above solution (obtained from expt. 3) add few copper turnings	
	Formation of white deposit (grey film of mercury becomes silver on rubbing)	Confirms the presence of mercury
	$Hg^{2+} + Cu \longrightarrow Cu^{2+}\downarrow + Hg$	
	If there is no characteristic reactions (confirms the absence of mercury), carry out the following tests To the part of filtrate add ammonium hydroxide (to make it alkaline), warm and filter. The appearance of the blue colour indicates the presence of copper.	
Confirmatory tests for copper (Group II)		
1	**Ferrocyanide test**: Acidify the part of filtrate/sample solution with acetic acid and add potassium ferrocyanide solution.	
	Appearance of chocolate red/pink precipitate	Confirms the presence of copper
	$2Cu^{+2} + K_4[Fe(CN)_6] \longrightarrow Cu_2[Fe(CN)_6]\downarrow + 4K^+$	

Table Contd...

2	To the solution of the substance add ammonium hydroxide.	
	Bluish white precipitate (copper hydroxide), turns black on warming	Confirms the presence of copper
	$Cu^{+2} + 2NH_4OH \longrightarrow Cu(OH)_2]\downarrow \xrightarrow{\Delta} CuO + H_2O$	
3	To the solution of the substance add potassium iodide solution.	
	Brown precipitate turns black on warming	Confirms the presence of copper
	$Cu^{+2} + 4KI \longrightarrow Cu_2I_2\downarrow + I_2 + 4K^+$	

Confirmatory Tests for Group III

Transfer the filtrate to a china dish and boil off H_2S (vapours give negative test with lead acetate paper). Add nitric acid, ammonium chloride and boil for 2 min. Add ammonium hydroxide to the hot solution collect the precipitate and use it for the further reactions. Preserve the filtrate for the analysis of next group.

Wash the precipitate with hot ammonium chloride solution (1%) and extract the precipitate with 10 ml of water and add sodium peroxide in small quantities. Stir and boil the contents for 5 min (to escape oxygen), cool and filter.

- Black precipitate indicates the presence of manganese
- Reddish-brown precipitate indicates the presence of ferric
- Yellowish precipitate indicates the presence of ferrous
- Yellowish filtrate indicates the presence of chromium

Confirmatory tests for aluminium (Group IIIa)		
1	Divide the filtrate into two parts	
2	To the filtrate/sample solution add solid ammonium chloride and boil.	
	Formation of white gelatinous precipitate	Confirms the presence of aluminium
3	**Lake test**: Acidify the filtrate/sample solution with dilute hydrochloric acid, add 2 drops of blue litmus solution and 2 drops of ammonium hydroxide, shake gently.	
	A crystalline blue precipitate floating in the clear solution	Confirms the presence of aluminium
4	To the solution of the substance add ammonium acetate and boil the mixture.	
	White gelatinous precipitate in hot, and disappears on cooling	Confirms the presence of aluminium
	$2Al^{3+} + 6CH_3COONH_4 \longrightarrow Al[CH_3COO]_3 + 6NH_4^+$	

Table Contd...

<table>
<tr><td colspan="3">Confirmatory tests for ferric (group IIIa)</td></tr>
<tr><td rowspan="4">1</td><td colspan="2">Dissolve the precipitate in dilute hydrochloric acid and use for the following tests</td></tr>
<tr><td colspan="2">To the part of precipitate/sample solution add potassium ferro cyanide solution.</td></tr>
<tr><td>Formation of prussian blue precipitate or colouration</td><td>Confirms the presence of ferric</td></tr>
<tr><td colspan="2">$4Fe^{3+} + 3K_4[Fe(CN)_6] \longrightarrow Fe_4[Fe(CN)_6]_3\downarrow + 12K^+$</td></tr>
<tr><td rowspan="2">2</td><td colspan="2">To the part of precipitate/sample solution add potassium sulpho cyanide solution.</td></tr>
<tr><td>Formation of blood red colouration</td><td>Confirms the presence of ferric</td></tr>
<tr><td rowspan="3">3</td><td colspan="2">To the solution of the substance add ammonium hydroxide.</td></tr>
<tr><td>Reddish brown gelatinous precipitate</td><td>Confirms the presence of ferric</td></tr>
<tr><td colspan="2">$Fe^{3+} + 3NH_4OH \longrightarrow Fe(OH)_3\downarrow + 3NH_4^+$</td></tr>
<tr><td colspan="3">Confirmatory tests for ferrous (group IIIa)</td></tr>
<tr><td rowspan="3">1</td><td colspan="2">To the solution of the substance add potassium ferricyanide solution.</td></tr>
<tr><td>Deep blue precipitate of ferrousferro cyanide (Turn's bull blue)</td><td>Confirms the presence of ferrous</td></tr>
<tr><td colspan="2">$3Fe^{2+} + K_3[Fe(CN)_6] \longrightarrow Fe_3[Fe(CN)_6]_2\downarrow + 3K^+$</td></tr>
<tr><td rowspan="3">2</td><td colspan="2">To the solution of the substance add ammonium hydroxide solution.</td></tr>
<tr><td>Formation of green gelatinous precipitate</td><td>Confirms the presence of ferrous</td></tr>
<tr><td colspan="2">$Fe^{2+} + 2NH_4OH \longrightarrow Fe(OH)_2\downarrow$</td></tr>
<tr><td colspan="3">Confirmatory tests for chromium (Group IIIa)</td></tr>
<tr><td rowspan="2">1</td><td colspan="2">Acidify the part of filtrate/sample solution with dilute acetic acid and add lead acetate solution.</td></tr>
<tr><td>Brilliant yellow precipitate (soluble in sodium hydroxide)</td><td>Confirms the presence of chromium</td></tr>
<tr><td rowspan="2">2</td><td colspan="2">Acidify the part of filtrate/sample solution with dilute hydrochloric acid. Add ether, add hydrogen peroxide solution and shake gently.</td></tr>
<tr><td>Appearance of blue colour on the ether layer</td><td>Confirms the presence of chromium</td></tr>
</table>

Pass the H_2S to the ammoniacal filtrate. Filter and add ammonia to make it alkaline, add H_2S until there is no further precipitation. Wash the precipitate with hot water and extract with 1N hydrochloric acid, shake and filter.

Boil the filtrate to remove sulphide and neutralize with sodium hydroxide solution. Add 10 ml sodium hydroxide, shake and filter.

White precipitate turning brown on the filter paper indicates presence of manganese.

Confirmatory tests for manganese (Group IIIb)		
1	Mix the part of precipitate with lead peroxide. Add concentrated nitric acid, heat, cool and dilute with water.	
	Formation of pink/violet colour (on settling insoluble matter)	Confirms the presence of manganese
	$2Mn^{2+} + 5PbO_2 + 5HNO_3 \longrightarrow 2HMnO_4 + 3Pb(NO_3)_2 + 2Pb^{2+} + 2H_2O$	
2	Dissolve the part of precipitate in nitric acid and add 0.5 ml of sodium bismuthate. Shake and allow it stand.	
	Formation of pink/purple colour (on settling insoluble matter)	Confirms the presence of manganese
	$2Mn^{2+} + 5BiO_3 + 14HNO_3 \longrightarrow 2HMnO_4 + 5Bi(NO_2)_3 + 7H_2O$	

Confirmatory tests for zinc (Group IIIb)		
1	Pass H_2S gas to the sample solution/filtrate.	
	Formation of white/dirty white precipitate	Confirms the presence of zinc
	$Zn^{2+} + H_2S \longrightarrow ZnS\downarrow$	
2	**Ferrocyanide test:** Acidify the sample solution/part of filtrate with hydrochloric acid and add potassium ferrro cyanide solution.	
	Formation of white/bluish white precipitate	Confirms the presence of zinc
	$2\,Zn^{2+} + K_4[Fe(CN)_6] \longrightarrow Zn_2[Fe(CN)_6]\downarrow + 4K^+$	
3	To the sample solution/part of filtrate add ammonium hydroxide solution.	
	Formation of white/bluish white precipitate	Confirms the presence of zinc
	$Zn^{2+} + 2NH_4OH \longrightarrow Zn(OH)_2\downarrow + 2NH_4^+$	

Confirmatory Tests for Group IV

Boil the filtrate with solid ammonium nitrate and evaporate the solution. Cool, make it alkaline with ammonia and add saturated solution of ammonium carbonate. Warm, filter and wash the precipitate with hot water and dissolve in dilute acetic acid. Boil off carbon dioxide, cool and add chromate solution drop wise, shake and filter.

Formation of yellow precipitate confirms the presence of barium.

Test the filtrate for calcium and strontium

Confirmatory tests for calcium (Group IV)		
1	To the sample solution/filtrate add ammonium oxalate solution.	
	Formation of white precipitate/suspension. [Sparingly soluble in dilute acetic acid and soluble in hydrochloric acid]	Confirms the presence of calcium
	$Ca^{2+} + (NH_4)_2C_2O_4 \longrightarrow CaC_2O_4\downarrow + 2NH_4^+$	
2	**Flame test:** Introduce the glass rod dipped in white precipitate formed above (or paste of sample made with hydrochloric acid) into the non-luminous flame.	
	Appearance of brick red flame	Confirms the presence of calcium
3	To the solution of the substance add ammonium carbonate solution.	
	Formation of white precipitate of calcium carbonate	Confirms the presence of calcium
	$Ca^{2+} + (NH_4)_2CO_3 \longrightarrow CaC_2O_4\downarrow + 2NH_4^+$	
4	To the solution of the substance add glacial acetic acid, ammonium chloride and potassium ferrocyanide solution.	
	Formation of white precipitate	Confirms the presence of calcium
	$Ca^{2+} + K_4[Fe(CN)_6] \longrightarrow K_2Ca[Fe(CN)_6]\downarrow + 2K^+$	

Confirmatory tests for barium (Group IV)		
1	To the solution of substance add dilute hydrochloric acid and dilute sulphuric acid.	
	Formation of white precipitate	Formation of white precipitate
	$Ba^{2+} + 2HCl \longrightarrow BaCl_2\downarrow + 2H^+$	

Table Contd...

2	To the solution of the substance add potassium chromate solution.	
	Formation of yellow precipitate of barium chromate (insoluble in water and dil. acetic acid)	Confirms the presence of barium
	$Ba^{2+} + 2K_2Cr_2O_7 \longrightarrow Ba(Cr_2O_7)_2\downarrow + 4K^+$	
3	To the solution of the substance add ammonium carbonate solution.	
	Formation of white precipitate of barium carbonate	Confirms the presence of barium
	$Ba^{2+} + (NH_4)_2CO_3 \longrightarrow CaCO_3\downarrow + 2NH_4^+$	
4	**Flame test:** Introduce the glass rod dipped in white precipitate formed above (or paste of sample made with hydrochloric acid) into the non-luminous flame.	
	Appearance of pale green flame	Confirms the presence of barium
5	To the solution of the substance add sodium hydroxide or ammonium hydroxide solution.	
	White precipitate of barium hydroxide	Confirms the presence of barium
	$Ba^{2+} + 2OH^- \longrightarrow Ba(OH)_2\downarrow + 2NH_4^+$	
6	To the solution of the substance add sodium phosphate solution.	
	White precipitate of barium phosphate	Confirms the presence of barium
	$Ba^{2+} + Na_2HPO_4 \longrightarrow BaHPO_4\downarrow + 2Na^+$	

Confirmatory Tests for Group V

Add 2 ml of ammonium oxalate to filtrate, boil, cool and filter (this treatment is not necessary if group IV is absent). Test the filtrate for the group V elements.

Add ammonium hydroxide and excess of ammonium phosphate solution to filtrate. Cool, shake and rub the walls of the test tube with glass rod. Allow it stand for 5 minutes and filter. Formation of white crystalline precipitate indicates the presence of magnesium.

Confirmatory tests for magnesium (Group V)		
1	**Magneson test:** To the precipitate suspended in water/to the sample solution add dilute hydrochloric acid and few drops of magneson reagent followed by an excess of sodium hydroxide.	
	Formation of flocculent blue precipitate/colouration	Confirms the presence of magnesium

Table Contd...

2	**Filter paper ash test:** To the filtrate/sample solution add drop of cobalt nitrate solution and dip a filter paper in it. Hold the paper in a flame until it burns away and leaves ash.	
	Formation of pink colour ash	Confirms the presence of magnesium
3	To the solution of the substance add sodium hydroxide solution.	
	Formation of white precipitate	Confirms the presence of magnesium
	$Mg^{2+} + 2NaOH \longrightarrow Mg(OH)_2\downarrow + 2Na^+$	
4	To the solution of the substance add ammonium hydroxide solution.	
	Formation of white gelatinous precipitate (soluble in excess of ammonium chloride)	Confirms the presence of magnesium
	$Mg^{2+} + 2NH_4OH \longrightarrow Mg(OH)_2\downarrow + 2NH_4^+$	
5	To the solution of the substance add hypoiodide solution.	
	Formation of reddish brown precipitate	Confirms the presence of magnesium
	$Mg^{2+} + 2\ HOI \longrightarrow Mg(OI)_2\downarrow + 2H^+$	

Confirmatory Tests for Group VI

Confirmatory tests for Potassium (Group VI)		
1	**Flame test:** Introduce the glass rod dipped in the paste of sample made with hydrochloric acid into the non-luminous flame	
	Formation of violet flame	Confirms the presence of potassium
2	To the solution of the substance add picric acid.	
	Formation of yellow precipitate	Confirms the presence of potassium
	$K^+ + C_6H_2(NO_2)_3OH \longrightarrow C_6H_2(NO_2)_3OK\downarrow + 2H^+$	
3	To the solution of the substance add perchloric acid.	
	Formation of white precipitate	Confirms the presence of potassium
	$K^+ + HClO_4 \longrightarrow KClO_4\downarrow + H^+$	
4	To the solution of the substance add sodium cobalt nitrite solution.	
	Formation of yellow precipitate	Confirms the presence of potassium
	$3K^+ + Na_3[Co(NO_2)_6] \longrightarrow K_3[Co(NO_2)_6]\downarrow + 3Na^+$	

Table Contd...

Confirmatory tests for ammonium (Group VI)		
1	**Nesslers test:** Boil the sample with sodium hydroxide. Pass the evolving ammonia gas to Nessler's reagent.	
	Formation of Reddish brown precipitate	Confirms the presence of ammonium
	$NH_4^+ + 4NaOH + 2K_2HgI_4 \longrightarrow NH_4HgOHgI\downarrow + 4KI + 3NaI + 3H_2O + Na^+$	
2	To the solution of the substance add picric acid.	
	Formation of Yellow precipitate of ammonium picrate	Confirms the presence of ammonium
	$NH_4^+ + C_6H_2(NO_2)_3OH \longrightarrow C_6H_2(NO_2)_3ONH_4\downarrow + H^+$	

CHAPTER 48

QUALITATIVE ANALYSIS OF ORGANIC COMPOUNDS

Qualitative organic analysis involves series of steps that help to establish the identity of the unknown organic compound. An unknown organic compound can be identified and reported for the following parameters.

1. Physical appearance
2. Solubility
3. Detection of melting point/boiling point
4. Detection of saturation
5. Detection of aromaticity
6. Detection of special elements
7. Detection of functional groups
8. Confirmatory tests for the functional groups
9. Preparation of derivatives and determination of their properties

Practical notes

- For most tests about 0.1 gm solid or 0.1-0.2 ml (2-3 drops) of liquid material (not more) should be used.
- Develop a general knowledge of the physical characteristics of common organic compounds.
- If in doubt about the expected result of a test between a compound and a reagent, carry out a trial test with a known compound and compare with the unknown.
- In general, 0.5-1 gm (or 0.5-1 ml) of substance gives the most satisfactory results in preparing derivatives.
- If a practical book instructs larger quantities (3-4 gm or more), the quantities should be scaled down to 1 gm or 1 ml of the unknown substance and corresponding quantities of reagents should be used.

48.1 Preliminary Tests

Experiment	Observation	Inference
Physical properties		
Physical state		
	1. Solid 2. Liquid	• Aromatic hydrocarbons, phenols, acids, amides may be present • Aromatic hydrocarbons, alcohols, aldehydes, ketones, esters, acids, anhydrides may be present
Colour		
	Solids • Pale yellow • Orange to red • Pink • Brown • Green • Colourless **Liquids** • Reddish brown • Orange to red	• Nitro compounds, iodoform • Nitro anilines, azo compounds • Naphthols • Diamines, aromatic amines and amino phenols • *P*-niroso phenols • Aldehydes, ketones, carbohydrates and alcohols • Amines • Nitro compounds
Odour		
	• Pleasant • Bitter almonds • Fishy • Pungent • Phenolic • Rotten eggs • Fruity smell • Garlic smell	• Ethers, aliphatic alcohols, chloroform • Benzaldehyde, nitro benzenes • Amines • Acetic acid, acetic anhydride, formic acid • Phenols and naphthols • Sulphur containing compounds • Esters • Thiophenols and thio alcohols
Physical Constants • Determine the boiling point in case of liquids or melting point in case of solids. • Distillation is recommended in case of liquids. (It serves the dual purpose of determining the boiling point as well as purification of the liquid for subsequent tests).		

48.2 Solubility Tests

Dissolve 0.1 gm/0.2 ml of sample in 3 ml of the solvent, shake thoroughly, if needed warm and cool, then check the solubility.			
1.	Sample + water	Completely soluble • Turns blue litmus red • Turns red litmus blue • Neutral to litmus	 • Water soluble acids and phenols • Water soluble bases (amines), alcohols, aldehydes and carbohydrates • Neutral compounds
	If acidic add a few drops 10% sodium- bicarbonate solution to the sample solution. • Evolution of carbon di-oxide by effervescence – indicates the presence of carboxylic acids and nitro phenols. • No evolution of carbon di-oxide - indicates the presence of phenols.		
2.	Sample + 5% HCl	Completely soluble in 5% HCl and reappears by addition of Conc. NaOH	Most amines (except 3° amines with only aromatic groups)
3.	Sample + Cold H_2SO_4	Completely soluble in Cold Conc. H_2SO_4	Aromatic nitro compounds, ethers, esters and anhydrides
4.	Sample + 5% NaOH	• Completely soluble in 5% NaOH and reappears by addition of Conc. HCl • Soluble in hot condition and turns into brown colour • Odour of alcohol	• Acids, anhydrides and phenols • Carbohydrates • Esters
5.	Sample + 5% $NaHCO_3$	• Compound dissolves with effervescence • No effervescence	• Most carboxylic acids • Phenols
6.	Insoluble in water, acid and alkali	Neutral	• Hydrocarbons, nitro-hydrocarbons • Alkyl/aryl halides, esters and ethers • Higher molecular weight alcohols, aldehydes and ketones

48.3 Test for Saturation

1.	**Baeyer test**		
	Dissolve 0.2 gm/0.2 ml of the sample in 2 ml of water or suitable solvent (acetone may also be used as solvent). Add 10 ml of Na_2CO_3 and then add 2% $KMnO_4$ drop by drop.	• Decolourisation • No decolourisation	• Unsaturated compound • Saturated compound
	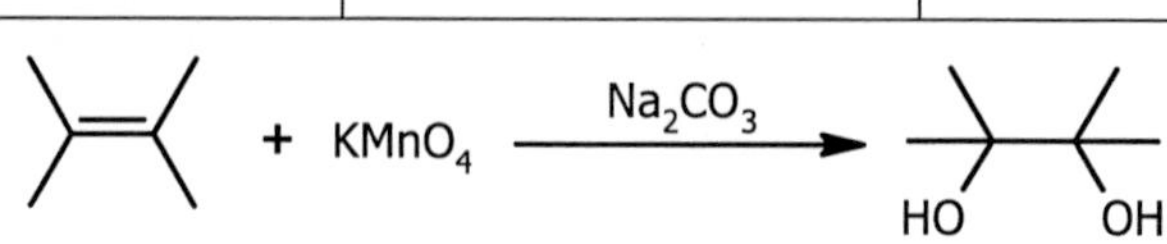		

Table Contd...

2.	**Bromine test**	
	Dissolve 0.2 gm (or 0.2 ml) of the compound in 2 ml of carbon tetrachloride or another suitable solvent and add the solution dropwise to 2 ml of 2% bromine solution in carbon tetrachloride and shake.	
	Rapid disappearance of the bromine colour (yellow)	Unsaturated compound
	No decolourisation with evolution of hydrogen bromide (HBr) gas	Saturated compound
	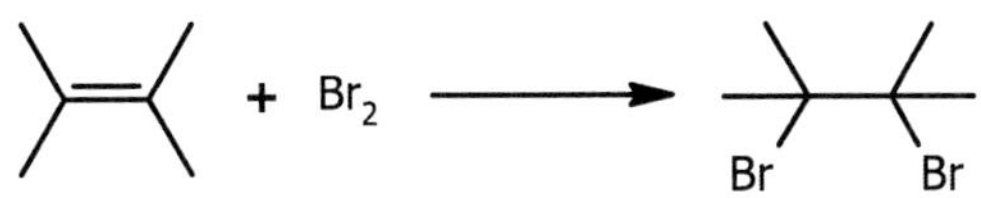	
	This reagent is very dangerous, hence protect skin, eyes and nose.	

48.4 Test for Aromaticity

1.	**Ignition test** Burn small amount of sample on metal spatula.	
	• Non-smoky flame • Smoky flame • Charing	• May be aliphatic • May be aromatic • May be carbohydrates
2.	**Nitration test**	
	Heat the sample with a mixture of equal volume of sulphuric acid and nitric acid (1:1). Pour this solution into beaker containing water.	
	Yellowish solution/precipitate	Indicates the presence of aromatic compounds

48.5 Qualitative Analysis for Elements

In organic compounds the elements commonly occurring along with carbon and hydrogen, are oxygen, nitrogen, sulphur, chlorine, bromine and iodine. The detection of these elements depends upon converting them to water-soluble ionic compounds (NaX, NaCN, Na_2S and NaCNS) and the application of specific tests.

Lassaigne's sodium fusion test: Place a piece of clean sodium metal and add a little of the given compound (50 mg or 2-3 drops) (For liquids, it is better to first melt the sodium and then add the liquid drop by drop). Heat the tube gently at first, allowing any distillate formed to drop back onto the molten sodium. When charring begins, heat the bottom of the tube to dull redness for about 3 minutes and finally plunge the tube, while still hot, into a clean dish containing cold distilled water (15 ml).

Stir the mixture, boil for 15 minutes and remove the glass splinters by filtration. The filtrate is called as "**sodium fusion extract**". The extract should be clear and colourless, and it is used for the specific tests described below:

Caution: The tube shatters, and any residual sodium metal reacts with water.

Test for Nitrogen		
	To a 1 ml of the extract add 2 drops of freshly prepared ferrous sulphate solution. Add 1 drop of NaOH solution and 2 drops of $FeCl_3$ solution. Boil the mixture for half a minute, cool and acidify by adding dilute sulphuric acid dropwise.	
	Formation of a bluish-green precipitate (prussian blue) or a blue solution	Indicates the presence of nitrogen
	If no precipitate appears, allow to stand for 15 minutes, filter and inspect filter paper.	
	It is essential to use excess of sodium, otherwise if sulphur and nitrogen both are present, sodium thiocyanate (NaSCN) may form (gives red-ferric ion). So it willn't respond to cyanide or sulphide ion tests. With excess of sodium the sodium thiocyanate, formed will be decomposed into sodium cyanide and sulphur.	
	$\underset{\text{Sodium thiocyanate}}{NaCNS} + \underset{\text{Sodium}}{2\,Na} \longrightarrow \underset{\text{Sodium cyanide}}{NaCN} + \underset{\text{Sodium sulphide}}{Na_2S}$	
	$\underset{\text{Sodium cyanide}}{6\,NaCN} + \underset{\text{Ferrous sulphate}}{FeSO_4} \longrightarrow \underset{\text{Sodium ferrocyanide}}{Na_4[Fe(CN)_6]} + Na_2SO_4$	
	$6\,Na_4[Fe(CN)_6] + 2\,Fe_2(SO_4)_3 \longrightarrow \underset{\text{Ferric ferro cyanate (prussian blue)}}{Na_4[Fe(CN)_6]_3} + 6\,Na_2SO_4$	

Test for Sulphur		
1.	**Sodium nitroprusside test**	
	To the 1 ml of the cold extract add 1 drop of dilute sodium hydroxide and 4 drops of cold, freshly prepared, dilute solution of sodium nitroprusside.	
	Development of deep red or purple colour	Indicates the presence of sulphur (This test is very sensitive. Only strong positive results are significant)
	$\underset{\text{Sodium sulphide}}{Na_2S} + \underset{\text{Sodium nitroprusside}}{Na_2[Fe(CN)_5]\,NO} \longrightarrow Na_4\,[Fe(CN)_5]\,NOS$	
2.	**Lead acetate test**	
	Acidify 1 ml of the extract with acetic acid and add 2 drops of lead acetate solution.	
	Formation of black precipitate	Indicates the presence of sulphur
	$Na_2S + 2\,CH_3COOH \longrightarrow 2\,CH_3COONa + H_2S$ $H_2S + (CH_3COO)_2\,Pb \longrightarrow PbS + 2\,CH_3COOH$	

Table Contd...

<table>
<tr><th colspan="3">Test for Halogens</th></tr>
<tr><td colspan="3">If a positive result is obtained: acidify the remaining portion of the 'fusion' filtrate with dilute sulphuric acid, boil and cool. Add carbon tetrachloride (1 ml) and a few drops of freshly prepared chlorine water. Shake the mixture.</td></tr>
<tr><td rowspan="8">Acidify 1 ml of the extract with 2 drops of dilute nitric acid, and if nitrogen and/or sulphur are present*, boil for 1-2 minutes to expel H_2S/HCN.
Cool and add 2 ml of aqueous silver nitrate, compare with a blank.</td><td>Formation of a white or yellow precipitate</td><td>Indicates the presence of halogens</td></tr>
<tr><td colspan="2">$$\underset{\text{Sodium halide}}{NaX} + \underset{\text{Silver nitrate}}{AgNO_3} \xrightarrow{\text{Dilute } HNO_3} \underset{\text{Silver halide}}{AgX} + NaNO_3$$</td></tr>
<tr><td>Formation of white precipitate, soluble in ammonium hydroxide</td><td>Indicates the presence of chlorides</td></tr>
<tr><td colspan="2">$$\underset{\text{Sodium Chloride}}{NaCl} + \underset{\text{Silver nitrate}}{AgNO_3} \xrightarrow{\text{Dilute } HNO_3} \underset{\text{Silver chloride}}{AgCl} + NaNO_3$$</td></tr>
<tr><td>Formation of pale yellow precipitate, soluble in excess of ammonium hydroxide</td><td>Indicates the presence of bromides</td></tr>
<tr><td colspan="2">$$\underset{\text{Sodium bromide}}{NaBr} + \underset{\text{Silver nitrate}}{AgNO_3} \longrightarrow \underset{\text{Silver bromide}}{AgBr} + NaNO_3$$</td></tr>
<tr><td>Formation of yellow precipitate, insoluble in ammonium hydroxide</td><td>Indicates the presence of iodides</td></tr>
<tr><td colspan="2">$$\underset{\text{Sodium iodide}}{NaI} + \underset{\text{Silver nitrate}}{AgNO_3} \xrightarrow{\text{Dilute } HNO_3} \underset{\text{Silver iodide}}{AgI} + NaNO_3$$</td></tr>
<tr><td colspan="3">* If nitrogen and/or sulphur are also present, the addition of silver nitrate to the acidified 'fusion' solution will precipitate silver cyanide and/or silver sulphide in addition to the silver halides. The removal of hydrogen cyanide and/or hydrogen sulphide is effected by boiling the 'fusion' solution.</td></tr>
</table>

<table>
<tr><th colspan="4">Test for Nitrogen and Sulphur</th></tr>
<tr><td>1</td><td>To the 1 ml of the extract add 2 drops of $FeCl_3$ solution</td><td>Formation of blood red colour</td><td>Indicates the presence of nitrogen and sulphur</td></tr>
<tr><td>2</td><td>To the 1 ml of the extract add 2 ml of alcohol and 2 drops of cobalt nitrate solution</td><td>Formation of blue colour</td><td>Indicates the presence of nitrogen and sulphur</td></tr>
</table>

Group classification tests

- It is strongly advised against carrying out unnecessary tests, since not only are they a waste of time but also increase the possibility of error.
- Thus it is pointless to first test for alcohol or ketone in a basic compound containing nitrogen.
- Instead tests for amines, etc., should be done on such a compound.
- It should be noted that whilst two substances with the same functional group may sometimes have very similar *b.p.* or *m.p.,* solid derivatives can usually be chosen from the literature, with *m.p.* differences of about 10 °C (or more), which distinguish between the two possibilities.

Example:

Compound	**b.p. (°C)**	**Derivatives [m.p (°C)]**	
		2,4-DNPH	**Semicarbazone**
Diethyl ketone	102	156	139
Methyl *n*-propyl ketone	102	144	112

2, 4-DNPH: 2,4-Dinitro Phenyl Hydrazine

48.6 Test for Functional Groups

Tests for Carboxylic Acids			
1.	**Litmus test**		
	Test the sample solution with litmus paper	Blue litmus paper turns into red colour	Indicates the presence of acids (COOH/OH)
2.	To the sample add 2 ml of 5% KI and KIO_3 in test tube. Boil for 2 minutes and then add drop of starch	Appearance of blue colour	Indicates the presence of acids
3.	**Test with 5% aqueous $NaHCO_3$**		
	To the sample add 1 ml of 5% aqueous $NaHCO_3$	Brisk effervescence of CO_2	Indicates the presence of acids (Distinction from phenol)
	$R{-}C({=}O){-}OH + NaHCO_3 \longrightarrow R{-}C({=}O){-}ONa + CO_2 + H_2O$ If the acid is insoluble in water and the reaction is sluggish, dissolve the acid in methanol and add carefully to a saturated sodium hydrogen carbonate solution, when a vigorous effervescence will be observed.		

Table Contd...

No.			
4.	**Distinguish tests among different carboxylic acid**		
	Dissolve 0.5 gm of sample in a little water, add dil. NaOH solution slowly until the solution is just alkaline to phenolphthalein. Then make just acidic with very dil. HCl and add slight excess of dil. HNO_3, boil to remove the excess of NH_3. Test with litmus paper for the neutral nature.		
	Add 3 drops of $FeCl_3$ solution to the neutralized solution, boil if necessary.	• Reddish brown precipitate • Violet colour • Faint yellow colour\ • Buff coloured precipitate	• Acetic acid • Salicylic acid • Oxalic acid • Benzoic acid/cinnamic acid
5.	**Distinction between aliphatic and aromatic acids**		
	Soda lime test		
	Mix sample and sodalime (1:2), heat in a dry a test tube	A gas burns with smoky flame	Indicates the presence of aromatic acid (distinction from aliphatic acids)
	Esterification test		
	Warm a mixture of 1 part of sample, 2 parts of absolute ethanol and 1 part of Conc. H_2SO_4 for 2 minutes. Cool and pour into beaker containing aqueous sodium hydroxide solution		
	• Sweet fruity odour • Gum like odour • Apple-fruity odour • Mint odour • Banana odour • Balmy odour	• Precence of COOH • Formic acid • Acetic acid • Benzoic acid • Acetic acid • Salicylic acid	
	$R\text{-}COOH + R\text{-}OH \xrightarrow[NaOH]{H^+} R\text{-}COO\text{-}C_2H_5$		
6.	**Test for sulphonic acid and sulphinic acid**		
	Dissolve a small amount of compound in cold Conc. H_2SO_4 and add 1 drop of phenetole or		
	Development of blue colour of *p*-substituted aromatic sulphoxide	Indicates the presence of aromatic sulphonic and sulphinic acids	
7.	**Fluorescein test**		
	Mix both substance and resorcinol and add 4 drops of sulphuric acid in a test tube. Heat for 5 minutes till deep red solution is obtained. Pour it into large volume of dilute sodium hydroxide		
	Intense greenish-yellow fluorescence	Presence of di-carboxylic acid	
	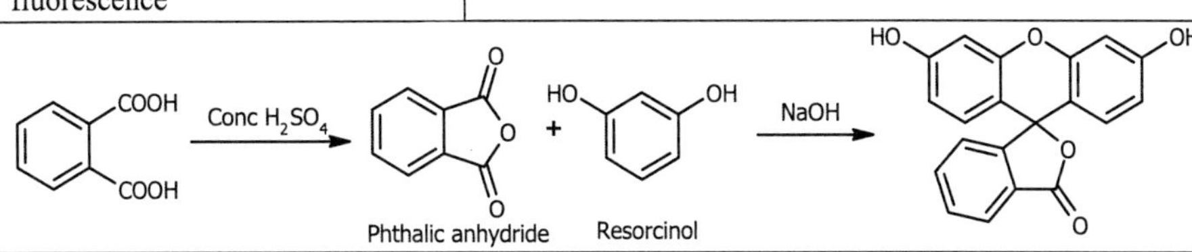		

	Tests for Alcohols		
1.	To the solution of the substance add piece of dry sodium metal and heat in a water bath	Brisk effervescence of H_2 gas evolves out	Indicates the presence of alcohol
2.	**Ceric ammonium nitrate test**		
	To the solution of the substance add 2 drops of ceric ammonium nitrate solution	• Red colour • Green colour	• Indicates the presence of alcohol • Indicates the presence of phenols
	$R\text{-}OH + (NH_4)_2Ce(NO_3)_6 \longrightarrow (NH_4)_2Ce(OR)(NO_3)_5 + HNO_3$ Ceric ammonium nitrate (Yellow) → (Red)		
3.	**Esterification test**		
	Warm 2 parts of sample, part of dilute acetic acid and 1 part of Conc. H_2SO_4 for 2 minutes. Cool and pour into beaker containing aqueous sodium carbonate solution and smell.		
	Sweet fruity odour	Indicates the presence of alcohol	
	$R\text{-}COOH + R\text{-}OH \xrightarrow[NaOH]{H^+} R\text{-}COO\text{-}C_2H_5$		
4.	**Xanthate test (carbon disulphide reagent test)**		
	To the solution of the substance add NaOH pellets, dissolve and cool. Then add 5 drops of carbon disulphide and shake	Forms yellow precipitate of thio carbamate	Indicates the presence of alcohol
	$R\text{-}OH + NaOH \longrightarrow R\text{-}OK + H_2O$ $R\text{-}OK + CS_2 \longrightarrow$ 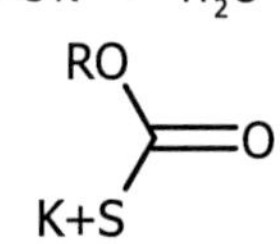 Yellow xanthate		
5.	**Oxidation test**		
	Heat copper wire in a non luminous flame and introduce into the solution of sample in fusion tube 2-3 times. Divide this into 2 parts.		
	To one part add few drops of Schiff reagent	Violet/purple colour	Methyl alcohol and ethyl alcohol
	To the 2nd part add few drops of 2,4-dinitro phenylhydrazine	Crystalline yellow precipitate • Odour of oil of bitter almond	Carbonyl group • Benzyl alcohol

Table Contd...

Differentiation between 1°, 2° and 3° alcohols		
6.	**Lucas test**	
	Lucas reagent (anhydrous $ZnCl_2$ + Conc. HCl) converts alcohols into the corresponding alkyl chlorides. Zinc chloride (a Lewis acid) increases the reactivity of alcohols towards acid. The test depends on the rate of reaction of primary, secondary, and tertiary alcohols with the reagent at room temperature.	
	To the sample add 2 ml of Lucas reagent (anhydrous $ZnCl_2$ + Conc. HCl). Shake the mixture and keep it aside for 5 minutes.	
	• Immediate separation of two phases • Cloudy in 3-5 minutes • Solution remains clear	• Tertiary alcohols • Secondary alcohols • Primary alcohols
	Primary alcohols - the aqueous phase remains clear. $R\text{-}CH_2OH \xrightarrow[HCl]{ZnCl_2}$ No reaction at room temperature	
	Secondary alcohols - very slow reaction. (1 hour or maybe longer) $R_2\text{-}CH\text{-}OH \xrightarrow[HCl]{ZnCl_2} R_2\text{-}CH\text{-}Cl + H_2O$	
	Tertiary alcohols - very fast reaction and droplets of the alkyl chloride formed almost immediately. $R_3\text{-}C\text{-}OH \xrightarrow[HCl]{ZnCl_2} R_3\text{-}C\text{-}Cl + H_2O$ (immediately)	
7.	**Jones Reagent (CrO_3-H_2SO_4 in H_2O)**	
	Place 1 ml of acetone in a test tube and dissolve one drop of a liquid or 10 mg of a solid in it. Add one drop of the reagent to the acetone solution and shake the tube to mix the contents.	
	Formation of a green or blue-green precipitate or emulsion	Indicates the presence of alcohols

Table Contd...

No.	Test / Observation	Inference
	Primary and secondary alcohols react within two seconds as indicated by the disappearance of the orange colour of the reagent and the formation of a green or blue-green precipitate or emulsion. $R\text{-}CH_2OH \xrightarrow{(O)} R\text{-}CHO \xrightarrow{(O)} R\text{-}COOH$ $R_2\text{-}CH\text{-}OH \xrightarrow{(O)} R_2\text{-}C{=}O$ Tertiary alcohols do not react even after 3 minutes. $R_3\text{-}C\text{-}OH \xrightarrow{(O)}$ No visible reaction	
	Test for Phenols	
1.	**Litmus paper test**	
	Test the sample solution with litmus paper.	
	Blue litmus paper turns into red colour	Indicates the presence of COOH/phenolic OH
2.	To the sample solution add 5 ml of sodium hydroxide solution.	
	Sodium hydroxide dissolves the sample	Indicates the presence of phenol OH group
3.	To the sample solution add 5 ml of sodium bicarbonate solution.	
	Evolution of no carbondi-oxide	Indicates the presence of phenol OH group
4.	**Neutral $FeCl_3$ test**	
	To 0.1% aqueous/alcoholic solution of sample add 1 drop of ferric chloride ($FeCl_3$) solution.	
	• Green/blue/violet colour • White to violet precipitate • Green to white precipitate • Bluish white precipitate • Greenish black/transient blue precipitate • Red precipitate	• Presence of phenols • Presence of α-naphthol • Presence of β-naphthol • Presence of resorcinol • Presence of hydroquinone • Presence of pyrogallol (presence of enol)
5.	**Bromine water test**	
	To the aqueous/acidic solution of the sample add excess of bromine dropwise until colour persists after shaking.	
	• White precipitate • Yellowish needle like crystals	• Presence of phenol/resorcinol • Presence of α-naphthol

Table Contd...

	Phenols are generally highly reactive towards electrophilic reagents and are readily brominated by bromine water. 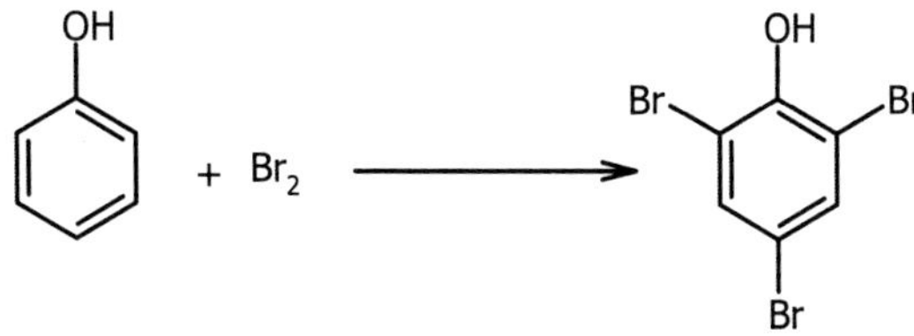	
6.	**Liebermann's nitroso test**	
	Heat sodium nitrite crystals and sample for 1 minute, cool and add 1 ml of sulphuric acid and shake.	
	• Green to blue colouration. – Dilution with water turns red – Basification gives original colour	• Indicates the presence of phenols
	• Deep blue/green colour	• *o*-Cresol/m-Cresol/Resorcinol/Catechol.
	• Dirty brown colour	• *p*-Cresol
	• Green colour	• α-Naphthol
	• Black colour	• β-Naphthol
	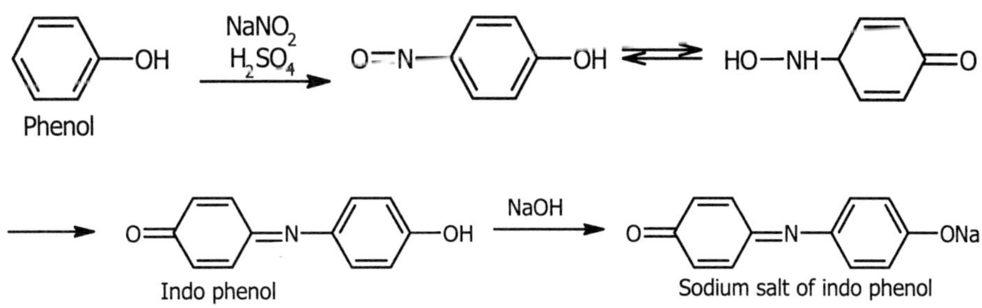	
7.	**Phthalein test**	
	Moisten phenol and phthalic anhydride with 3 drops of Conc. H_2SO_4. Heat it and cool, then add to a beaker containing 10% NaOH.	
	• Red colouration	• Phenol/*o*-Cresol
	• Blue colouration	• *m*-Cresol/Pyrogallol
	• Yellowish red with green fluroscence	• Resorcinol
	• Green colouration	• α-naphthol
	• Faint green with fluroscence	• β-naphthol

Table Contd...

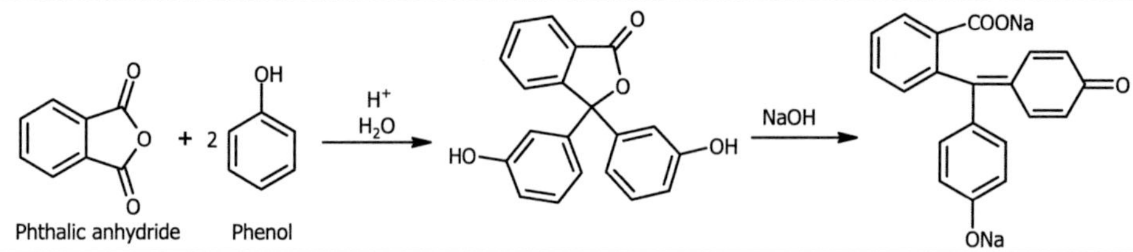

	Tests for Carbonyl Compounds	
1.	**2,4-Dinitrophenylhydrazine test (Brady's reagent)**	
	2,4-Dinitrophenylhydrazine gives sparingly soluble yellow or red 2,4-dinitrophenylhydrazones with aldehydes and ketones.	
	Add 3 ml of the reagent to 2 drops of the compound in a test tube and shake. If no precipitate forms immediately, warm and allow to stand for 5-10 minutes.	
	Yellow/reddish crystalline precipitate	Indicates presence of carbonyl compounds
	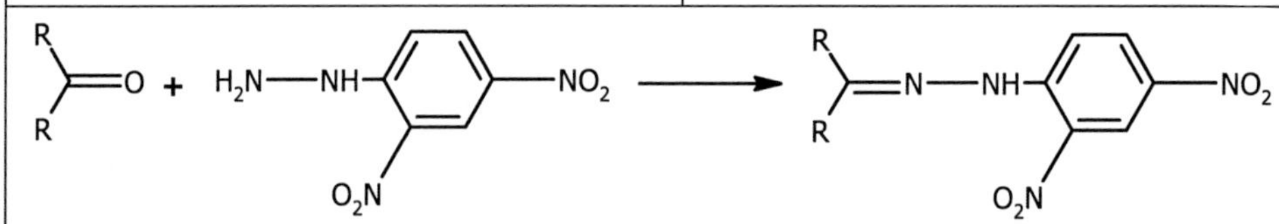	
	Tests for aldehydes	
1.	**Tollen's reagent test (silver mirror test)**	
	Place 2 ml of Tollen's reagent (ammoniacal silver nitrate, $[Ag(NH_3)_2NO_3]$ and 2 ml sample solution in a test tube and boil for 5-6 minutes and cool.	
	Appearance of black color or silver mirror on the sides of the test tube	Indicates the presence of aldehyde
	Aldehydes are readily oxidized to carboxylic acids and will reduce tollen's reagent to produce a silver mirror on the sides of the clean test tube.	

Table Contd...

<table>
<tr><td>2.</td><td colspan="3">Fehling's test</td></tr>
<tr><td></td><td>To 2 drops/0.05 gm of the compound, add 3 ml of equal volume of fehling's solution I and II. Warm the solution in water bath</td><td>Yellowish red precipitate of cuprous oxide-Cu_2O</td><td>Indicates the presence of aldehyde</td></tr>
<tr><td></td><td colspan="3">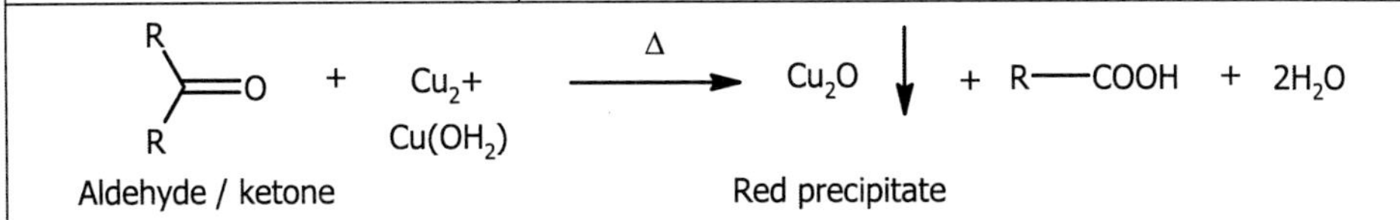
</td></tr>
<tr><td>3.</td><td colspan="3">Schiff's reagent test- Caro's test (majenta solution decolourized by SO_2)</td></tr>
<tr><td></td><td colspan="3">Schiff's reagent is a solution of p-rosaniline hydrochloride dissolved by passing SO_2</td></tr>
<tr><td></td><td>To 0.2 gm of the compound, add 3 ml of schiff's reagent and shake the mixture in the cold condition</td><td>• Pink or purple colour instantaneously, due to the displacement of sulphurous acid
• Pink or purple colour on standing</td><td>• Indicates the presence of aliphatic aldehyde
• Indicates the presence of aromatic aldehyde</td></tr>
<tr><td></td><td colspan="3">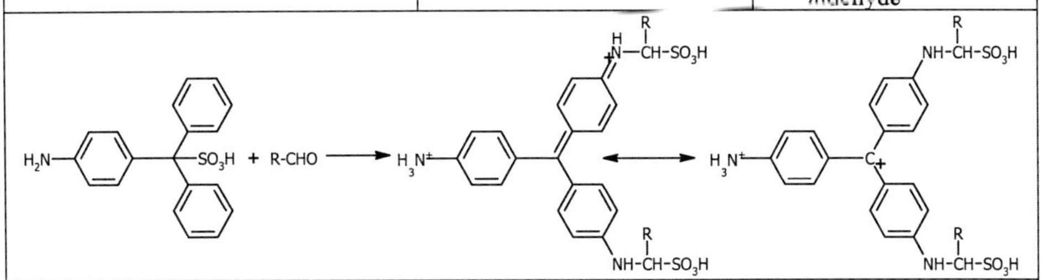
</td></tr>
<tr><td>5.</td><td colspan="3">Benedict's test</td></tr>
<tr><td></td><td>Boil the sample with 2 ml of benedict's reagent for few minutes</td><td>Yellowish red precipitate of cuprous oxide-Cu_2O is formed</td><td>Indicates the presence of aldehyde</td></tr>
<tr><td></td><td colspan="3">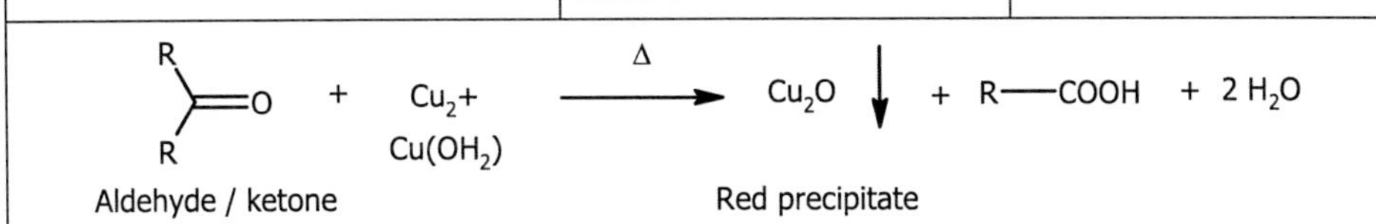
</td></tr>
<tr><td>6.</td><td colspan="3">Jones reagent (CrO_3-H_2SO_4 in H_2O)</td></tr>
<tr><td></td><td colspan="3">Place 1 ml of acetone in a test tube and dissolve one drop of a liquid or 10 mg of a solid sample in it. Add one drop of the reagent to the acetone solution and shake the tube to mix the contents.</td></tr>
<tr><td></td><td colspan="2">Formation of a green or blue-green precipitate or emulsion</td><td>Indicates the presence of aldehydes</td></tr>
</table>

<table>
<tr><th colspan="3">Test for Ketones</th></tr>
<tr><td rowspan="4">1.</td><td colspan="2">Iodoform test</td></tr>
<tr><td colspan="2">To 0.2 gm of the substance add 2 ml of water/methanol/dioxane, 1 ml of 10% sodium hydroxide (NaOH) and I_2-KI reagent drop wise, until dark colour of iodine persists. Allow to stand for 2-3 minutes; if no iodoform separates at room temperature, warm the test tube in a beaker of water at 60 °C. Add few more drops of the iodine reagent if the faint iodine colour disappears. Continue the addition of the reagent until a dark colour is not discharged after 2 minutes heating at 60 °C. Remove the excess of iodine by the addition of a few drops of dilute sodium hydroxide solution with shaking, dilute with an equal volume of water, and allow to stand for 10 minutes.</td></tr>
<tr><td>Yellow precipitate of iodoform</td><td>Indicates the presence of ketone (methyl ketone)</td></tr>
<tr><td colspan="2">The reaction is given by acetaldehyde and simple methyl ketones. Alcohols containing the CH_3-CH-R-OH group will be oxidized under the reaction conditions and also give a positive test.
$$NaOH + I_2/KI \longrightarrow NaOI$$
$$\underset{\text{Ketone}}{R-\overset{O}{\overset{\|}{C}}-CH_3} + NaOI \longrightarrow R-\overset{O}{\overset{\|}{C}}-CI_3 + NaOH$$
$$R-\overset{O}{\overset{\|}{C}}-CI_3 + NaOH \longrightarrow R-\overset{O}{\overset{\|}{C}}-OH + \underset{\text{yellow precipitate}}{CHI_3}$$</td></tr>
<tr><td rowspan="3">2.</td><td colspan="2">Sodium nitroprusside test</td></tr>
<tr><td colspan="2">To the sample add sodium nitroprusside solution and excess of 30% sodium hydroxide</td></tr>
<tr><td>Orange colour changes to pink/purple colour (formation of complex)</td><td>Indicates the presence of ketone (methyl ketone)</td></tr>
<tr><td rowspan="3">3.</td><td colspan="2">Zimmerman reaction (m-di-nitro benzene test)</td></tr>
<tr><td colspan="2">To the sample add powdered m-dinitro benzene and add excess of NaOH solution and shake.</td></tr>
<tr><td>• Red colour develops immediately
• Light red colour on standing
• Yellow turns to red on standing</td><td>• Methyl ketone
• Benzophenone/benzaldehyde
• Salicyladehyde</td></tr>
</table>

<table>
<tr><th colspan="4">Test for Esters And Anhydrides</th></tr>
<tr><td rowspan="5">1.</td><td colspan="3">Hydroxamic acid test</td></tr>
<tr><td colspan="3">Esters react with hydroxylamine in the presence of sodium hydroxide to form the sodium salt of the corresponding hydroxamic acid. On acidification and addition of ferric chloride the magenta-coloured iron (III) complex of the hydroxamic acid is formed.</td></tr>
<tr><td colspan="3">To 0.2 gm of the sample add 2 gm of hydroxylamine hydrochloride ($NH_2OH.\ HCl$) in ethanol. Then add saturated solution of KOH in alcohol until mixture becomes alkaline to litmus. Boil for 2 minutes, cool and acidify with HCl then add 2 drops of $FeCl_3$.</td></tr>
<tr><td>Violet/wine colour</td><td colspan="2">Indicates the presence of anhydrides and esters</td></tr>
<tr><td colspan="3">$R{-}C({=}O){-}OR + NH_2OH \longrightarrow R{-}C({=}O){-}NH{-}OH + R{-}OH$</td></tr>
<tr><td>2</td><td>To the solution of the sample add a drop of phenolphthalein and add NaOH drop by drop, then heat</td><td>Odour and colour vanishes</td><td>Indicates the presence of anhydrides and esters</td></tr>
</table>

<table>
<tr><th colspan="4">Test for Ethers</th></tr>
<tr><td>1.</td><td>To 1 ml of iodine in carbon disulphide (CS_2) add 0.5 ml of the substance</td><td>Purple colour changes to tan colour</td><td>Indicates the presence of ethers</td></tr>
</table>

<table>
<tr><th colspan="3">Test for Polyhydric Alcohols And Carbohydrates</th></tr>
<tr><td rowspan="4">1.</td><td colspan="2">Molisch's test (General test for carbohydrates)</td></tr>
<tr><td colspan="2">Place 1 ml of the carbohydrate solution in a clean test tube. Add 2 drops of molisch reagent (alcoholic α-naphthol, 10%) and mix well. Incline the test tube and slowly add 0.5 ml of concentrated sulfuric acid in drops along the sides of the test tube.</td></tr>
<tr><td>A red colouration, changing to dark purple forms at the interface</td><td>Indicates the presence of carbohydrates</td></tr>
<tr><td colspan="2">The concentrated sulphuric acid dehydrates pentoses and hexoses to form furfural and 5-hydroxymethyl furfural. These furfural derivatives condense with alcoholic α-naphthol (molisch reagent) to form violet colored products.</td></tr>
</table>

Table Contd...

	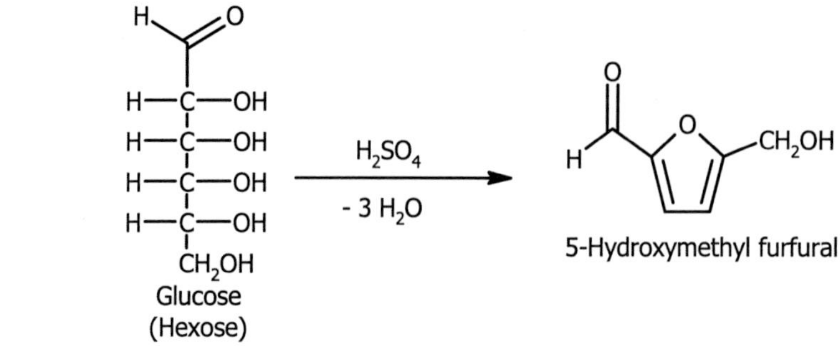 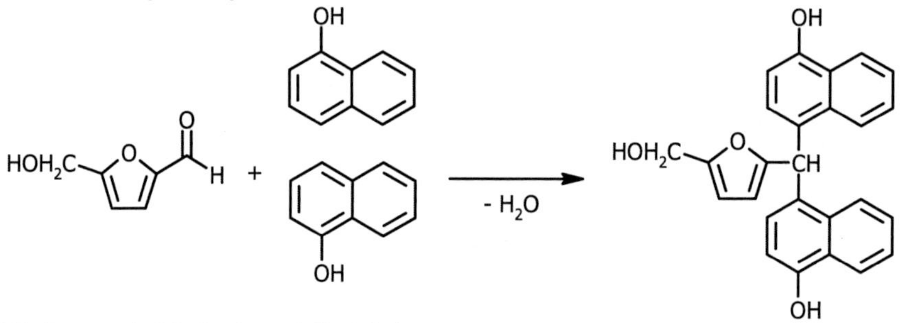 5 Hydroxymethyl furfural 1 Naphthol Furfuryl-diphenyl-methane dye **Note:** Concentrated solution of organic compounds may give a red colour instead violet due to the charing action of the sulphuric acid. In case of doubt the experiment should be repeated on a more diluted solution of the substance to be tested.		
2.	**Fehling's solution test**		
	To 2 drops/0.05 gm of the compound, add 3 ml of equal volume of Fehling's solution I and II. Warm the solution in water bath		
	Yellowish red precipitate of cuprous oxide-Cu_2O is formed	Indicates the presence of aldehyde (reducing sugars)	
	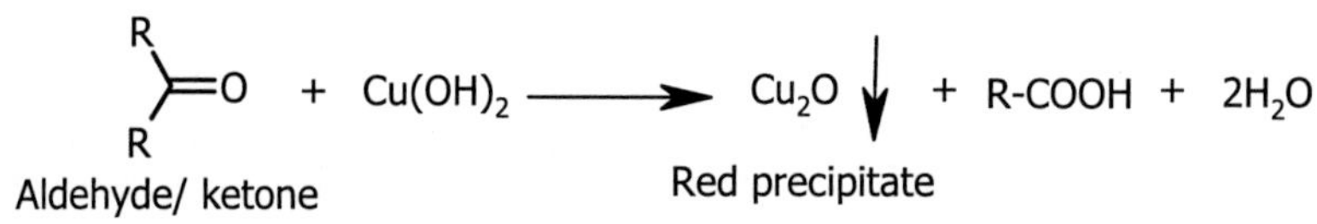		
3.	**Benedict's test**		
	Boil the sample with 2 ml of Benedict's reagent for few minutes	Yellowish red precipitate of cuprous oxide-Cu_2O is formed	Indicates the presence of aldehyde (reducing sugars)

Table Contd...

	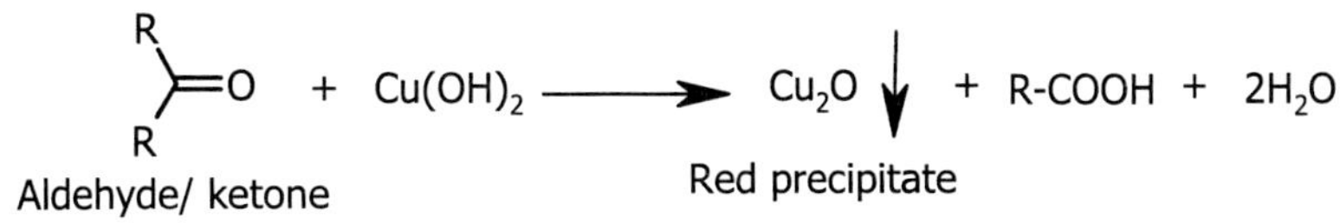	
4.	**Tollen's reagent test (silver mirror test)**	
	Place 2 ml of Tollen's reagent (ammoniacal silver nitrate, $[Ag(NH_3)_2NO_3]$) and 2 ml sample solution in a test tube and boil for 5-6 minutes and cool.	
	Appearance of black color or silver mirror on the sides of the test tube	Indicates the presence of aldehyde (carbohydrates)
	Aldehydes are readily oxidized to carboxylic acids and will reduce Tollen's reagent to produce a silver mirror on the inside of a clean test tube.	
	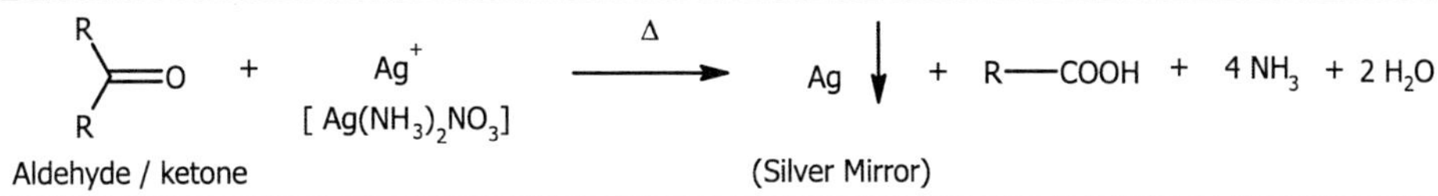	

Tests for Compounds Containing Nitrogen

	Tests for Amines		
1.	**Conc. HCl test**		
	To the sample add concentrated hydrochloric acid	Formation of white precipitate	Indicates the presence of amines
2.	**Diazotization test**		
	To the sample add 3 ml of concentrated hydrochloric acid, cool thoroughly to 0 - 5 °C in an ice-bath for 5 minutes and add few drops cold solution (kept in ice-bath) of sodium nitrite (0.5 gm sodium nitrite in 2 ml of water) from a dropper. Add the above solution to ice cold solution of 0.2 gm of β-naphthol in 2 ml of sodium hydroxide with swirling keeping the test tube in ice-bath.		
	Formation of an orange-red azo dye	Indicates the presence of amines	
	1° Aliphatic	$R{-}NH_2 + HNO_2 \longrightarrow R{-}OH + N_2 + H_2O$ Orange red dye (Nitrogen gas (N_2) evolves)	

Table Contd...

	1° Aromatic	$Ar\text{-}NH_2 + HNO_2 \xrightarrow{HCl} Ar\text{-}N{=}N^+Cl^- + H_2O$ Diazonium salt 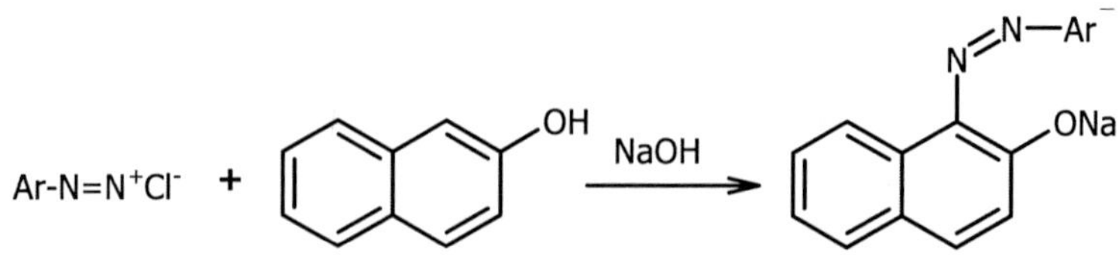 Orange red dye
	2° Aliphatic	$R_2NH + HNO_2 \longrightarrow R_2N\text{-}N{=}O$ Yellow oily nitrosamines
	2° Aromatic	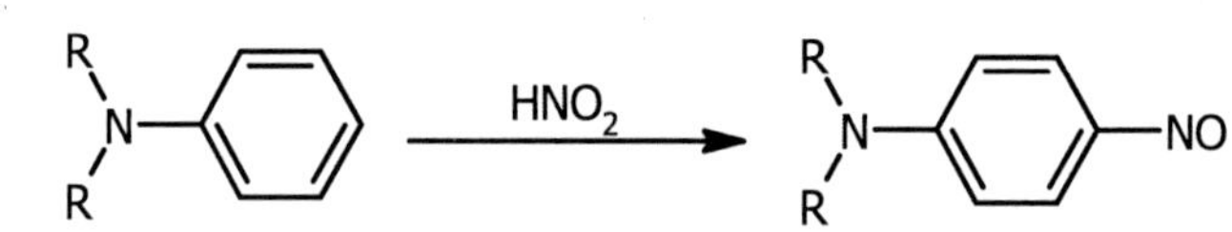 Dialkyl anilines yield green solid of *p*-nitroso compounds
	3° Aliphatic	Red – green solid on NaOH addition (if para position is unsubstituted)
	3° Aromatic	No colouration, but gives blue colour with starch iodide paper (nitrous acid liberates iodine from starch iodide and produces blue colour)
3.	**Carbylamine test**	
	Dissolve 0.2 gm of the sample in alcoholic sodium hydroxide and add 4 drops of chloroform warm gently.	
	Foul odour of isocyanide	Indicates the presence of amines
	Carbylamines formed will be readily identified by its extremely nauseating odour	
	$R\text{—}NH_2 + CHCl_3 + 3\ NaOH \longrightarrow R\text{—}CN + 3\ NaCl + 3\ H_2O$	
4.	**Reaction with benzene sulphonyl chloride**	
	Place 0.5 ml/0.5 gm of the compound, 10 ml of 5% NaOH and 1 ml of benzenesulphonyl chloride in a test tube, stopper the tube and shake until the odour of the sulphonyl chloride has disappeared. The solution must be kept alkaline.	

Table Contd...

	• If there is no precipitate, acidify it cautiously to congo red with concentrated hydrochloric acid • If a precipitate appears in the alkaline solution, dilute with about 10 ml of water and shake • If no reaction has occurred	• Indicates the presence of primary amine • The precipitate does not dissolve, a secondary amine is indicated • The substance is probably a tertiary amine
	Benzene sulphonyl chloride reacts with primary and secondary amines but not with tertiary amines to yield substituted sulphonamides.	

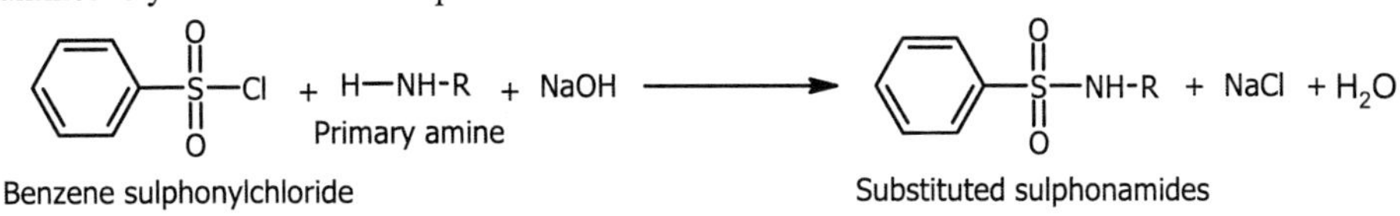

The substituted sulphonamide formed from a primary amine dissolves in the alkali medium.

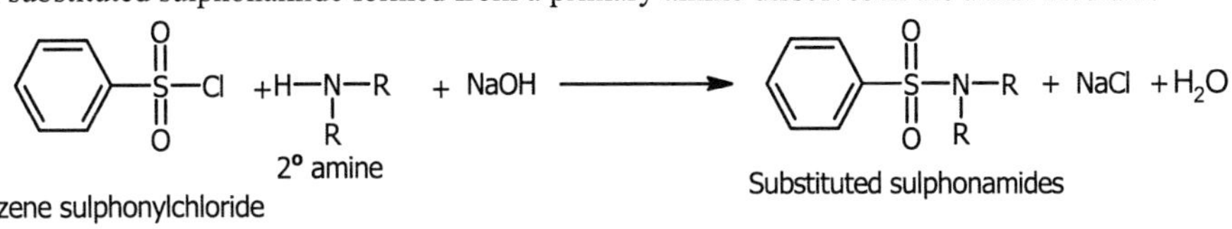

The substituted sulphonamide formed from a secondary amine is insoluble in alkali.

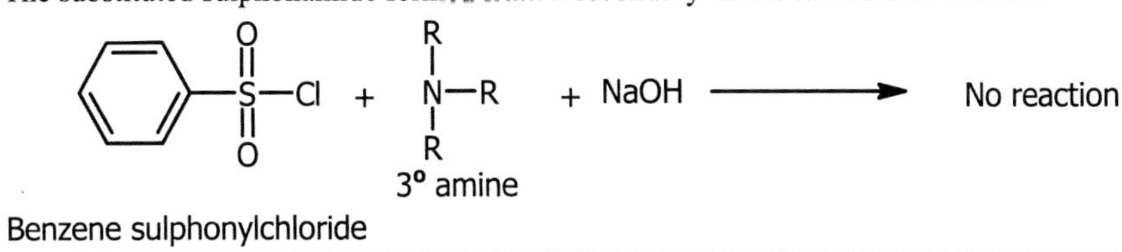

	Tests for Amides (R-CO-NH$_2$)	
1.	**Hydroxamic acid test**	
	To 0.2 gm of the sample add 2 gm of hydroxylamine hydrochloride (NH_2OH. HCl) in ethanol. Then add 5 ml of 10% NaOH and boil for 2 minutes. Cool and add 2 drops of $FeCl_3$.	
	Formation of deep red colour	Indicates the presence of amides

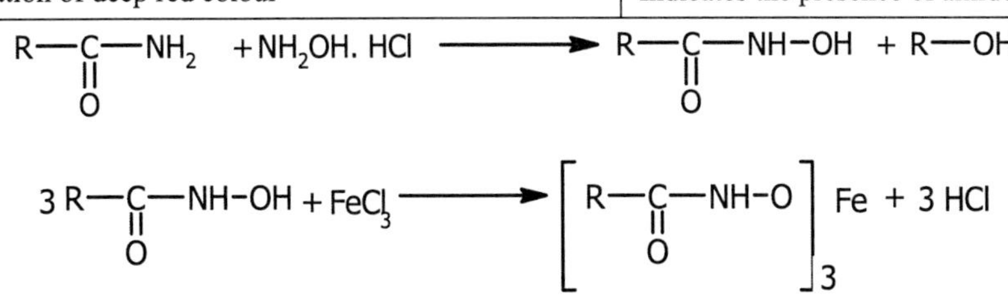

Table Contd...

2.	Add 1 ml of 10% NaOH to the sample and boil	Smell of NH_3/white fumes	Indicates the presence of amides
	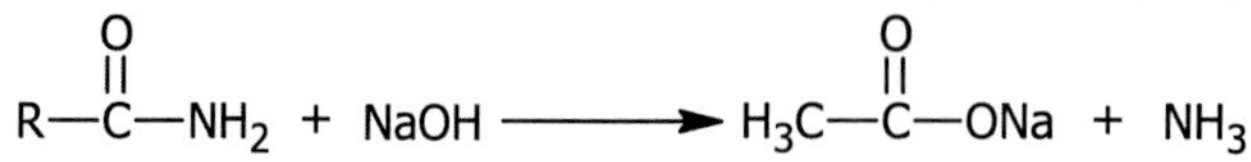		
	Add 5 drops of concentrated HCl to the above solution	Formation of white precipitate	Indicates the presence of aromatic amides
3.	To 1 ml of the ice-cold $NaNO_2$ solution add 1 ml of acetic acid and add ice-cold solution of sample	Brisk effervescence of nitrogen (N_2)	Indicates the presence of amides
4.	**Biuret test**		
	Heat 0.2 gm of the sample (urea) to melt, cool, dissolve it in 10% sodium hydroxide and add 1 drop of 1% copper sulphate solution	Develops violet/purple colour	Indicates the presence of di-amides
	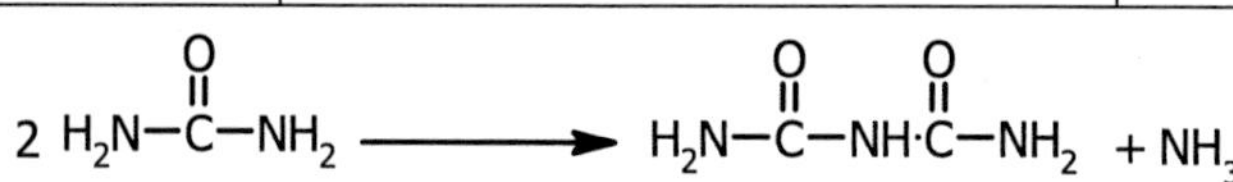 Biuret		
5.	**Action of hypobromide**		
	To a solution of the substance add sodium hypobromide solution and note the brisk effervescence.		

	Tests for Anilides	
1.	**Tafel's test**	
	To 0.2 gm of the sample add 5 ml of sulphuric acid, shake and add 0.2 gm potassium dichromate solution.	
	Red/violet colour (changes to green colour)	Indicates the presence of anilides

Table Contd...

2.	**Hydrolysis test**	
	To 0.2 gm of the sample add 5 ml of dilute hydrochloric acid, boil, cool and then add sodium nitrite ($NaNO_2$) solution followed by β-napthol in sodium hydroxide.	
	Formation of orange-red azo dye due to the cleavage of anilide into aniline	Indicates the presence of anilides

	Tests for Thiourea	
1.	**Ferric chloride test**	
	Heat the sample to melt, cool and add ferric chloride ($FeCl_3$) solution.	
	Formation of deep-red colour	Indicates the presence of thiourea
2.	**Potassium ferrocyanide test**	
	To the hot solution of substance in acetic acid add 2 ml of potassium ferro cyanide.	
	A green colour changes to blue	Indicates the presence of thiourea

	Tests for Nitro Group		
1.	**Mulliken Barker test**		
	To 0.2 gm of sample in 2 ml of ethanol add 3 drops of calcium chloride ($CaCl_2$) solution and pinch of zinc dust. Boil, cool and filter. Divide the filtrate into 2 parts.		
	To one part add Tollen's reagent	Black/gray precipitate	Indicates the presence of nitro group
	$R\text{-}NO_2 \xrightarrow{Zn,\ CaCl_2} R\text{-}NH\text{-}OH + H_2O$ $R—NH\text{-}OH + [Ag(NH_3)_2NO_3] \longrightarrow Ag^+ + R—NO + H_2O + 2NH_3$		
	To the 2nd part add equal volumes of feehling's solutions A and B and heat	Red brown precipitate	Indicates the presence of nitro group
2.	To the sample add 2 ml of 5% Ferric ammonium sulphate and drop of dilute sulphuric acid, followed by 1 ml of methanolic potassium hydroxide (KOH). Shake vigorously.		
	Red brown precipitate	Indicates the presence of nitro group	
	$R\text{-}NO_2 + [Fe(NH_3)_2SO_4] \longrightarrow R\text{-}NH_2 + 6\ Fe(OH)_3$		

Table Contd…

3.	**Azo dye test**		
	Heat 0.1 gm compound with 0.5 ml of hydrochloric acid (HCl). Add metal tin and boil for 3-4 minutes	Smell of aniline	Indicates the presence of nitro group
	Filter the above solution and add 3 ml of aqueous sodium nitrite ($NaNO_2$) solution, pour into a beaker containing β-naphthol	Brick-orange dye	Indicates the presence of nitro group

48.7 Preparation of Derivatives

1. The final characterization of the unknown is made by the preparation of suitable solid derivatives.
2. The derivative should be carefully selected and its melting point should preferably be between 90-150 °C for ease of crystallization and melting point determination.
3. Preparation of one derivative should be attempted.
4. The derivative should be purified by recrystallization, dried and the melting point should be determined.

The preparation of derivatives of organic compounds: The preliminary examination and group classification tests indicate the particular class (functional group) to which an unknown organic compound may belong.

Further characterization and identification depends on the selection and preparation of a suitable solid derivative and accurate determination of its melting point (best, between 90 - 150 °C).

List of some of the derivatives of organic compounds

Class of compound	Derivatives
1. Acids	Anilide, amide, p-toluidide
2. Alcohols	3,5-Dinitrobenzoate
3. Phenols	Benzoate, acetate, bromo-derivative
4. Aldehydes and ketones	Semicarbazone, 2,4-dinitrophenyl-hydrazone, oxime
5. Amines	Benzoyl, acetyl and sulphonamide derivatives

Methods for the Preparation of Derivatives

1. **Acid derivatives**

1A) ***Amides, anilides and p-toluidides:*** The acid (0.5 gm) is refluxed with thionyl chloride (2-3 ml) in a fume cupboard for about 30 mins*. It is advisable to place a plug of cotton wool in the top of the reflux condenser to exclude moisture. The condenser is removed and the excess of thionyl chloride is distilled off (b.p. 78 °C). The acid chloride thus produced is treated with concentrated ammonia solution (5 ml) or aniline (0.5-1 ml) or p-toluidine (0.5-1 gm), when the solid derivative separates out. It is collected and recrystallized from alcohol adding decolourising charcoal if found necessary.

* Alternately use PCl_5 to form the acid chloride.

2. Alcohol Derivative

2A) ***3,5-Dinitrobenzoates:*** 3,5-Dinitrobenzoyl chloride is usually partially hydrolysed and should be prepared in the pure state by heating gently a mixture of 3,5- dinitrobenzoic acid (1 gm) and phosphorus pentachloride (1.5 gm) in a dry test tube, until it liquifies (5 min) *. The liquid is poured on a dry watch glass and allowed to solidify. The phosphoryl chlorides are removed by pressing the solid with a spatula on a wad of filter paper. The residual acid chloride is suitable for immediate use in the preparation of the derivatives.

* Work under fume hood. Fumes are irritating to the eyes and nose.

The 3,5-dinitrobenzoyl chloride is mixed with the alcohol (0.5-1 ml) in a loosely corked dry test tube and heated on a steam bath for about 10 minutes. Secondary and tertiary alcohols require up to 30 minutes. On cooling add 10 ml of sodium hydrogen carbonate solution, stir until the ester crystallizes out, and filter at the pump. Wash with a little carbonate solution, water and suck dry. Recrystallize from the minimum hot ethanol or light petroleum. Cool slowly to avoid the formation of oily droplets of ester.

3. Phenol Derivative

3A) ***Benzoates (Schotten-Baumann method):*** Dissolve phenol (0.5 gm) add 5% sodium hydroxide (10 ml) in a small conical flask. Add benzoyl chloride (2 ml) in small quantities at a time, and shake the mixture vigorously with occasional cooling under the tap or in ice-water. After 15 minutes the solid benzoate separates out. The solution should be alkaline at the end of the reaction; if not alkaline, or if the product is oily, add a solid pellet of sodium hydroxide and shake again. Collect the benzoate, wash thoroughly with cold water, and recrystallize from alcohol or light petroleum.

3B) ***Acetates:*** Acetates is a suitable derivative for polyhydric and substituted phenols. Dissolve phenol (0.5 gm) in 10% sodium hydroxide solution and add an equal quantity of crushed ice, then add acetic anhydride (2 ml). Shake the mixture vigorously in a test tube until the acetate separates out. Filter the product and recrystallize from alcohol.

3C) ***Bromo derivatives:*** Suspend phenol (0.3 gm) in dilute hydrochloric acid (10 ml) and add bromine water dropwise until no more decolourisation occurs. Filter the bromo derivative which precipitates out and recrystallize from alcohol.

4. Aldehydes and Ketones

4A) ***Semicarbazones:*** Dissolve semicarbazide hydrochloride (1 gm) and sodium acetate (1.5 gm) in water (8-10 ml), add the aldehyde or ketone (0.3 ml) and shake. Shake the mixture for a few minutes and then cool in ice-water. Filter off the crystals, wash with a little cold water and recrystallize from methanol or ethanol.

4B) ***2,4-Dinitrophenylhydrazones:*** Suspend 0.25 gm of 2,4-dinitrophenylhydrazine in 5 ml of methanol and add 0.5 ml of concentrated sulphuric acid cautiously. Filter the warm solution and add a solution of 0.2 gm of the carbonyl compound in 1 ml of methanol. Recrystallize the derivative from methanol, ethanol or ethyl acetate.

4C) ***Oximes:*** Dissolve hydroxylamine hydrochloride (0.5 gm) in water (2 ml). To this add solution of 10% sodium hydroxide (2 ml) and the carbonyl compound (0.2-0.3 gm) in alcohol (1-2 ml). Warm on a steam bath for 10 minutes and then cool in ice. Induce the crystallisation by scratching the sides of the test tube with a glass rod. Dry and recrystallize from alcohol.

5. Amine Derivatives

5A) ***Acetyl derivatives (acetamides):*** Reflux gently in a small dry flask under a dry condenser the amine (1 gm) with acetic anhydride (3 ml) for 15 minutes. Cool the reaction mixture and pour into 20 ml cold water. Boil to decompose the excess acetic anhydride. Cool and filter by suction the insoluble derivative. Recrystallize from ethanol.

5B) ***Benzoyl derivatives (benzamides):*** Suspend 1 gm of the amine in 20 ml of 5% aqueous sodium hydroxide in a well-corked flask, and add 2 ml benzoyl chloride (fume hood), about 0.5 ml at a time, with constant shaking. Shake vigorously for 5-10 min until the odour of the benzoyl chloride has disappeared. Ensure that the mixture remains alkaline. Filter off the solid derivative, wash with a little cold water and recrystallize from ethanol.

5C) ***Benzenesulphonamides:*** To 1 gm of the amine in 20 ml of 5% sodium hydroxide solution in a well-corked flask add 1 ml benzenesulphonyl chloride (fume hood). Shake the mixture until the odour of the sulphonyl chloride disappears. Check that the solution is alkaline. Acidify if necessary to obtain the precipitated derivative. Concentrated hydrochloric acid added dropwise should be used. Filter the product, wash with a little cold water and dry. Recrystallize from ethanol.

48.8 Recording of Results

The results should be recorded in a systematic manner.

1. Results should be recorded in the practical book at the time (not written up afterwards).
2. A record should be made of every test carried out, no matter whether a **negative result has been obtained**.
3. Test, observation and inference should be given.
4. At the conclusion of the analysis a brief summary of results should be included, giving the following details
   - (i) Saturation
   - (ii) Aromaticity
   - (iii) Special element(s)
   - (iv) Functional group(s)
   - (v) Derivatives
   - (vi) Observed melting point of derivative
   - (vii) Literature melting point of derivative

CHAPTER 49

QUALITATIVE ANALYSIS OF CARBOHYDRATES

Carbohydrates are naturally occurring organic compounds. Chemically they are polyhydroxy aldehydes or polyhydroxy ketones or the compounds that produce these compounds on hydrolysis. Carbohydrates present in foods serve as a primary source of energy. Two types of carbohydrates are known

1. Simple carbohydrates (sugars)
2. Complex carbohydrates (starch and dietary fiber)

Simple Carbohydrates

Simple carbohydrates (sugars) are present in fruits, milk and other foods. The two main types of simple sugars are

1. **Monosaccharides:** The most common monosaccharides found in the human diet are aldoses (glucose and galactose) and ketoses (fructose).
2. **Disaccharides:** Disaccharides consists of two monosaccharides joined by a process of condensation. The important disaccharide sugars present in human nutrition are sucrose (table sugar), lactose (sugar in milk) and maltose (a product of starch breakdown).

Complex Carbohydrates

Complex carbohydrates are chains of more than two sugar molecules. They may be oligosaccharides or polysaccharides depending upon the chain length.

Oligosaccharide: It is a short chain composed of 3 to 10 sugar molecules.

Polysaccharides: They are long chain carbohydrates, composed of more than 10 sugar molecules. Fruits and vegetables are particularly rich in polysaccharides. The two common polysaccharides are storage polysaccharides [starch, glycogen, inulin (homoglycan) and glucosamine glycans] and dietary fibers (cellulose, hemicellulose, pectin, gums, chitin and mucilage etc.) Dietary fibers (non-starch polysaccharides) provides structure to plant cell wall and is found inside plant cells. The fibrous carbohydrates promote the elimination of digestive waste from the gut and lower the absorption of both glucose and bile salt in humans. Starch is digestible in gastrointestinal tract but dietary fiber is not digestible.

49.1 General Test for Carbohydrates

1. Molisch test

Place 1 ml of the carbohydrate solution in a clean test tube, add 2 drops of molisch reagent 10% (alcoholic α-naphthol) and mix well. Incline the test tube and slowly add 0.5 ml of sulfuric acid in drops along the sides of the test tube.	
Violet/purple colour ring develops at the junction of two liquids	Indicates the presence of carbohydrates

Mechanism: The concentrated sulphuric acid dehydrates pentoses to form furfural and dehydrates hexoses to form 5-hydroxymethyl furfural. These furfural derivatives condense with alcoholic α-naphthol (molisch reagent) to form violet colored products. In case of oligo and polysaccharides, they are first hydrolysed to monosacchrides by acid, which undergoes dehydration to form furfural and its derivatives.

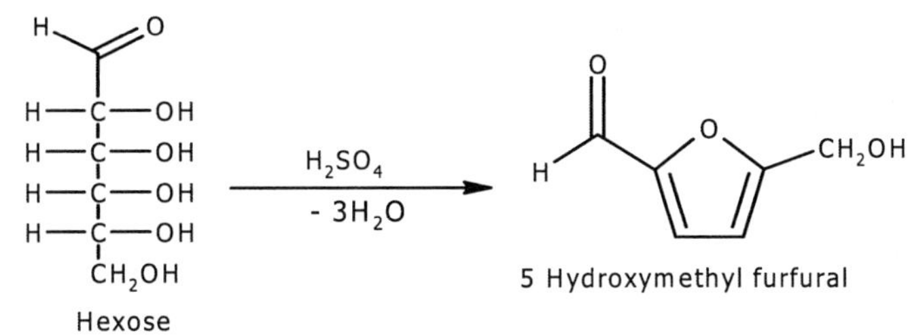

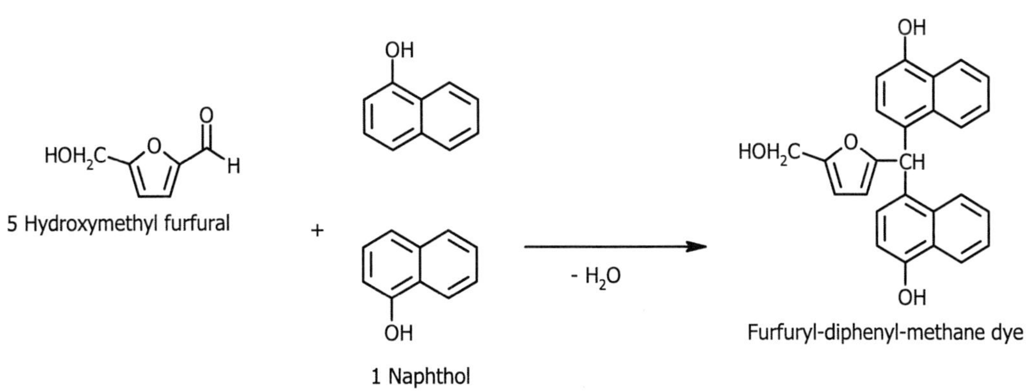

***Note*:** Concentrated solution of organic compounds may give a red colour instead of violet due to the charing action of the sulphuric acid. In case of doubt the experiment should be repeated on a more diluted solution of the substance to be tested.

2. Solubility test

Dissolve the substance in distilled water	
Soluble	May be mono and disaccharides
Insoluble	May be polysaccharides

49.2 Tests for Reducing Sugars

Carbohydrates (monosaccharides and many disaccharides) with free aldehyde or ketone groups have reducing properties and reduce weak oxidizing agents like Cu^{2+} ion in alkaline medium.

Fehling's test

Place 2 ml of the carbohydrate solution in a test tube, add 2 ml of Fehling's solution A, 2 ml of Fehling's solution B and mix well. Place the test tube into the water bath for five minutes. The colour of Fehling's solution changes from blue to red/brown in the presence of reducing sugars	
Yellow or brick red precipitate	Indicates the presence of reducing sugars

Fehling's solution A: *7* gm Copper sulphate ($CuSO_4.5H_2O$) dissolved in distilled water containing 2 drops of dilute sulphuric acid in 100 ml of distilled water.

Fehling's solution B: 35 gm of potassium tartrate and 12 gm of NaOH in 100 ml of distilled water.

Mechanism: The colour of Fehling's solution (contains Cu^{2+}) changes from blue to red/brown in the presence of reducing sugars. The alkali converts glucose to enediol, which reduces copper sulphate to cupric hydroxide. Cupric hydroxide is unstable and decomposes to cuprous oxide on boiling. The precipitated cuprous oxide will have different shades of colour depending upon the glucose concentration. The purpose of using potassium tartarate (Rochelle salt) is to dissolve the precipitate of cupric hydroxide. Fehling's A and Fehling's B should be used separately to avoid the formation of black precipitate of cupric oxide. The test is sensitive enough that even 1 mg of glucose will produce the characteristic red colour of the compound.

$$CuSO_4 + 2NaOH \longrightarrow Cu(OH)_2 + Na_2SO_4$$

$$Cu(OH)_2 \longrightarrow Cu^{2+} + 2OH^-$$

$$\underset{}{Cu^{2+}} + \underset{\text{(Reducing sugars)}}{\text{Aldose/ ketose}} \xrightarrow{\Delta} \underset{\text{Red precipitate}}{Cu_2O} + H_2O + \underset{\text{Oxidises sugars to acids}}{(O)}$$

Fructose (ketose monosaccharides) also gives positive result, due to conversion to aldoses by the base present in the reagent. For this reason, Fehling's test is referred as a general test for monosaccharides.

Benedict's test

Benedict's test is based on the reduction of alkaline copper sulphate. Benedict's test is more specific than Fehling's test because strong alkali in Fehling's solution may destroy small amounts of reducing sugar.

Place 8 drops of the carbohydrate solution in a test tube and add 5 ml of Benedict's reagent. Place the test tube in the water bath for 10 minutes.	
Colour change from blue to • Green (Low glucose present) • Yellow • Orange • Red • Brick red/brown (high glucose present)	Indicates the presence of reducing sugars

Mechanism: The bluish Cu^{2+} ion present in the reagent will be reduced to reddish cuprous oxide by reducing sugars. Benedict's reagent contains copper sulphate (gives cupric ions in the solution), sodium carbonate (alkaline medium) and sodium citrate (prevents the precipitation of cupric hydroxide by forming cupric sodium citrate complex). Sodium citrate also improves the shelf life of the reagent by preventing an interaction between sodium carbonate and copper sulphate that may otherwise get precipitated as cupric carbonate.

$$CuSO_4 \longrightarrow Cu^{2+} + SO_4^-$$

$$\underset{\text{(Reducing sugars)}}{Cu^{2+} + \text{Aldose / Ketose}} \longrightarrow \underset{\text{Red precipitate}}{Cu_2O} + \text{Oxidized product of Aldose / Ketose}$$

Sucrose does not react under normal conditions (heating 4-10 min in water bath). Fructose, an alpha hydroxyl ketone (not strictly reducing sugar) gives positive test, because it gets converted to aldoses (glucose and mannose) by the reagent.

Tollen's test (Silver mirror test)

Place 2 ml of Tollen's reagent and 2 ml sugar solution in a test tube and boil for 5-6 minutes and cool.	
Silver mirror will appear on the sides of test tube	Indicates the presence of reducing sugars

Tollen's reagent: Ammoniacal silver nitrate, $[Ag(NH_3)_2NO_3]$

Preparation: Mix aqueous silver nitrate and drop of dilute sodium hydroxide (forms brown prepciptate of silver oxide) then add ammonia to dissolve the precipitate.

Mechanism: Ag^+ ions present in ammoniacal silver nitrate react with OH^- of aqueous ammonia and produces AgOH (brown precipitate), excess of aqueous ammonia dissolves this and forms silver diamine complex $[Ag(NH_3)_2]^+$. The silver diamine complex is reduced by reducing sugar to metallic silver. The reducing sugar is oxidized to corresponding acid.

$$\underset{[Ag(NH_3)_2NO_3]}{Ag^+} + \underset{\text{(Reducing sugars)}}{\text{Aldose / ketose}} \xrightarrow{\Delta} \underset{\text{(Silver Mirror)}}{Ag\downarrow} + 2H_2O + \text{Oxidized product of sugar}$$

Safe note:

1. The reaction mixture should be acidified with dilute acid before disposal to prevent the formation of highly explosive silver azide.

Tommer's test

To 2 ml of Tommer's reagent (5% copper sulphate in water and equal volume of sodium hydroxide solution) add 3 ml of sugar solution, boil for 2 minutes and cool	
Formation of yellow or red precipitate	Indicates the presence of reducing sugars

Mechanism: The reagent consists of blue Cu^{2+} ions that reacts with the carbohydrate to reduce to cuprous oxide, this changes color from blue to red/yellow.

$$CuSO_4 + 2NaOH \longrightarrow Cu(OH)_2 + Na_2SO_4$$

$$Cu(OH)_2 \longrightarrow Cu^{2+} + OH^-$$

$$Cu^{2+} + \underset{\text{(Reducing sugars)}}{\text{Aldose/ ketose}} \xrightarrow{\Delta} \underset{\text{Red precipitate}}{Cu_2O} + H_2O + \underset{\text{Oxidises sugars to acids}}{(O)}$$

49.3 Test to Distinguish Monosaccharides from Disaccharides

Barfoed's test

Barfoed test is used for detecting the presence of monosaccharides. Monosacchrides (reducing sugars) reduces of copper (II) acetate to brick red colour copper (I) oxide (Cu_2O). Disacchrides responds to the reaction much slower.

Place 8 drops of the carbohydrate solution in a test tube. Add 5 ml of Barfoed's reagent. Place the test tube into the water bath.	
Formation of light red precipitate • Precipitate within 2-3 minutes • Precipitate within 4-10 minutes	Indicates the presence of • Monosaccharides • Disaccharides

***Barfoed's reagent*:** Barfoed's reagent is copper (II) acetate in acetic acid and not as reactive as Benedict's reagent.

***Mechanism*:** Copper ion present in the solution oxidizes reducing monosaccharides and forms carboxylic acid and brick red copper (I) oxide (Cu_2O) within 3 minutes. The aldehyde group of monosaccharides will be converted in to hemiacetal and then oxidized to carboxylate. Reduction of cupric ions will effect only in mild acidic condition (difference from benedict's test). Disaccharides may give positive result on prolonged boiling.

$$(CH_3COO)_2Cu + 2H_2O \longrightarrow Cu(OH)_2 + 2CH_3COOH$$

$$\underset{\text{(Reducing sugars)}}{\text{Monosaccharide}} + Ca(OH)_2 \longrightarrow \underset{\text{(Red precipitate)}}{Cu_2O} + \underset{\text{(Corresponding acid)}}{\text{Oxidized product of sugar}} + H_2O$$

49.4 Test to Distinguish Ketoses and Aldoses

Seliwanoff's test

Seliwanoff' test is used to distinguish aldoses from ketoses. Fructose and sucrose gives positive result.

Place 1 ml of the carbohydrate solution in a test tube. Add 3 ml of Seliwanoff's reagent, mix well. Place the test tube into the water bath for 30 seconds and cool.	
Formation of deep red precipitate	Indicates the presence of fructose (ketose)

Seliwanoff' reagent: It consists of resorcinol and hydrochloric acid.

Mechanism: The hydrochloric acid dehydrates ketoses more rapidly in to 5-hydroxy methyl furfural than aldoses. 5-Hydroxy methyl furfural reacts with resorcinol and forms red complex within 2 minutes. Ketoses will form deep red colour while aldoses will show light pink colour (takes longer time than the ketoses). Acid hydrolysis of poly and oligo saccharides gives deep cherry red colour.

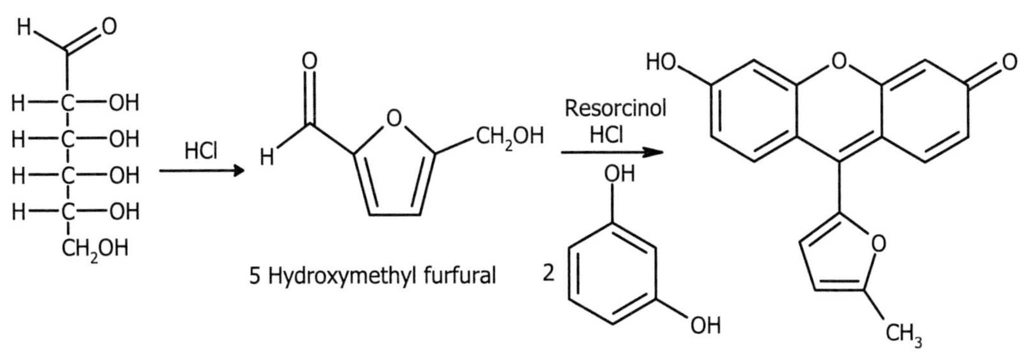

Rapid furfural test

Place 1 ml of the carbohydrate solution in a clean test tube. Add 2 drops of Molisch reagent (1% alcoholic α-naphthol) and mix well. Add 5 ml of concentrated hydrochloric acid to the test tube. Boil the solution for few minutes.	
Development of deep purple colour	Indicates the presence of ketoses (fructose and sucrose)

49.5 Test to Distinguish Pentoses and Hexoses

Bial's test

Place 2 ml of a sample solution in a test tube. Add 2 ml of Bial's reagent (a solution of orcinol, HCl and ferric chloride). The solution is then heated gently in a hot water bath. If the color is not obvious, more water can be added to the tube.	
• Formation of bluish-green colour within five minutes • Formation of green, red and brown colour	• Indicates the presence of pentoses • Indicates the presence of hexoses

Formation of a bluish-green color within 5 minutes without the formation of a precipitate is due to the presence of pentoses. All other colors indicate a negative result for pentoses. Hexoses generally react to form green, red, or brown products.

This test is used to distinguish pentoses from hexoses. Pentoses give a positive test with Bial's reagent. The test reagent dehydrates pentoses to form five member ring, furfural. Furfural further reacts with orcinol and the iron ion present in the test reagent to produce a blue-green color. A positive test is the formation of a bluish-green color within 5 minutes without the formation of a precipitate.

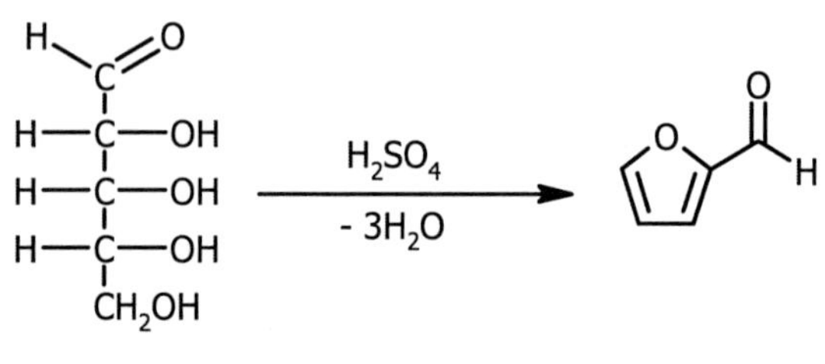

Taubers's benzidine test

To 3- 4 drops of sugar solution add 1 ml of benzidene reagent. Boil the reaction mixture vigorously on open flame and immediately place the test tube on an ice bath for cooling. A red/violet color forms, indicates the presence of pentose sugar.

Benzidine reagent: Contains benzidine in acetic acid.

Benzidine reacts with pentose in presence of acetic acid to give intense red color indicating the presence of pentose sugar.

49.6 Confirmatory Tests for Reducing Sugars

Osazone test: This test is given only by reducing sugars like glucose, fructose, lactose and maltose. Reducing sugars when treated with phenylhydrazine hydrochloride in the presence of acetate buffer, pH 5.0 give characteristic yellow colored crystals of osazones. All these reducing sugars give characteristic yellow crystalline structures of the osazone when observed under microscope.

Place 0.2 gm of the unknown sample in a test tube and add 0.4 gm of phenylhydrazine hydrochloride, 0.6 gm of crystallized sodium acetate, and 4 ml of distilled water. Place the test tube in a beaker of boiling water. After 20 min, remove the test tube from bath and set it aside to cool. Pour a small amount of the liquid and solid on a watch glass. Tap the watch glass from side to side to spread out the crystals, and absorb some of the mother liquid with a piece of filter paper, taking care not to crush or break up the clumps of crystals. Examine the crystals under a low-power microscope (about 80-100 X).	
• Broomed shape crystals within 5 minutes	• Indicates the formation of glucosazone
• Broomed shape crystals within 2 minutes	• Indicates the formation of fructosazone
• Spherical shape crystals within 10-15 min	• Indicates the formation of maltosazone
• Ball of prickles shape crystals within 10-12 min	• Indicates the formation of lactosazone

Note: The formation of tarry products due to oxidation of the phenylhydrazine hydrochloride may be prevented by the addition of 0.5 ml of saturated sodium bisulfite solution. This should be done before heating, if it is desired to isolate the osazone and determine its melting point.

Mechanism: Osazone formation involves hydrazone formation at C-1 of an aldose (or C-2 of a ketose) and oxidation of C-2 (or C-1) of an alcohol group to a ketone (or an aldehyde). The new carbonyl group is also converted to a hydrazone. It has been suggested that the reaction stops here (rather than further oxidation at C-3, etc.,) because of hydrogen-bonding stabilization of the osazone.

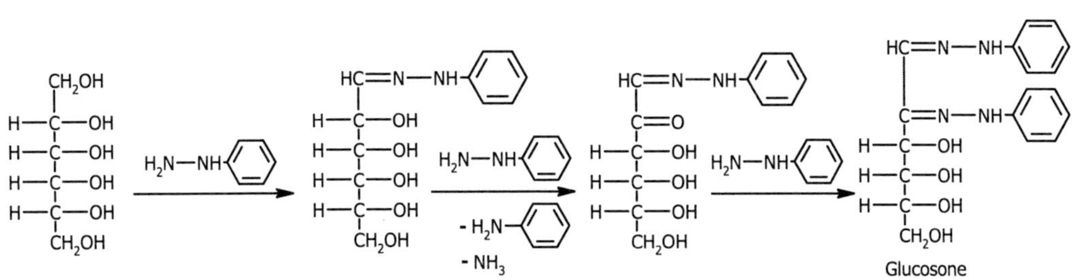

49.7 Test for Starch and Other Polysaccharides

Iodine test: Polysaccharides such as starch and glycogen forms coloured adsorption complexes.

Add 2 drops of dilute iodine solution to 3 ml of dilute starch/dextrin/glycogen solution.	
• Formation deep blue colour solution	• Indicates the presence of starch
• Formation deep red colour solution	• Indicates the presence of dextrin
• Formation deep reddish brown colour solution	• Indicates the presence of glycogen

Iodine solution: Iodine dissolved in an aqueous solution of potasssium iodide.

Starch gives blue colour with iodine due to formation of complex between iodine and amylase fraction of starch. The linear amylase molecule coils into a spiral (held by hydrogen bonds between hydroxyl groups of adjacent loop), while the iodine molecule aligns itself in the center of spiral. This arrangement allows interaction of electron orbital of iodine atoms to cause light absorption giving the complex the characteristic blue colour. The amylopectin or branched portion of starch, forms much shorter helices and iodine molecules are unable to assemble, leading the color to be of an orange/yellow colour.

Iodine solution will also react with glycogen although the brown color produced is and much less intense.

Precipitation reactions of polysaccharides

> Take 5 ml of saturate solution of ammonium sulphate in a test tube and add 5 ml of sample solution to make the solution saturated with ammonium sulphate. Mix well by inverting the test tube several times and allow it to stand for 5 minutes.
>
> Filter the flocculent precipitate of sugar (polysaccharide) formed and perform iodine test with the residue as well as with the filtrate.

Sugar	Residue	Filter	Inference
Starch	Blue	No blue	Complete precipitation
Glycogen	Brown	No brown	Incomplete precipitation
Dextrin	Red	Red	Incomplete precipitation

Generally the bigger molecules are easily precipitated by salts. The starch is precipitated when its solution is half-saturated with ammonium sulphate. Glycogen is completely precipitated only when its solution is fully saturated with ammonium sulphate, whereas dextrin is not completely precipitated even on full saturation.

49.8 Test for Non-reducing Sugar (Sucrose)

Inversion test

Add 2 drops of hydrochloric acid and one drop of thymol blue to 5 ml of sucrose solution. The development of pink colour indicates the solution acidic. Divide it into two equal parts. Boil one portion for about one minute and then cool it under the tap. Neutralize both portions with 2% sodium carbonate solution by adding it drop by drop. Upon neutralization blue colour will develop.

- The boiled sucrose gives positive results for Benedict's test and Seliwinoff's test.
- The unboiled sucrose does not reduce Benedict's reagent.

This test is confirmation test for of sucrose. Sucrose is a non-reduced disaccharide and it gives negative results for Fehling's, Benedict's, Tommer's and Barfoed's tests. Sucrose on hydrolysis with hydrochloric acid it forms the reducing mono saccharides, glucose and fructose. These monosaccharides (reducing sugars) can be detected by reducing sugar tests.

$$C_6H_{22}O_{11} + H_2O \xrightarrow[\text{Boil}]{\text{Conc. HCl}} \underset{\text{Glucose}}{C_6H_{12}O_6} + \underset{\text{Fructose}}{C_6H_{12}O_6}$$

Results of Benedict's, Seliwanoff's and Iodine Tests

Carbohydrates	Benedict's Test	Seliwanoff's Test	Iodine Test
Glucose	Red	Peach	Orange-Red
Fructose	Red	Deep Red	Orange-Red
Sucrose	Blue	Deep Red	Orange-Red
Lactose	Red	Red	Orange-Red
Starch	Blue	Pinkish	Black
Water	Blue	Almost clear	Orange-Red

Tests of foods

Food Sample	Benedict's Test	Seliwanoff's Test	Iodine Test
Honey	Red	Deep Red	Orange-Red
Brown Sugar	Red	Deep Red	Orange-Red
Teddy Grahams	Red	Deep Red	Black

49.9 Scheme for Identification of Unknown Sugar

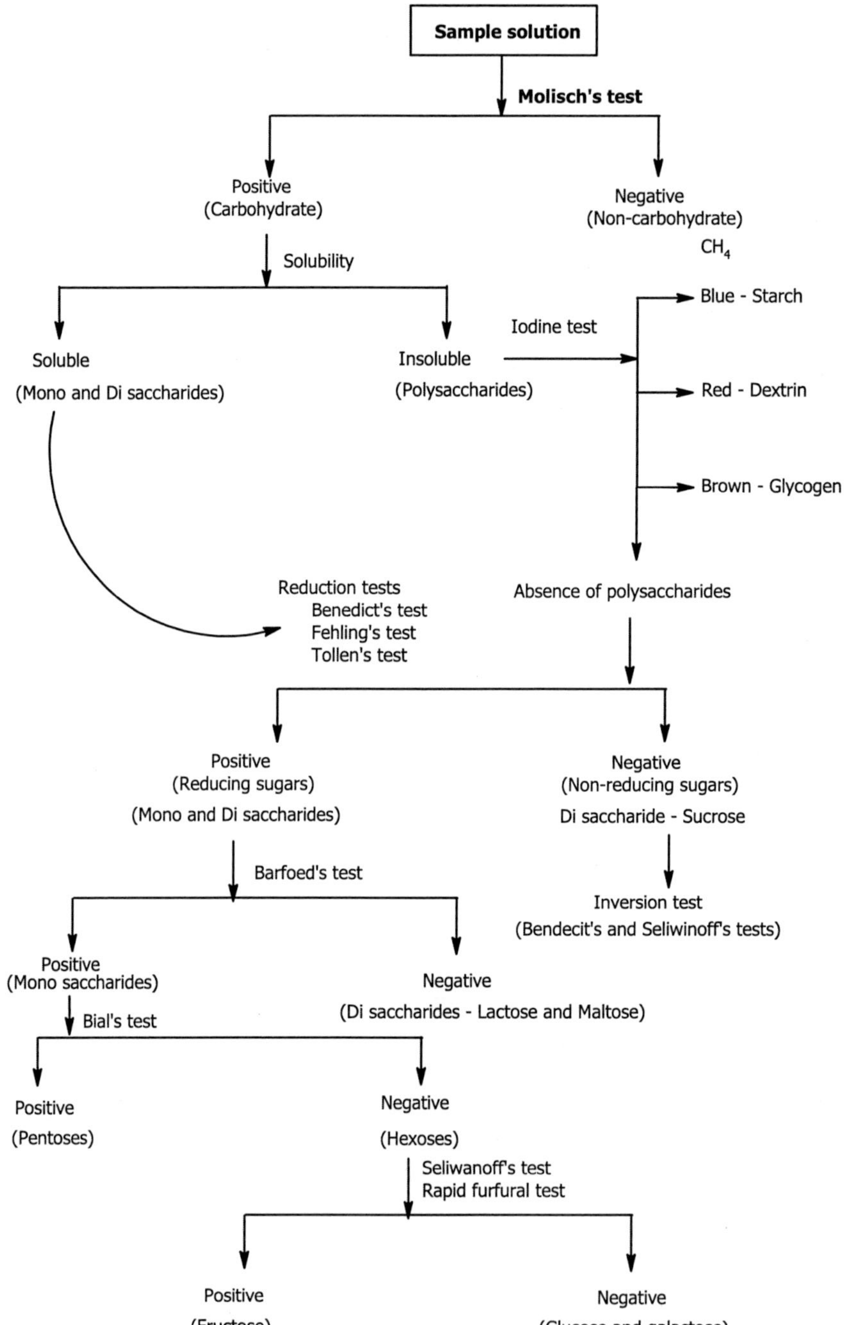

CHAPTER 50

QUALITATIVE ANALYSIS OF AMINO ACIDS AND PROTEINS

Amino acids

Amino acids are the compounds having one or more amino groups and one or more carboxyl groups in the same molecule. They are usually classified into α, β and γ amino acids. Because of the fact that only α–amino acids constitute the protein molecules, the term amino acid is generally used instead of the α–amino acid in reference to proteins. The general structure for an α–amino acid is:

$$R-\underset{NH_2}{\underset{|}{CH}}-COOH$$

Amino acids are zwitterion's contain both acidic (carbocyclic acid) and basic (amino) functional groups. At particular pH, number of protonated amine (ammonium) and deprotonated carboxylic groups (carboxylate) exist in equal and are known as 'zwitterion'.

Body synthesized certain amino acids and are called as non-essential aminoacids. But some amino acids can't be synthesized in the body and hence they must be supplied through diet. The first categories of amio acids are called dispensible or non-essential acids, while those of the second category are known as indespensible or essential amino acids.

Essential amino acids	Non- essential amino acids
1. Threonine	1. Glycine
2. Valine	2. Alanine
3. Leucine	3. Serine
4. Isoleucine	4. Tyrosine
5. Lysine	5. Aspartic acid
6. Methionine	6. Glutamic acid
7. Phenylalanine	7. Glutamine
8. Tryptophan	8. Lysine
9. Arginine	9. Hydroxylysine
10. Histidine	

Classification of amino acids

Nature	No. of groups		Amino acids
	NH_2	COOH	
Neutral	1	1	Glycine, alanine, serine, threonine, valine, leucine, isoleucine
Acidic	1	2	Aspartic acid, asparagine, glutamic acid and glutamine
Basic	2	1	Arginine, lysine and hydroxylysine

Proteins

The proteins are the most abundant macro molecules composed of C, H, O, N, S and P elements. Chemically proteins are polymers of α-amino acids, where α-amino acids are polymerized through peptide linkages. Proteins consists of very long polypeptide chains (100 to 1000 amino acid units) joined by peptide linkages. Cells contain hundreds or thousands of different proteins with a different function or biological activity.

Properties:

1. Proteins are colloidal in nature hence don't diffuse through intact membrane.
2. Proteins posses free amino (basic) and carboxyl (acidic) groups and are amphoteric in nature. Due to this amphoteric property, they act as buffers in maintaining the desired pH of body fluids and tissues.
3. Proteins are optically active molecules.
4. Metaproteins, proteases and peptones are the intermediate products of protein hydrolysis.

Appearance	• Turbid solution • Clear solution • Milky opalescent	• Albumin, globulin, casein • Gelatin, peptone • Albumin, casein
Colour	• Faint yellow • Faint brown	• Peptone • Gelatin
Smell	• Egg smell • Milk like smell • Meat like smell	• Albumin • Casein • Peptone, gelatin
Litmus test	• Neutral • Acidic • Alkaline	• Albumin, gelatin • Peptone • Casein and metaproteins

Colour reactions of proteins

Proteins react with variety of reagents and gives coloured products. Each amino acid differ with respect to functional groups and produce different colour reactions.

50.1 General Tests for Proteins

Biuret test

Biuret test is the characteristic test for detecting the peptide bonds (-CO-NH-) of protein and their hydrolytic products (peptides and metaproteins).

To 2 ml of protein solution add 1 ml of 40% sodium hydroxide and 1 drop of 1% copper sulphate solution	
• Development of violet colour • Development of rosy pink colour	• Indicates the presence of peptide linkage of the protein. • Indicates the presence of peptide linkage of the peptone.
Control: To 2 ml of distilled water add 1 ml of 40% sodium hydroxide and 1 drop of 1% copper sulphate solution	
• Development of blue/no violet colour	• Indicates the absence of peptide linkage

Mechanism: A copper (II) ion is reduced to cuprous and produce violet colour by forming complex with nitrogen of the peptide bond in alkaline medium. The colour of the reaction varies from pink to purple to violet, which depends upon the number of peptide groups.

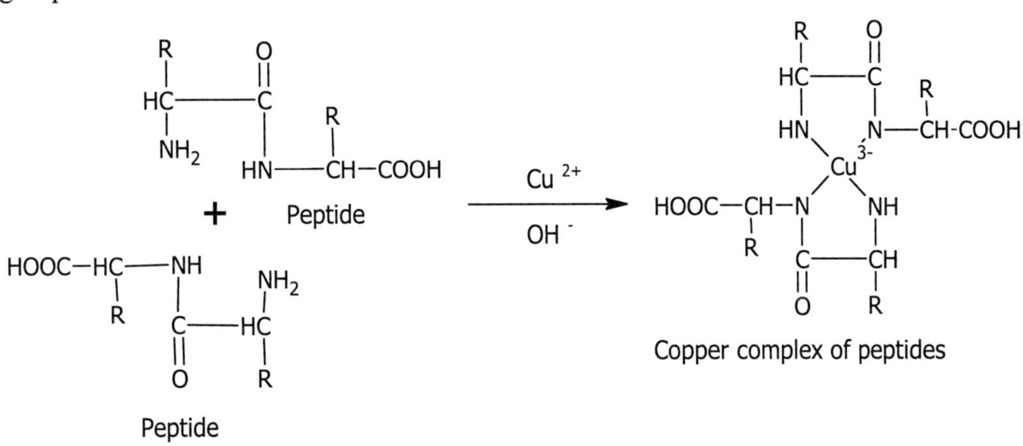

Biuret test is also positive for compounds containing two carbamoyl groups joined directly or through a single atom of nitrogen or carbon and similar substances which contain -$CSNH_2$, -C(NH) NH_2 and –CH_2NH_2.

Note: Excess of copper sulphate (i.e., > 2 drops) will produce blue colour instead of violet colour.

Ninhydrin test (Triketohydrindene hydrate test)

This test is characteristic for all the amino acids containing free amino group and carboxylic group. It also gives positive results for other primary amines.

Place 2 ml of protein solution in a test tube and add 2 drops of 0.2% ninhydrin solution. Boil the mixture for 2 min and then cool.	
• Development of purple colour • Development of yellow colour	• Indicates the presence of peptides • Indicates the presence of proline and hydroxyl proline

Mechanism: The triketohydrindene hydrate (ninhydrin) forms a complex with the amino or carboxylic group of the amino acids or other primary amino groups containing compounds. This complex upon reduction gives diketo amino derivative which further complex with another molecule of ninhydrin and gives purple colour. Proline and hydroxyl proline gives yellow colour due to the presence of secondary amino group.

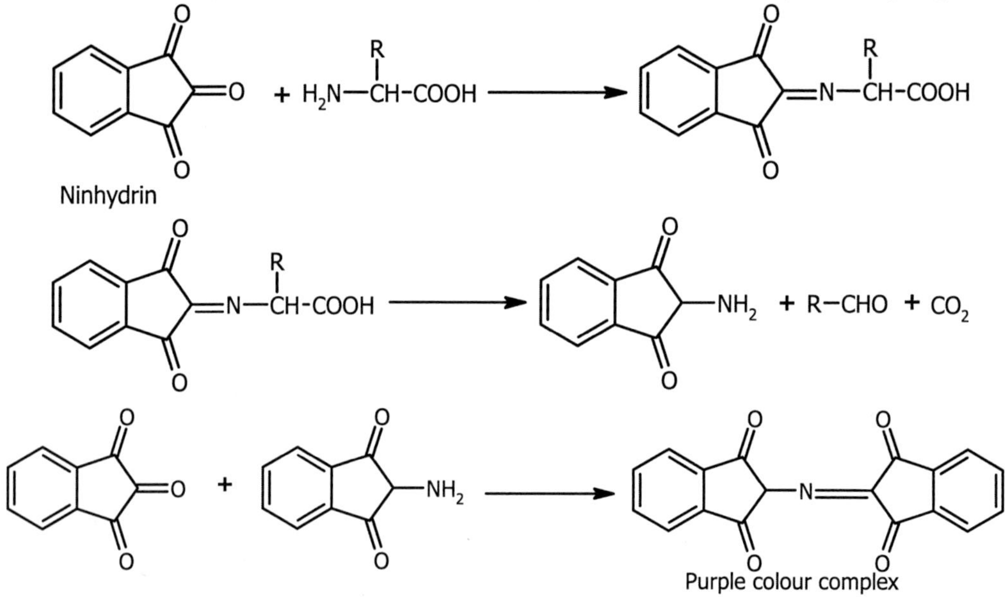

50.2 Test for Aromatic Amino Acids

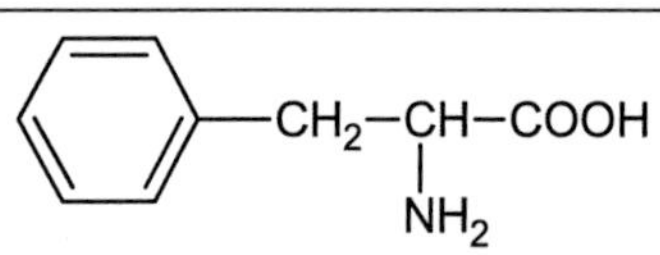 Phenyl alanine	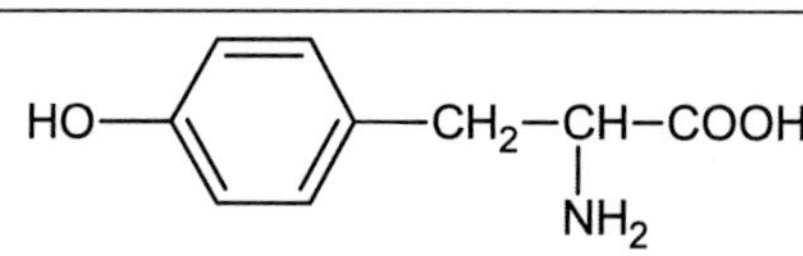 Tyrosine
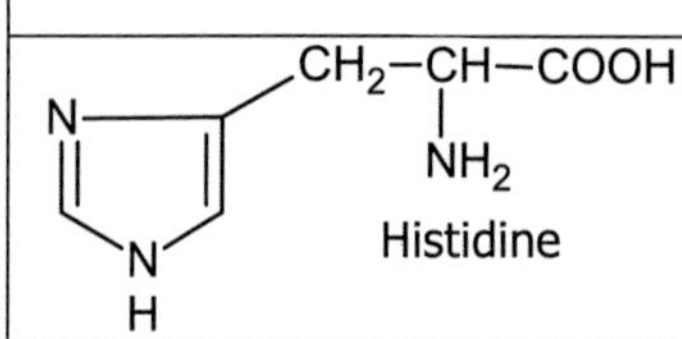 Histidine	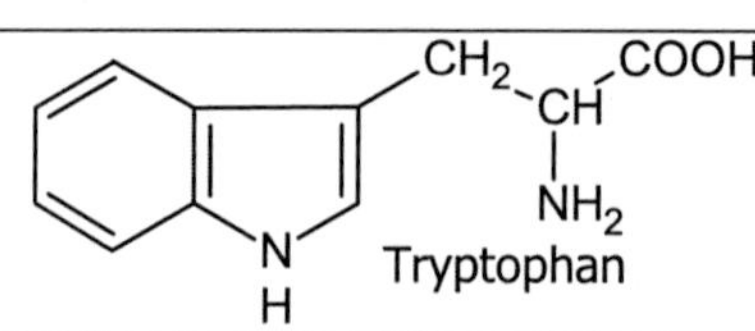 Tryptophan

Xanthoprotic test

Place 2 ml of protein solution in a test tube and add 1 ml of nitric acid. A white precipitate forms and boil the solution in water bath, the colour changes to yellow. Cool the test tube and add 2 ml of 40% sodium hydroxide/ammonia drop wise to make it alkaline.	
• Development of orange colour	• Indicates the presence of aromatic amino acids

Mechanism: Drop of nitric acid turns the skin yellow, due to the reaction with benzene ring of amino acid unit of the protein (tyrosine, tryptophan) and nitric acid. Nitric acid converts protein into white precipitate of metaprotein (insoluble). Aromatic group gets converted to yellow coloured aromatic nitro derivative on treatment with nitric acid (nitration). Alkaline pH (sodium hydroxide) changes yellow colour to orange colour, due to ionization.

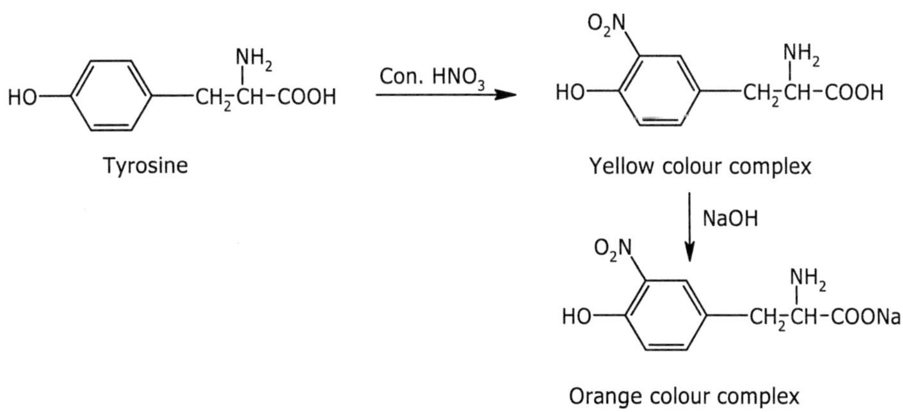

Test for Tyrosine

Millon's test:

Proteins containing tyrosine amino acids such as albumin, globulin, casein and peptone alone gives positive result for this reaction.

Place a few drops of the reagent in the test tube and add 1 ml of the acidified protein solution. Boil the mixture by placing in water bath for 10 minutes at 100 °C. Cool, add few drops of 1% sodium nitrite solution and warm gently.	
• Development of red colour	• Indicates the presence of phenolic group

Millon's reagent: 10% mercuric sulphate in 10% sulphuric acid, 1% sodium nitrite solution.

Solution of mercuric and mercurous nitrates in nitric acid contains little nitrous acid.

Mechanism: The neutral protein solution reacts with acidified mercuric nitrate to form a yellow precipitate of mercury-protein complex. This complex forms a nitrite complex with nitrite solution on warming and develops red colour. Phenols or phenolic substances such a salicylicacid (phenolic hydroxyl group) also gives positive results.

$$Hg + \text{Con. } HNO_3 \longrightarrow Hg(NO_3) + 2\,NO_2 + 2\,H_2O$$

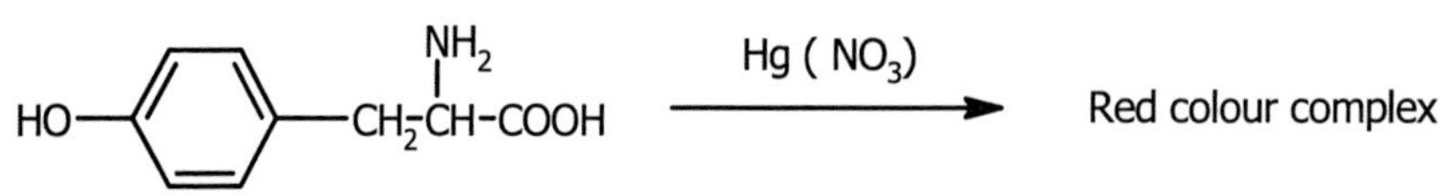

Tests for Tryptophan

Aldehyde test: (Hopkins-Cole Adam-MicKiewicz reaction)

To 1 ml protein solution in a test tube add 1 drop of 10% mercuric sulphate solution in 10% sulphuric acid and 1 drop of 40% formaline. Mix well, incline the test tube and add 1 ml sulphuric acid from the sides of the test tube.	
• Development of purple ring at the juction of two liquids	• Indicates the presence of tryptophan

This test is given by proteins containing amino acid tryptophan (e.g., peptone). In this test protein is precipitated by mercuric sulphate and the amino acid tryptophan condenses with formaldehyde in presence of H_2SO_4 to form violet coloured compound.

Glyoxalic Acid Reation (Hopkin's cole test)

To 2 ml of the protein solution add 2 ml of glyoxalic acid reagent. Pour this mixture to another test tube containing 4 ml of sulphuric acid from the sides of the tube. Rotate the test tube gently.	
• A purple ring develops at the junction of two liquids	• Indicates the presence of tryptophan

Glyoxalic acid reagent: Exposing glacial acetic acid to sunlight for a few minutes offers glyoxalic acid.

Note: Sulphuric acid must be highly pure; otherwise, the impurities may act as oxidizing agent. Oxidizing agents like nitrates and chlorates will inhibits the reaction.

Echrlich's test

Echrlich's test is specific for indole containig amino acid tryptophan, which gives red colour with the reagent.

To 1 ml of the echrlich's reagent, add 1 ml of the protein solution.	
• Development of red colour	• Indicates the presence of tryptophan

Echrlich's reagent: *para*-Dimethyl amino benzaldehyde (10%) in hydrochloric acid (10%).

Pauly's test

Mix 1 ml of sulphanilic acid (1% solution in 10% hydrochloric acid) with 2 ml of test solution and cool in ice. Add 1 ml of 5% sodium nitrite solution to cool solution and add 2 ml of 1% Na_2CO_3 solution after 5 minutes.	
• Development of red colour • Development of pink colour • Development of orange red colour	• Indicates the presence of histidine • Indicates the presence of tyrosine • Indicates the presence of tryptophan

50.3 Test for Guanidine Containing Amino Acids

Test for Arginine

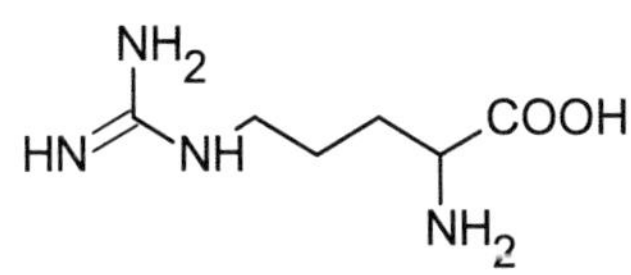

Sakaguchi test (Arginine test)

Sakaguchi test is specific test for guanidine group of arginine. Gelatin containing arginine also gives positive result. Non-aminoacids creatinine guanidine group also answers this test.

To 3 ml of the protein solution in a test tube add 1 ml of 10% NaOH, 5 drops of 1% alcoholic α-naphthol and mix. Then add 5 drops of bromine water/5 drops of freshly prepared sodium hypobromide solution.	
• Development of bright orange red colour	• Indicates the presence of arginine

Mechanism: Guanidine group of arginine reacts with α-naphthol in presence of bromine solution/sodium hypobromide and gives red orange condensation product.

50.4 Test for Sulphur Containing Amino Acids

Sulphur test

This test is given by sulphur containing amino acids cysteine, cystine and methionine.

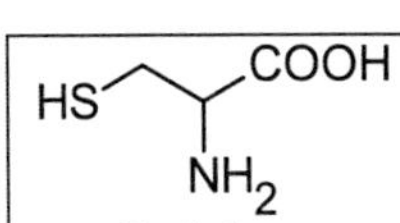

Cysteine

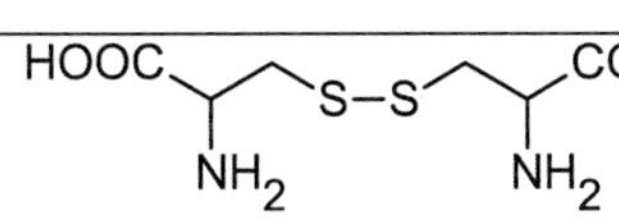

Cystine

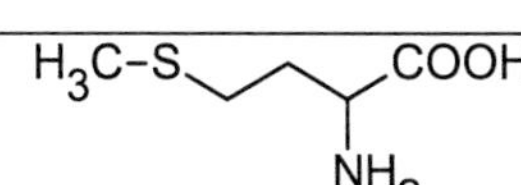

Methionine

Boil 1 ml of the protein solution with 1 ml of 40% NaOH for 1 minute and add a drop of lead acetate solution.	
• Formation of black/brown precipitate (insoluble in dilute hydrochloric acid)	• Indicates the presence of sulphur containing amino acids

Sulphur (present in amino acids) reacts with sodium hydroxide and forms sodium sulphide. Sodium sulphide upon reaction with lead acetate produces black/brown precipitate of lead suphide, which is insoluble in dilute hydrochloric acid.

50.5 Test for Carbohydrates Present in Proteins

Molish's test: Proteins containing carbohydrate (glycoprotein), can be detected by Molish's test.

Place 1 ml of the protein solution in a clean test tube, add 2 drops of Molisch's reagent (alcoholic α-naphthol, 10%) and mix well. Incline the test tube and add 0.5 ml of sulfuric acid drop wise along the sides of the test tube.	
• Formation of purple colour at the interface of two liquids	• Indicates the presence of glycoprotein (carbohydrate in protein)

Test for Phosphorus Present in Casein

Neumann's test

To the dry casein taken in dry test tube add 2 drops of nitric acid and 10 drops of sulphuric acid. Heat the contents strongly over a flame in a fume cupboard with constant stirring to avoid spurting. The organic material chars and nitrous fumes evolved. Allow the contents to cool, add 2 ml of distilled water and 2 ml of ammonia. Mix, add 2 ml of ammonium molybdate solution and warm.	
• Development of lemon yellow/Shining yellow/canary yellow colour precipitate of ammonium phosphomolybdate	• Indicates the presence of phosphorous present in protein

50.6 Coagulation Reactions

Denaturation of protein structure by heat or acids (nitric acid) is called as coagulation of protein. Albumin and globulin are coagulating proteins.

Heat coagulation

Take protein solution about $2/3^{rd}$ of the test tube and heat the upper portion of the solution.	
• Appearance of opalescent (becomes deep with few drops of 2% acetic acid)	• Indicates the presence of heat coagulating proteins (albumin and globulin)

Coagulation with nitric acid

Take 2 ml of nitric acid in a test tube and add the protein solution by the side of the test tube.	
• Appearance of white ring at the junction of the two liquids	• Indicates the presence of coagulating proteins

50.7 Precipitation Reactions

Proteins are precipitated from their solutions by variety of reagents and these precipitates are useful in isolation, identification and derivatization of proteins. Heavy metals (mercuric chloride, silver nitrate and copper sulphate), alkaloidal reagents (picric acid, phosphotungstic acid and tannic acid), acids (nitric acid and metaphosphoric acid) and concentrated solutions of ammonium sulphate and sodium chloride precipitates the proteins by 'salting out' methods.

Dehydrating agents such as alcohols and acetone also can precipitate by converting them into suspensoids (which flocculate by few drops of salt solution). Alcohols bring protein solution into isoelectric point at which it is precipitated.

Precipitation of mineral acids (Heller's test)

Mineral acids precipitates the protein by changing the pH and/inducing the protein denaturation (coagulation).

To 3 ml of the nitric acid in a test tube add 3 ml of protein solution along the sides of test tube.	
• Appearance of white precipitate at the junction of the two liquids (excess of acid dissolves the precipitate)	• Indicates the precipitation of proteins

Precipitation by heavy metals

Protein solution on treatment with solution of salts of heavy metals forms the precipitates and are soluble in excess of salt solution. So precipitation is done by adding the reagent drop by drop. The reagent salt solutions include 5% copper sulphate solution, 5% zinc sulphate solution, 2% lead acetate solution, 5% mercuric nitrate solution, 5% ferric chloride solution and diluted solution of mercuric chloride, silver nitrate and barium chloride.

Alkalify the protein solution by adding 1% sodium carbonate solution. To the 2 ml of alkaline protein solution add few drops of 5% solution of salts of heavy metals.
• Copper sulphate solution gives bluish white precipitate (soluble in excess of copper sulphate, reappears on further addition of copper sulphate) • Zinc sulphate solution gives white precipitate • Lead acetate solution gives white precipitate • Mercuric nitrate solution gives brown precipitate • Ferric chloride solution opalescence (disappears with excess of ferric chloride)

Precipitation by alkaloidal reagents

Alkaloidal reagents (complex organic acids) such as 2% tannic acid solution, 20% sulphosalicylic acid solution, 10% trichloro acetic acid and Esbach's reagent cause precipitation of protein solutions. In alkaline medium precipitation will not occur (may dissolve).

Sulphosalicylic acid test: To 2 ml of the dilute protein solution add a drop of 20% solution of salicylsulphonic acid (a solution of salicylic acid in sulphuric acid).	
• Formation of white precipitate/turbidity	• Indicates the precipitation of protein
Esbach's test: To 1 ml of the dilute protein solution add 1 ml Esbach's reagent (mixture of picric acid and citric acid).	
• Formation of yellow precipitate	• Indicates the precipitation of protein
Trichloroacetic acid test: To 2 ml of the dilute protein solution add 2 ml of 10% trichloroacetic acid.	
• Formation of white precipitate	• Indicates the precipitation of protein
Tannic acid test: To 2 ml of the dilute protein solution add 3-5 drops of 2% tannic acid.	
• Formation of white precipitate	• Indicates the precipitation of protein

Precipitation by neutral salts

Full saturation test: To 5 ml protein solution taken in a test tube add solid ammonium sulphate until it becomes saturated.	
• Formation of gelatinous precipitate	• Indicates the presence of albumin
Half saturation test: To 3 ml saturated solution of ammonium sulphate add 3 ml of saturated protein solution and shake vigrously.	
• Formation of gelatinous precipitate	• Indicates the presence of albumin

Precipitation by alcohol

To 1 ml of protein solution add 3 ml of alcohol.	
• Formation of white opalescence (doesn't dissolve in water).	• Indicates the presence of protein

50.8 General Scheme for the Analysis of Proteins

Test	Observation	Inference
A General Test for Proteins		
Biuret test	• Violet or purple colour • Rosy pink colour	• Indicates the presence of proteins • Indicates the presence of proteins
Control	• Blue colour	• Indicates the absene of proteins
Ninhydrin test	• Blue colour • Yellow colour	• Proteins present • Proline and hydroxy proline

Table Contd...

Test for Aromatic Amino Acids		
Xanthoprotic test	Colour changes to yellow and to orange	Indicating the presence of aromatic amino acids (tyrosine, phenyl alanine and tryptophan).
Test for Tyrosine		
Millon's test	Red precipitate and solution becomes red	Presence of tyrosine
Tests for Tryptophan		
Aldehyde test: (Hopkins-Cole- Adam-MicKiewicz reaction)		
A purple violet ring develops at the junction of two liquids		Presence of tryptophan.
Glyoxalic Acid Reation (Hopkin's cole test)		
A purple ring develops at the junction of two liquids		Presence of tryptophan
Echrlich's test	Formation of red colour	Indicates the presence of tryptophan
Paulys test	• Formation of red colour • Formation of pink colour • Formation of orange-red colour	• Indicates the presence of histidine • Indicates the presence of tyrosine • Indicates the presence of tryptophan
Test for Arginine		
Sakaguchi test (arginine test)	A development of bright red-orange colour	Indicates the presence of arginine
Test for sulphur containing amino acids		
Sulphur test	Formation of black/brown precipitate (insoluble in dilute hydrochloric acid)	Indicates the presence of methionine, cysteine and cystine (sulphur containing amino acids)
Test for Carbohydrates Present in Proteins		
Molish's test	A purple color develops at the interface	Indicates the presence of glycoprotein.
Coagulation Reactions		
Heat coagulation	Appearance of opalescence (becomes deep with few drops of 2% acetic acid)	Indicates the presence of heat coagulating proteins (albumin and globulin)
Coagulation with Nitric acid	Appearance of white ring at the junction of two liquids	Indicates the presence of heat coagulating proteins (albumin and globulin)
PRECIPITATION OF REACTIONS		
Precipitation of Mineral Acids (Heller's test)		
Appearance of white precipitate at the junction of the two liquids.		Indicates the presence of albumin

Table Contd...

Precipitation by Heavy Metals		
Copper sulphate test	Formation of bluish white precipitate (Soluble in excess of copper sulphate but reappears further addition)	Indicates the presence of protein
Zinc sulphate test	Formation of white precipitate	Indicates the presence of protein
Lead acetate test	Formataion of white precipitate	Indicates the presence of protein
Ferric chloride test	Appearance of opalescence (disappears on the addition of excess ferric chloride)	Indicates the presence of protein
Mercuric nitrate test	Formation of brown colour precipitate	Indicates the presence of protein
Precipitation by Alkaloidal Reagents		
Sulphosalicylic acid test	Formation of white precipitate/turbidity	Indicates the presence of protein
Esbach's reagent test	Formation of yellow precipitate	Indicates the presence of protein
Trichloroacetic acid test	Formation of white precipitate	Indicates the presence of protein
Tannic acid test	Formation of white precipitate	Indicates the presence of protein
Precipitation by Neutral Salts		
Full saturation test	Formation of gelatinous white precipitate	Indicates the presence of albumin
Half saturation test	Formation of gelatinous white precipitate	Indicates the presence of albumin or globulin

Precipitation by Alcohol		
Protein solution with alcohol	A white opalescent forms (doesn't dissolve in water)	Indicates the presence of protein
Test for Phosphorus Present in Casein		
Neumann's Test	Development of lemon yellow/shining yellow/canary yellow colour precipitate of ammonium phosphomolybdate	Indicates the presence of phosphorous

CHAPTER 51

QUALITATIVE ANALYSIS OF ALKALOIDS

Alkaloids are the physiologically active basic compounds (alkali-like) obtained from plant kingdom, in which at least one nitrogen atom forms part of a cyclic system. Alkaloids (cyclic organic compounds) contain nitrogen (one/two) in a negative state of oxidation.

Physical Properties

Physical state: Alkaloids occurs as crystalline solids and amorphous solids (emetine). Some are obtained as liquids, which includes volatile (nicotine and coniine) and non-volatile (pilocarpine and hyoscine) liquids.

Color: The majority of alkaloids are colorless, but colchicine and berberine appears yellow, canadine as orange and salts of sanguinarine appears as copper-red colour.

Solubility: The alkaloidal bases are soluble in organic solvents and insoluble in water. Salts are usually soluble in water and insoluble (sparingly soluble) in organic solvents.

- Alkaloids insoluble (sparingly soluble) in certain organic solvents: Morphine in ether, theobromine and theophylline in benzene.
- Water soluble alkaloids: Caffeine, ephedrine, codeine, colchicine, pilocarpine and quaternary ammonium bases.
- Liquid alkaloid soluble in water: Coniine.
- Water soluble alkaloidal salts: Quinine monosulphate.
- Alkaloidal salts soluble in organic solvents: Lobeline and apoatropine hydrochlorides in chloroform.

The alkaloids form insoluble precipitates with solutions of phospotungstenic acid, phosphomolybdic acid, picric acid and potassium mercuric iodide. Many of these precipitates have definite crystalline shapes and used in the identification of an alkaloid.

Most alkaloids are optically active (laevorotatory) compounds and produce bitter taste.

Functions of alkaloids

- Protective against insects and herbivores due to their bitterness and toxicity.
- Nitrogen reservoirs for protein synthesis and in case of nitrogen deficiency.
- Detoxication by methylation, condensation and cyclization of the compounds (whose accumulation might otherwise damage to the plants).
- Plant stimulants or regulators involve in growth, metabolism and reproduction.
- Source of energy in case of deficiency in carbon dioxide assimilation.

Classification

Type	Source	Example
True (typical) alkaloids	Derived from amino acids and have nitrogen in a heterocyclic ring.	Atropine
Proto alkaloids	Derived from amino acids and do not have nitrogen in a heterocyclic ring.	Ephedrine
Pseudo alkaloids	Not derived from amino acids but have nitrogen in a heterocyclic ring.	Caffeine
False alkaloids	Non alkaloids give false positive reaction with alkaloidal reagents.	

Alkaloids are also classified according to the amino acid from which they are derived

- ***Histidine:*** Pilocarpine, pilosine (imidazole alkaloids)
- ***Tryptophan:*** Ajmalin, ajmalicin, yohimbine, reserpine, vincristine and strychnine (indole alkaloids)
- ***Phenylalanine (α-amino acid)*****:** Phenylethylamines and tetrahydroisoquinoline alkaloids, mescaline
- ***Tyrosine (α-amino acid):*** Morphine, codeine and thebaine (phenanthrene alkaloids), berberine, papaverine, emetine and tubocurarine (tetrahydroisoquinoline alkaloid), mescaline (phenyl alkyl amine)
- ***Lysine (α-amino acid):*** Piperidine, lobelanine, lobeline, pelletierine, isopelletierine, pseudopelletierine, quinolizidine, and indolizidine alkaloids
- ***Proline (α-amino acid):*** Nicotine
- ***Nicotinic acid (β-amino acid):*** Nicotine
- ***Anthranilic acid (β-amino acid):*** Ajmalin, ajmalicin, ergotamine, yohimbine, reserpine, vincristine and strychnine (indole alkaloids)
- ***Ornithine (dibasic amino acid):*** Senecionine (pyrrolizidine alkaloids) and cocaine, scopolamine (tropane alkaloids), hygrine, nicotine (pyridine and pyrazolidine alkaloid).

It is most satisfactory to classify them according to the chemical nature of the nucleus present in alkaloids.

Chemical nature	Alkaloids
Phenylethylamine	Ephedrine
Tropane	Atropine, cocaine
Tropolone	Colchicine
Pyrrolidine	Hygrine
Pyrrolizidine	Retronecine, senecionine
Imidazole (glyoxaline)	Pilocarpine, pilosine
Indole	Ergotamine, reserpine, strychnine, vincristine and physostigmine
Quinoline	Quinine, cinchonine
Isoquinoline	Papaverine, emetine, berberine and tubocurarine
Quinolizidine (lupin alkaloids, nor-lupinane)	Sparteine, cystisine, lupanine, angustifoline
Phenanthrene	Morphine, codeine and thebaine
Pyridine and piperidine	Nicotine, coniine, arecoline, piperine, pelletierine
Purine (xanthine alkaloids)	Caffeine, theophylline, theobromine
Steroid	Ceveratum, jeveratum

Alkaloids derived from animal kingdom

- *Salamandra maculosa*: Samandarone and cycloneosamandarine
- *Phyllobates terribilis, Phyllobates aurotaenia and Phyllobates bicolor*: Batrachotoxin A
- *Bufo alvarius*: Bufotenine
- *Castor fiber*: Deoxy-nuphradine, castoramine
- *Moschus moschiferus*: Muscopyridine

51.1 General Tests for Alkaloids

To the substance add dilute hydrochloric acid, shake well and filter. With the filtrate perform the following general tests.

1. **Mayer's test**

Mayer's reagent: Potassio-mercuric iodide solution.

To 2 ml of the filtrate add 2 drops of Mayer's reagent	
Development of creamy white/yellow precipitate	Indicates the presence of alkaloids

Ephedrine, caffeine and theobromine will give no precipitate with Mayer's reagent.

2. **Dragendorff's test**

Dragendorff's reagent: Potassium bismuth iodide solution

To 2 ml of the filtrate add 2 drops of Dragendroff's reagent	
Development of orange or orange red precipitate	Indicates the presence of alkaloids

3. **Wagner's test**

Wagner's reagent: Solution of iodine in potassium iodide

To 2 ml of the filtrate add 2 drops of Wagner's reagent	
Development of reddish brown precipitate	Indicates the presence of alkaloids

4. **Hager's test**

Hager's reagent: 1 gm of picric acid in 100 ml of water

To 2 ml of the filtrate add 2 drops of Hager's reagent	
Development of yellow precipitate	Indicates the presence of alkaloids

5. **Tannic acid test**

Tannic acid reagent: 10 gm of tannic acid in 10 ml of alcohol and dilute with water to 100 ml

To 2 ml of the filtrate add 2 drops of Tannic acid reagent	
Development of yellowish white precipitate	Indicates the presence of alkaloids

6. **Mercuric chloride test**

To 2 ml of the filtrate add 2 drops of Mercuric chloride	
Development of white precipitate	Indicates the presence of alkaloids

51.2 General Colouring Agents

A simple colour change produced by the following reagents indicates the presence of alkaloids.

- ***Mandelin's reagent (sulphovanadic acid):*** 1 gm of ammonium vanadate in 100 ml of concentrated sulphuric acid (Sulphovanadic acid).
- ***Froehde's reagent:*** 5 gm of molybdic acid or sodium molybdate in 5 ml of hot concentrated sulphuric acid.

- ***Marquis reagent:*** 3 ml of concentrated sulphuric acid containing 3 drops of 40% formaldehyde.
- ***Erdmann's reagent:*** A mixture of 10 drops of concentrated nitric acid and sulphuric acid.
- ***Mecke's reagent (selenious-sulphuric acid):*** 1 gm of selenious acid in 200 ml of concentrated sulphuric acid.
- ***Modified Dragendroff's reagent (Kraut's reagent):*** Dissolve 1.6 gm of bismuth subnitrate in 60 ml of 20% glacial acetic acid and add 5 ml of each 40% potassium iodide solution and glacial acetic acid. Make up the volume with 100 ml water.
- ***Rosenthaler's reagent (selenious-sulphuric acid):*** 1 gm of potassium arsenate in 200 ml of concentrated sulphuric acid.

51.3 Identification of Phenylethylamine Alkaloids

Ephedrine, hordenine, mescaline and narceine are the structural examples of phenylethylamine alkaloids.

Ephedrine and Pseudo ephedrine (*Ephedra sinica, Ephedra vulgaris*) 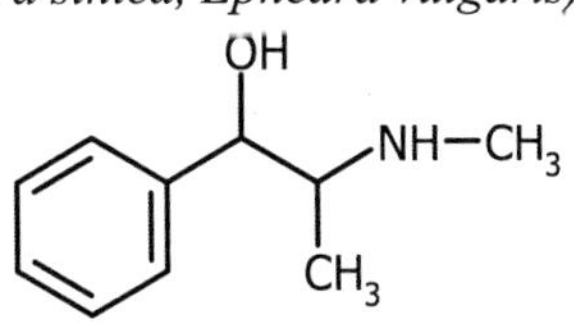	Hordenine (*Ariocarpus fissuratum, Hordeum vulgare*) 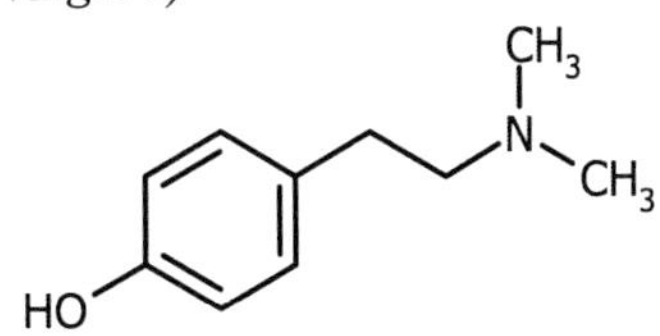
Mescaline (*Lophophora williamsii, Echianopsis pachanoi, Echianopsis peruviana*) 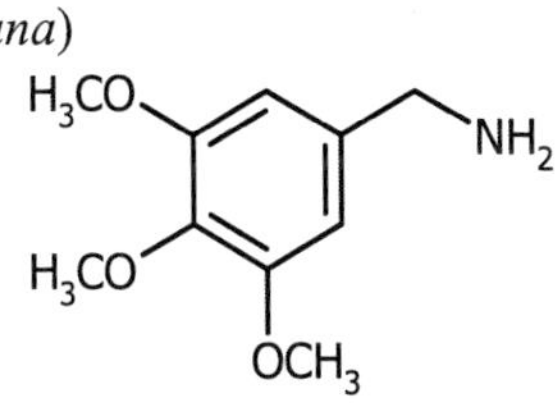	Narceine (*Papaver somniferum*) 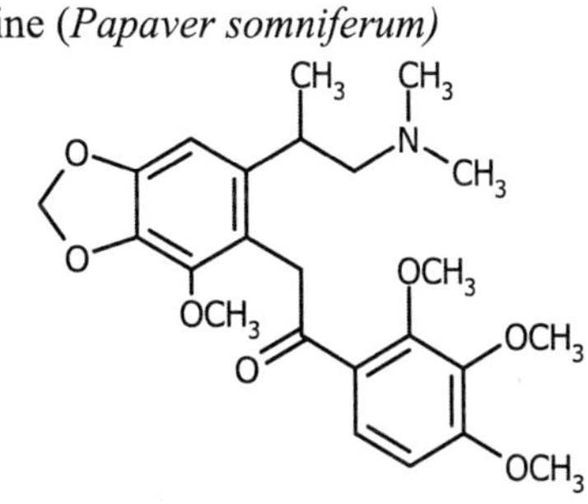

Test for ephedrine	
Chen's test: To 2 ml of alkaloid solution, add few drops of dilute hydrochloric acid, 1 ml 5% copper sulphate, 1 ml 20% sodium hydroxide and 1 ml petroleum ether. Develops reddish colour and shake well the contents.	
Formation of purple colour in organic layer and blue colour in aqueous layer	Indicates the presence of ephedrine (phenylethylamine alkaloid)

Table Contd...

Dissolve alkaloid in 30 ml of chloroform in a stoppered flask. Shake the contents vigorously and allow it stand at room temperature for 12 hours. Evaporate the chloroform layer.	
Ephedrine hydrochloride crystals are separates out	Indicates the presence of ephedrine (phenylethylamine alkaloid)
Triturate ephedrine with few crystals of potassium ferricyanide and add few drops of water. Heat on a water bath.	
Odour of bitter almonds (distinct odour of benzaldehyde)	Indicates the presence of ephedrine (phenylethylamine alkaloid)

51.4 Identification of Tropane Alkaloids

Tropane alkaloids contain tropane ring (bicyclic [3.2.1] ring). These are abundant in solanaceae family plants and acts as anticholinergic.

Atropine (*Atropa belladonna*) 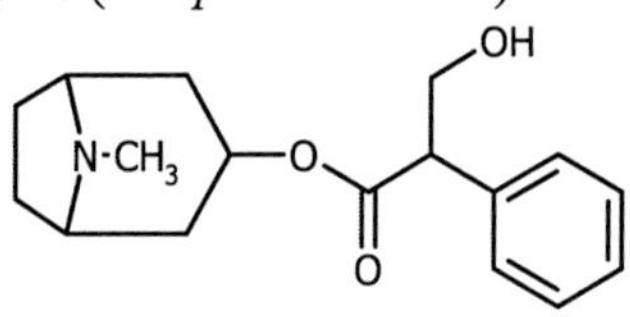 Hyoscyamine Levo form of atropine	Hyoscine (*Atropa belladonna, Hyoscyamus niger Datura stromanium)* 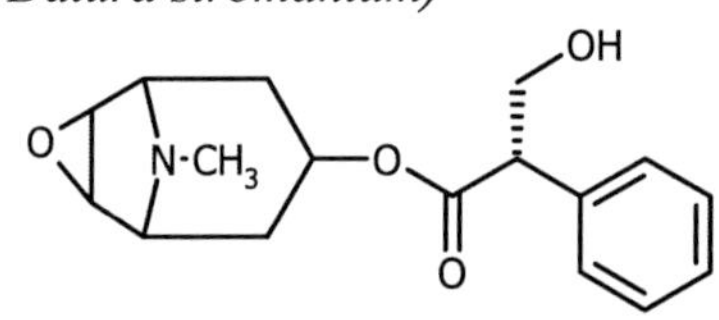
Cocaine (*Erythroxylon coca)* 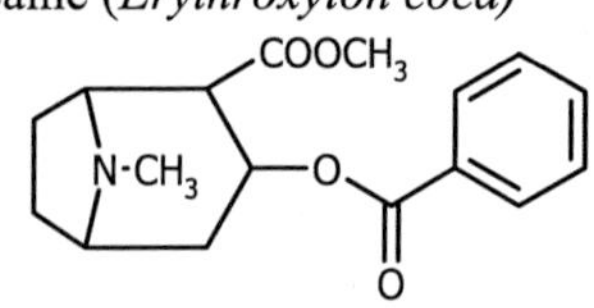	Ecgonine (*Erythroxylon coca)* 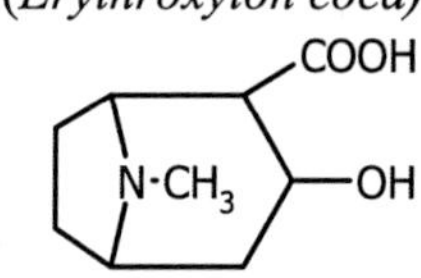

Test for Tropane Alkaloids	
Test for atropine and cocaine	
Vitali-Morin's Test: Evaporate 1 ml of alkaloid solution to dryness and cool. Add 2 drops of fuming nitric acid, evaporate this mixture to dryness, cool and add 1 drop of 10% alcohlic potassium hydroxide.	
Development of violet colour	Indicates the presence of tropane alkaloids (atropine, cocaine)
***p*-Dimethylamino benzaldehyde (DMAB) test:** To 1 ml alkaloid solution add few drops of *p*-dimethylamino benzaldehyde and heat the mixture by placing in hot plate with continuous stirring.	
Development of cherry red colour	Indicates the presence of tropane alkaloids (atropine, cocaine)
Test for atropine	
Gerrard reaction: To the 10 mg of atropine add 2% w/v ethanolic mercuric chloride solution	
Produces instant red colour	Indicates the presence of atropine

Table Contd...

Test for cocaine	
Permanganate test: To the sample solution prepared in the saturated alum add few drops of potassium permnganate.	
Formation of violet crystalline precipitate (cocaine permanganate)	Indicates the presence of cocaine
Test for hyoscyamine	
Schaer's test: To the sample add drop of hydrogen peroxide (30%) and 1 ml of sulphuric acid.	
Formation of distinct green colour	Indicates the presence of hyoscyamine
Gerrard reaction: To the 10 mg of atropine add 2% w/v ethanolic mercuric chloride solution.	
Produces red colour after warming	Indicates the presence of hyoscyamine
Vitali-Morin's Test:	
Violet colour changes into red and becomes colourless	Indicates the presence of hyoscyamine
Test for hyoscine	
Silver nitrate test: To the sample solution add few drops of silver nitrate solution.	
Formation of yellowish white precipitate (Insoluble in nitric acid and soluble in dilute ammonia)	Indicates the presence of hyoscine
Gerrard reaction: To the 10 mg of atropine add 2% w/v ethanolic mercuric chloride solution.	
• Produces white colour	• Indicates the presence of hyoscine

51.5 Identification of Pyrrolizidine Alkaloids

Pyrrolizidine alkaloids are also known as necine bases. They are produced by the plants as a defensive mechanism against insect herbivores.

Retronecine *(Senecio vulgaris, Heliotropium europaeum)* 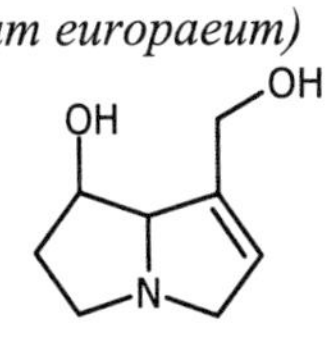	Senecionine *(Senecio aureus, Tussilago farfara)* 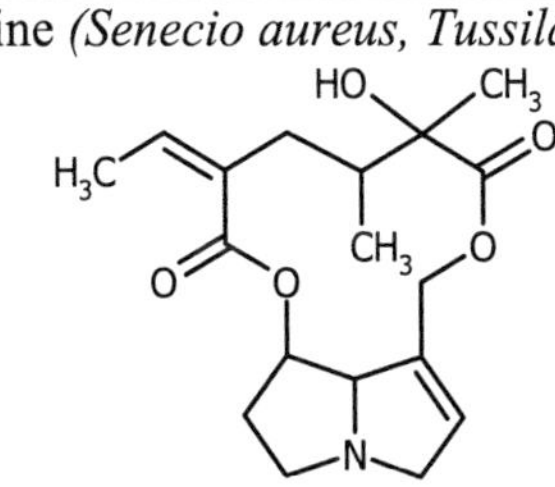
Test for pyrrolizidine alkaloids	
Ehrlich test: Mix powdered plant (1.5 grams) with 5% ascorbic acid (40 mg) and with small amount of sand, separately. Shake each of them and after 5 min filter and divide into sample and blank. To the sample tube add 12 ml of alkaline sodium nitroprusside (5%) reagent. To both the tubes (test and control) add ehrlich reagent and heat by placing in water bath for 5 minutes.	
Ehrlich regent: Mixture of 5 gm of 4–dimethylamino benzaldehyde (DMAB) (dissolved in 3 ml of water), 10 ml of perchloric acid 60% and 60 ml of acetic acid.	
Formation of red colour	Indicates the presence of pyrrolizidine alkaloids

51.6 Identification of Imidazole Alkaloids

Pilocarphus genus produces imidazole alkaloids from amino acid histidine.

Pilocarpine (*Pilocarpus jaborandi*)	Pilosine (*Pilocarpus microphyllus)*	Pilocarpidine (*Pilocarpus microphyllus)*
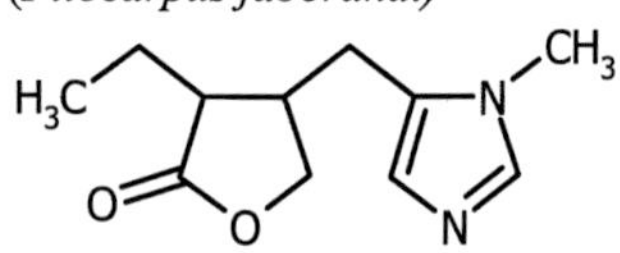	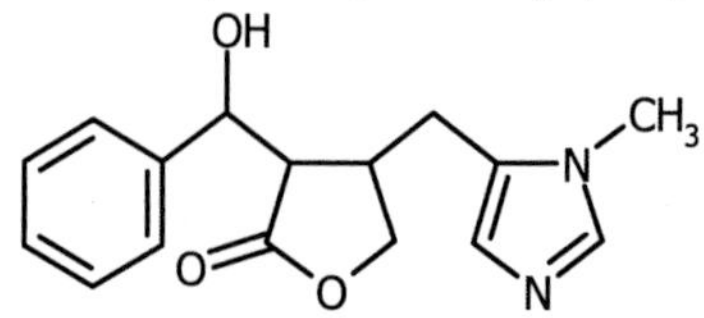	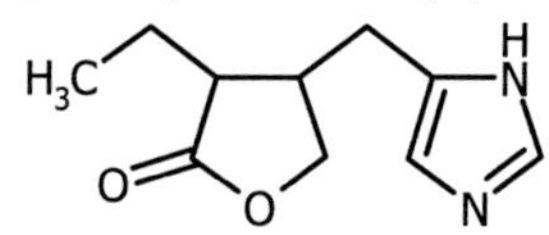

Test for imidazole alkaloids	
Helch's test: To the aqueous solution add 3 drops of hydrogen peroxide and few drops of potassium dichromate.	
Development of violet colour	Indicates the presence imidazole alkaloids
Ekkerts test: To the aqueous solution of sample add sodium nitroprusside, few drops of dilute sodium hydroxide. After few minutes add dilute hydrochloric acid (acidification).	
Development of wine red colour (addition of sodium thiosulphate changes red into green)	Indicates the presence imidazole alkaloids

51.7 Identification of Indole Alkaloids

Indole alkaloids are derived from amino acid tryptophan.

Ergotamine (*Claviceps purpurea*)	Reserpine (*Catharanthus roseus, Rauwolfia serpentina*)
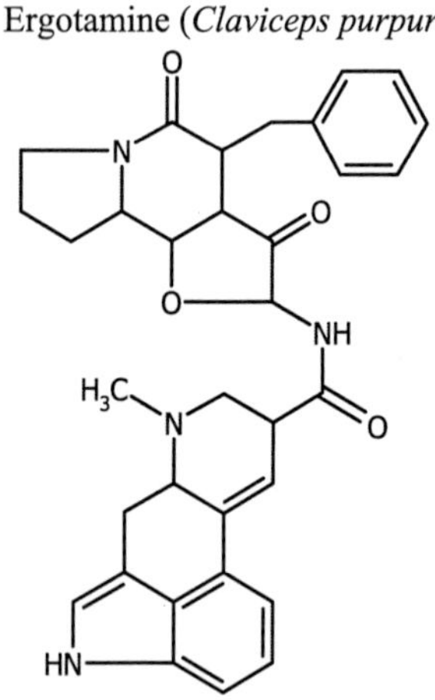	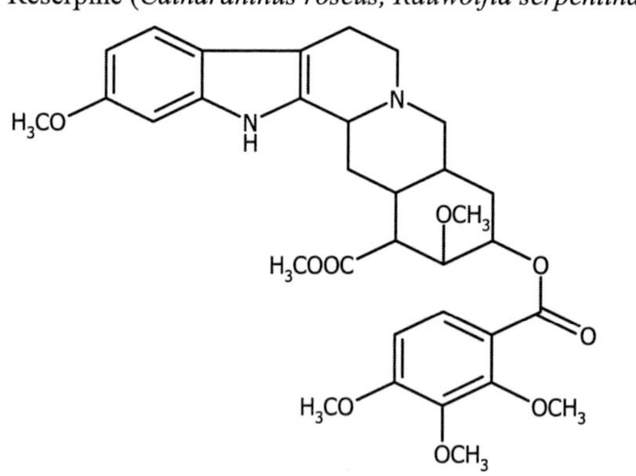

Strychnine and Brucine (*Strychnus nuxvomica*)	Vincristine (*Vinca rosea*)
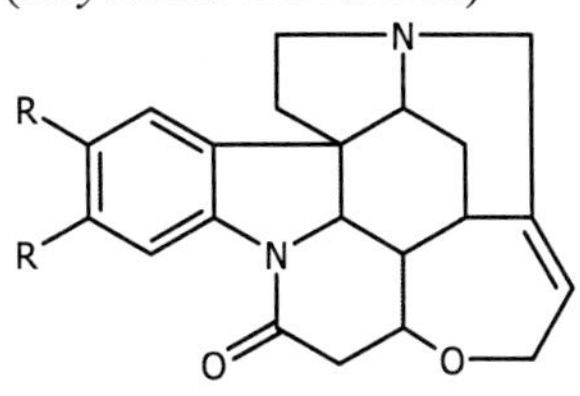 Strychnine R= H, Brucine R= OCH_3	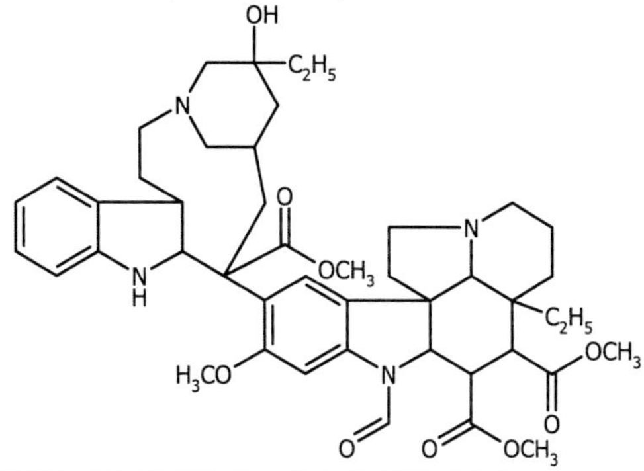
Physostigmine (*Physostigma venenosum*) 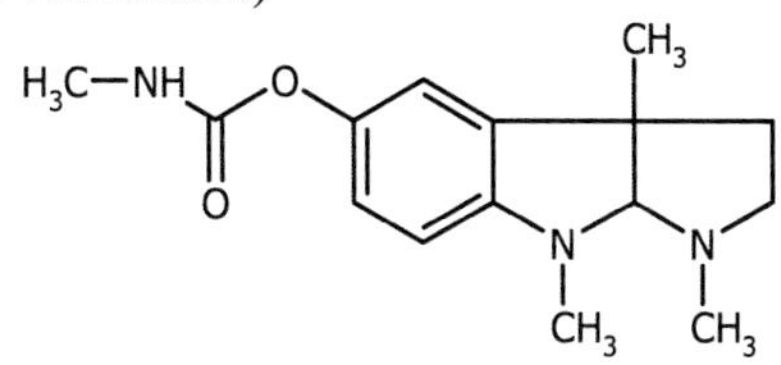	

Test for Ergot alkaloids	
Ehrlich test: To 2 ml of alkaloid solution add few drops of Ehrlich reagent.	
Ehrlich *regent:* 20 mg of *p*-dimethylamino benzaldehyde (DMAB) in 15% H_2SO_4, containing trace of $FeCl_3$.	
Formation of deep blue colour	Indicates the presence of ergot and polypeptide alkaloids.
Van-Urk's test: To 2 ml of alkaloid solution add few drops of Van-Urk's reagent.	
Van-Urk's *regent:* Mixture of 5 gm of 4–dimethylamino benzaldehyde (DMAB) (dissolved in 3 ml of water), 10 ml of perchloric acid 60% and 60 ml of acetic acid.	
Formation of red colour	Indicates the presence of pyrrolizidine alkaloids
Keller's test: To the solution of the alkaloid in acetic acid with traces of ferric chloride add sulphuric acid along the wall of the test tube.	
Formation of intense blue colour between two phases	Indicates the presence of ergot and polypeptide alkaloids

Test for Vinca alkaloids	
Vanillin-hydrochloric acid test: To the powdered sample add vanillin in hydrochloric acid.	
• Development of pink colour • Development of orange-yellow colour	• Indicates the presence vinblastine • Indicates the presence vincristine
Van-Urk's test: To 2 ml of alkaloid solution add few drops of Van-Urk's reagent.	
Development of reddish brown colour	Indicates the presence of vinca alkaloids

Test for Physostigma alkaloids	
Eserine blue test: To 2 ml of alkaloid solution add strong ammonia solution.	
Development of yellowish red colour [gives bluish residue (eserine blue) on evaporation]	Indicates the presence of physostigmine
Alkali hydrolysis test: To 2 ml of physostigmine (eserine) solution add strong alkali.	
• Ammonium hydroxide produces red colour • Potassium hydroxide produces yellow colour	Indicates the formation of eserine hydroxides
Fuming nitric acid test: To the 1 ml physostigmine (eserine) add five drops of potassium ferricyanide and two drops of ferric chloride (1%)	
Develops blue colour	Indicates the formation of chloroeserine
Potassium ferric cyanide test: To the small quantity of physostigmine (eserine) solution in a porcelain dish add few drops of fuming nitric acid.	
Develops yellow colour [which upon evaporation produces green residue]	Indicates the presence of physostigmine

Test for Nux-vomica alkaloids	
Nitric acid test: To few crystals of alkaloid add drops of nitric acid.	
• Development of faint yellow colour (turns to yellow on evaporation) • Development of intense red colour (turns to violet on evaporation and addition of stannous chloride)	• Indicates the presence of strychnine • Indicates the presence of brucine
Tests for strychnine	
Sulfuric acid-dichromate test: To few crystals of alkaloid add 3 drops of sulphuric acid and few crystals of potassium dichromate.	
Development of deep blue colour (which gradually changes to violet to purplish red to orange and to yellow)	Indicates the presence of strychnine
Mandelin's reagent: To few crystals of alkaloid add Mandelin's reagent.	
• Development of deep violet colour (changes to red to cherry red on addition of water)	• Indicates the presence of strychnine
Tests for brucine	
Transfer 5 ml of the alkaloidal solution to separatory funnel, add 2 ml of ammonia and 5 ml of chloroform. Shake well and separate the chloroform layer. Evaporate 1 ml of chloroform layer to dryness, which gives residue after evaporation.	
Erdmann's reagent test: To the part of residue add 2 drops of Erdmann's reagent.	
Development of orange red colour	Indicates the presence of brucine
Nitric acid test: To another part of residue add 2 drops of nitric acid.	
Development of orange red colour	Indicates the presence of brucine
Stannous chloride test: To the above residue add freshly prepared stannous chloride solution.	
Development of violet colour	Indicates the presence of brucine

Test for Rauwolfia alkaloids	
Vanillin-hydrochloric acid test: To the powdered sample add vanillin in hydrochloric acid.	
• Development of red colour • Development of orange-yellow colour	• Indicates the presence rauwolfia alkaloids

51.8 Identification of Quinoline Alkaloids

Alkaloid	Structure	Source	Functions
(-)-Quinine, (+)-Quinidine	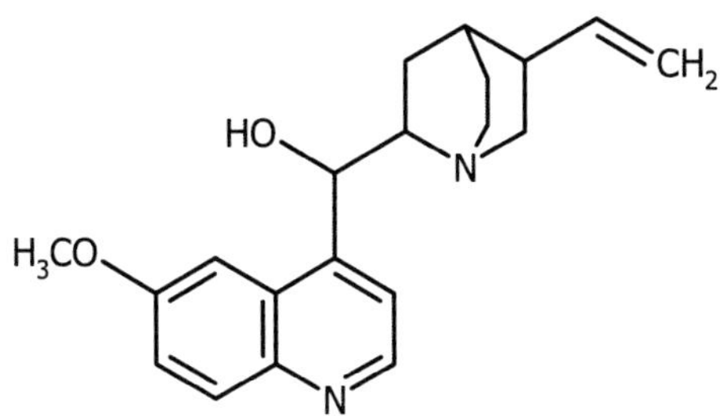	*Cinchona officinalis*	Antimalarial, oxytocic, skeletal muscle relaxant
(+)-Cinchonine, (-)-Cinchonidine	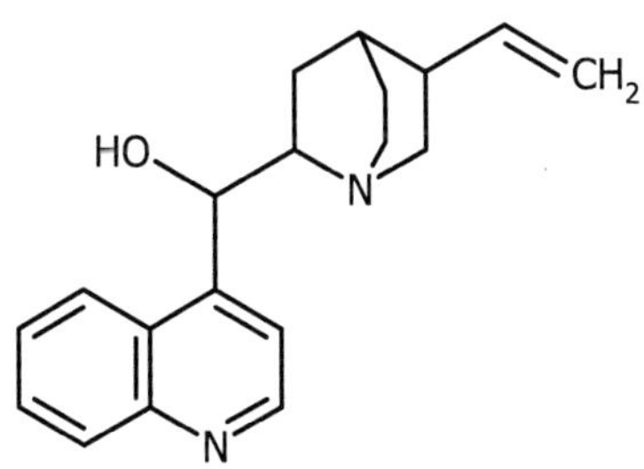		

Test for quinine	
Fluorescence test: To the sample solution (1 ml) add sulphuric acid (1 ml) and 5 ml of water.	
Appearance of blue fluorescence under ultra violet light	Indicates the presence of quinine
Perform blank (without sample solution) and observe the difference.	
Thalleoquin test: Quinine sulphate with chlorine/bromine and excess of ammonia produces greenish resinous mass, which is called as Thalleoquin.	
To the 1 ml of sample solution add 5 drops of bromine water and 1 ml of ammonia	
Development of emerald green colour	Indicates the presence of quinine
Herapathite test: To the boiling mixture of quinine (0.3 gm), glacial acetic acid (7.5 ml), ethanol (3 ml) and 5 drops of concentrated sulphuric acid, add ethanolic iodine (1%, 3.5 ml).	
Formation of crystals on standing [appears dark in reflected light and green in transmitted light]	Indicates the formation of iodosulphate of quinine

Table Contd...

Glacial acetic acid test: To the 10 mg of powdered sample in a test tube add few drops of glacial acetic acid.	
Development of purple colour	Indicates the presence of quinine
Silver nitrate test: To the 1 ml of solution of the sample add 5 drops of silver nitrate solution	
Formation of white precipitate (soluble in nitric acid)	Indicates the presence of quinine
Erythroquinine test (Rosequin/Ferrocyanide test): Dissolve few mg of quinine in dilute acetic acid and add few drops of bromine water (fresh). To the above mixture add drop of 10% potassium ferrocyanide and concentrated ammonium hydroxide.	
Formation of instant red colour	Indicates the presence of quinine (formation of erythroquinine)
Shake the above mixture with 2 ml of chloroform, organic layer becomes red colour.	

51.9 Identification of Isoquinoline Alkaloids

Papaverine (*Papaver somniferum*)	Emetine (*Cephaelis ipecacuanha*)
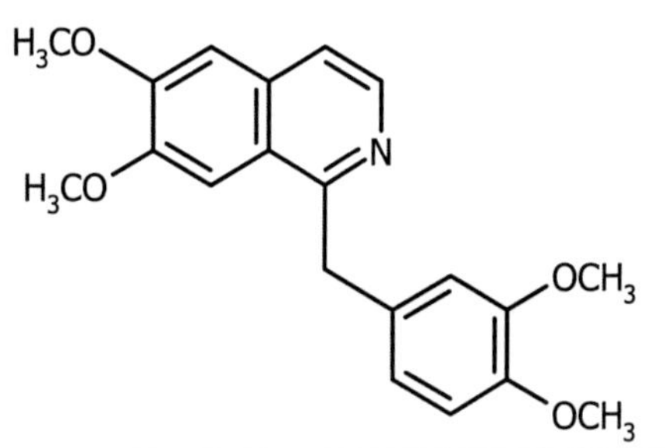	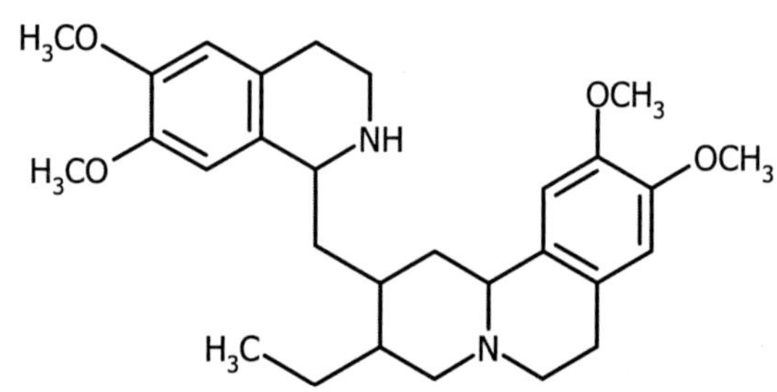
Berberine (*Berberis vulgaris*)	Tubocurarine (*Chondrodendron tomentosum*)
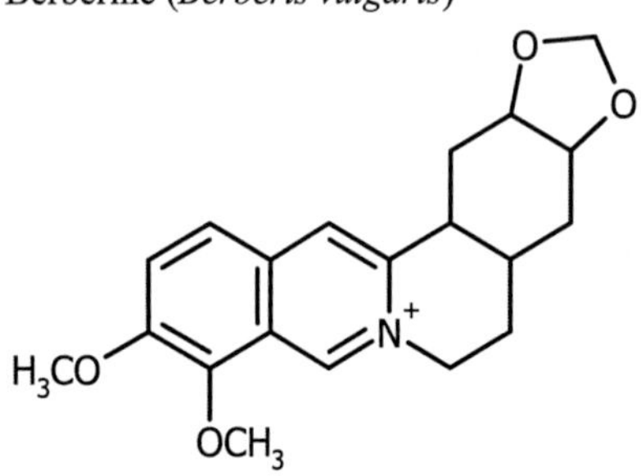	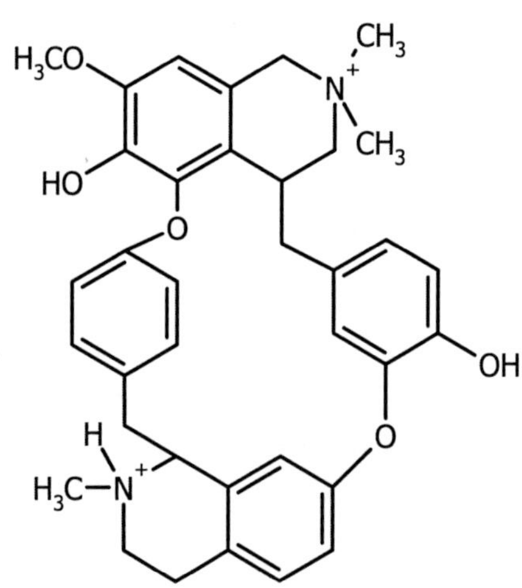

Test for isoquinoline alkaloid	
Transfer 5 ml of the alkaloidal solution to separating funnel, add 2 ml of ammonia and 5 ml of chloroform. Shake well and separate the chloroform layer. Evaporate 1 ml of chloroform layer to dryness, which gives residue after evaporation.	
Marquis test: To the part of residue add 2 drops of Marquis reagent.	
Formation of greenish-violet colour	Indicates the presence of isoquinoline alkaloid
Froehd's test: To another part of residue add 2 drops of Froehd's reagent.	
• Formation of green colour	• Indicates the presence of papaverine (isoquinoline alkaloid)
• Development of dirty green colour	• Indicates the presence of emetine
Warren's test: To the part of residue present in porcelain dish add few crystals of potassium permanganate and crush them. To the mixture add 3 drops of Marquis reagent.	
Formation of purple colour which changes to bluish green and then to violet blue	Indicates the presence of papaverine (isoquinoline alkaloid)
Schaer's test: To the 10 mg of the substance add 2 drops of hydrogen peroxide (30%) and 1 ml of sulphuric acid. Heat the mixture for 2 min.	
• Development of orange colour	• Indicates the presence of emetine

51.10 Identification of Pyridine Alkaloids

Nicotine ***(Nicotiana tabacum)***

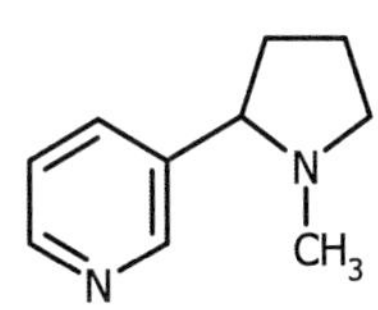

Test for pyridine alkaloids	
***p*-Dimethylamino benzaldehyde (DMAB) test:** To 1 ml alkaloid solution add few drops of *p*-dimethylamino benzaldehyde and heat the mixture by placing in hot plate with continuous stirring.	
Development of red colour (changes to violet)	Indicates the presence of nicotine
Vanillin-hydrochloric acid test: To the powdered sample add vanillin in hydrochloric acid.	
Development of red colour	Indicates the presence nicotine

51.11 Identification of Purine Alkaloids

Caffeine, theophylline and theobromine are the structural examples of purine alkaloids.

Caffeine (1,3,7-trimethylxanthine) *Coffea Arabica, Camellia sinensis* 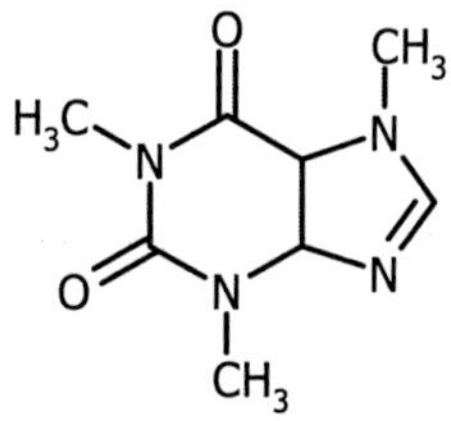 CNS stimulant	Theophylline (3,7-dimethylxanthine) *Theobroma cacao, Theobroma bicolor, Theobroma angustifolium* 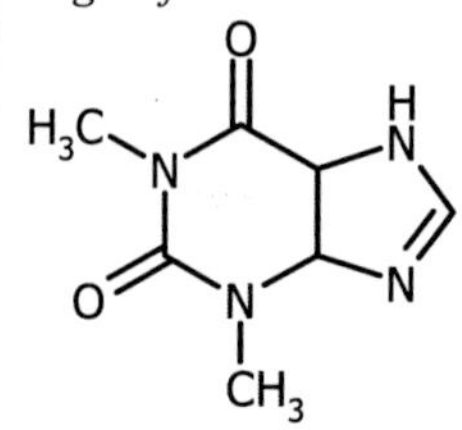 Respiratory stimulant	Theobromine (1,3-dimethylxanthine) *Theobroma cacao, Theobroma bicolor, Theobroma angustifolium, Coffea Arabica* 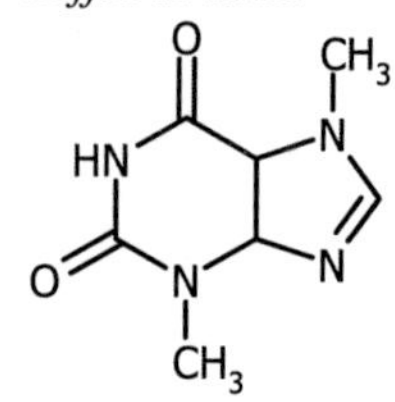 Vasodilator, diuretic, cardiac stimulant

Test for purine alkaloids	
Murexide Test: Transfer 5 ml of the alkaloidal solution to separating funnel, add 2 ml of ammonia and 5 ml of chloroform. Shake well and separate the chloroform layer. Evaporate 1 ml of chloroform layer to dryness, which gives residue after evaporation.	
To the residue add 0.5 ml of hydrochloric acid and 0.5 ml of potassium chlorate solution. Evaporate the mixture to dryness.	
Formation of yellow residue, which becomes crimson on exposure to few drops of ammonia and finally to purple	Indicates the presence of purine alkaloids (caffeine, theophylline)
Schaer's test: To the sample add 1 drop of hydrogen peroxide (30%) and 1 ml of sulphuric acid	
Formation of orange residue (changes into violet after alkalifying with ammonia)	Indicates the presence of purines
Silver nitrate test: To the sample add 1 ml of hot water and 3 drops of silver nitrate	
Formation of jelly like floccules	Indicates the presence of purines
Test for caffeine, theophylline and theobromine	
Silver nitrate test: To the sample add 1 ml of hot water and 3 drops of silver nitrate. To the jelly like floccules add ammonia.	
• Formation of purple colour solution (dispappears by the addition of sodium hydroxide)	• Indicates the presence of caffeine
• Formation of violet red colour solution	• Indicates the presence of theophylline
• Colourless solution	• Indicates the presence of theobromine

Test Name	Ephedrine	Atropine	Cocaine	Reserpine	Brucine	Strychnine	Quinine	Papaverine	Theo-bromine/ Caffeine
Mayer's reagent	- -	Creamy white ppt	Creamy white ppt	Creamy white ppt	Creamy white ppt	Creamy white ppt	Creamy white ppt	Creamy white ppt	No ppt
Wagner's reagent	Brown ppt	Brown ppt	Brown ppt	Brown ppt	Brown ppt	Brown ppt	Brown ppt	Brown ppt	Brown ppt for caffeine only
Tannic acid	- -	- -	-	White ppt	White ppt	White ppt	White ppt	White ppt	White ppt
Mandelin's reagent	-	-	-	Violet turns into orange red on adding water	-	Violet turns into orange red on adding water	-	-	-
Froehde's reagent	- -	- -	-	-	-	-	-	Green	-
Marquis reagent	- -	-	-	-	-	-	-	Greenish to violet	-
Erdmann's reagent	- -	-	-	-	Orange red color	-	-	-	-
Confir-matory test	**Chen's test** Purple color in $CHCl_3$ layer, Blue color in the H_2O layer	**Vitali-Morin's test** Purple color to red to colorless	**Permangan ate test** Violet crystalline precipitate	**Vanillin-hydrochlori c acid test** Red or orange yellow color	**Stannous chloride** Violet color	**Dichromate test** Intense violet streaks appear	**Thalle-oquine test** emerald green color is produced **Blue** fluorescence under UV light	**Warren's test** purple color to blue green to intense violet-blue to brown	**Murexide test** a crimson or reddish residue

Note: *ppt: Precipitate*

CHAPTER 52

QUALITATIVE ANALYSIS OF STEROIDS

A steroid is a terpenoid lipid contains cyclopentano phenanthrene (CPP) skeleton (four fused rings) in 6-6-6-5 pattern. Steroids are found in plants, animals and fungi. The precursor of animal and fungi steroids is lanosterol, where as plant sterols derived from cycloartenol. Both of these precursors are cyclization product of triterpene squalene.

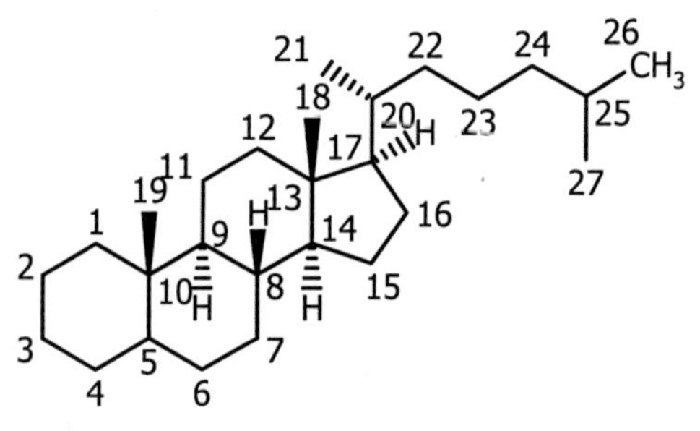

Steroid skeleton with numbering

All the steroids on dehydrogenation with selenium at 360 °C give 'diels hydrocarbon'.

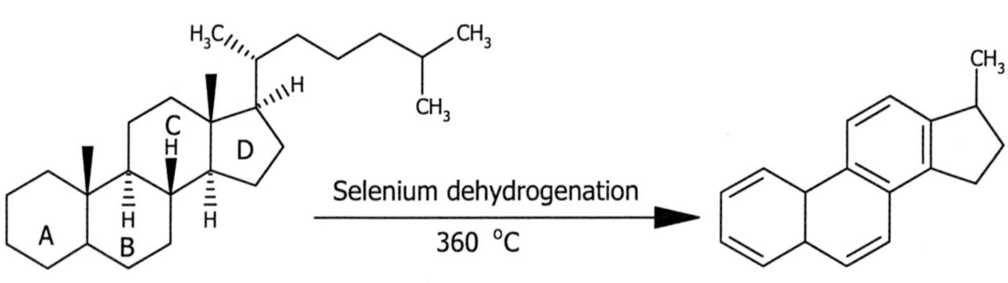

Classification

I. Functional

1. ***Sterols:***

(a) *Zoo sterols:* Cholesterol,

(b) *Phyto sterols:* Stigmasterol,

(c) *Myco sterols:* Ergosterols, zymosterol

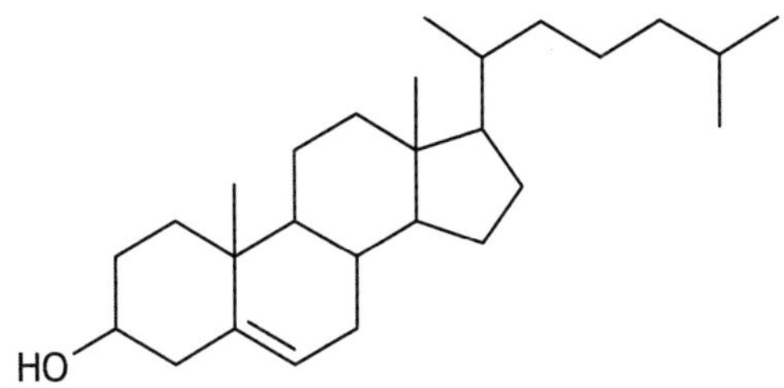

Cholesterol

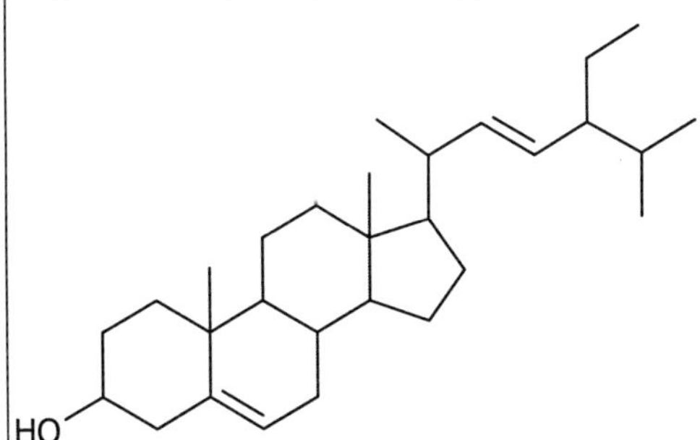

Stigmasterol

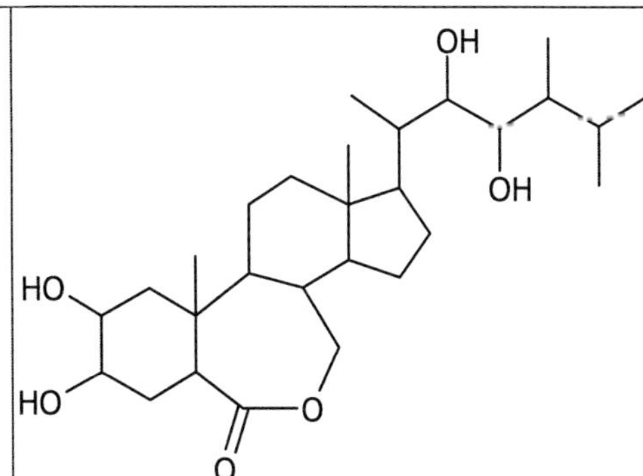

Brassinosteroid

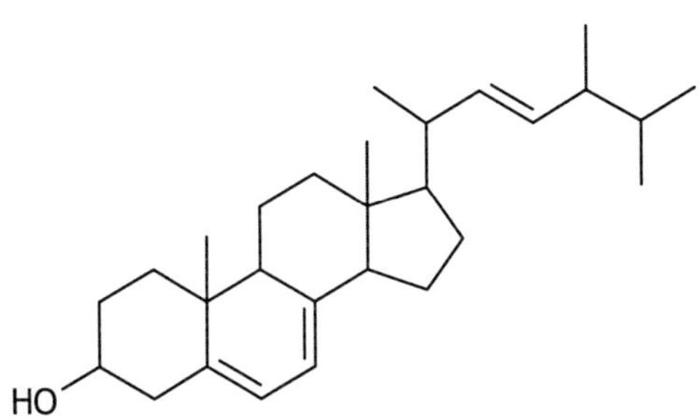

Ergosterol

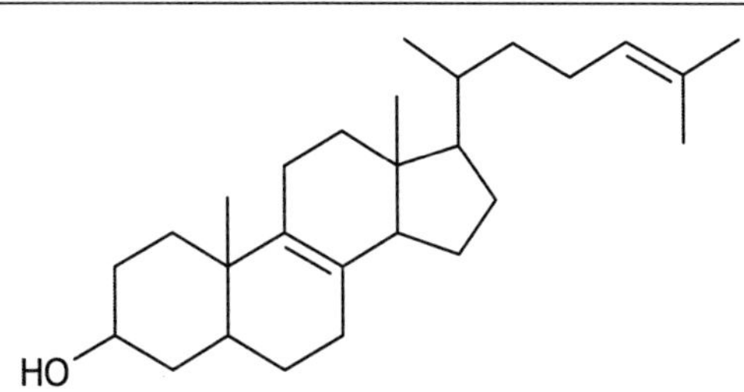

Zymosterol

2. Vitamin D

(a) Vitamin D_2

(b) Vitamin D_3

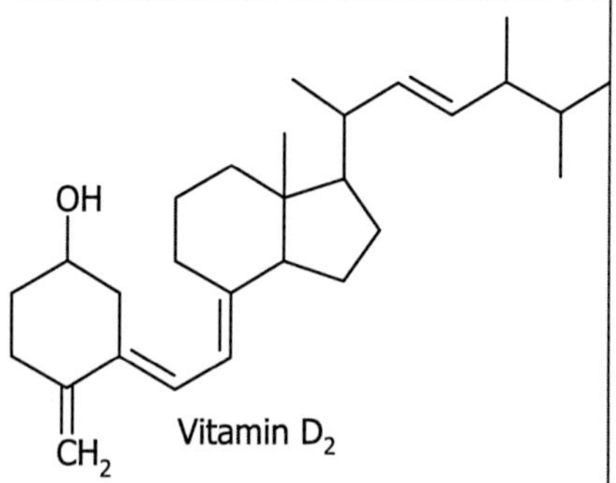

Vitamin D_2

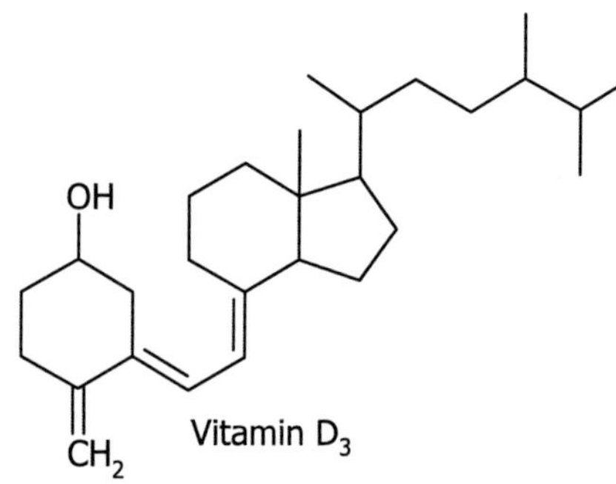

Vitamin D_3

3. Bile acids

(a) Cholic acid

(b) Deoxycholic acid

(c) Chenodeoxycholic acid

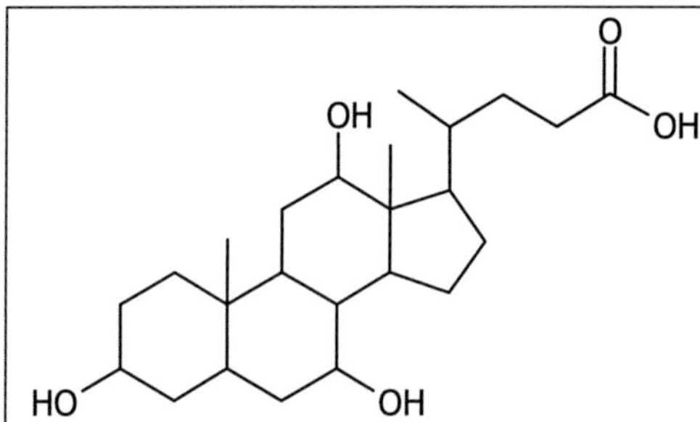

Cholic acid

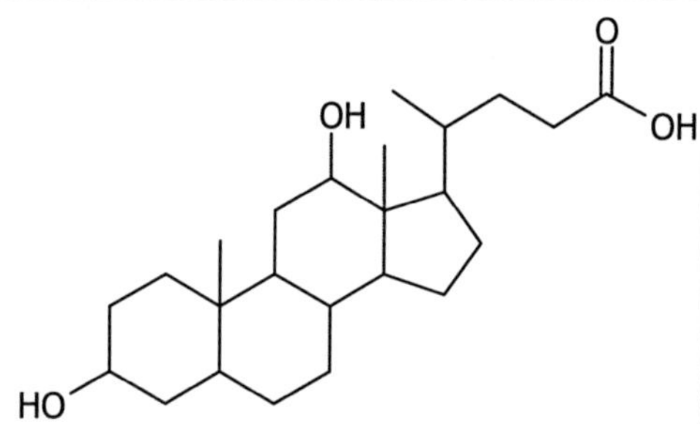

Deoxycholic acid

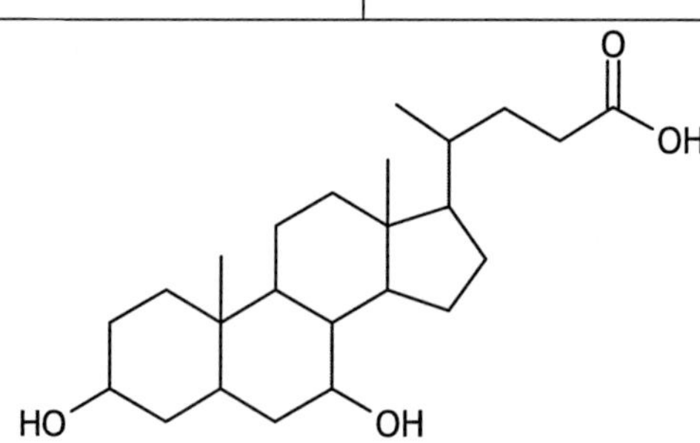

Chenodeoxycholic acid

4. Sex hormones

(a) ***Estrogens (female hormones):*** Estrone, estradiol and estriol

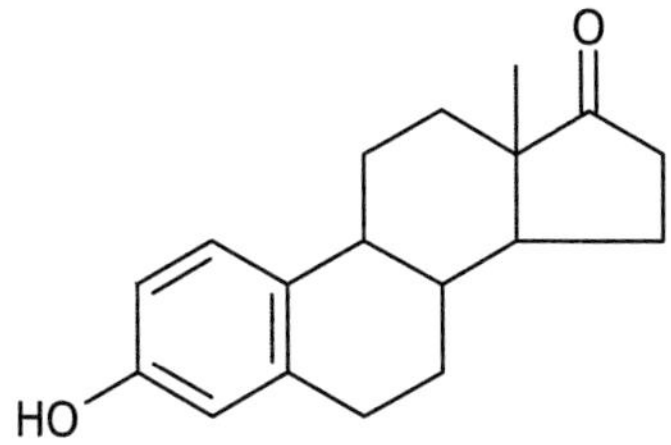

Estrone

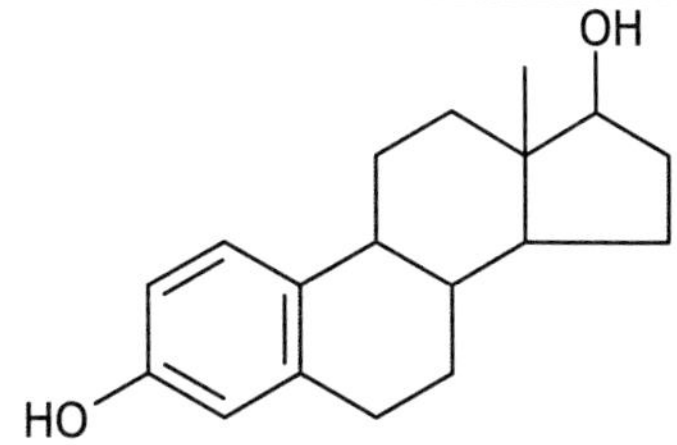

Estradiol

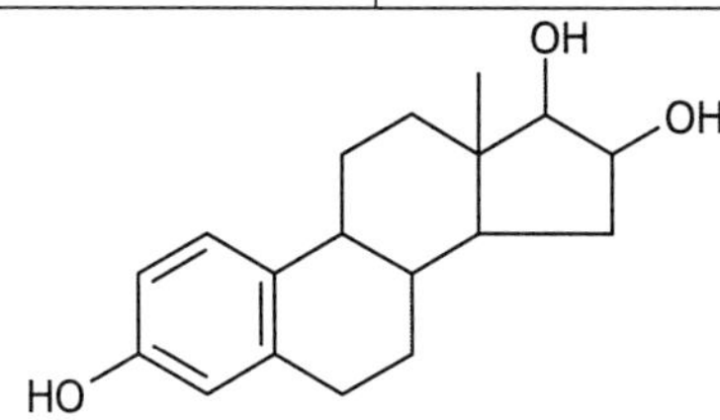

Estriol

(b) ***Gestogens:*** Progesterone

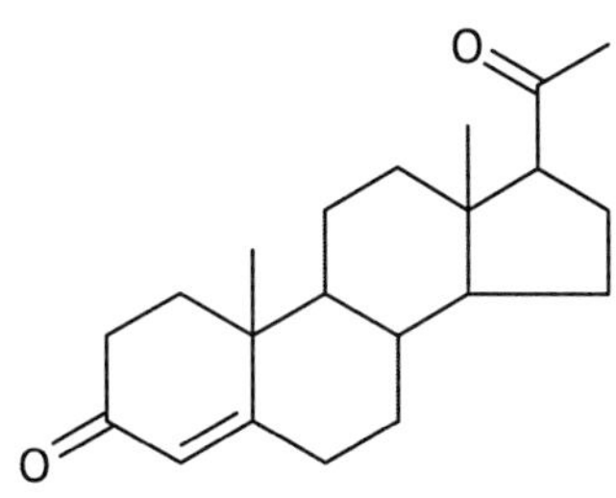

Progesterone

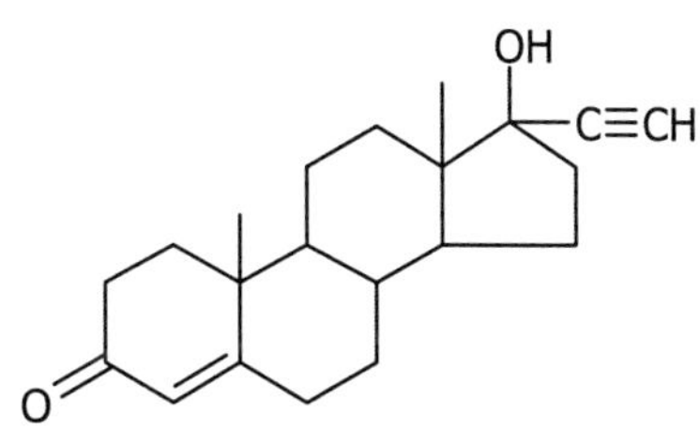

Ethisterone

(c) ***Androgens (male hormones):*** Androsterone, dihydroepiandrosterone and testosterone

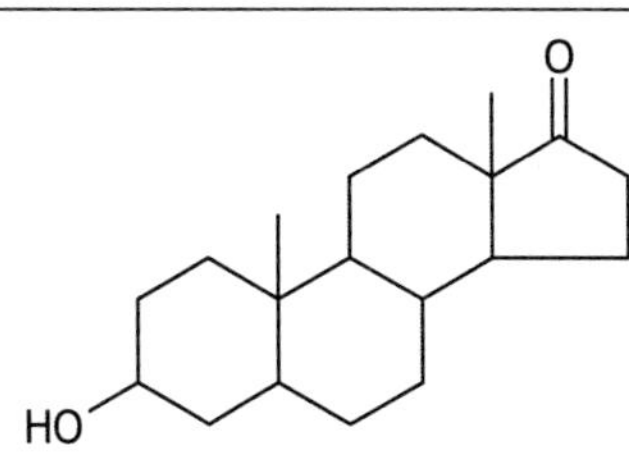

Androsterone

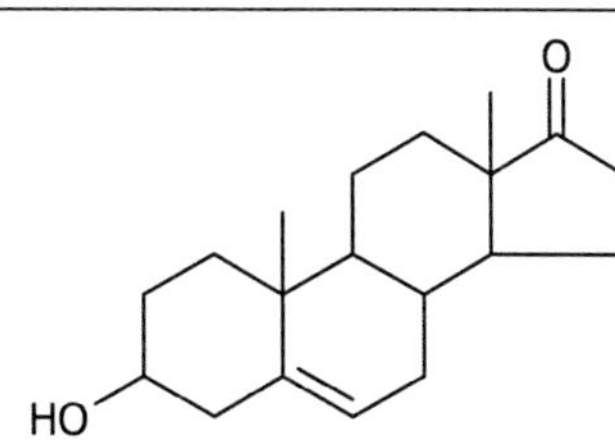

Dihydroepiandrosterone

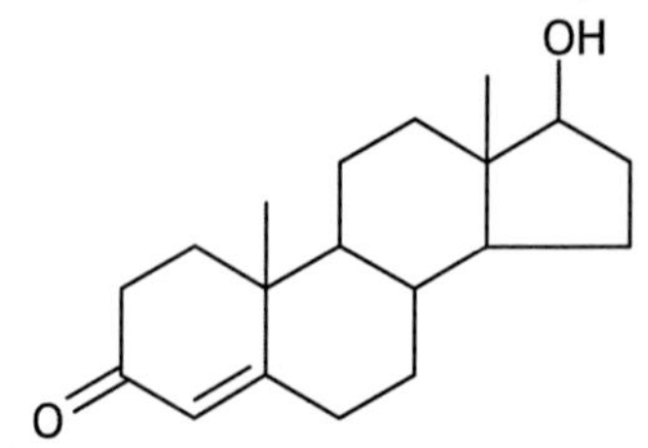

Testosterone

5. Adrenocorticol hormones:

(a) ***Glucocorticoids:*** Corocosterone, cortisone and cortisol

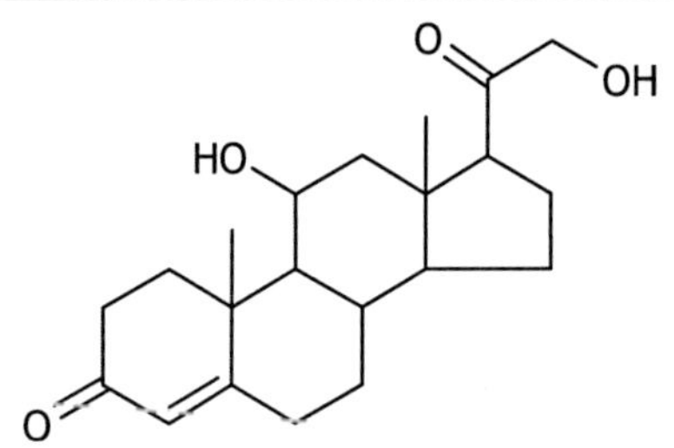

Corticosterone

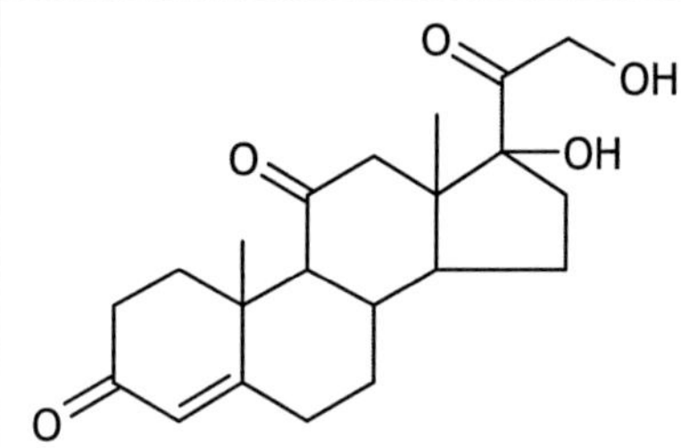

Cortisone

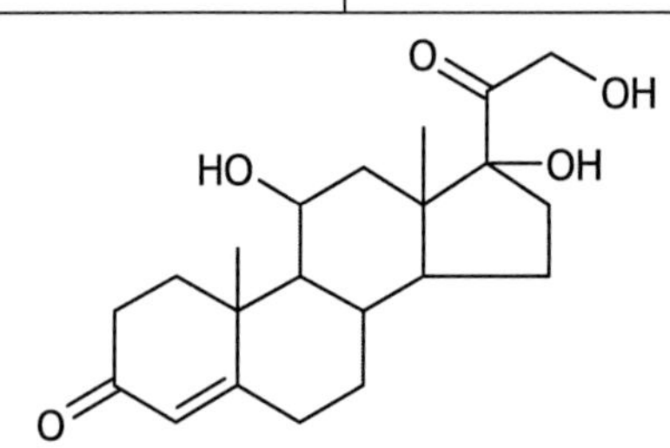

Cortisol

(b) ***Mineralo corticoids:*** Aldosterone

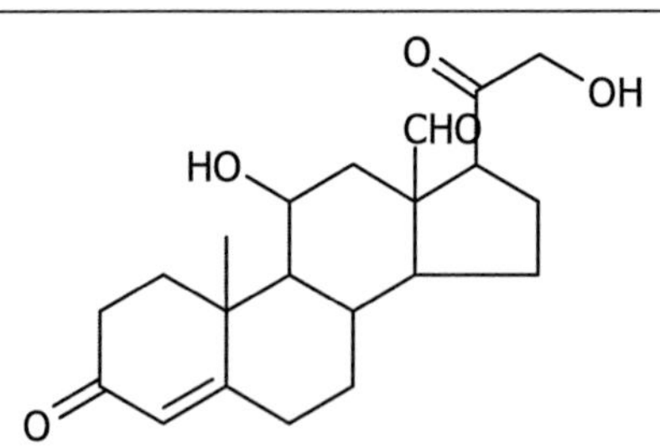

Aldosterone

II. Structural

Steroids can be classified based on the number of carbon atoms present in it.

Class	Number of carbons	Example
Cholestanes	27	Cholesterol
Cholanes	24	Cholic acid
Pregnanes	21	Progesterone
Androstanes	19	Testosterone
Estranes	18	Estradiol

General tests for the identification of steroids

Salkowski reaction (sulphuric acid test)

<table>
<tr><td colspan="2">Dissolve a little chlosterol in 2 ml of chloroform contained in a dry test tube. Add 2 ml of sulphuric acid and shake gently.</td></tr>
<tr><td>• Chloroform layer (upper) appears as bluish red which changes to cherry red and then to purple.
• Sulphuric acid layer (lower) shows yellowish-green colour with green fluorescence.</td><td>Indicates the presence of steroids</td></tr>
</table>

Liebermann- Burchard reaction (Acetic anhydride-sulphuric acid test)

This test is widely used in the determination of blood cholesterol.

<table>
<tr><td colspan="2">Dissolve a few crystals of cholesterol in 2 ml of chloroform contained in a dry test tube. Add 10 drops of acetic anhydride, 3 drops of sulphuric acid and mix well.</td></tr>
<tr><td>• Formation of rose red colour, which quickly changes to blue then to green</td><td>Indicates the presence of steroids</td></tr>
</table>

Zak's test

<table>
<tr><td colspan="2">Evaporate 1 ml of cholesterol solution in chloroform to dryness. Add 3 ml of 0.05% ferric chloride in glacial acetic acid and 1 ml of sulphuric acid.</td></tr>
<tr><td>Formation of red colour</td><td>Indicates the presence of steroids</td></tr>
</table>

Formaldehyde- H_2SO_4 test

<table>
<tr><td colspan="2">To the 2 ml of sample in chloroform add 2 ml of formaldehyde - sulphuric acid solution (1 part of 40% formaldehyde to 50 parts of the sulphuric acid).</td></tr>
<tr><td>Formation of cherry red colour in chloroform layer</td><td>Indicates the presence of steroids</td></tr>
<tr><td colspan="2">Decant the chloroform layer into test tube containing 3 drops of acetic anhydride</td></tr>
<tr><td>Formation of blue colour</td><td>Indicates the presence of steroids</td></tr>
</table>

Rosenheim test

To the 1 ml of sample in chloroform add 1 ml of 90% trichloroacetic acid.	
Appearance of blue or pink colour	Indicates the presence of steroids

Lifschutz test

Heat 1 ml of sample and few mgs of perbenzoic acid in glacial acetic acid. Add 0.5 ml of concentrated sulphuric acid.	
Appearance of intense blue or green colour	Indicates the presence of steroids

Fieser test

To 1 ml of sample in benzene add few mgs of selenium dioxide in benzene. Maintain the temperature at 20°C.	
Appearance of yellow colour	Indicates the presence of 5α series sterols (the 5β series does not give any colour).

Tortelli-Jaffe test

To 1 ml of sample in acetic acid add 1 ml of 2% bromine in chloroform.	
Appearance of green colour at the interface	Indicates the presence of sterols

Zlatkis test (This test is used in the determination of plasma cholesterol)

To 1 ml of sample in acetic acid add 1 ml of solution of ferric chloride in concentrated sulphuric acid and glacial acetic acid.	
Appearance of purple colour.	Indicates the presence of sterols

52.1 Colour Reactions of Cholesterol

Cholesterol occurs as free or fatty esters in higher animals. It was first isolated from gallstones. It is optically active crystalline compound with melting point 149 °C and specific rotation $[\alpha]_n$ = 39°. Higher concentration of cholesterol in blood circulation raises blood pressure and leads to arteriosclerosis.

S.No.	Test	Observation	Inference
	General Tests for Steroids		
1	**Salkowski reaction**	Chloroform layer (upper) appears as bluish red which changes to cherry red and then to purple Sulphuric acid layer (lower) shows yellowish-green colour with green fluorescence	Presence of cholesterol

Table Contd...

S.No.	Test	Observation	Inference
2	**Liebermann-Burchard reaction (Acetic anhydride-H_2SO_4 test)**	Formation of rose red colour, which quickly changes to blue then to green	Presence of cholesterol
3	**Zak's test**	Formation of red colour	Presence of cholesterol
4	**Formaldehyde-H_2SO_4 test**	Formation of cherry red colour in chloroform layer Formation of blue colour in chloroform layer after the addition of acetic anhydride	Presence of cholesterol

52.2 Colour Reactions of Vitamin D

Vitamin D possess anti-rachitic properties. Vitamin D_2 (ergocalciferol) and vitamin D_3 (cholecalciferol) are the two major types of vitamin D. Natural vitamin D_3 is more active than vitamin D_2.

S. No	Test	Observation	Inference
1	**Antimony trichloride test:** To 1 ml of sample add 1 ml of solution of antimony trichloride ni chloroform.		
		Appearance of intense yellow colour	Presence of vitamin D
2	**Diclorohydrin test:** To 1 ml of sample add 1 ml of solution of diclorohydrin and acetyl chloride in chloroform.		
		Appearance of green colour	Presence of vitamin D
3	To 1 ml of sample add 1 ml of solution of aluminum chloride in ethanol.		
		Appearance of violet colour	Presence of vitamin D
4	To 1 ml of sample in chloroform add 1 ml of acetic anhydride and sulphuric acid		
		Appearance of bright red colour [red colour changes through violet and green]	Presence of vitamin D

52.3 Colour Reactions of Bile Acids

Bile acids are the alkaline secretions of the gall bladder. These are the end products of the cholesterol metabolism and play major role in fat metabolism. These acids occurs as conjucates of choline / taurine.

S. No	Test	Observation	Inference
1	**Mylius reaction [for cholic acid]:** Dissolve cholic acid in ethanol add few drops of iodine solution and warm.		
		Appearance of blue colour in transmitted light and needles, which appear yellow in reflected light.	Presence of cholic acid

Table Contd...

S. No	Test	Observation	Inference
2	**Hammersten test:** Add 1 ml of 25% hydrochloric acid to 1 ml of cholic acid.		
		Appearance of violet colour, [which changes to yellow on heating]	Presence of cholic acid
3	**Pettenkofer reaction:** To the 1 ml of solution of bile acid add 1ml of sucrose or furfural. Further add few drops of concentrated sulphuric acid and warm.		
		Appearance of red colour	Presence of cholic acid

52.4 Colour Reactions of Estrogens

Estrogens stimulate the growth and development of secondary sexual characteristics in female. Hyperestrogenism is responsible for the menstrual dysfunction in women and gynaecomastia in men.

Kober test

This reaction is useful in the determination of urinary estrogens (pregnancy test).

Heat 1 ml of sample and 60% sulphuric acid containing phenol, hydroquinone / β-naphthol.	
Appearance of yellow colour with green fluorescence. [On dilution with water and reheating produces pink colour]	Presence of estrogens

Diazo coupling test

To 1 ml of sample add diazotized *p*-nitroaniline / sulphanilic acid.	
Development of orange or red colour [This is due to the coupling of diazo compound with estrogen, which reacts as a phenol]	Presence of estrogens

David test for estriol

To 1 ml of sample add sulphuric acid and arsenic acids.	
Development of blue colour	Presence of estrogens

Bachmann test for estriol

To 1 ml of oestriol add 1 ml of mixture of sodium p-toluene sulphonate and phosphoric acid	
Development of violet colour	Presence of estriol

Zimmermann reaction (17-ketosteroids)

To 1 ml of alcoholic solution of the keto steroid sample add 1 ml of *m*-dinitrobenzene and 1 ml of potassium hydroxide	
Development of coloured complex	Presence of keto steroid.

52.5 Colour Reactions of Androgens

Testosterone is responsible for the spermatogenesis and maintenance of secondary sexual characteristics in men. Androsterone (metabolite of testosterone) possess anti-convulsant effect.

S. No	Test	Observation	Inference
1	**Zimmermann test** (Test for 17-keto steroids and not specific for androgens): Add m-dinitrobenzene and potassium hydroxide to the sample solution.		
		• Brown colour • Purple colour	• Testosterone • Androsterone
2	**Pinus reaction:** To the solution of androsterone add a solution of antimony trichloride in acetic acid.		
		Produces blue colour	Presence of androsterone
3	**Pettenkofer reaction:** Tp the solution of dehydroepiandrosterone add solution furfural in acetic acid. Further add sulphuric acid and warm.		
		Appearance of red colour	Presence of dihydroepiandro-sterone

CHAPTER 53

QUALITATIVE ANALYSIS OF TERPENOIDS

The terpenoids (isoprenoids) are a large and diverse class of naturally-occurring organic chemicals similar to terpenes, derived from five-carbon isoprene units. The term terpene is applied to a mixture of isomeric hydrocarbons of molecular formula $C_{10}H_{16}$ occurring in turpentine and many other essential oils. The essential oils are complex mixtures of hydrocarbons and their oxygenated derivatives, oil of bitter almond and oil of wintergreen (contain mainly benzaldehyde and methyl salicylate). The oxygenated derivatives of terpenes, like alcohols, aldehydes, ketones, etc., were called as camphors. The compounds with empirical formula C_5H_8 of terpenoids, represent the hydrocarbons as well as the oxygenated hydrocarbons having $(C_5)_n$.

Most terpenoids contain multicyclic structures that differ from each other in terms of functional groups and basic carbon skeletons. Plant terpenoids possess insecticidal, anthelmintic, antibacterial and antineoplastic functions. Terpenoids contribute to the scent of eucalyptus, the flavors of cinnamon, cloves and ginger and the color of yellow flowers. Citral, menthol, camphor, salvinorin A and cannbinoids are the important terpenoids.

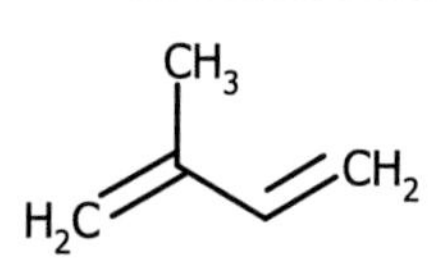

Terpene

Due to their pleasant-smelling nature, the mono and sesqui-terpenoids are useful in perfume industry.

Properties

Terpenoids are mostly colourless, pleasant smelling liquids or solids, insoluble in water and readily volatile in steam. Most terpenoids are optically active compounds. All of the terpenoids have one or more double bonds and they add on halogens, hydrogen halides and Tilden reagent (nitrosyl chloride). Some of the terpenoids having conjugated system of double bonds undergo Diel's - Alder reaction, and due to the presence of olefinic bonds easily undergo oxidization by all the oxidizing agents.

Classification

Terpenoid hydrocarbons are classified on the basis of the number of isoprene units which they contain.

Terpenoid	Number of isoprene unit(s)	Examples
Hemiterpenes	1	Prenol, isovaleric acid
Monoterpenes	2	Geraniol, terpineol, citral, limonene, camphor
Sequiterpenes	3	Farnesol, zingiberene, cadinene
Diterpenes	4	Retinol, retinal, phytol, abietic acid
Sesterpenes	5	Geranylfarnesol
Tritepenes	6	Squalene, amyrins
Tetraterpenes	8	Carotenoids
Polyterpenes	n	Rubber

Some of the important essential oils with their terpenoid constituents are given below.

Essential oils	Terpenoids		Essential oils	Terpenoids
Turpentine	Pinene		Jasmine	Linalool
Caraway	Carvone, ά-Limonene		Lemon	D-Limonene, citral
Coriander	Linalool, Pinene		Sandalwood	Santalol
Eucalyptus	Cineole		Camphor	Cineole, eugenol, terpineol

General Tests for Terpenoids		
1	To the 1 ml of sample solution add 1 ml of alcoholic Sudan III.	
	Development of red colour	Indicates the presence of terpenoids.
2	To the 1 ml of sample solution add few drops of tincture alkane.	
	Development of red colour	Indicates the presence of terpenoids
3	**Salkowski test:** To the 1 ml of the sample add few drops of chloroform and add 3 ml of concentrated sulphuric acid.	
	Greyish or reddish brown colour appears	Indicates the presence of terpenoids

Tests for camphor

Vanillin test: To the powdered camphor add drop of vanillin in hydrochloride (1:100 fresh) and drop of concentrated sulphuric acid.	
• Development of instant yellow colour • Colour changes into red to violet and blue	Indicates the presence of camphor (natural)

Tests for menthol

To 10 mg of crystals add 4 drops of concentrated sulphuric acid and drop of vanillin sulphuric acid reagent	
• Development of orange yellow colour • Colour changes into violet on the addition of water	Indicates the presence of menthol
To 10 mg of crystals add 1 ml glacial acetic acid and drop of concentrated sulphuric acid and concentrated nitric acid	
No colour change	Indicates the presence of menthol

Tests for thymol

To 10 mg of crystals add 1 ml glacial acetic acid and drop of concentrated sulphuric acid and concentrated nitric acid	
• No colour change • Appearance of green/bluish colour	• Indicates the presence of menthol • Indicates the presence of thymol
To the thymol sample add phthalic anhydride	
• Development of violet red colour • Addition of dilute alkali forms blue colour (thymolphthalein)	Indicates the presence of thymol

Tests for eucalyptus oil

To the drop of oil present in microslide add drop of 5% alcoholic hydroquinone	
• Appearance of colourless prisms/rhomboids	• Indicates the presence of eucalyptus oil
To the drop of oil add drop of 50% resorcinol	
• Appearance of leaf like crystals	• Indicates the presence of eucalyptus oil

Tests for asafoetida

Triturate the sample with water.	
• Formation of milky white emulsion	• Indicates the presence of umbelliferone
Place drop of sulphuric acid on the freshly fractured surface.	
• Appearance of reddish brown colour	• Indicates the presence of umbelliferone
Place drop of nitric acid on the freshly fractured surface.	
• Appearance of green colour (turns into violet after washing with water)	• Indicates the presence of umbelliferone
Umbelliferone test: Boil 1 gm of sample with 50% hydrochloric acid, filter and alkalify with ammonium hydroxide.	
• Produces blue fluorescence	• Indicates the presence of umbelliferone

Tests for cineole

To the drop of oil present in microslide add drop of 5% alcoholic hydroquinone	
• Appearance of colourless prisms/rhomboids	• Indicates the presence of cineole
To the drop of oil add drop of 50% resorcinol	
• Appearance of leaf like crystals	• Indicates the presence of cineole

Tests for colophony resin

To the powdered resin add 10 ml of acetic anhydride and drop of sulphuric acid. Shake well the mixture.	
• Appearance of purple colour	• Indicates the presence of colophony resin
Test the alcoholic resin for the acidity with litmus paper	
• Turns blue litmus red	• Indicates the acidic property of resin
To the test solution add tollens reagent	
• Formation of silver mirror	• Indicates the presence of furfuraldehyde of resin.

Tests for carotenoids

Stratify the ether/chloroform solution of sample with 85% sulphuric acid.	
Appearance of blue colour between the junction of two layer	Indicates the presence of carotenoids
Carr-price test: To the sample add antimony trichloride in chloroform	
Appearance of blue colour	Indicates the presence of carotenoids

Tests for curcumin

To the powdered sample add drops of sulphuric acid.	
Appearance of crimson red colour	Indicates the presence of curcumin
To the aqueous solution of sample add boric acid and few drops of alkali solution.	
Appearance of reddish brown colour (changes in to greenish blue on the addition of alcohol)	Indicates the presence of curcumin
Fluorescence test: To the sample powder add drop of acetic anhydride and drops of sulphuric acid	
• Appearance of violet colour • Produces red fluorescence	Indicates the presence of curcumin

Tests for eugenol

Bromine test: To the 1 ml of sample solution add 1 ml of bromine in dichlromethane.	
Decolurization of yellow colour of bromine	Indicates the presence of carbon-carbon double bond (C=C) in eugenol
Potassium permanganate test: To the 1 ml of sample solution add 1 ml of potassium permanganate	
Decolurization of purple colour of bromine	Indicates the presence of carbon-carbon double bond (C=C) in eugenol
Ferric chloride test: To the 1 ml of sample solution add 1 ml of 5% alcoholic ferric chloride	
Development of blue colour	Indicates the presence of phenolic group of eugenol

CHAPTER 54

QUALITATIVE ANALYSIS OF GLYCOSIDES

Glycosides are composed of both sugar moiety (glycone, genin) and non-sugar moiety (aglycone, agenin). Sugar moieties establishes link with non-sugar moieties through glycosidic bond. The sugar can be mono- (glucose, L-rhamnose, L-fructose, l-arabinose and D-xylose), di(galactose, mannose and gentiobiose), and poly saccharide. When an aglycone part (non-sugar moiety) becomes terpene the compounds are named as flavanoids. Glycosides possessing soap properties are known as saponins.

Properties

Glycosides are colourless (flavanoid: yellow, anthraquinone: red/orange), solid, amorpous, non-volatile and bitter (except glycyrrhizin, stevioside and populin) compounds. They are non-reducing sugars gives positive test with Molisch's test and Fehling's test (after hydrolysis). But the aglycone portion of K-strophanthoside gives positive test. Most compounds are water soluble and insoluble in organic solvents.

Classification

Depending upon type of functional group to which sugar moiety bond (glycosidic bond) they are named as

Glycosides	Functional group of non-sugar (aglycone)	Examples
O-Glycoside	Hydroxyl	Salicin
N-Glycoside	Amino	Adenine
S-Glycoside	Thiol	Sinigrin
C-Glycoside	Carbon	Barbaloin

Glycosides are further classified depending upon the basic nucleus/functionality of aglycone present in it.

Glycosides	Functionality/ nucleus of aglycone	Examples
Alcoholic glycosides	Alcohol	Salicin
Aldehyde glycosides	Aldehyde	Glucovanillin
Isothiocyanate glycosides	Isothiocyanates	Sinigrin, sinalbin
Anthraquinone glycosides	Anthraquinone	Sennoside, barbaloin, cascaroside
Coumarin glycosides	Coumarin	Apetrin
Cyanogenic glycosides	Nitrile	Amigdalin, prunasin
Flavanoid glycosides	• Flavanol, • Flavone, • Flavanone	• Quercitrin • Vitexin • Naringin
Phenolic glycosides	Phenol	Arbutin, Sennoside, barbaloin, cascaroside
Saponin glycosides		Diosgenin
Steroidal glycosides (cardiac glycosides)	Cyclopentano perhydrphenathrene	Lannatoside A, digitoxin, gitoxin, quabain
Steviol glycosides	Diterpene (steviol)	Stevioside
Thioglycosides (Sulphur/thioglycoside)	Thio	Allicin

54.1 General Tests for Glycosides

1. Legal's test

To the 1 ml solution of the sample add few drops of pyridine and sodium nitroprusside 92%, w/v) and sodium hydroxide (20%, w/v).	
Development of pink/red colour	Indicates the presence of glycosides

2. Baljet test

To the fractured part of sample add alkaline picric acid.	
Development of yellow to orange colour	Indicates the presence of glycosides

3. Raymond test

To the solution of glycoside in 1 ml of ethanol (50%) add 0.1 ml of Raymond reagent and 3 drops of sodium hydroxide (20%, w/v).	
Development of violet colour (changes into blue)	Indicates the presence of glycosides
Raymond reagent: 1% w/v *m*-Nitro benzene in ethanol.	

4. Tollen's test (Ammoniacal silver nitrate, $[Ag(NH_3)_2NO_3]$)

To the 1 ml solution of the sample in pyridine and 1 ml of Tollen's reagent and heat. Sodium hydroxide (20%, w/v).	
Formation of silver mirror	Indicates the presence of sugar portion of glycosides

5. Antimony trichloride test

To the 1 ml solution of the sample add 1 ml of antimony trichloride and 1 ml of trichloroacetic acid and heat the mixture.	
Development of blue/violet colour	Indicates the presence of glycosides

6. Kedde test

To the 1 ml solution of the sample add equal volume of Kedde reagent	
Development of blue/violet colour (1-2 hr)	Indicates the presence of glycosides
Kedde reagent: Equal volume of 2% w/v 3,5-Dinitro benzoic acid in methanol and 75% w/v of aqueous potassium hydroxide.	

54.2 Identification of Aldehyde Glycosides

Glucovanillin obtained from vanilla represents this class of glycosides.

Glycoside name	Structure	Source	Function
Glucovanillin (vanilloside)	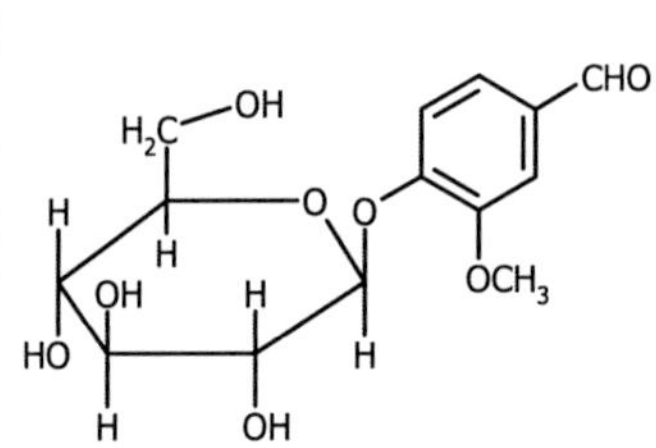	*Vanilla fragrans*	Flavouring and perfuming agent

Vanillin test

To 0.5 gm of sample add 10 ml of water and dissolve by warming. Add 3 drops of ferric chloride solution to the mixture.	
Development of blue-purple colour (changes to brown and white to grey-white precipitate)	Indicates the presence of vanillin
To 1 g of sample add 5 ml of sodium hydrogensulphate and dissolve by warming. Add 10 ml of dilute sulphuric acid and warm the mixture at 60-70 °C for 5 minutes.	
Formation of crystals	Indicates the presence of vannilin

54.3 Identification of Cardiac Glycosides

Digitalis, strophanthidin and squill contain cardiac glycosides. All are cardiotonic agents.

Glycoside	Structure
Lanatoside A *(Digitalis lanata)*	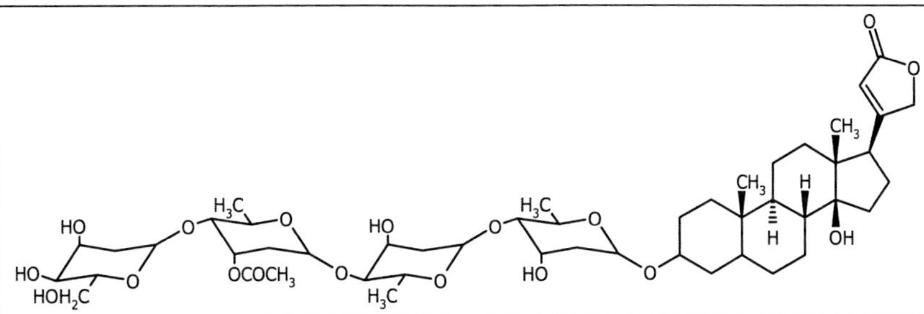
Digitoxin *(Digitalis pupurea)*	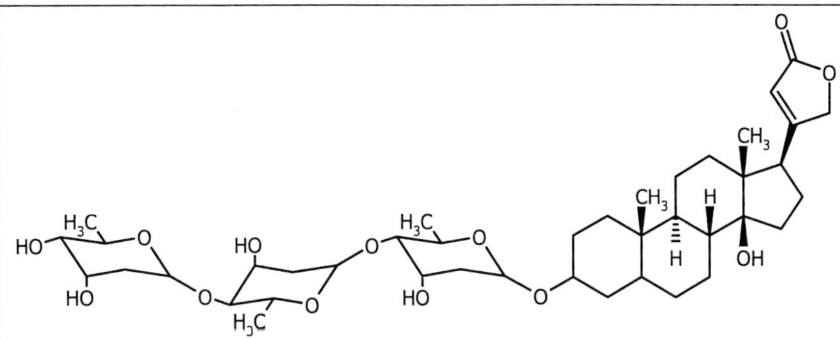
Digoxin *(Digitalis pupurea)*	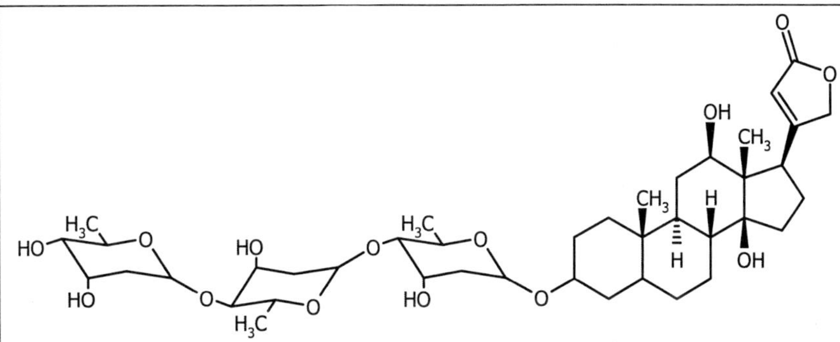
Ouabain *(Strophanthus kombe)*	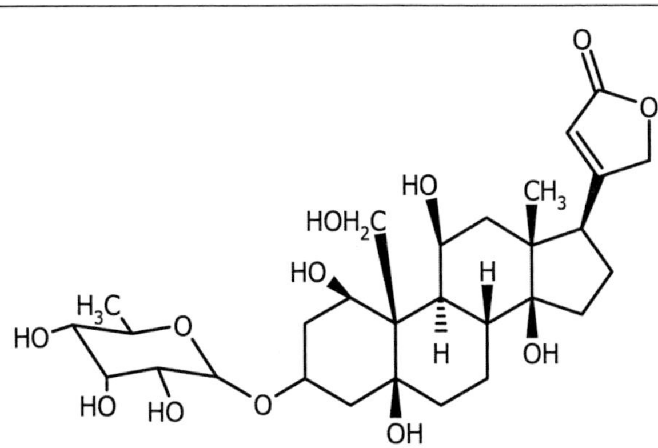

Table Contd...

Glycoside	Structure
Scillaren A *(Urginea maritama,. Scilla maritama)*	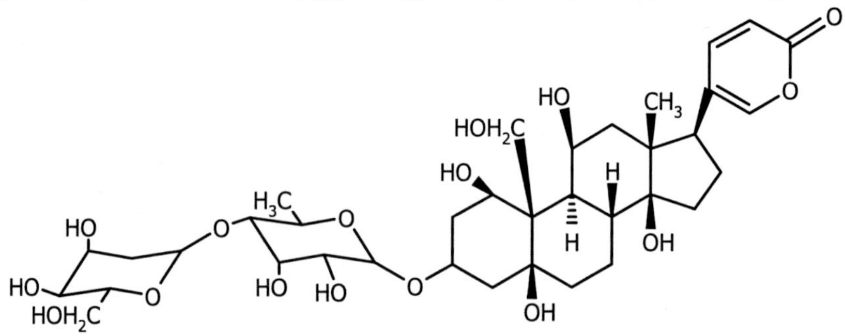
Proscillaridin A	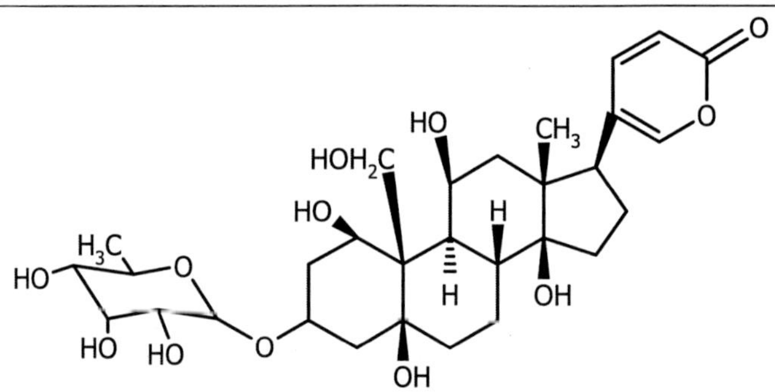

Keller-Killiani test (for deoxy sugar)

To the 2 ml of glycoside solution add glacial acetic acid and trace of ferric chloride in sulphuric acid along the sides of the test tube.	
Development of reddish brown to bluish brown colour at the junction of the two liquids (2-5 minutes)	Indicates the presence of cardiac glycosides (deoxy sugars)

Cardiac glycosides are steroid in nature, hence give positive results for the following steroidal identification tests.

Salkowski reaction (sulphuric acid test)

Dissolve a little sample in 2 ml of chloroform in a dry test tube, add an equal volume of sulphuric acid and shake gently.	
• Chloroform layer (upper) appears as bluish red which changes to cherry red and then to purple • Sulphuric acid layer (lower) shows yellowish-green colour with green fluorescence	Indicates the presence of steroids

Liebermann- Burchard reaction (Acetic anhydride-sulphuric acid test)

Dissolve a little sample in 2 ml of chloroform in a dry test tube. Add 10 drops of acetic anhydride, 3 drops sulphuric acid and mix well.	
Formation of rose red colour, which quickly changes to blue then to green	Indicates the presence of steroids [Addition of sulphuric acid to steroidal nucleus containing compound in the presence of acetic anhydride gives a green (red absorbing) chromophore]

54.4 Identification of Anthraquinone Glycosides

Cascara, aloes and senna plants

Glycoside	Structure	Source	Function
Sennoside	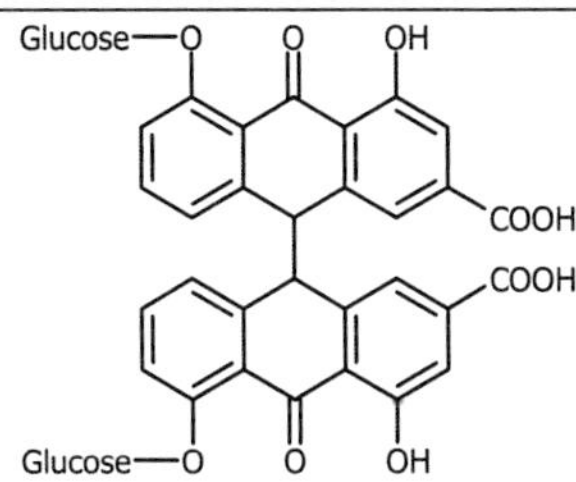	*Ephedra sinica,* *Ephedra vulgaris,* *Ephedra equisetina*	Sympathomimetic, bronchodilator
Barbaloin	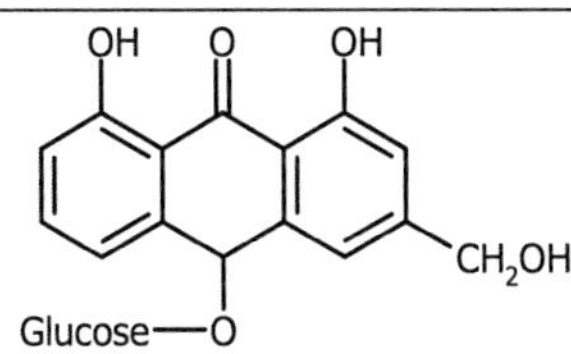	*Aloe vera*	Laxative
Cascaroside	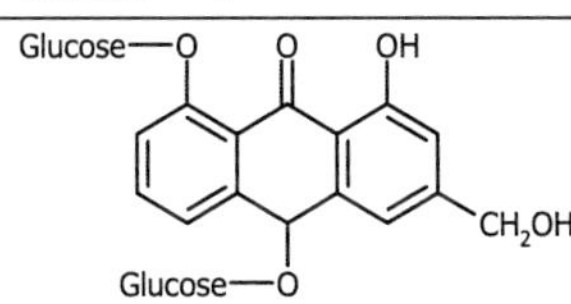	*Rhamnus purshiana*	Laxative

Borntrager test

To the sample add 2 ml of dilute sulphuric acid and heat (hydrolysis). Filter, cool, add benzene/carbon tetrachloride and shake the mixture. Separate the organic layer and add equal volume of ammonia solution to it.	
• Development of pink/cherry-red colour in ammoniacal layer • Development of yellow-orange colour (rapidly changes to red)	• Indicates the presence of anthraquinone glycosides • Indicates the presence of anthranol (Addition of hydrogen peroxide produces rapid result)

Borax test

Preparation of 0.1% aqueous solution: Heat aqueous solution of aloe by gentle heating and add 0.5 gm Kieselguhr and filter.	
To the 5 ml of filtrate add 0.2 gm of borax and heat to dissolve. Pour few drops of the liquid into test tube containing water (full).	
Produces green colour fluorescence	Indicates the presence of aloe glycosides

Bromine test

To the 2 ml of filtrate add 2 ml of bromine solution (fresh).	
Forms yellow colour precipitate	Indicates the presence of aloe glycoside tetra bromaloin

Nitric acid test

To 5 ml of filtrate add 2 ml of concentrated nitric acid.	
• Brown to green	• Indicates the cape aloe
• Deep brownish red	• Indicates the barbados
• Pale brownish to yellow	• Indicates the socotrine
• Yellow to brown	• Indicates the zanzibar

Klunge's test

To the 20 ml of filtrate add drop of saturated copper sulphate solution, 1 gm of sodium chloride and 10 ml of ethanol (90%).	
• Formation of wine red colour	• Indicates the presence of barbados
• No colour develops	• Indicates zanzibar and socotrine

Nitrous acid test

To the 1 ml of filtrate add few crystals of sodium nitrite and 1 ml of acetic acid.	
• Produces rich pink to carmine colour	• Indicates the presence of barbados
• Produces lemon pink colour	• Indicates the presence of cape aloe

Modified Borntrager test (for C-glycosides)

To the sample add 5 ml of dilute sulphuric acid and 2 ml of aqueous ferric chloride (5%). Boil the mixture for 5 minutes in water bath. Filter, cool, add benzene/carbon tetrachloride and shake the mixture. Separate the organic layer and add equal volume of ammonia solution to it.	
• Development of yellow to orange colour • Development of pink/red colour in ammoniacal layer	• Indicates the presence of anthraquinone glycosides • Indicates the presence of anthraquinone glycosides

54.5 Identification of Phenolic Glycosides

Arbutin, sennoside, barbaloin and cascaroside contains phenolic group.

Glycoside	Structure	Source	Function
Arbutin	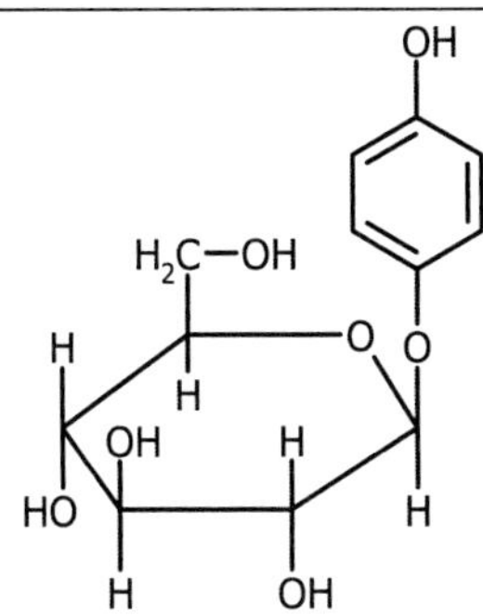	*Arectostaphyllos uva ursi*	Diuretic, antibacterial

Sulphuric acid test: To the 1 ml of sample add few drops of sulphuric acid.	
Produces red colour (disappears by the addition of few drops of water)	Indicates the presence of phenolic glycosides
Mandelin's test: To the 1 ml of sample add few drops of Mandelin's reagent.	
Mandelin's reagent (sulphovanadic acid): 1 gm of ammonium vanadate in 100 ml of concentrated sulphuric acid (Sulphovanadic acid).	
Produces purple red colour	Indicates the presence of phenolic glycosides
Froehde's test: To the 1 ml of sample add few drops of Froehde's reagent.	
Froehde's reagent: 5 gm of molybdic acid or sodium molybdate in 5 ml of hot concentrated sulphuric acid.	
Produces violet colour	Indicates the presence of phenolic glycosides
Erdmann's test: To the 1 ml of sample add few drops of Erdmann's reagent.	
Erdmann's reagent: A mixture of 10 drops of concentrated nitric acid and sulphuric acid.	
Produces bright red colour	Indicates the presence of phenolic glycosides

54.6 Identification of Cyanogenic Glycosides

Amigdalin and prunasin contains cyanogens group.

Glycoside	Structure	Function
Amygdalin *(prunas amygdalus, Prunas communis)*	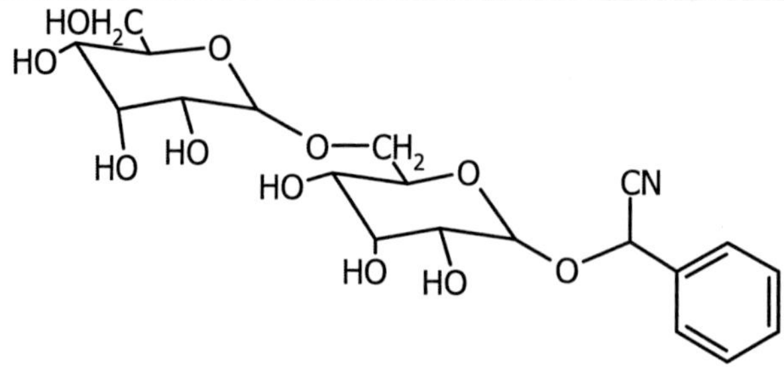	Sedative, demulscent
Prunasin (*Prunas japonica*)	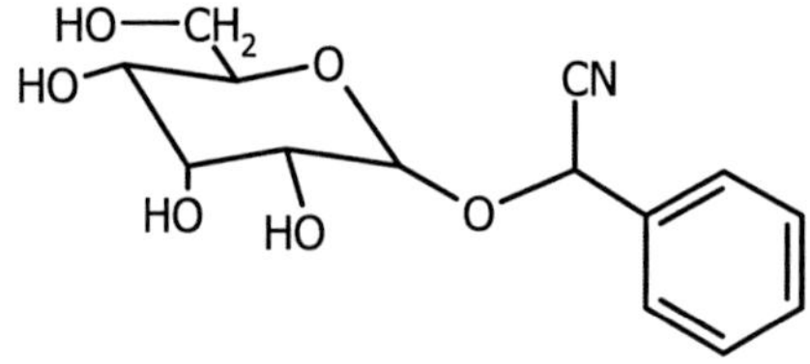	Diuretic, demulscent

Ferri ferrocyanide test: Macerate the 1 gm of the sample add 5 ml of alcoholic potassium hydroxide (5%, w/v) for 5 minutes. Transfer this solution into aqueous solution of ferrous sulphate (2.5%) and ferric chloride (91%). Allow to stand the mixture at 60-70 °C for about 10 minutes. Transfer the contents into test tube containing hydrochloric acid (20%).	
Appearance of prussian blue colour	Indicates the liberation of HCN
Grignard test (Sodium picrate test): Soak the filter paper with picric acid (10%) and drain the excess reagent. Dip this paper (sodium picrate paper) into solution of sodium carbonate (10%), drain and dry. Now place the paper in the flask containing moistened sample and place the flask in warm condition for 1 hr.	
Filter paper turns into brick red/maroon (liberated HCN reacts with sodium picrate and produces colour)	Indicates the formation of sodium isopurpurate

54.7 Identification of Coumarin Glycosides

Apterin belongs to coumarin class of glycosides contains furanocoumarin ring.

Glycoside	Structure	Function
Apterin *(Garden angelica, Zizia aptera)*	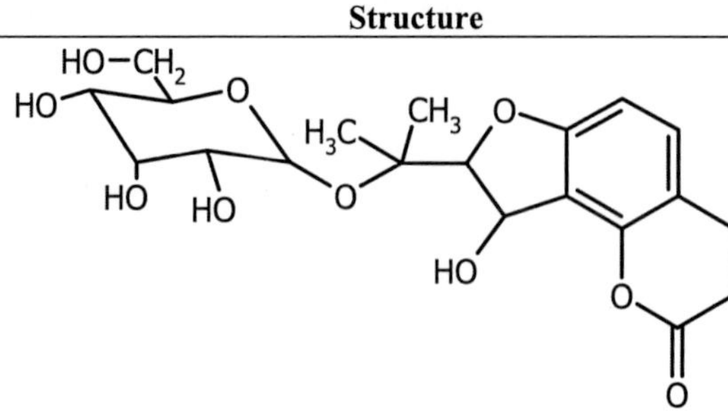	Calcium channel blocker

Alkali test: To the alcoholic solution add alkali hydroxide.	
Development of blue/green fluorescence	Indicates the liberation of coumarin glycoside
Guaicum test/Benzidine test: To the sample add few drops of alcoholic guaicum resin (1%) or alcoholic benzidine (1%) and few drops of hydrogen peroxide.	
Development of blue colour	Indicates the liberation of coumarin glycoside

CHAPTER 55

COLOUR TESTS FOR THE IDENTIFICATION OF DRUGS

Colour tests are the important tool for the presumptive identification of pharmaceuticals, biological fluids and poisons.

Identification tests for functional groups

Functionality	Reagent/test
Alkaloids	Dragendorff's reagent
Antidepressants (tricyclic amines)	Marquis test
Barbiturates	• Koppanyi-Zwikker reagent • Mercurous nitrate • Vanillin reagent
Benzodiazepines	Formaldehyde-sulphuric acid reagent
Chlorinated phenols	Fuming nitric acid
Ergot alkaloids (indole alkaloids)	*p*-Dimethylamino benzaldehyde reagent
Opiates	• Froehde's reagent • Marquis test
Phencyclidine	*p*-Dimethylamino benzaldehyde reagent
Phenothiazines	• Ferric chloride test • Formaldehyde-sulphuric acid reagent • Forrest reagent • FPN reagent • Liebermann reagent
Quaternary amines	Dragendorff's reagent
Quinines	Thalleioquin test
Quinones	Methanolic potassium hydroxide test
Reducing sugars	• Benedicts reagent • Tollen's reagent
Salicylates	• Ferric chloride test • McNallys test

Table Contd...

Functionality	Reagent/test
Steroids	Naphthol-sulphuric acid reagent
Sulphonamides	• Copper sulphate test • Koppanyi-Zwikker reagent • Nitrous acid test • Mercurous nitrate test
Tertiary amines	Dragendorff's reagent
Tetracyclines	• Benedicts reagent • Libermann reagent • Mandelin violet reagent • Meckes reagent • Sulphuric acid test

Identification tests for pharmaceuticals

Pharmaceuticals	Reagent/test
Amantadine	Ninhydrin test
Amiloride	Diazotization test
Amitriptyline	• Mandelin reagent • Marquis test
Ampicillin	Ferric chloride test
Antazoline	Libermann's reagent
Aspirin	Ferric chloride test
Azathioprine	Diazotization test
Benserazide	*p*-Dimethylamino benzaldehyde reagent
Bisacodyl	• Libermann's reagent • Marquis test • Mandelin reagent
Bumetanide	Libermann's reagent
Bupivacaine	• Formaldehyde-sulphuric acid test • Hinsberg test
Busulphan	Potassium permanganate test
Carbidopa	• Ferric chloride test • Libermann's reagent • Methanolic potassium hydroxide test • Nessler's reagent • Potassium dichromate test • Tollen's reagent
Carbamazepine	Formaldehyde-sulphuric acid test
Chlorambucil	Potassio-mercuric iodide test

Table Contd...

Pharmaceuticals	Reagent/test
Chloramphenicol	• Fujiwara reagent • Naphthol-sulphuric acid test
Chlorpheniramine	Ferric chloride test
Chlorpromazine	• Ferric chloride test • Forrest reagent • FPN test • Froehde's reagent • Mandelin's test • Mecke's test
Chlorpropamide	Koppanyi-Zwikker reagent
Chloroquine	Libermann's reagent
Cimetidine	• Hydroxyl amine test • Nessler's reagent • Steyn test
Clofibrate	• Libermann's reagent • Marquis test
Clomipramine	• Forrest reagent • FPN reagent • Libermann reagent • Mandelin's reagent
Clonidine	Libermann's reagent
Cyclophosphamide	Phosphorous test
Diazepam	Formaldehyde-sulphuric acid test
Diclofenac	• Libermann's reagent • Mandelin's reagent
Diethylstilbesterol	Ferric chloride test
Diphenhydramine	• Froehde's reagent • Libermann's reagent • Mecke's reagent
Dipyridamol	Mandelin's test
Ethambutol	Diazotization test
Ethchlorvynol	Tollens reagent
Ethinyl estradiol	• Libermann's reagent • Naphthol-sulphuric acid test • Tollen's reagent
Ephedrine	Diazotization test
Fenclofenac	Libermann's reagent
Flurazepam	• Formaldehyde-sulphuric acid test • Froehde's reagent • Mandelin's test

Table Contd...

Pharmaceuticals	Reagent/test
Folic acid	Diazotization test
Furosemide	• Formaldehyde-sulphuric acid test • Koppanyi-Zwikker reagent • Libermann's reagent
Glibenclamide	Libermann's reagent
Glutethimide	• Koppanyi-Zwikker reagent • Libermann's reagent • Vannilin reagent
Glyclazide	Libermann's reagent
Haloperidol	Libermann's reagent
Hydrochlorthiazide	Libermann's reagent
Ibuprofen	Libermann's reagent
Imipramine	• Forrest reagent • Libermann's reagent
Indomethacin	• Libermann's reagent • Mandelin's test
Isocarboxazid	Libermann's reagent
Isoniazid	• Nessler's reagent • Silver nitrate test
Ketamine	Forrest reagent
Ketoprofen	Sodium nitroprussside test
Labetalol	• Ferric chloride test • Libermann's reagent • Mandelin's reagent • Nessler's reagent
Levamizole	• Libermann's reagent • Mandelin's reagent
Levodopa	• Ferric chloride test • Potassium dichromate test • Tollen's reagent
Mebendazole	Copper sulphate test
Mefanemic acid	Libermann's reagent
Methadone	• Mandelin's reagent • Marquis test
Methyldopa	• Ferric chloride test • Mandelin's reagent • Marquis test • Methanolic potassium hydroxide test • Potassium dichromate test • Tollen's reagent
Metronidazole	Methanolic potassium hydroxide test

Table Contd...

Pharmaceuticals	Reagent/test
Methotrexate	Mandelin's test
Metoclopramide	Libermann's reagent
Naproxen	Liberman's reagent
Nifedipine	Methanolic potassium hydroxide test
Nitrazepam	Diazotization test
Orciprenaline	• *p*-Dimethylamino benzaldehyde reagent • Potassium dichromate test
Oxytetracycline	Formaldehyde-sulphuric acid test
Paracetamol	• Ferric chloride reagent • Libermann's reagent • Nessler's reagent
Pethidine	• Froehde's reagent • Formaldehyde-sulphuric acid test • Libermann's reagent
Phenformin	Mandelin's test
Pheniramine	Libermann's reagent
Phenobarbitone	Cobalt chloride test
Phensuximide	Libermann's reagent
Phenytoin	• Cobalt chloride test • Copper sulphate test • Libermann's reagent • Vanillin's test
Pindolol	• Libermann's reagent • Mandelin's test • Marquis test
Piroxicam	Libermann's reagent
Prazocin	Libermann's reagent
Procaine	• *p*-Dimethylamino benzaldehyde reagent • Mandelin's test • Mecke's test
Promazine	• Formaldehyde-sulphuric acid test • Libermann's reagent • Mandelin's test
Promethazine	• Forrest reagent • FPN reagent • Marquis test • Mandelin's reagent • Mecke's test
Propoxyphene	• Froehde's reagent • Libermann's reagent • Mandelin's reagent

Table Contd…

Pharmaceuticals	Reagent/test
Propyl thiouracil	Copper sulphate test
Pyrazinamide	Nessler's reagent
Rimantadine	Ninhydrin test
Salbutamol	• Libermann reagent • Mandelin's reagent • Marquis test
Sulindac	• Libermann's reagent • Mandelin's reagent
Thiobendazole	Copper sulphate test
Timolol	Libermann reagent
Terbutaline	• *p*-Dimethylamino benzaldehyde reagent • Potassium dichromate test
Tranylcypromine	• Libermann's reagent • Mandelin's reagent
Valproic acid	• Cobalt chloride test • Ferric chloride test
Verapamil	Libermann's reagent
Warfarin	Libermann's reagent
Zomepirac	• Formaldehyde-sulphuric acid test • Libermann's reagent • Marquis's test
Xylometazoline	Mandelin's test

Aromaticity

Add two or three drops of nitric acid to the sample, heat in a water bath at 100 °C for 1 min, cool the mixture, dilute three to four times with water and make the solution alkaline by the addition of 40% (w/v) solution of sodium hydroxide.	
• Change of yellow (or colourless) in to darker colour	• Indicates the presence of benzene ring (formation of nitro compound)
• Orange colour	• Indicates the non-aromatic corticosteroids (cortisone) and nifursol

Benedict's test

Add 0.5 ml of the reagent to the sample and heat in a water-bath at 100 °C for 3 min.	
Tollen's reagent: Dissolve 1.73 gm of copper sulphate in 10 ml of water. Dissolve 1.73 gm of trisodium citrate and 10 gm of anhydrous sodium carbonate in 80 ml of water with the aid of heat. Pour this solution into copper sulphate solution and dilute the mixture to 100ml.	
• Red precipitate (due to cuprous oxide)	• Indicates the presence of ascorbic acid, glucose, tetracyclines, dithionate • Cardiac glycosides and clindamycin produces no colour

Citric acid test

Dissolve 200 mg of sample in a mixture of sodium hydroxide (2 ml) and water (2 ml). Add 3 ml of citric acid and shake vigorously.	
• White precipitate	• Indicates the presence of allopurinol

Copper sulphate test

Dissolve the sample in a minimum volume of 0.1M sodium hydroxide and add a 1% (w/v) solution of copper sulphate, drop by drop, until the colour change is complete.	
• Green, blue and brown colours	• Indicates the presence of sulphonamides
• Green solution with grayish precipitate	• Indicates the presence of propyl thiouracil
• Greenish precipitate	• Indicates the presence of mebendazole and thiobendazole
• Blue-green precipitate with pink needles	• Indicates the presence of phenytoin

Cobalt chloride test

To the 20 mg of sample add 4 drops of cobalt chloride solution and 1 drop of ammonia	
• Formation of blue colour	• Indicates the presence of phenytoin
• Formation of violet colour	• Indicates Phenobarbital, ethosuximide and valproic acid

Diazotization test

Dissolve the sample in 2M hydrochloric acid and cool in ice (< 10 °C). Add 4 drops of sodium nitrite (1%) and one drop of 2-naphthol (4%) in sodium hydroxide (2M).	
• Bright red/orange-red colour	• Indicates the presence of ergot alkaloids, psilocybin, psilocybin and dimethyltryptamine
• Violet colour	• Indicates the aminonitrothiazoles
• Reddish-brown (by adding few drops of hydrochloric acid turns into greenish yellow)	• Indicates amiloride
• Dark blue with violet tinge	• Indicates the presence of ethambutol
• Violet colour	• Indicates the presence of ephedrine
• Pink colour precipitate	• Indicates azathioprine
• Pale colour	• Indicates folic acid
• Dark colour	• Indicates nitrazepam

Dragendorff test

Dissolve the sample in three drops of 2M hydrochloric acid, add 2 to 3 ml of the reagent and dilute to 10 ml with water.	
Dragendorff reagent: Dissolve 1 gm of bismuth subnitrate in 3 ml of 10 ml hydrochloric acid with the aid of heat. Dilute to 20 ml with water and dissolve 1 gm of potassium iodide in the mixture. If black bismuth triiodide separates, add 2M hydrochloric acid and more potassium iodide to dissolve it.	
• Orange, red-orange or brown precipitate	• Indicates the presence of alkaloidal base • Primary, secondary, tertiary and quaternary amines also produces positive result

Diphenylamine test

Apply the reagent to the sample on a white tile or in a test-tube.	
Diphenylamine reagent: Mix 0.5 gm of diphenylamine in 20 ml of water and dilute to 100 ml with concentrated sulfuric acid.	
• Blue colour	• Indicates the presence of oxidizing agents (bromated, chlorated, chromate, dichromate, iodate, nitrite, permanganate, vanadate)
• Violet colour	• Indicates the aminonitrothiazoles

Diphenylamine test (Modified)

Mix the blood sample with 1 ml acetone in a tube and centrifuge. To the 50 µl of supernatant add 50 µl of diphenylamine reagent and 25 µl of chloroform. Centrifuge the mixture and allow to stand. A pink colour that develops in the chloroform layer indicates ethchlorvynol.	
Diphenylamine reagent: Mix 0.5 gm of diphenylamine in 20 ml of water and dilute to 100 ml with concentrated sulfuric acid.	
• Pink colour in chloroform layer	• Indicates the presence of ethchlorvynol.

Erdmann's test

Dissolve the sample in 1 ml of ethanol, add a pellet of potassium hydroxide and evaporate to dryness (100 °C in a water bath). To the residue add 0.5 ml of water and 1 ml of carbon tetrachloride layer and shake it with 1 ml of reagent.	
Erdmann's reagent: Mix 1 ml of nitric acid with 30 ml of sulfuric acid.	
• Red colour (in acid layer) (changes to orange and green)	• Indicates the presence of clofenotane, dichlorodiphenyldichloroethylene (DDE) • Dichlorodiphenyldichloroethane (DDD) and mitotane produces no colour change from red.
• Pink colour	• Aldrin, dialdrin and endrin

Ferric chloride test

To the 2 drops of ferric chloride solution add the aqueous or ethanolic solution of the sample.	
Ferric chloride reagent: Dissolve 5 gm of anhydrous ferric chloride, or 8.25 gm of ferric chloride hexahydrate, in 100 ml of distilled water.	
• Red, orange, green, blue/brown colour	• Indicates the presence of phenolic compounds, fatty acids and phenyl pyrazoline (in aqueous medium phenols produce negative result) • Indicates the salicylates (aspirin require alkali hydrolysis)
• Violet colour	• Phenothiazines in higher concentrations gives positive result
• Yellow-orange	• Indicates the presence of chlorpheniramine
• Orange-brown	• Indicate the presence of amino caproic acid
• Yellow solution	• Indicates ampicillin
• Greenish-brown	• Indicates carbidopa
• Blue-greenish	• Indicates diethylstibesterol

Ferrous sulphate test

To the sample add 2 drops of ferrous sulphate reagent.	
Ferrous sulphate reagent: To one volume of a 10% (w/v) solution of ferrous sulphate ($FeSO_4.7H_2O$) add five volumes of sulphuric acid with cooling.	
• Red/pink colour	• Indicates the presence of nitrates and nitrites (glyceryl trinitrate)

Folin-Ciocalteu test

Add the sample to 0.5 ml of reagent.	
Folin-Ciocalteu reagent: ***Stock solution*:** Dissolve 100 gm of sodium tungstate and 25 gm of sodium molybdate in 800 ml of water in a 1500 ml flask, add 50 ml of phosphoric acid and 100 ml of hydrochloric acid, and reflux for 10 hr cool, add 150 gm of lithium sulphate, 50 ml of water and four to six drops of bromine, cool, filter and dilute to 1000 ml with water. This stock solution should be stored at a temperature not exceeding 4 °C and used within 4 months of its preparation; it has a yellow colour and must not be used if any traces of green colour is present. ***Working solution:*** Dilute one volume of the stock solution with two volumes of water.	
• Blue colour	• Indicates the presence of phenolic compounds (halogenations on phenolic ring inhibits this reaction)

Formaldehyde-sulphuric acid test

Mix the sample with reagent and heat at 100 °C for 1 min.	
Formaldehyde-sulphuric acid reagent: To four volumes of sulfuric acid add six volumes of formaldehyde solution (using a pipette with the tip below the surface of the acid) with stirring and adequate cooling. When the reagent is warm it remains clear for about 1 hr. If turbidity develops, this may be dispelled by heating in a water-bath at 100 °C for about 1 min.	
• Orange colour	• Indicates the presence of benzodiazepines, betamethasone
• Yellow colour	• Indicates bromazepam
• Pink colour	• Indicates flurazepam
• Brown colour	• Indicates tryptamine, betamethasone
• Red colour	• Indicates zomepirac, bupivacaine
• Deep yellow with brownish tinge	• Indicates furosemide
• Violet red (red fluorescence)	• Indicates pethidine
• Purple-brown	• Indicates clomifene
• Yellow-orange	• Indicates carbamazepine
• Immediate dark violet	• Indicates codeine

Forrest test

Dissolve the sample in a minimum volume of 2M hydrochloric acid and an equal volume of the reagent. To test urine, add 1 ml of reagent to 0.5 ml of urine.	
Forest reagent: Mix together equal volumes of a 0.2% (w/v) solution of potassium dichromate, a 30% (v/v) solution of sulfuric acid, a 20% (w/w) solution of perchloric acid and a 50% (v/v) of nitric acid.	
• Red, pink, orange, blue or violet colour	• Indicates the presence of phenothiazines (Dibenzazepines produce blue colour, phenothiazines inhibits this colour development)

FPN test

Dissolve the sample in a minimum volume of 2M hydrochloric acid and add an equal volume of reagent.	
FPN reagent: Mix together 5 ml of 5% (w/v) ferric chloride solution, 45 ml of a 20% (w/w) solution of perchloric acid and 50 ml of a 50% (v/v) solution of nitric acid.	
• Red, pink, orange, blue or violet colour	• Indicates the presence of phenothiazines

Fujiwara test

Mix together 2 ml of reagent and 1 ml of pyridine. Add the sample and heat in a water-bath at 100 °C for 2 min with shaking.	
Fujiwara reagent: Freshly prepared 20% (w/v) sodium hydroxide solution.	
• Red colour (in pyridine layer)	• Indicates the presence of halogen containing hydrocarbons (chloramphenicol, dichloralphenazone, trichloroethane, chloral hydrate).
• Yellow colour	• Indicates the 2,2,2-trichloro ethanol.

Furfuraldehyde test

Dissolve the sample in ethanol, place one drop of solution on a filter paper, add one drop of reagent and expose the paper to hydrochloric acid fumes for 2 to 3 min.	
Furfuraldehyde reagent: A 10% (v/v) solution of furfuraldehyde in ethanol.	
• Black spot	• Indicates the presence of non-aromatic carbamates

Hinsberg test

Dissolve 5 mg in 1 ml of pyridine and add 3 drops of sodium hydroxide and 2 drops of benzene sulphonyl chloride	
• Development of yellow-orange colour • Development of dark red-violet colour	• Indicates the presence of bupivacaine • Indicates the presence of rifampicin

Hydrogen peroxide test

To the10 mg of the substance add 2 drops of hydrogen peroxide and 1 ml of hydrochloric acid. Heat the mixture for 2 min.	
• Development of orange colour	• Indicates the presence of emetine

Hydroxylamine test

Dissolve 5 mg sample in 3 ml water, add 100 mg of hydroxyl amine hydrochloride and 1 ml of sodium hydroxide. Allow it stand for 5 min, add 2 ml of hydrochloric acid and 10 drops of ferric chloride.	
• Development of violet-red (changes to violet-brown)	• Indicates the presence of penicillins
• Development of violet colour	• Indicates clofibrate

Iodine test

Mix the sample with an equal volume of manganese dioxide and heat the mixture carefully to dull redness over a small flame. Repeat the test by heating the sample alone.	
• Violet vapours	• Indicates the presence of iodine in the molecule
Better results can be obtained by omitting manganese dioxide (in selective cases, e.g.: amiodarone)	

Koppanyi-Zwikker test

Dissolve the sample in 1 ml of ethanol, add one drop of the reagent followed by drop of pyrrolidine and agitate the mixture.	
Koppanyi-Zwikker reagent: A 1% (w/v) solution of cobalt nitrate in ethanol.	
• Development of violet colour	• Indicates the presence of imide group (barbiturates, glutethimide, saccharin, oxyphenisatine, theophylline and paramethadione)
• Pink or red-violet colour	• Indicates the sulphonamides
Hydrochlorides produces blue colour before the prior to the addition of pyrrolidine	

Lead nitrate test

To the 2 ml of sample solution add 2 ml of lead nitrate (2 gm in 10 ml water).	
• Formation of yellow precipitate	• Indicates the presence of dimercaprol

Libermann test

Add two or three drops of the reagent to the sample on a white tile. Occasionally it is necessary to carry out the test in a tube and heat in a water-bath at 100 °C.	
Libermann reagent: Add 1 gm of sodium or potassium nitrite to 10 ml of sulfuric acid with cooling and swirling to absorb the brown fumes.	
• Orange colour	• Indicates the presence of mono substituted benzene
• Brown colour	• Indicates the two mono substituted benzene
Phenols produce intense colour. Compounds with hydroxyl, alkoxy, methylene oxy groups produce wide range of colours.	

Marquis test

Add a drop of reagent to the sample on a white tile.	
Marquis reagent: Carefully mix 100 ml of concentrated sulphuric acid with 1 ml of 10% (v/v) formaldehyde solution (stable for several weeks if protected from light).	
• Green to orange red	• Indicates the presence of morphine, amphetamine and diamorphine.

McNally's test

Dissolve the sample (1 mg) in a few drops of acetone, and add 1 to 2 ml water. Add three drops of solution 1 and an equal volume of solution 2. Shake and heat in a water-bath at 100 °C for 3 min.	
McNally's reagent: 1. A 0.5% solution of copper sulfate in 10% acetic acid. 2. A freshly prepared 2% (w/v) solution of sodium nitrite.	
• Red colour	• Indicates the presence of free salicylic acid
• Brown precipitate	• Indicates the amino salicylic acid
• Violet colour	• Indicates diflunisal
Phenols produce intense colour. Compounds with hydroxyl, alkoxy, methylene oxy groups produce wide range of colours.	

Methanolic potassium hydroxide test

Add a few drops of reagent to a solution of the sample in methanol and heat if necessary to boiling point to develop the colour.	
Methanolic potassium hydroxide reagent: A 20% (w/v) solution of potassium hydroxide in methanol.	
• Colourless to pale red, orange, yellow, green blue colour	• Indicates the presence of phenols with adjacent hydroxyl groups and nitro benzenes.

Millons reagent test

Add a few drops of reagent to a solution of the sample in methanol and heat if necessary to boiling point to develop the colour.	
Millons reagent: Dissolve 3 ml of mercury in 27 ml of fuming nitric acid and add an equal volume of water with stirring.	
• Red to orange colour	• Indicates the presence of phenols
Combination of Folin-Ciocalteu reagent is advised for basic phenolic compounds	
Phenols with halogen substitution produce negative result	
Primary arylamines produce negative result	
Phenols with more than one hydroxyl group produce red colour	

Naphthol-sulphuric acid test

<table>
<tr><td colspan="2">Mix the sample with 1 ml of reagent, heat in a water bath at 100 °C for 2 min and note any colour produced. Cool, add 1 ml of water and note the colour again.</td></tr>
<tr><td colspan="2">Naphthol-sulfuric acid reagent: Mix 1 gm of 2-naphthol with 40 ml sulfuric acid and heat in a water bath at 100 °C, with occasional stirring until the naphth-2-ol is dissolved.</td></tr>
<tr><td>• Red, orange, Brown, yellow to green</td><td>• Indicates the presence of steroids</td></tr>
<tr><td>• Brown-yellow colour</td><td>• Indicates the chloramphenicol</td></tr>
<tr><td>• Green colour</td><td>• Indicates starch and tartaric acid</td></tr>
<tr><td colspan="2">Combination of Folin-Ciocalteu reagent is advised for basic phenolic compounds</td></tr>
</table>

Nessler's test

<table>
<tr><td colspan="2">Add the reagent (three drops) to the sample (three drops), agitate and heat the mixture to 100 °C in a water bath, examining it every minute for 10 min. A blank solution should be treated similarly at the same time.</td></tr>
<tr><td colspan="2">Nessler's reagent:
1. Dissolve 50 gm of mercuric chloride and 35 gm of potassium iodide in 200 ml of water and cool.
2. Dissolve 50 gm of sodium hydroxide in 50 ml of water and cool.
Add the cold solution 2 to the cold solution 1 and make up to 100 ml. Allow the mixture to stand and decant the clear supernatant (stable for many months) for use. Store in dark brown bottles away from light.</td></tr>
<tr><td>• Brown-orange colour</td><td>• Indicates the presence of aliphatic amides and thioamides (chloramphenicol, nicotinamide)</td></tr>
<tr><td>• Immediate black colour</td><td>• Indicates compounds with hydroxyl group at ortho and para position
• Indicates compounds with NH-NH and NH-NH_2 group (nicotinamide)</td></tr>
<tr><td>• Orange colour</td><td>• Indicates acebutalol and methotrexate</td></tr>
<tr><td>• Black colour</td><td>• Indicates carbidopa
• Ascorbic acid, hydralazine, iproniazide, isoniazid, adrenaline gives immediate colour
• Cimetidine, gentamycin, labetalol, meprobamate gives at 100 °C.</td></tr>
<tr><td>• Yellow colour</td><td>• Indicates dihydrostreptomycin</td></tr>
</table>

Ninhydrin test

Dissolve the sample in methanol, place one drop of solution on a filter paper, add one drop of reagent and dry in a current of hot air.	
Ninhydrin reagent: Dissolve 0.5 gm of ninhydrin in 40 ml of acetone.	
• Violet colour	• Indicates the presence of aliphatic primary amines and amino acids. • Gentamycin produces after 4 hr.
• Pink-orange colour	• Indicates amphetamine
• Yellow colour	• Indicates procainamide and proxymetacaine
• Weak pink-violet	• Indicates amantadine, rimantadine

Nitric acid test

1. Mix the sample with three drops of fuming nitric acid, heat at 50 °C for 30 s and observe any colour produced (observation 1). Cool the mixture. 2. Add two drops of cool mixture to 2 ml of sulfuric acid and observe the colour (observation 2). 3. To the remainder of the cool mixture, add 2 ml of water followed by 2M sodium hydroxide, dropwise, until pH 8 is reached (use an indicator paper) (observation 3).	
• Orange-red colour (Observation 1) • Orange colour (Observation 2) • Orange-brown colour (Observation 3)	• Indicates the presence of chlorinated phenol (hexachlorophene)
• Red colour (Observation 1) • Red colour (Observation 2) • Brown-violet colour (Observation 3)	• Indicates the presence of chlorinated phenol (pentachlorophenol)

Nitrous acid test

Dissolve the sample in a minimum volume of water, and add an amount of solid sodium nitrite equal in volume to the sample followed by a few drops of 2M hydrochloric acid.	
• Orange	• Indicates sulphafurazole
• Yellow	• Indicates sufadoxine, sulfachlorpyridazine, sulfadimidine, sulfaethidole, sulfamethizole, sulfamethoxazole, sufamethoxydiazine, sulphamethoxypyridazine, sulphametopyrazine, sulfamoxole, sulfaphenazole, sulfapyridine, sulfasomidine, sulfathiazole, sulfinpyrazone
• Green colour	• Indicates phenazone
• Blue colour	• Indicates dipyrone
• Violet colour	• Indicates amidopyrine

Phosphorus test

To the sample add 0.5 ml of nitric acid and 0.2 ml of sulfuric acid and heat at 100 °C in a water bath for 30 min. Cool add 1 ml of a 10% (w/v) solution of ammonium molybdate and replace in the water bath at 100 °C for 5 min. A blank solution should be treated at the same time.	
• Bright yellow solution/ precipitate	• Indicates the presence of phosphorous and oraganophosphorous pesticide • Chloroquine phosphate, cyclophosphamide and triclofos also give positive result.

Potassio-cupric tartrate test

Dissolve 10 mg of the sample in 1 ml of water and add 2 ml of potassio-cupric tartrate solution.	
• Red violet colour	• Indicates the presence of ampicillin
• Reddish-yellow precipitate	• Indicates betamethasone

Potassium dichromate test (Jone's test)

Dissolve the sample by shaking in 0.5 ml of 2M hydrochloric/sulphuric acid and add a few drops of potassium dichromate.	
• Immediate brown/green colour	• Indicates the presence of phenols (with two/more adjacent hydroxyl group) and amino phenol • Cyclophosphamide and triclofos also give positive result. • Monophenols, halogenated phenols and phenols with hydroxyl groups *meta* to each other react more slowly or not at all.
• Wine red colour	• Indicates griseofulvin
• Violet	• Indicates paracetamol
• Brownish-red	• Indicates chloroquine
• With acid orange-red (turns into dark brown with potassium dichromate)	• Indicates amitriptyline

Potassium permanganate test

Heat 100 mg with 10 ml water. Add 2.5 ml of sodium hydroxide and 3 drops of potassium permanganate	
• Violet colour turns to blue and finally into emerald green	• Indicates the presence of busulfan
• An intense red colour	• Indicates folic acid
• Brown precipitate	• Indicates ascorbic acid

Potassio-mercuric iodide test

Dissolve 50 mg of sample in sodium hydroxide (5 ml) and add alkaline potassio-mercuric iodine. Boil the mixture and allow to stand.	
• Light beige colour precipitate	• Indicates the presence of chlorambucil
• Yellow flocculent precipitate	• Indicates allopurinol
• White curdy precipitate	• Indicates chloroquine
• White turbidity	• Indicates ergotamine

Schiff's test

Add the sample to 1 ml of reagent.	
Schiff's reagent: Dissolve 0.2 gm of basic magenta (fuchsin, CI 42510) in 120 ml of hot water, cool, add 20 ml of 10% (w/v) solution of sodium hydrogen sulfite and 2 ml of hydrochloric acid and dilute to 200 ml. Store at 4 °C and protect from light.	
• Violet colour	• Indicates the presence of aliphatic aldehyde • Longer carbon chain containing compounds respond weak.

Sodium nitroprusside test

Add the sample to 2 ml of reagent followed by one drop of 2M sodium hydroxide.	
Sodium nitroprusside reagent: A 1% (w/v) solution of sodium nitroprusside.	
Orange colour	Indicates the presence of ketones (ketoprofen)

Simon's test (modified sodium nitroprusside test)

Add one drop of solution 1 to the sample, followed by drops of solution 2.	
Simon's reagent: **Solution 1:** Dissolve 1 gm of sodium nitroprosside in 100 ml of water and add 2 ml of acetaldehyde to the solution with thorough mixing. **Solution 2:** Freshly prepared 2% sodium carbonate in distilled water.	
• Dark blue colour • Pink to cherry-red colour	• Indicates the presence of secondary amine (ephedrine) • Indicates primary amines such as amphetamine, methylenedioxyamphetamine (MDA)

Silver nitrate test

Dissolve 20 mg in water (freshly boiled and cooled, 50 ml). Add 0.5 ml of nitric acid and 3 drops of silver nitrate.	
• White precipitate (dissolves in excess of ammonia)	• Indicates the chloride salts of basic drugs
• Metallic silver (bubbles of nitrogen evolves)	• Indicates isoniazid
• Dark grey precipitate	• Indicates ascorbic acid

Steyn test (sodium picrate test)

Mix the sample with few drops of chloroform and sulfuric acid to hasten the reaction while holding a piece of filter paper, impregnated with the reagent, in the vapours that issue from the tube, and heating the contents to 30 °C.	
Steyn reagent: Prepare a solution of 5 gm sodium bicarbonate and 0.5 gm picric acid in 100 ml of water.	
• Yellow colour in filter paper (changes to orange, brown-orange, orange-red and red)	• Indicates primary cyanide group containing compounds such as cimetidine, diphenoxylate, isoaminile.

Sulfuric acid test

Dissolve small quantity in 1 ml of sulphuric acid and allow to stand for 5 minutes.	
Steyn reagent: Prepare a solution of 5 gm sodium bicarbonate and 0.5 gm picric acid in 100 ml of water.	
• Orange/yellow colour (produce fluorescence under UV light)	• Indicates steroids
• Red or orange colour (produce fluorescence under UV light)	• Indicates thioxanthenes, diethylstilbesterol
• Purplish brown colour	• Indicates beclomethasone

Thalleoquin test

Dissolve the sample in a minimum volume of 2M hydrochloric acid, add two drops of bromine solution, place one drop of the mixture on a piece of filter and expose the paper to ammonia fumes.	
• Green colour	• Indicates the presence of quinine (hydroquinine, hydroquinidine, quinidine, quinine)
	• Cinchonidine and cinchonine will not react.

Tollen's test (Silver mirror test)

Dissolve the sample in a minimum amount of water, with the addition of ethanol if necessary, add an equal volume of the reagent and note any colour that develops. Heat the mixture in a water bath at 100 °C for 30 s.	
Tollen's reagent: To 20 ml of 0.1M silver nitrate add sufficient strong ammonia solution to dissolve the initial brown precipitate.	
• Red, yellow, brown, black colours • White precipitate (turns yellow on heating) • Silver mirror (on heating)	• Indicates the presence of reducing sugars • Indicates the presence of ethchlorvynol and ethinylestradiol • Indicates the presence of carbidopa

Vanillin test

Add two drops of reagent to the sample, heat in a water bath at 100 °C for 30 s and note any colour that is produced. Dilute the cooled mixture by adding a few drops of water and note any change of colour.	
Vanillin reagent: Dissolve 1 gm of vanillin in 20 ml of sulfuric acid, warming if necessary	
• Dark violet, blue, green colour (on dilution)	• Indicates the presence of cycloalkane ring

Wasicky test (Van Urk test)

Add the reagent to the sample in a test tube, warming if necessary. Observe any colour produced, then carefully dilute with water or spray dried spots on filter paper and heat.	
Wasicky reagent (Van Urk reagent): Dissolve 2 gm of *p*-Dimethylamino benzaldehyde (*p*-DMAB) to 50 ml of 95% ethanol and 50 ml of concentrated hydrochloric acid. The reagent should be freshly prepared.	
• Violet colour	• Indicates the presence of primary aromatic amines (diphenyl amine gives negative result)
• Red colour (changes to violet)	• Indicates the cannabinols and indole nucleus
• Red to orange colour (changes to violet upon dilution)	• Indicates phenols and phenolic amines

CHAPTER 56

IDENTIFICATION OF ADULTERANTS

56.1 Identification of the Invert Sugar (adulterant) Present in Honey

Principle: Honey is natural sugar produced by honeybees. Fructose corn syrup is the most common adulterant of honey. The reducing property of fructose is utilized in the determination.

Method 1: Lane-Eynon titrimetric method

This is the most accurate method for the estimation of reducing sugars (invert sugars). This method is based on the determination of the volume of sample solution required to reduce copper present in Fehling's solution. Invert sugars reduce copper present in Fehling's solution into cuprous oxide (red in colour).

Requirements: Fehling's solutions A and B, honey samples, glucose (0.5%, standard), methylene blue indicator (0.2%), neutral lead acetate solution (45%), potassium oxalate solution (22%).

Preparation of sample solution

1. Weigh 25 gm of sample into volumetric flask (250 ml capacity), add 10 ml of neutral lead acetate solution and make up the volume to 250 ml with water.
2. Pipette and transfer 25 ml of this solution into volumetric flask (500 ml capacity), containing 100 ml of water.
3. Add potassium oxalate solution in small portions until there is no precipitation occurs.
4. Make up the volume with water to 500 ml and filter.
5. Transfer the filtrate into 50 ml burette.

Preparation of standard invert sugar solution

1. Weigh accurately 9.5 gm of sucrose into 1000 ml volumetric flask.
2. Add 100 ml of water and 5 ml of concentrated hydrochloric acid.
3. Allow the mixture to stand at room temperature for 3 days.

4. Pipette 25 ml of standard solution into 100 ml volumetric flask and add 50 ml water.
5. Neutralize the solution with sodium hydroxide (20%) using phenolphthalein as indicator.
6. Acidify the mixture with hydrochloric acid (1N).
7. Make up the volume to 100 ml with water.
8. Adjust the volume such that each ml contains 2.5 mg of invert sugar.

Standardization of Fehling's solution

1. Transfer 10 ml of Fehling's solution A and B (each of 5 ml) into conical flask.
2. Add 25 ml of water and add 15 ml of standard invert sugar loaded in the burette.
3. Add 3 drops of methylene blue indicator and titrate the mixture with standard invert sugar solution.
4. The colour change from blue to brick-red indicates the end point.

$$\text{Factor for Fehling's solution} = \text{Titre} \times 2.5/1000$$

Determination of invert sugar

1. Transfer 10 ml of Fehling's solution A and B (each of 5 ml) into conical flask.
2. Add 50 ml of water and boil the mixture.
3. Add 3 drops of methylene blue indicator.
4. Titrate the mixture with sample solution loaded in the burette.
5. The colour change from blue to brick-red indicates the end point.

$$\%\ \text{Reducing sugars} = \frac{\text{Fehling factor} \times \text{Dilution} \times 100}{\text{Titer} \times \text{Weight of sample}}$$

Method 2: ***Fiehe's test***

1. Dissolve 10 gm of honey in 10 ml of cold water and shake it with 20 ml of diethyl ether.
2. Separate the ether layer and evaporate it.
3. Dissolve this residue in 10 ml of ether.
4. To the 2 ml ether solution add 2 ml of Seliwanoff reagent (fresh).

   [Seliwanoff reagent: 1% resorcinol in concentrated hydrochloric acid]
5. Development of pink colour in the acid layer indicates the presence of invert sugar (it turns into cherry red at the junction of the layers).

Method 3: Aniline acetate test

1. Evaporate the remaining ether extract in a porcelain dish and to the residue add 2 ml of fresh aniline acetate solution.

2. Keep this solution for 15 min.
3. Appearance of pink or orange colour confirms the presence of invert sugar.

56.2 Identification of the Adulterants Present in Milk

Principle: Milk adulteration is of two types

- Synthetic milk: Blend of urea, cooking oil, detergent, caustic soda, sugar, salt and skimmed milk powder.
- Dilution of milk: Diluting milk with water and addition of urea, fat, neutralizers, sugar, soap, starch and formalin to increase the viscosity.

Adulteration of milk reduces the quality of the milk and produce hazardous effect to the body. The food industry preservatives benzoic acid and salicylic acid are also the adulterants. These adulterants increase the shelf life of milk.

Procedure:

Test for microorganism

Methylene blue reduction test: Microbes (if any) present in the milk reacts with the indicator (dye) and converts blue colour into white colour.

Test for table sugar (sucrose)

Sucrose increases the density of the milk.

Ketone sugar test: Sucrose (adulterant) present in the milk produce red colour with resorcinol.

Test for starch

Iodine test: The development of blue colour with iodine indicates the formation of starch-iodo complex.

Test for acids

Ferric chloride test: Milk with drop of ferric chloride in presence of drop of sulphuric acid develops buff colour and indicates the presence of benzoic acid (adulterant). The development of violet colour indicates the presence of salicylic acid.

Test for soap

Soaps increase the foaming of milk.

Phenolphthalein test: The development of pink colour with phenolphthalein indicates the basic medium due to the presence of soaps.

Test for formalin

Ferric chloride test: Milk with drop of ferric chloride in presence of drop of sulphuric acid develops purple colour and indicates the presence of formalin.

Test for ammonium sulphate

Ammonium sulphate increases the density of milk.

Milk with sodium hydroxide, sodium hypochlorite and phenol produce blue colour.

56.3 Determination of the Adulteration of Lead Chromate with Turmeric

Principle: Turmeric is polished with lead chromate and metanil yellow. Lead chromate cause anaemia, abortion, paralysis, mental retardation and brain damage.

Procedure:

1. Ash about 2 gm of the sample and dissolve it in 10 ml of sulphuric acid.
2. Boil the mixture for about 5 min and filter.
3. To the 5 ml of filtrate add 0.5 ml of alcoholic diphenyl carbazide (2%).
4. Development of pink colour indicates the presence of lead chromate.

56.4 Identification of the Vanaspati (adulterant) Present in Ghee

Principle: The presence of vanaspati in ghee may be detected by testing for sesame oil (content of vanaspati). The test is called 'Baudoiun test'.

Procedure:

1. Transfer 10 ml of melted ghee into test tube.
2. Dissolve 0.2 gm of sucrose in 10 ml of dilute hydrochloric acid (in another test tube).
3. Mix the two solutions and shake thoroughly.
4. Heat the mixture until it becomes colourless.
5. Keep it aside for 10 min.
6. Development of pink or red colour indicates the presence of vanaspati.

56.5 Identification of the Tea Seed Oil (adulterant) Present in Vegetable Oil

Principle: The detection of tea seed oil in vegetable oil is based on the formation of deep green colour by reflected and transmitted light in Fitelsons test for tea seed oil.

Procedure:

1. Mix 1 ml of acetic anhydride and 1.5 ml of chloroform with 2 drops of concentrated sulphuric acid in a test tube.
2. Cool the mixture by placing in ice bath and add 10 drops of given oil.

   [If the liquid appears cloudy make it clear by adding drop wise acetic anhydride]
3. Keep the test tube in an ice bath for 10 min.

4. Development of a green colour in a reflected and transmitted light indicates the presence of tea seed oil.
5. Add 10 cm^3 of dry ether and mix the contents by shaking the tube.
6. Brown colour is formed which changes to red within 1 min and finally fades.

56.6 Identification of the Rhodamine B (adulterant) Present in Foods

Principle: Rhodamine B produce artificial red colour and is used as a colouring agent in food industry (confectioneries). It is carcinogenic and also causes damage to kidney, liver and spleen.

Procedure:

1. Extract the red colour with ether/alcohol.
2. To the 5 ml of extract add 5 ml of concentrated hydrochloric acid.
3. Development of strong pink colour indicates Rhodamine B and light pink colour indicates absence of Rhodamine B.

56.7 Identification of the Chicory (adulterant) Present in Coffee

Requirements: Aqueous lead acetate (10%), concentrated hydrochloric acid, Seliwanoff reagent.

Principle: Chicory contains inulin, which upon hydrolysis produces fructose. Coffee does not contain inulin. Hence chicory alone can give positive results with Seliwanoff reagent.

***Seliwanoff reagent*:** 50 mg of resorcinol in 100 ml of dilute hydrochloric acid (1:2)

Procedure:

1. To the aqueous extract of sample add neutral lead acetate solution.
2. Filter the solution and to the 5 ml of filtrate add 5 ml Seliwanoff reagent and 1 ml concentrated hydrochloric acid.
3. Development of distinct red colour indicates the presence of fructose (hydrolysis product of inulin).

ISOLATION AND DETERMINATION OF ACTIVE PRINCIPLES FROM NATURAL PRODUCTS

CHAPTER 57

ISOLATION OF ACTIVE PRINCIPLES FROM NATURAL PRODUCTS

57.1 Isolation of Caffeine from Tea Leaves

Requirements: Tea leaves, Sodium carbonate/calcium carbonate, Tea bags.

Principle: Caffeine belongs to alkaloids, basic nitrogenous compounds found in plants. Caffeine produces stimulant effects in central nervous, respiratory and cardiac systems. It causes nervousness, insomnia and headache. Tea leaves, coffee seeds and cocoa seed contains caffeine as one of the major ingredient. Tea leaves consists of cellulose, caffeine, tannins and small amount of chlorophyll.

Isolation of caffeine involves the extraction of water soluble caffeine present in the leaves into hot water. The solubility of caffeine in water is 22 mg/ml at 25 °C, 1080 mg/ml at 80 °C and 670 mg/ml at 100 °C. The caffeine present in the aqueous layer will be separated by the extraction with dichloromethane. Caffeine is more soluble in dichloromethane (140 mg/ml) than water (22 mg/ml) at 25 °C. Addition of sodium carbonate to the mixture forms sodium salt of phenolic compounds (tannins). The sodium salt of tannins are more water soluble and remain in the water.

Procedure:

1. Place the tea bags containing tea powder/tea leaves into beaker, add 30 ml of distilled water and 2 gm of sodium carbonate.
2. Gently boil the mixture for 20 minutes by placing in a water bath (push the tea bag into bottom whenever necessary).
3. Decant the hot liquid into flask, add 20 ml of distilled water to the beaker, boil and decant again into the flask.
4. Press tea bags to collect maximum amount of extract.
5. Cool the tea extract to room temperature and transfer into separatory funnel.
6. Add 5 ml of dichloromethane to the separatory funnel and swirl the contents.
7. Allow the contents to settle down and to form two distinct layers (If any emulsion forms stir it with glass rod to break the emulsion).

8. Collect the lower organic layer and extract the remaining aqueous layer with dichloromethane.
9. Combine the organic layers and add 0.5 gm of anhydrous sodium sulphate and swirl the mixture (anhydrous sodium sulphate will absorb the water if any present).
10. Decant the liquid into beaker containing boiling chips and place it in a hot plate.
11. Add petroleum ether when volume reduces into 3 ml.
12. Allow the beaker to cool to room temperature when cloudiness appears in the beaker.
13. Crystals of caffeine will appear in the solution and isolate it by vacuum filtration.
14. The melting point of the pure caffeine is 238 °C.

57.2 Isolation of Pectin from Apples

Requirements: Apple, Oxalic acid.

Principle: Pectin (derived from Greek meaning 'congealed and curled') is a structural hetero polysaccharide. The higher concentrations of pectin are found in the middle lamella of cell wall of terrestrial plants. During ripening, enzymes pectinase and pectin esterase breaks the pectin and converts the fruit soft. Pectin acts as a soluble fiber and reduces blood cholesterol level. It increases the viscosity of the intestinal tract, there by decreases the absorption of cholesterol from bile and food. Pectin in large intestine and colon gets converted to short chain fatty acids, which have prebiotic effect.

The natural sugars found in pectin are D-galactose, L-arabinose and D-xylose. Pectin consists mainly of D-galacturonic acid (GalA) units joined by α-(1-4) glycosidic linkage. The uronic acids have carboxyl groups, which are naturally present as methyl esters and others. Which on treatment with ammonia produces carboxamide groups. In some cases galacturonic acid is replaced with (1-2)-linked L-rhamnose and this type of pectin is called as rhamnogalacturonan-I. Commercial pectins are derived from citrus peel (20-35%) or apple pomace (10-15%). Citrus pectins are light cream/light tan in colour where as apple pectins are darker. Pectins also can be obtained from sugarbeet waste, mango waste and sunflower heads. Pectin, monovalent salt of pectinic acid and pectic acid are soluble in water, where as di and tri valent salts are weakly soluble.

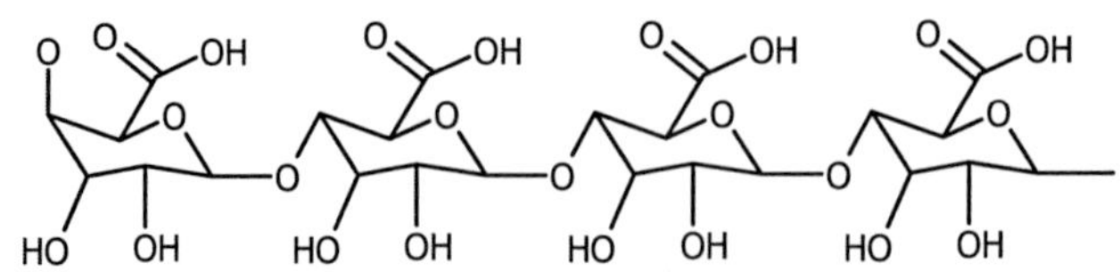

Applications

1. Gelling agent: Pectins are mainly used as gelling agent, thickener, water binder and stabilizer.
2. Pectin increases viscosity and volume of stool, used against constipation.
3. Pectin is an active ingredient of lozenges, demulcent.
4. Pectin is used in wound healing preparations and medical adhesives (colostomy devices).
5. Pectin is useful in cosmetics as stabilizer.
6. Pectin is an effective antidote for heavy metal poisoning.
7. Bactericidal action of pectin is useful during diarrhoea especially for infants and children.
8. Pectin due its large water binding capacity reduces the food intake.
9. Pectin beads are useful in sustained release drug delivery systems.
10. Pectin and calcium pectin will be degraded by colonic pectinolytic enzyme. It offers colon-specific drug delivery, which is useful in ulcerative colitis, Crohn's disease.

Conventional extraction of pectin

- Direct boiling is a conventional method of pectin extraction and gives relatively good yield after two hours of extraction.
- Longer heating may degrade the pectin (thermal degradation). It involves extraction of pectin with acidified water (pH up to 2) at temperature not more than 70 °C.
- Mineral acids - sulphuric acid, hydrochloric acid and phosphoric acids can be used for extraction.
- Oxalic acid, ammonium oxalate, tartaric acid and polyphosphates also can be used. Zeolites, Zeocarb also assists the extraction procedure.
- Double extraction at 85-90 °C after one hour using cationic resin gives higher yields with better gel strength of pectin.

57.3 Isolation of Casein, Albumin and Lactose from Milk

Requirements: Acetic acid (10%), Powdered milk, Calcium carbonate, Nylon hose.

Principle: Milk contains water, vitamins, minerals, proteins, sugars and lipids. Three major proteins present in the milk are casein, lactalbumin and lactoglobulin. Thiamine, riboflavin, pantothenate, vitamin A, vitamin D and vitamin K are the vitamins present in milk. It contains calcium, potassium, sodium, phosphorous and trace metals. The precipitation of casein from water during destruction is called as protein denaturation. Lactose is the main carbohydrates present in it.

Composition	Human	Cow	Goat	Sheep
Water	87.4	87.1	87.0	82.6
Protein	1.4	3.4	3.3	5.5
Fats	4.0	3.9	4.2	6.5
Carbohydrates	7.0	4.9	4.8	4.5
Minerals	0.2	0.7	0.7	0.9

Casein is a phosphoprotein attached to serine and threonine. Casein exists as calcium salt and composed of α, β and κ caseins. α and β caseins are soluble in water individual as well as in combination. Addition of κ caseins to the above mixture results in micelle. α and β casein are phosphoproteins and are precipitated by calcium ions. Calcium caseinate has its isoelectric point at pH 4.6, therefore it is insoluble in solution of pH less than 4.6. Casein bears negative charge at milk pH 6.6, but when acid is added the negative charges gets neutralized and becomes precipitate.

$$\text{Calcium caseinate} + 2HCl \longrightarrow \text{Casein} + CaCl_2$$

Albumins are globular proteins soluble in water and dilute salt solutions. The second most abundant protein present in milk is lactalbumin and the other one is lactoglobulins.

Procedure:

Isolation of casein

1. Place about 4 gm of powdered non-fat milk and add 42 ml of water into 100 ml beaker.
2. Heat the milk to 40 °C by placing in a hot plate.
3. Add 5 drops of acetic acid (10%) dropwise to the warm milk (after it reaches 40 °C).
4. Maintain the temperature after every addition of acetic acid (5 drops) and stir the solution (on stirring casein precipitates out).
5. Isolate the casein precipitated and add dilute acetic acid to complete the precipitation (excess of acid hydrolyses lactose to glucose and galactose).
6. Remove the milk solution present in the precipitate by squeezing.
7. Dry the casein by pressing between filter paper.

Isolation of albumin

1. Neutralize the milk present in beaker (obtained in above reaction) by adding sufficient calcium carbonate.
2. Heat the mixture to 75 °C for 5 minutes on a sand bath.
3. Separate the precipitate of albumin by filtration in hot condition.

Isolation of lactose

1. Cool the liquid (obtained in above step) to room temperature and centrifuge for 5 min.
2. Decant the liquid into beaker and add 95% ethanol to the beaker.
3. Solid get precipitates and heat the mixture on a sand bath to about 60 °C to dissolve the soild.
4. Centrifuge the solution for 2 min and remove the supernatant, and collect the solid lactose.

57.4 Isolation of Cinnamon Oil from Cinnamon

Requirements: Cinnamon, Ethyl acetate, anhydrous Calcium chloride/magnesium sulphate.

Principle: Cinnamon is dried barks of *Cinnamomum zylanicaum*. Essential oils are volatile organic substances that can be isolated from odoriferous plants. The oils are generally concentrated in seeds, flowers and also in other parts. Essential oils can be separated from plants by grinding, maceration, solvent extraction, distillation and concentration. Combination of all these techniques results in effective and efficient isolation.

***Procedure*:**

1. Weigh 5 gm of dry cinnamon bark and grind them into a coarse powder with the aid of mortar and pestle.
2. Transfer the grinded mass into round bottom flask loaded with boiling chips.
3. Add 35 ml of ethyl acetate to the flask and reflux the mixture for 20 min.
4. Cool the mixture to room temperature and decant the solution to flask.
5. Rinse the residue with ethyl acetate and combine the organic layers.
6. Add anhydrous magnesium sulphate/calcium chloride and swirl the flask for about 30 seconds.
7. Filter the mixture into round bottom flask and remove the ethyl acetate by distillation.
8. The oil remains in the flask is cinnamon oil.

57.5 Isolation of Clove Oil from Clove

Requirements: Cloves, Dichloro methane.

Principle: Cloves are dried flower buds of *Eugenia caryophyllata*. Clove oil used as flavouring agent and perfume. Eugenol is the major component (85-90%) along with eugenol acetate (9-10%). Clove bud oil posses antibacterial, antifungal, insecticidal and antioxidant properties. Clove oil also useful in the treatment of asthma and allergic disorders.

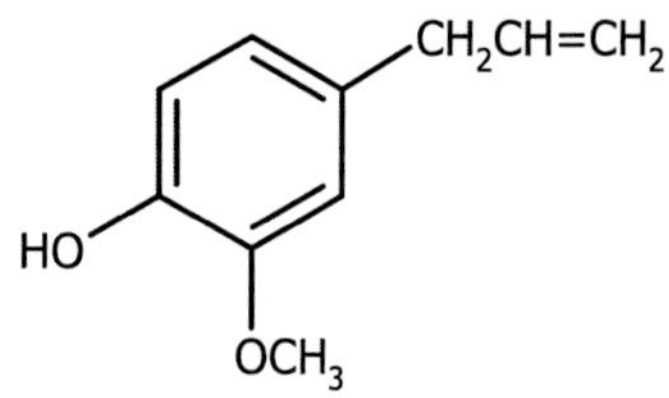

Eugenol

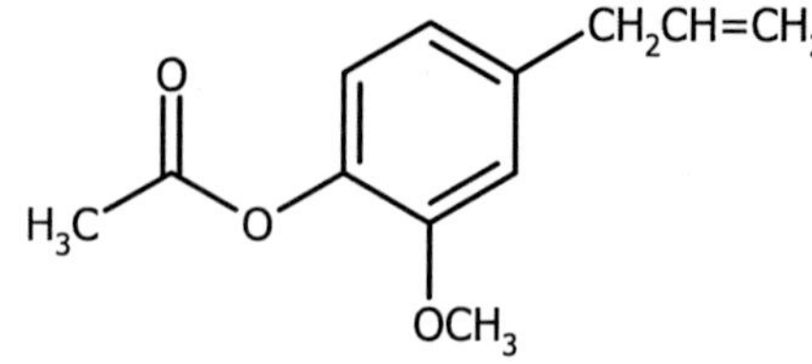

Eugenol acetate

Procedure:

Isolation of clove oil by organic extraction

Clove oil can be isolated by the process which is similar and described for cinnamon.

Isolation of clove oil by hydro distillation

1. Weigh 5 gm of dry cloves and grind them into a coarse powder with the aid of mortar and pestle.
2. Transfer the grinded mass into round bottom flask, loaded with boiling chips and add 50 ml of distilled water to the flask.
3. Distil the mixture ($3/4^{th}$) and allow the distillate to cool to room temperature.
4. Transfer the distillate into separating funnel and add 10 ml of saturated sodium chloride solution.
5. Rinse the condenser and flask with 10 ml of dichloromethane (DCM) and pour into separatory funnel.
6. Swirl the contents of separating funnel gently, and allow the two layers to separate.
7. Collect the DCM layer into flask and extract the aqueous layer with two portions of DCM.
8. Combine the organic layer and dry with anhydrous sodium sulphate and decant the DCM layer and evaporate the DCM by placing in water bath.
9. Clove oil remains in the evaporation dish after complete evaporation.

Isolation of clove oil by distillation in clevenger

1. Place 5 gm of clove previously grinded into coarse powder in the round bottom flask.
2. Add boiling chips to the flask and attach Clevenger and distil the contents of the flask for 12 hr (until no oil drop comes out).
3. Wash the distillate with saturated sodium chloride and separate the hydro layer.
4. The oil remains in the separatory funnel.

57.6 Isolation of Starch from Potatoes

Requirements: Potatoes, Muslin cloth.

Principle: Starch is an important polysaccharide, which is insoluble in water. It occurs as granules.

Procedure:

1. Peel the raw potato and cut into small pieces.
2. Grind the pieces with water using mortar and pestle (mixer can be used).
3. Collect the homogenate into beaker and add water.
4. Filter the homogenate through muslin cloth to remove particles.
5. Allow the filtrate to settle and decant the supernatant liquid.
6. Wash the residue 3 times and decant the supernatant.
7. The solid remains in the flask is polysaccharide starch.

57.7 Isolation of Carotenoids from Tomato and Carrot

Requirements: Tomatoes, carrots, ethanol, dichloro methane.

Principle: Carotenoids are natural phytochemicals of the tetraterpenoid family, which means that they are comprised of 8 isoprene units, hence the 40 carbon atoms in each molecule. Carotenoids naturally occur in the chloroplasts of plants and other photosynthetic organisms. Carotenoids are of interest because they are great antioxidants, stimulate the immune system, and are required for the production of provitamin A. β-Carotene and lycopene are the most studied carotenoids since they are the most abundant in human diet.

β-Carotene is yellow-orange in color and can be isolated from the roots of carrots. The most common side effect of excessive β-carotene consumption is carotenodermia (skin turns orange), a harmless condition. High doses of β-carotene have been associated with increased risk of lung cancer in people who smoke or were exposed to asbestos.

However, dietary levels of β-carotene have been known to reduce the risk of many types of cancer. Lycopene is a red-orange color; excess amounts of lycopene can accumulate in the adipose tissue under the skin, producing lycopenemia, a variant of carotenodermia. Lycopene is two times more effective antioxidant than β-carotene; and it is associated with the reduced risk of prostate, lung and stomach cancers.

Procedure:

1. Place the 10 gm tomato paste/carrot paste into a beaker containing 30 ml of 95% ethanol.
2. Stir the solution for about 6 minutes, isolate the solid by vacuum filtration and place into a 25 ml round-bottom flask loaded with 10 ml of dichloromethane (DCM).
3. Reflux the solution for about 5 minutes and collect the solid by vacuum filtration (repeat the operation for two times).
4. Combine the DCM layers, wash with brine solution and dry with sodium sulfate.

5. Evaporate the organic layer and add 1 ml of hexane.
6. TLC was used to determine if β-carotene was present in the organic solution.
7. Separate the mixture on an alumina gel column using three different mobile phases.
8. Use hexane to separate the color bands, 2% DCM in hexanes elute the yellow color band (β-carotene) and DCM to elute the scarlet color band (lycopene).

57.8 Isolation of Trimyristin and Myristin from Nutmeg

Requirements: Ground nutmeg, Diethyl ether, Ethanol (95%), Sodium hydroxide, Sodium chloride solution (35%), Distilled water, Vegetable oil, Ferric chloride solution (1%), Calcium chloride solution (1%).

Principle: Nutmeg is the fruit of *Myristica fragrans* houtt, when the fruit of the tree matures, it splits into two, exposing a single seed sheathed in a crimson aril. The aril, a fibrous covering, constitutes the spice known as mace, and the seed is a whole nutmeg. Nutmeg oil is obtained by steam distillation of kernels of the fruit and consists of approximately 90% of terpene hydrocarbons. Major components are sabinene and α-, β-pinene. Nutmeg oil is used primarily for food flavoring. Its characteristic odour is due to safrole derivatives like myristicin and elemicin. Hot extraction of nutmeg with diethyl ether will produce trimyristin (glyceryl trimyristate) along with a small quantity of myristicin. Nutmeg oil generally consists of over 80% trimyristin. Pure trimyristin can easily be isolated by recrystallization from ethanol. Myristicin is toxic and has got narcotic activity which may also cause convulsions. Nutmeg has carminative and stimulative properties.

Trimyristin is an ester with the chemical formula $C_{45}H_{86}O_6$. It is a saturated fat which is the triglyceride of myristic acid. Trimyristin is found naturally in many vegetable fats and oils. Trimyristin is a white to yellowish-gray solid that is insoluble in water, but soluble in ethanol, benzene, chloroform, dichloromethane, and ether. Trimyristin makes up between 20-25% of the overall mass of dried, ground nutmeg.

Separation is generally carried out by steam distillation and purification uses extraction from ether followed by distillation or rotary evaporator to remove the volatile solvent. The extraction of trimyristin can also be done with diethyl ether at room temperature, due to its high solubility in the ether. The experiment is frequently included in curricula, both for its relative ease and to provide instruction in these techniques.

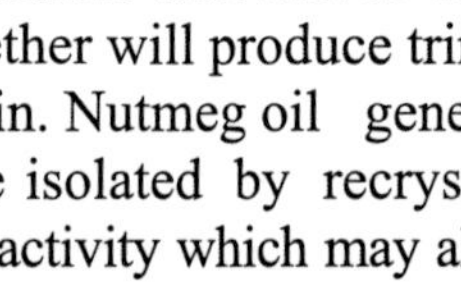

Trimyristin

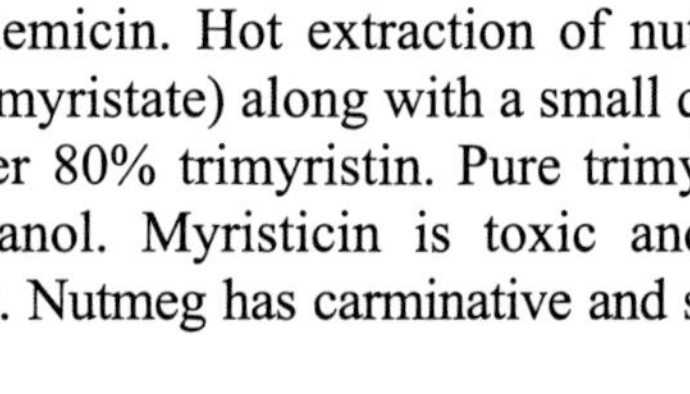

Myristin

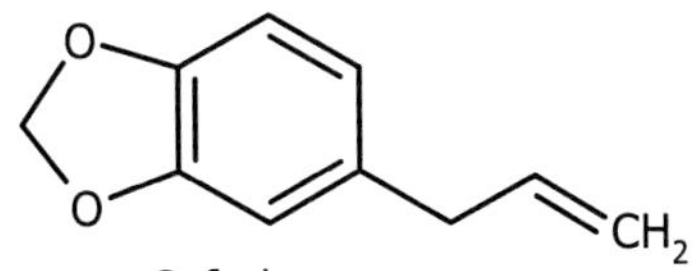

Safrole

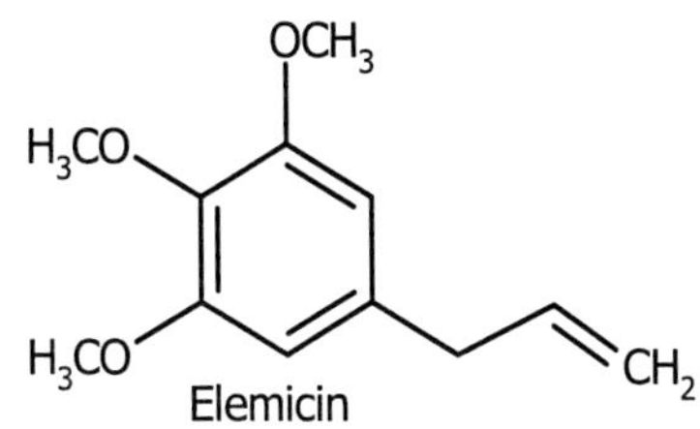

Elemicin

Saponification

Saponification of trimyristin produces the sodium salt of myristic acid, $CH_3(CH_2)_{12}CO_2Na$, which is a soap with a linear aliphatic chain and glycerol as a side-product. Sodium myristate on further acidification produces a fatty acid, myristic acid.

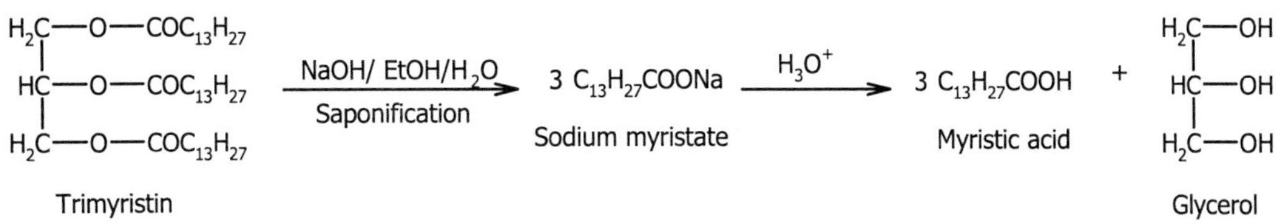

Myristicin is a phenylpropene, a natural organic compound present in small amounts in the essential oil of nutmeg and to a lesser extent in other species, such as parsley and dill. Myristicin is a naturally occurring insecticide and acaricide (pesticides that kill members of the acari group, which includes ticks and mites) with possible neurotoxic effects on neuroblastoma cells. It has psychoactive properties at doses much higher than used in cooking. Myristicin is a week inhibitor of monoamine oxidase. Raw nutmeg produces anticholinergic like symptoms which is attributed to myristicin and elemicin.

Procedure:

Isolation of trimyristin

1. Weigh 20 gm of ground nutmeg, and place in a 250 ml round bottom flask.
2. Add 50 ml of diethyl ether to the above flask and reflux the mixture on water bath at 45 °C in a hood for 30 minutes.
3. Cool the flask to room temperature, and decant the supernatant liquid containing the impure trimyristin into a clean 100 ml beaker.
4. Wash the solid residue with an additional 20 ml of diethyl ether, and add that liquid to the beaker.
5. Evaporate the liquid ether to recover the yellow solid residue which melts at 56 °C (If the product melts during evaporation of the solvent, solidify in an ice bath before adding ethanol).

6. Dissolve the residue in small portion of 95% ethanol at room temperature. If necessary, decant to separate solution from insoluble solids.
7. Cool in ice bath to recrystallize, and vacuum filter. Rinse crystals with cold ethanol.
8. Determine the yield as mass of trimyristin extracted per mass of nutmeg used.

Saponification of trimyristin

1. Weigh the solid trimyristin.
2. Add the trimyristin, along with 20 ml of ethanol (95%) and 0.20 gm of solid sodium hydroxide to a 100 ml round bottom flask.
3. Reflux the mixture on a water bath at 85-90 °C for 15 minutes.
4. Cool to room temperature, and then add 20 ml of water, followed by 20 ml of sodium chloride solution (35%).
5. Sodium myristate forms as white voluminous solid (if not place in an ice bath) which is vacuum filtered and washed with 30 ml of sodium chloride solution (35%).
6. Allow product to dry and recrystallize from ethanol.
7. The melting range of pure trimyristin is 51-52 °C.

57.9 Isolation of Naringin from Grapes

Requirements: Grapes, Ethanol.

Principle: Naringin is a flavone glycoside abundant in grapes and oranges also contains traces of naringine gives the grape juice bitter taste. It is metabolized to the flavone naringenin in humans, which is aglycone. In the commercial production of grapefruit juice the enzyme naringinase removes the bitterness of naringin. Naringin inhibit metabolizing enzymes cytochrome P450 (CYP3A4 and CYP1A2) and results in drug-drug interactions.

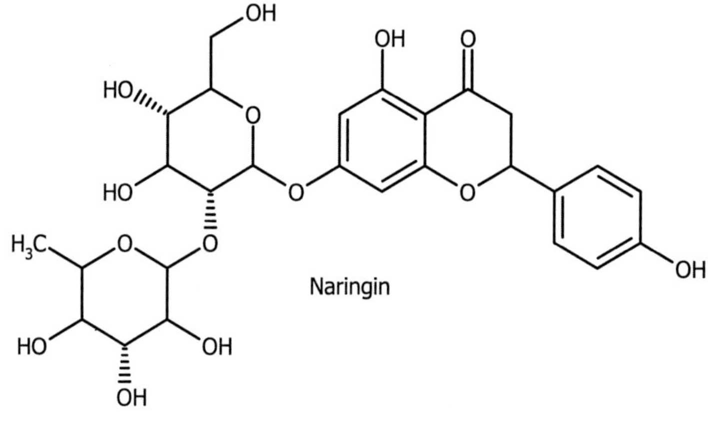

Naringin

Naringin causes anti-angiogenesis effect by inhibiting vascular endothelial growth factor (VEGF) release. Naringin also shown nueroprotective in diabetes induced neuropathy, protective effects on cognition and antioxidanat effects.

Procedure 1

1. Remove the grapefruit rind with cork borer and blend 40 gm with 100 ml of ethyl alcohol.
2. Filter the mixture and dry the residue to remove alcohol.
3. Divide the dry residue into two equal portions and use in different extraction procedures.

Soxhlet extraction:

- To the one portion of the residue taken in an extraction thimble add 50 ml of ethyl alcohol.
- Extract the mixture for 3 hr and filter the contents of flask.
- Make up the volume of filtrate to 250 ml with ethyl alcohol and dilute to 1 to 100 for analysis.

Kesterson (Hendrickson) extraction:

- Allow the combined mixture of portion of filtrate and residue to stand for 16 hr with occasional stirring.
- Extract the residue for 2 hr with water containing calcium oxide and then with water heated to 95 °C immediately and allowed to stand for 2 hr.
- Combine the three filtrates make up the volume to 500 ml, and diluted 1 to 50 for analysis.

Analysis

- The analysis depends upon the production of a yellow color on the addition of alkali to the mixture in presence of diethylene glycol.
- Measure the absorbance of the yellow colour.

Procedure 2

1. Heat one part of chopped grape peel and four parts of water to 90 °C, and allow it stand at this temperature for 5 min.
2. Filter the water extract and add two parts of water to the solid.
3. Repeat the extraction at 80 °C and filter immediately.
4. Boil the combined extracts with 1% celite, filter and concentrate *in vacuo* to approximately $1/9^{th}$ of original volume.

5. Crystallize the concentrated extract in a refrigerator and filter as an octa-hydrate needles (naringin, m.p. 83 °C).
6. Dissolve the naringin in 100 ml of boiling isopropanol and filter while hot.
7. Heat the filtrate to its boiling point to initiate crystallization, cool, filter and wash with cold isopropanol.
8. The melting point of pure naringin is 171 °C (Alternatively, naringin can also be recrystallized from hot water).

57.10 Isolation of Hesperidin from Oranges

Requirements: Oranges, Calcium hydroxide, Hydrochloric acid.

Principle: Hesperidin was identified from oranges of Hesperides family, hence given the name Hesperidin. Later its presence in citrus fruits also identified. Neohesperidin, an isomer of hesperidin also found in these fruits. Hesperidin is non-bitter compound, whereas neohesperidin is bitter compound. Hesperidin decreases the fragility of blood capillaries.

Hesperidin can be isolated by two different methods

1. **First method:** In this method hesperidin is isolated by extracting dried citrus peel with petroleum ether followed by methanol. Petroleum ether removes the essential oils present in the peel and the methanol extracts the glycoside-hesperidin.
2. **Second method:** In this method, alkaline extraction of chopped orange peel followed by acidification gives crystals of hesperidin.

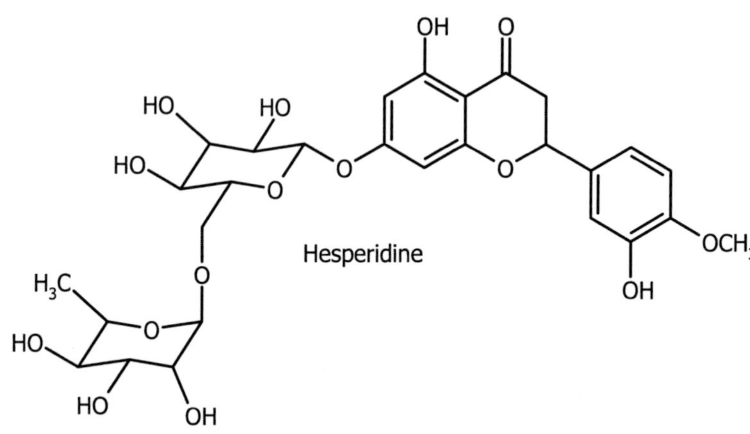

Hesperidine

Procedure:

1. Place chopped orange peel (20 gm) and calcium hydroxide (10%, 75 ml) in an Erlenmeyer flask and allow it to remain in room temperature for overnight.
2. Filter the mixture through celite placed on the Buchner funnel.

3. Acidify the yellow-orange filtrate obtained with hydrochloric acid to bring the pH 4 to 5.
4. Filter the solid, wash with water and dry.
5. Purify the hesperidin by recrystallization with aqueous formamide solution (20 to 25% v/v).
6. The melting point of the pure hesperidin is 252 to 254 °C
   [Concentrate the solution under reduced pressure, in case if there is no crystals formed after acidification].

CHAPTER 58

DETERMINATION OF ACTIVE PRINCIPLES OF NATURAL PRODUCTS

58.1 Determination of Acid Value

Requirements: Fat or oil, Potassium hydroxide solution (0.1M), Alcohol-ether mixture (Equal volumes of ethanol (95%) and ether, previously neutralized with 0.1M KOH to phenolphthalein), Phenolphthalein solution.

Principle: Acid value (AV) is an important indicator of vegetable oil quality. Acid value is the number which expresses in milligrams the amount of potassium hydroxide necessary to neutralize the free acids present in 1 gm of the substance. The amount of free acid present in most of the fixed oils is generally low and the acid value is used to eliminate low-grade and rancid oils, which tend to have higher acid values. The free acids present in the sample reacts with base and gives the salt form of the acid. The complete neutralization of the acid by base is indicated by the appearance of faint pink colour by the phenolphthalein.

$$\text{R-COOH} + \text{KOH} \longrightarrow \text{R-COOK} + H_2O$$

A high acid value is indicative of a stale oil or fat stored under improper conditions.

Procedure:

1. Accurately weigh about 10 gm of the substance being examined or the quantity specified in the monograph into a 250 ml flask.
2. Add 50 ml of the mixture of equal volume of ethanol (95%) and ether, previously neutralized with 0.1M potassium hydroxide to phenolphthalein solution.
3. Heat, if necessary, until the sample has completely dissolved.
4. Cool and add 1 ml of phenolphthalein solution and titrate against 0.1M potassium hydroxide until the solution remains faintly pink colour after shaking for 30 seconds.

5. Note the volume of potassium hydroxide consumed (**a**) and calculate the acid value from the expression

$$\text{Acid value} = \frac{a \times 0.00561 \times 1000 \times \text{Molarity of KOH used}}{\text{Weight of the substance} \times \text{Molarity of KOH}}$$

a = Volume of 0.1M potassium hydroxide consumed (in ml)

Note: If the oil has been saturated with carbon dioxide for the purpose of preservation, gently reflux the solution of the oil in ethanol (95%) and ether for 10 minutes before titration. The oil may be freed from the carbon dioxide by exposing it in a shallow-dish in a vacuum desiccator for 24 hr before weighing the sample.

Preparation of 0.1M potassium hydroxide: Weigh accurately about 5.61 gm of potassium hydroxide and dissolve in about 50 ml of distilled water and make the volume up to 1000 ml.

Standardization of 0.1M potassium hydroxide:

1. Weigh accurately about 1.26 gm of oxalic acid and dissolve in 100 ml of distilled water in a volumetric flask.
2. Take 25 ml of this solution into a conical flask and titrate against 0.1M potassium hydroxide using phenolphthalein as indicator.
3. Appearance of pale pink colour indicates the end point.
4. The molarity of potassium hydroxide is calculated by using formula

$$M_1V_1 = M_2V_2$$

M_1: Molarity of oxalic acid V_1: Volume of oxalic acid

M_2: Molarity of potassium hydroxide V_2: Volume of potassium hydroxide

58.2 Determination of Saponification Value

Requirements: Fat or oil, Alcoholic potassium hydroxide solution (0.5M), Hydrochloric acid (0.5M), Phenolphthalein solution.

Principle: The saponification value is the number of milligrams of potassium hydroxide necessary to neutralize the free acids and to saponify the esters present in 1 gm of the substance. The saponification value is a measure of both free and combined acids.

$$R\text{-}COO\text{-}R^1 + KOH \longrightarrow R\text{-}COOK + R^1\text{-}OH$$

It can be determined by refluxing a known amount of the sample with excess of standard alcoholic potassium hydroxide solution, and then titrating the unused alkali against a standard acid solution. Since each molecule of fat, regardless of its size, requires three molecules of KOH to saponify it, it is evident that the saponification number really indicates the number of fat molecules per gram of the fat. The larger the fat molecule, the fewer number will be per gram of fat. The saponification number, therefore, becomes a measure of the size of the fat molecule, or more specifically the size or molecular weight of the fatty acids in the fat. A high saponification number indicates the low molecular weight and vice-versa.

***Procedure*:**

1. Accurately weigh about 2 gm of the substance being examined or the quantity specified in the monograph into a 200 ml flask fitted with reflux condenser.
2. Add 25 ml of 0.5M ethanolic potassium hydroxide and a little pumice powder.
3. Boil under reflux on a water bath for 30 minutes.
4. Add 1 ml of phenolphthalein solution and titrate immediately against 0.5M hydrochloric acid, note the volume of acid consumed (**a**).
5. Repeat the operation omitting the sample; note the volume of acid consumed (**b**).
6. Calculate the saponification value from the expression.

$$\text{Saponification value} = \frac{(\text{b} - \text{a}) \times 0.02805 \times 1000 \times \text{Molarity of HCl used}}{\text{Weight of the substance} \times \text{Molarity of HCl}}$$

a = Volume of 0.5M hydrochloric acid consumed by the sample (in ml)

b = Volume of 0.5M hydrochloric acid consumed by blank (in ml)

Note: If the oil has been saturated with carbon dioxide for the purpose of preservation, gently reflux the solution of the oil in ethanol (95%) and ether for 10 minutes before titration. The oil may be freed from the carbon dioxide by exposing it in a shallow-dish in a vacuum desiccators for 24 hr before weighing the sample.

Preparation of 0.5M hydrochloric acid: Take about 44.5 ml of concentrated hydrochloric acid (11N) and slowly add 900 ml of distilled water. Stir the contents and make the volume up to 1000 ml and keep in a stoppered flask.

Standardization of 0.5M hydrochloric acid:

1. Weigh accurately about 5.3 gm of sodium carbonate and dissolve in 100 ml of distilled water in a volumetric flask.
2. Take 25 ml of this solution into a conical flask and titrate against hydrochloric acid using methyl orange as indicator.
3. Appearance of orange red colour indicates the end point.
4. The molarity of hydrochloric acid is calculated by using formula

$$\mathbf{M_1V_1 = M_2V_2}$$

M_1: Molarity of sodium carbonate **V_1:** Volume of sodium carbonate

M_2: Molarity of hydrochloric acid **V_2:** Volume of hydrochloric acid

58.3 Determination of Peroxide Value

Requirements: Fat or oil, Glacial acetic acid, Chloroform, Saturated potassium iodide solution, Sodium thiosulphate (0.01M), Starch solution.

Principle: The autoxidation of lipids is major cause of their deterioration and hydroperoxides formed by the reaction between oxygen and unsaturated fatty acid is the primary product of this reaction. The peroxide concentration usually expressed as peroxide value, gives early measure of lipid oxidation. The peroxide value is the number of milli equivalents of active oxygen that expresses the amount of peroxide contained in 1000 gm of the sample. Peroxide value can be determined based on iodometric principles. The peroxide present in the sample acts as an oxidizing agent and oxidizes potassium iodide present in the mixture under acidic conditions and liberates the iodine. This liberated iodine is titrated against 0.01M sodium thiosulphate using starch mucilage as an indicator.

Procedure:

1. Accurately weigh about 5 gm of the substance being examined or the quantity specified in the monograph into a 250 ml glass stoppered conical flask.
2. Add 30 ml of a mixture of 3 volumes of glacial acetic acid (18 ml) and 2 volumes of chloroform (12 ml).
3. Swirl until dissolved and add 0.5 ml of saturated potassium iodide solution.
4. Allow to stand for 10 minutes, with occasional shaking.
5. Add 30 ml of water and titrate gradually, with continuous and vigorous shaking with 0.01M sodium thiosulphate ($Na_2S_2O_3$) until the yellow colour becomes faint.
6. Add 0.5 ml of starch solution and continue the titration, shaking vigorously until the blue colour disappears (**a**).
7. Repeat the operation omitting the sample, note the volume of sodium thiosulphate consumed (**b**).
8. Calculate the peroxide value from the expression

$$\text{Perioxide value} = \frac{10(a-b) \times \text{Molarity of } Na_2S_2O_3 \text{ used}}{\text{Weight of the substance} \times \text{Molarity of } Na_2S_2O_3}$$

a = Volume of 0.01M sodium thiosulphate consumed by the sample (in ml)

b = Volume of 0.01M sodium thiosulphate consumed by blank (in ml)

Preparation of 0.01M sodium thiosulphate:

Dissolve about 2.5 gm of sodium thiosulphate in freshly boiled and cooled water and make the volume up to 1000 ml. Keep the solution aside and filter if necessary (if cloudy).

Standardization of 0.01M sodium thiosulphate:

1. Weigh accurately about 0.294 gm of potassium dichromate into conical flask containing 100 ml of recently boiled and cooled water.

2. Take 25 ml of this solution into an iodine flask; add 5 ml of hydrochloric acid and 2 gm of potassium iodide.
3. Swirl the flask, cover it with watch glass and allow to stand for 10-15 minutes in the dark.
4. Dilute the solution to about 20 ml with water and titrate immediately against 0.01M sodium thiosulphate until the yellowish color becomes faint.
5. Add 0.5 ml of starch solution, a deep violet colour will appear.
6. Continue the titration until the last drop of sodium thiosulphate solution removes blue colour of the starch iodide complex.
7. The resulting solution will be light green in colour due to chromium ions.
8. The molarity of sodium thiosulphate is calculated by following formula

$$M_1V_1 = M_2V_2$$

M_1: Molarity of potassium dichromate **V_1:** Volume of potassium dichromate

M_2: Molarity of sodium thiosulphate **V_2:** Volume of sodium thiosulphate

Note: Starch should not be added to a solution containing a large excess of iodine. The starch may get coagulated and its complex with iodine may not easily break-up.

58.4 Determination of Iodine Value

Requirements: Fat or oil, Carbon tetrachloride, Iodine mono chloride, Potassium iodide solution, Sodium thiosulphate (0.01M), Starch solution.

Principle: The iodine value is the number of grams of iodine taken up by 100 gm of fat or oil. Iodine itself reacts too slowly, instead either iodine monochloride or bromine in pyridine may be used.

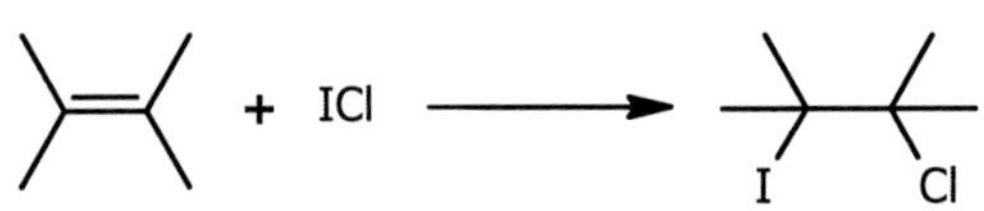

The iodine value of a fat or oil may be regarded as a measure of its degree of unsaturation and gives an idea about its drying character. It also helps in finding the adulteration in a fat or oil and judging its suitability for making soap. The unreacted iodine is titrated against 0.01M sodium thiosulphate using starch mucilage as an indicator.

Iodine value of

Non-drying oils (ground nut, mustard) = 85-105

Semi drying oils (cotton, seed, sesame) = 105-120

Drying oils (linseed, soyabean) = ≥ 120

Iodine monochloride method or Wijs method:

1. Transfer accurately weighed quantity of the substance being examined (the quantity specified in the monograph) into a 500 ml glass stoppered conical flask.
2. Add 10 ml of carbon tetrachloride and 20 ml of iodine mono chloride solution, insert the stopper and allow to stand in the dark at a temperature between 15 °C to 25 °C for 30 minutes.
3. Place 15 ml of potassium iodide solution in the cup top, carefully remove the stopper, rinse the stopper and the sides of the flask with 100 ml water.
4. Shake and titrate with 0.1M sodium thiosulphate until the yellow colour becomes faint.
5. Add 0.5 ml of starch solution and continue the titration, shaking vigorously until the blue colour disappears (**a**).
6. Repeat the operation omitting the sample, note the volume of sodium thiosulphate consumed (**b**).
7. Calculate the iodine value from the expression

$$\text{Iodine value} = \frac{(a - b) \times 0.0127 \times 100 \times \text{Molarity of } Na_2S_2O_3 \text{ used}}{\text{Weight of the substance} \times \text{Molarity of } Na_2S_2O_3}$$

a = Volume of 0.1M sodium thiosulphate consumed by the sample (in ml)

b = Volume of 0.1M sodium thiosulphate consumed by blank (in ml)

Preparation of Iodine Monochloride Solution

Dissolve iodine trichloride (8 gm) in about 200 ml of glacial acetic acid and dissolve iodine (9 gm) in carbon tetrachloride (300 ml). Mix the two solutions and adjust to 1000 ml with glacial acetic acid. This operation should be carried out in fume cup-board. Iodine monobromine in glacial acetic acid, may be used in place of Iodine monochloride.

Preparation of 0.1M sodium thiosulphate:

Dissolve about 25 gm of sodium thiosulphate in freshly boiled and cooled water and make the volume up to 1000 ml. Keep the solution aside and filter if necessary (if cloudy).

Standardization of 0.1M sodium thiosulphate

1. Weigh accurately about 2.94 gm of potassium dichromate into conical flask containing 100 ml of recently boiled and cooled water.
2. Take 25 ml into an iodine flask, add 10 ml of hydrochloric acid and 2 gm of potassium iodide.
3. Swirl the flask, cover it with watch glass and allow to stand for 10-15 minutes in the dark.

4. Dilute the solution to about 100 ml with water and titrate immediately against 0.1M sodium thiosulphate until the yellowish color becomes faint.
5. Add 5 ml of starch solution; a deep violet colour will appear.
6. Continue the titration until the last drop of sodium thiosulphate solution removes blue colour of the starch iodide complex.
7. The resulting solution will be light green in colour due to chromium ions.
8. The molarity of sodium thiosulphate is calculated by following formula

$$M_1V_1 = M_2V_2$$

M_1: Molarity of potassium dichromate **V_1:** Volume of potassium dichromate

M_2: Molarity of sodium thiosulphate **V_2:** Volume of sodium thiosulphate

Note: Starch should not be added to a solution containing a large excess of iodine. The starch may get coagulated and its complex with iodine may not easily break-up.

58.5 Determination of Alcohol Number

Requirements: Alkaline phenolphthalein, Sodium hydroxide, Potassium carbonate, Petroleum ether and Sodium chloride.

Principle: Alcohol number represents the number of milliliters of ethanol at 15 °C obtained from 10 ml of tincture or alcohol containing preparations.

Pretreatment

For the following samples pretreatment is necessary

1. ***Glycerin:*** Add sufficient water to the sample.
2. ***Iodine:*** Decolourize with zinc powder.
3. ***Volatile substances [Chloroform, ether, camphor]:***
    - Mix 10 ml of the sample with 10 ml of saturated sodium chloride present in separating funnel.
    - Add 10 ml of petroleum ether, shake and collect the aqueous layer.
    - Extract the organic layer with two 5 ml volumes of saturated sodium chloride.
    - Combine the aqueous layers and distilled.
4. Convert volatile acids slight alkaline by adding sodium hydroxide.
5. ***Others:*** Render them slightly acidic with dilute acid.

Amount of distillate after pre-treatment

Ethanol content (vol %)	Distillate (ml)
> 80%	13
80-70	12
70-60	11
60-50	10
50-40	9
40-30	8
< 30	7

Procedure:

1. Transfer 10 ml of sample to the distillation flask and maintain the temperature at 15 ± 2 °C.
2. Add 5 ml of water and add boiling chips.
3. Distill alcohol carefully into the glass stopper volumetric cylinder.
4. To the distillate add potassium carbonate (5 g), alkaline phenolphthalein solution (2 drops) and shake vigorously.
    - *If the aqueous layer shows no white turbidity agitate the distillate with additional potassium carbonate.*
    - *If there is no clear boundary surface between these two layers add few drops of water and shake vigourously.*
5. Allow it to stand in water bath for 30 min and maintain the temperature at 15 ± 2 °C.
6. Read the volume of upper reddish ethanol layer in ml.
7. The volume measured indicates the alcohol number.

Preparation of alkaline phenolphthalein solution: To phenolphthalein (1 g) add sodium hydroxide (7 ml) and make up the volume upto 100 ml with distilled water.

Note: Render the sample acidic with phosphoric acid or sulphuric acid to prevent bumping during distillation.

58.6 Determination of Ester Value

Requirements: Fat or oil, Alcoholic potassium hydroxide solution (0.5M), Hydrochloric acid (0.5M), Phenolphthalein solution, Potassium hydroxide solution (0.1M), Alcohol-ether mixture (equal volumes of ethanol (95%) and ether, previously neutralized with KOH to phenolphthalein).

Principle: The ester value is the number of milligrams of potassium hydroxide required to saponify the esters present in 1 gm (1000 mg) of the substance. This can be calculated by using the formula

$$\text{Ester value} = \text{Saponification value} - \text{Acid value}$$

Determination of saponification value

1. Accurately weigh about 2 gm of the substance being examined or the quantity specified in the monograph into a 200 ml flask fitted with reflux condenser.
2. Add 25 ml of 0.5M ethanolic potassium hydroxide and a little pumice powder.
3. Boil under reflux on a water bath for 30 minutes.
4. Add 1 ml of phenolphthalein solution and titrate immediately against 0.5M hydrochloric acid, note the volume of acid consumed (**a**).
5. Repeat the operation omitting the sample; note the volume of acid consumed (**b**).
6. Calculate the saponification value.

$$\text{Saponification value} = \frac{(b-a)\times 0.02805\times 1000\times \text{Molarity of HCl used}}{\text{Weight of the substance}\times \text{Molarity of HCl}}$$

Determination of acid value:

1. Accurately weigh about 10 gm of the substance being examined or the quantity specified in the monograph into a 250 ml flask.
2. Add 50 ml of the mixture of equal volume of ethanol (95%) and ether, previously neutralized with 0.1M potassium hydroxide to phenolphthalein solution.
3. Heat if necessary, until the sample has completely dissolved.
4. Cool and add 1 ml of phenolphthalein solution and titrate against 0.1M potassium hydroxide until the solution remains faintly pink colour after shaking for 30 seconds.
5. Note the volume of potassium hydroxide consumed (**a**) and calculate the acid value.

$$\text{Acid value} = \frac{a\times 0.00561\times 1000\times \text{Molarity of KOH used}}{\text{Weight of the substance}\times \text{Molarity of KOH}}$$

Preparation of 0.5M hydrochloric acid: Take about 44.5 ml of concentrated hydrochloric acid (11N) and slowly add 900 ml of distilled water. Stir the contents and make the volume up to 1000 ml and keep in a stoppered flask.

Standardization of 0.5M hydrochloric acid:

1. Weigh accurately about 5.3 gm of sodium carbonate and dissolve in 100 ml of distilled water in a volumetric flask.

2. Take 25 ml of this solution into a conical flask and titrate against hydrochloric acid using methyl orange as indicator.
3. Appearance of pale orange red colour indicates the end point.
4. The molarity of hydrochloric acid is calculated by following formula

$$M_1V_1 = M_2V_2$$

M_1: Molarity of sodium carbonate **V_1:** Volume of sodium carbonate
M_2: Molarity of hydrochloric acid **V_2:** Volume of hydrochloric acid

Preparation of 0.1M potassium hydroxide: Weigh accurately about 5.61 gm of potassium hydroxide and dissolve in about 50 ml of distilled water and make the volume up to 1000 ml.

Standardization of 0.1M potassium hydroxide: Weigh accurately about 1.26 gm of oxalic acid and dissolve in 100 ml of distilled water in a volumetric flask.

1. Take 25 ml of this solution into a conical flask and titrate against 0.1M potassium hydroxide using phenolphthalein as indicator.
2. Appearance of pale pink colour indicates the end point.
3. The molarity of potassium hydroxide is calculated by following formula

$$M_1V_1 = M_2V_2$$

M_1: Molarity of oxalic acid **V_1:** Volume of oxalic acid
M_2: Molarity of potassium hydroxide **V_2:** Volume of potassium hydroxide

58.7 Estimation of Quinine Sulphate

Requirements: Quinine sulphate powder, Perchloric acid (0.1M), Chloroform, Glacial acetic acid, Crystal violet solution.

Principle: Quinine sulphate is a weak base. According to bronsted theory anion of weak acid is the weak base. Glacial acetic acid is an amphoteric solvent, which levels the salt into strong base. Therefore this base can be titrated against perchloric acid using crystal violet as indicator.

Procedure:

1. Accurately weigh about 0.2 gm of the substance or the quantity specified in the monograph into a 250 ml flask.
2. Add 10 ml of chloroform and 20 ml of acetic anhydride; shake well.
3. Titrate the contents of the flask with 0.1M perchloric acid using crystal violet as indicator.
4. Each ml of 0.1M perchloric acid = 0.0249 gm of $C_{40}H_{50}N_4O_8S$.

Preparation of 0.1M perchloric acid: Slowly add perchloric acid (72%, 85 ml) to glacial acetic acid (900 ml) with continuous stirring. Add acetic anhydride (30 ml) to the above mixer, stir and adjust the volume to 1000 ml with glacial acetic acid. Allow the solution stand for 24 hr.

Standardization of 0.1M perchloric acid:

1. Weigh accurately about 0.5 gm of the potassium hydrogen phthalate into a flask.
2. Attach reflux condenser (fitted with drying tube) and add glacial acetic acid (25 ml). Warm until the solution is dissolved.
3. Cool and titrate with 0.1M perchloric acid using crystal violet as indicator till the green colour persists at least for 30 seconds.
4. Perform blank titration omitting the sample.

Each ml of 0.1M $HClO_4$ = 0.2041 gm of potassium hydrogen phthalate

$$\text{Molarity of perchloric acid} = \frac{\text{Weight of potassium hydrogen phthalate}}{\text{Volume of perchloric acid consumer} \times 0.2041}$$

58.8 Estimation of Ascorbic Acid

Chemicals: Citrus fruits, Sulphuric acid (1M), Iodine (0.05M), Starch solution.

Principle: Ascorbic acid (vitamin C) is an essential antioxidant plays essential role in the protection of body from infections and diseases. It is not synthesized in the body, hence should be supplied through diet. The amount of ascorbic acid can be determined by redox-titration using iodine (iodimetric titration). Ascorbic acid reduces iodine into iodide and gets oxidized to dehydroascorbic acid. Iodine reduction to iodide will continue as long as ascorbic acid is present. The excess of iodine can be detected by the reaction with starch indicator, which gives blue colour with iodine. This method is applicable for ascorbic acid present in fruits, food and pharmaceutical formulations.

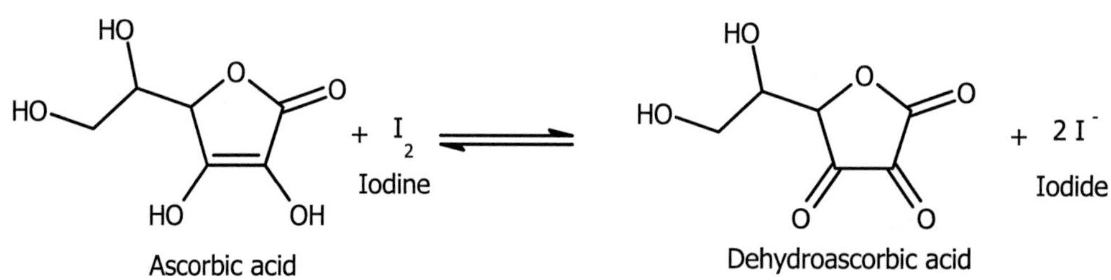

Sample preparation: Ascorbic acid is susceptible to oxidation by atmospheric oxygen, hence should be prepared immediately before the titrations. It can be minimized by the addition of small amount of oxalic acid.

Fruits and vegetables: Cut into small pieces and grind in a mortar and pestle (blending fruits and vegetables also can be done). Add distilled water and decant the liquid into flask (repeat this operation for several times). Strain the decanted portion through cheese cloth and dilute it with distilled water.

Fresh fruit juice: Strain the juice through cheese cloth to remove seeds and pulp (otherwise blocks the pipette).

Packaged fruit juice: Strain through cheese cloth, if it contains pulp and seeds.

Procedure:

1. Pipette 20 ml of sample solution into conical flask.
2. Add distilled water (100 ml) and starch paste (1 ml).
3. Titrate the mixture with 0.005M iodine solution.
4. The appearance of persistent dark blue-black colour due to the starch iodine complex is the end point.
5. Repeat the titration for correction.

Preparation of 0.005M iodine: Dissolve 2 gm of iodate-free potassium iodide in 30 ml of distilled water in a glass-stoppered flask. Add 1.27 gm of resublimed iodine to above flask and shake well until iodine dissolves. Make the volume up to 1000 ml and stored in cool and dark place.

Standardization of iodine:

1. Weigh accurately about 0.15 gm of arsenic trioxide and dissolve in 20 ml of 1N sodium hydroxide (warm if necessary).
2. Add 40 ml of water, 2 drops of methyl orange indicator followed by dil. hydrochloric acid until pink colour is obtained.
3. Add 2 gm of sodium bicarbonate, dilute with 50 ml water followed by few drops of starch solution and titrate with iodine solution until a permanent blue colour is produced.

$$\text{Molarity of iodine} = \frac{\text{Weight of arsenic trioxide}}{\text{Volume of iodine consumed} \times 0.04946}$$

Note: Any excess of sodium hydroxide added to dissolve arsenic trioxide is neutralised with hydrocloric acid. Iodine may react with NaOH to form sodium hypoiodide (NaOI). Sodium thiosulphate which is recently standardized can be used as an alternative to arsenic trioxide.

PREPRATION OF SOLUTIONS AND REAGENTS

CHAPTER 59

PREPARATION OF SOLUTIONS

59.1 Molar Solutions

Number of moles of solute per one liter of solution is known as molarity (ratio between solute and liters of solution).

Molarity, M = Number of moles/Number of liters

Molar solutions of solids: To prepare the molar solution, mass of the solid can be calculated from the following formula

$$g = MMs \times M \times V$$

MMs = Molar mass (molecular weight)

M = Molarity of solution (need to be prepared)

V = Volume of solution (in liter)

E.g. 1: Preparation of 1000 ml of 1M sodium hydroxide

The mass of sodium hydroxide required $= MMs \times M \times V$

$= 40 \times 1 \times 1$

= 40 gm of sodium hydroxide

Similarly for 1000 ml of 0.5M sodium hydroxide

The mass of sodium hydroxide required $= MMs \times M \times V$

$= 40 \times 0.5 \times 1 = 20$ gm of NaOH

Molar solutions of liquids: In case of preparation of molar solutions from concentrated solutions

The volume of concentrated solution required, $V_1 = M_2 \times V_2/M_1$

V_1 = Volume of concentrated solution required

M_1 = Molarity of concentrated solution

V_2 = Volume of the final solution

M_2 = Molarity of the final solution

E.g. 2: Preparation of 1000 ml of 1M sulphuric acid

The volume of concentrated solution required, $V_1 = M_2 \times V_2/M_1$

$V_1 = 1 \times 1000/18$

$V_1 = 55.5$ ml of concentrated sulphuric acid

Similarly Preparation of 1000 ml of 0.5 M sulphuric acid

$V_1 = 0.5 \times 1000/18$

$V_1 = 27.8$ ml of concentrated sulphuric acid

E.g. 3: Preparation of 100 ml of 1M hydrochloric acid from concentrated hydrochloric acid (12.1M)

The volume of concentrated solution required, $V_1 = M_2 \times V_2/M_1$

$V_1 = 1 \times 100/12.1$

$V_1 = 8.26$ ml of concentrated hydrochloric acid

59.2 Normal Solutions

An equivalent weight of solute present in one liter (1000 ml) of solution is called as normality.

Normality, N = Number of equivalents/Number of liters

Equivalent weight is equal to its molecular weight divided by number of replaceable hydrogens (incase of acid)/replaceable hydroxyl group (incase of bases).

Acid	Formula	Number of replaceable hydrogens
Hydrochloric acid	HCl	1
Hydrofluric acid	HF	1
Nitric acid	HNO_3	1
Phosphoric acid	H_3PO_4	3
Sulphuric acid	H_2SO_4	2

Base	Formula	Number of replaceable hydroxyl ions
Ammonium hydroxide	NH_4OH	1
Calcium hydroxide	$Ca(OH)_2$	2
Potassium hydroxide	KOH	1
Sodium hydroxide	$NaOH$	1

E.g. 4: Preparation of 1000 ml of 1N sulphuric acid

Normality, N = Number of equivalents/Number of liters

Equivalent weight of sulphuric acid is equal to its molecular weight divided by number of replaceable hydrogens.

= 98/2 = 49

Hence, 49 gm of sulphuric acid in 1000 ml water produces 1N solution of sulphuric acid.

The number of equivalent of a solute is equal to grams of the solute divided by its weight

Equivalents = Grams of solute/Equivalent weight of solute

E.g. 5: Number of equivalents of 33.26 gm of phosphoric acid (H_3PO_4)

Equivalent weight = Molecular weight/Number of replaceable hydorgens

= 98/3 = 32.6

Hence, equivalents = Grams of solute/Equivalent weight of solute

= 33.26/32.6

= 1

The number of liters is equal to the number of milliliters of solution divided by 1000.

Liters = Milliliters/1000

Now normality can be calculated from the formula

$$\text{Normality} = \frac{\text{grams of solute/equivalent weight of solute}}{\text{ml of solution/1000}}$$

The equation can be rewritten as

$$\text{Normality} = \frac{(1000)\ (\text{grams of equivalent})}{(\text{equivalent weight of solute})\ (\text{ml of solution})}$$

Grams of solute can be calculated by rewriting the above equation as

$$\text{Grams of solute} = \frac{(\text{N})\ (\text{equivalent weight of solute})\ (\text{ml of solution})}{1000}$$

E.g. 6: Calculate the normality of solution containing 5 gm of NaOH in 200 ml of water

Normality, N = Number of equivalents/Number of liters

The number of equivalent of NaOH

Equivalents = molecular weight/number of replaceable hydroxide ions

= 40/1

= 40

Now normality can be calculated from

Normality, N = Number of equivalents/Number of liters

$$\text{Normality} = \frac{(1000)\ (\text{grams of equivalent})}{(\text{equivalent weight of solute})\ (\text{ml of solution})}$$

$$\text{Normality} = \frac{(1000)\ (5)}{(40)\ (200)} = 0.625\ \text{N}$$

E.g. 7: Calculate the number of gram of calcium hydroxide [$Ca(OH)_2$]required to prepare 150 ml of 0.15N $Ca(OH)_2$ solution

Equivalents = Molecular weight/Number of replaceable hydroxide ions

= 74/2

= 37

$$\text{Grams of solute} = \frac{(\text{N})\ (\text{equivalent weight of solute})\ (\text{ml of solution})}{1000}$$

$$\text{Grams of solute} = \frac{(0.15)(37)(150)}{1000} = 0.83\ \text{gm}$$

Concentrated reagents	**Formula**	**Formula weight**	**Density**	**Molarity**	**Normality**	**Volume (ml) required for 1000 ml**	
						1M	**1N**
Acetic acid (99.8%)	CH_3COOH	60.05	1.05	17.4	17.4	57.5	57.5
Formic acid (90%)	HCOOH	46.03	1.13	23.6	23.6	42.5	42.5
Hydrochloric acid (37.2%)	HCl	36.46	1.18	12.1	12.1	82.5	82.5
Nitric acid (69.6%)	HNO_3	63.01	1.41	15.6	15.6	64	63
Perchloric acid (70%)	$HClO_4$	100.46	1.67	11.6	11.6	86.2	86.2
Perchloric acid (60%)	$HClO_4$	100.46	1.54	9.2	9.2	108.7	108.7
Phosphoric acid (85%)	H_3PO_4	98.00	1.70	15.2	45.6	65.8	21.9
Sulphuric acid (98%)	H_2SO_4	98.08	1.84	18	36.8	54.7	27.2
Ammonium hydroxide (25%)	NH_4OH	35.05	0.91	13.4	13.4	74.6	74.6
Ammonium hydroxide (35%)	NH_4OH	35.05	0.88	18.1	18.1	55.3	55.3

Solid	Formula	Formula weight	Grams required for 1000 ml	
			1M	1N
Potassium hydroxide	KOH	56.11	40	40
Sodium hydroxide	NaOH	40.00	56	56

59.3 Percent Solutions

***Mass percent solution (w/v)*:** Weight of solute present in 100 gm of solution.

E.g.:

- **20% w/v NaOH solution:** 20 gm of NaOH in 100 gm of solution.
- **20% w/v KOH solution:** 20 gm of KOH in 100 gm of solution.
- **35% w/v NaCl solution:** 35 gm of NaCl in 100 gm of solution.

***Volume percent solution (v/v)*:** Milliliters of solute present in 100 ml of solution.

E.g.:

- **10% v/v C_2H_5OH:** 10 ml of C_2H_5OH in (90 ml water) 100 ml solution.
- **40% v/v HCHO:** 40 ml of HCHO in (90 ml water) 100 ml solution.

***Mass volume percent solution (w/v)*:** weight of solute present in 100 ml of solution.

E.g.:

- **1% w/v ethanolic phenolphthalein solution:** 1 gm of phenolphthalein in 100 ml of 95% ethanol solution.
- **5% w/v Ferric chloride solution:** 5 gm of ferric chloride in 100 ml of water.

CHAPTER 60

PREPARATION OF REAGENTS

Reagents

A chemical substance, which promotes and confirms the occurrence of chemical reaction, is known as reagent. It includes mixture of inorganic/organic compounds.

1. **Synthetic (organic) reagent:** It is useful in the chemical transformation of substrates in to products.

   E.g.: Grignard reagent.

2. **Analytical reagent:** It is useful in the conformation of any chemical reactions.

   E.g.: Tollen's reagent.

Preparation of fewer important reagents are given below

- **Acetic acid solution (dilute, 6% w/w):** Dilute 57 ml of glacial acetic acid into 1000 ml with water.
- **Ammonia solution (10% w/v)***:* Dilute 425 ml of strong ammonia into 1000 ml with water.
- **Ammonium acetate (0.1M):** Dissolve 7.71 gm of ammonium acetate in 200 ml of water. Add 1 ml of glacial acetic acid and dilute to 1000 ml with water.
- **Ammonium carbonate:** Dissolve 5 gm of ammonium carbonate in a mixture of 75 ml of dilute ammonia and 50 ml of water. Dilute to 100 ml with water, filter of necessary.
- **Ammonium oxalate:** 4% w/w solution of ammonium oxalate in water.
- **Chlorinated lime solution:** Mix chlorinated lime (100 gm) with water (1000 ml) transfer the mixer to a stoppered bottle, set aside for three hours, shaking occasionally and filter through calico. Chlorinated lime must be prepared freshly.
- **Ammonium thiocyante:** 10% w/w solution of ammonium thiocyanate in water.
- **Anisaldehyde acetic acid reagent:** Anisaldehyde (0.5 ml) is mixed with 98% acetic acid (10 ml).

- **Anisaldehyde sulphuric acid reagent:** Anisaldehyde (0.5 ml) is mixed with glacial acetic acid (10 ml) followed by methanol (85 ml) and sulphuric acid (5 ml).
- **Antimony chloride reagent:** 20% solution of antimony (III) chloride in chloroform.
- **Barfoed's reagent:** Dissolve crystalline copper acetate (6.5 gm) in 1% acetic acid solution (100 ml).
- **Barium chloride solution:** 10% w/v solution of barium chloride in water.
- **Barium hydroxide solution:** 3% w/v solution of barium hydroxide in water.
- **Bayer's reagent (potassium permanganate solution, 1% w/v):** Dissolve potassium permanganate (1 gm) in water (100 ml) and add few drops potassium hydroxide solution to have a clear solution of pink colour.
- **Benedict's reagent:** Pour Benedict's reagent A slowly into Benedict's reagent B. Mix well, and adjust the volume to **100** ml.

  ***Benedict's reagent A*:** Dissolve 1.73 gm of copper sulphate in 10 ml of water.

  ***Benedict's reagent B*:** Dissolve 17.3 gm of trisodium citrate and 10 gm of anhydrous sodium carbonate in 80 ml of water with the aid of heat. Pour this solution into copper sulphate solution and dilute the mixture to 100 ml.
- **Benzidine reagent:** Benzidine (0.5 gm) is dissolved in glacial acetic acid (10 ml) and the volume adjusted to 100 ml with ethanol. Bromine (9.6 ml) and potassium bromide (30 gm) in sufficient water to produce 100 ml.
- **Brady's reagent:** Dissolve 1 gm of powdered 2,4-dinitrophenyl hydrazine in 2 ml of concentrated sulphuric acid. Pour this solution into 20 ml of methanol and warm if necessary. Add 5 ml of water and mix thoroughly by mixing filter if necessary.
- **Bromine solution (aqueous):** Dissolve 9.6 ml of bromine and 30 gm of potassium bromide in sufficient water to produce 100 ml.
- **Bromine solution (acetic):** Dissolve 100 gm of potassium acetate in glacial acetic acid and add 4 ml of bromine and sufficient glacial acetic acid to produce 1000 ml.
- **Calcium chloride solution:** 10% w/v solution of calcium chloride in water.
- **Chromic acid:** Dissolve 84 gm of chromium trioxide in 700 ml of water and add 400 ml of sulphuric acid [slowly with stirring].
- ***para*-Dimethylamino benzaldehyde [*p*-DMAB] solution:** Dissolve 0.2 gm of DMAB in 20 ml of ethanol (95%) and add 0.5 ml of hydrochloric acid. Shake the solution with activated charcoal and filter.

- ***Para*-Dimethylamino benzaldehyde [*p*-DMAB] reagent:** Dissolve 0.125 gm of DMAB in a cooled mixture of 65 ml of sulphuric acid and 35 ml of water. Add 0.1 ml of ferric chloride solution and allow to stand for 24 hr in the dark before use.
- **2,4-Dinitrophenyl hydrazine reagent:** *see Brady's* reagent.
- **Diphenylcarbazide:** A 0.2% w/v solution of diphenylcarbazide in a mixer of glacial acetic acid (10 ml) and alcohol (90 ml).
- **Dragendorff's reagent:**

  ***Solution 1*:** Dissolve 0.85 gm of bismuth subnitrate in 10 ml of glacial acetic acid and 40 ml of water with the aid of heat and filter if necessary.

  Solution 2: Dissolve 8 gm of potassium iodide in 30 ml of water.

  Mix equal volume of solution 1 and 2 to obtain Dragendorff's reagent.
- **Diphenylamine reagent:** Mix 0.5 gm of diphenylamine in 20 ml of water and dilute to 100 ml with concentrated sulphuric acid.
- **Erdmann's reagent:** Mix 1 ml of nitric acid with 30 ml of sulphuric acid.
- **Fehling reagent:**

  ***Fehling reagent A*:** Dissolve crystalline copper sulphate (34.64 gm) in water and made the solution to 500 ml with water containing few drops of sulphuric acid.

  ***Fehling reagent B*:** Dissolve sodium potassium tartarate (Rochelle's salt, 173 gm) and sodium hydroxide (70 gm) in water. Make up the solution to 500 ml with water.
- **Ferric chloride reagent (5% w/v):** Dissolve 5 gm of anhydrous ferric chloride [or 8.25 gm of ferric chloride hexahydrate] in 100 ml of distilled water.
- **Ferric chloride reagent (neutral):** Dissolve ferric chloride in water and add ammonium hydroxide [till it precipitates]. Filter to remove the precipitate and use the filtrate as reagent.
- **Ferrous sulphate reagent:** To one volume of a 10% (w/v) solution of ferrous sulphate ($FeSO_4.7H_2O$) add five volumes of sulphuric acid with cooling.
- **Folin-Ciocalteu reagent:**

  ***Stock solution*:** Dissolve 100 gm of sodium tungstate and 25 gm of sodium molybdate in 800 ml of water in a 1500 ml flask, add 50 ml of phosphoric acid and 100 ml of hydrochloric acid, and reflux for 10 hr. Cool, add 150 gm of lithium sulphate, 50 ml of water and four to six drops of bromine, cool, filter and dilute to 1000 ml with water. [This stock solution should be stored at a temperature not exceeding 4 °C and used within 4 months of its preparation; it has a yellow colour and must not be used if any traces of green colour is present]

Working solution: Dilute one volume of the stock solution with two volumes of water.

- **Formaldehyde-sulphuric acid reagent:** To four volumes of sulphuric acid add six volumes of formaldehyde solution (using a pipette with the tip below the surface of the acid) with stirring and adequate cooling. When the reagent is warm it remains clear for about 1 hr. If turbidity develops, this may be dispelled by heating in a water-bath at 100 °C for about 1 min.
- **Forest reagent:** Mix together equal volumes of a 0.2% (w/v) solution of potassium dichromate, a 30% (v/v) solution of sulphuric acid, a 20% (w/w) solution of perchloric acid and a 50% (v/v) of nitric acid.
- **FPN reagent:** Mix together 5 ml of 5% (w/v) ferric chloride solution, 45 ml of a 20% (w/w) solution of perchloric acid and 50 ml of a 50% (v/v) solution of nitric acid.
- **Fujiwara reagent:** Freshly prepared 20% (w/v) sodium hydroxide solution.
- **Furfuraldehyde reagent:** A 10% (v/v) solution of furfuraldehyde in ethanol.
- **Hydroxyl amine hydrochloride reagent:** Dissolve 34.75 gm of hydroxyl amine hydrochloride in 950 ml of ethanol (60%) and add 5 ml of 0.2% w/v solution of methylene orange in ethanol (60%). Add 0.5M of potassium hydroxide in ethanol (60%) until the yellow colour is produced. Add sufficient ethanol to produce 1000 ml.
- **Hydroxyl amine hydrochloride solution:** Dissolve 1 gm of hydroxyl amine hydrochloride in 50 ml of water and 1 ml of bromo phenol blue solution. Add sodium hydroxide (0.1M) until the solution turns in to green.
- **Iodine-potassium iodide reagent:** Dissolve iodine (2 gm) and potassium iodide (4 gm) in water (10 ml). Adjust the volume into 100 ml with water.
- **Koppanyi-Zwikker reagent:** A 1% (w/v) solution of cobalt nitrate in ethanol.
- **Kedde reagent:** Freshly prepared ethanolic 3,5-dinitrobenzoic acid solutions (5 ml) are mixed with 2M NaOH solution (5 ml).
- **Lead acetate solution:** 10% w/v solution of lead acetate in carbon-dioxide free water.
- **Liebermann-Burchard reagent:** Acetic anhydride (5 ml) and Conc. sulphuric acid (5 ml) are added carefully to absolute alcohol (5 ml), while cooling in ice. The reagent must be prepared in freshly.
- **Liebermann reagent:** Add 1 gm of sodium or potassium nitrite to 10 ml of sulphuric acid with cooling and swirling to absorb the brown fumes.
- **Marquis reagent:** Carefully mix 100 ml of concentrated sulphuric acid with 1 ml of 10% (v/v) formaldehyde solution (stable for several weeks if protected from light).

- **McNally's reagent:** A 0.5% solution of copper sulfate in 10% acetic acid (solution 1). A freshly prepared 2% (w/v) solution of sodium nitrite (solution 2). [don't mix]
- **Mercuric acetate solution:** A 5% solution of mercuric acetate (5 gm) in glacial acetic acid (100 ml).
- **Methanolic potassium hydroxide reagent:** A 20% (w/v) solution of potassium hydroxide in methanol.
- **Millon's reagent (mercury nitrate solution):** Dissolve mercury (3 ml) in cold fuming nitric acid (27 ml), dilute the solution with equal volume of water with stirring.
- **Molisch's reagent:** Dissolve α-naphthol (10 gm) in rectified spirit (100 ml).
- **Naphthol-sulphuric acid reagent:** Mix 1 gm of 2-napthol with 40 ml sulphuric acid and heat in a water bath at 100 °C, with occasional stirring until the napth-2-ol is dissolved.
- **Nessler's reagent:**
  1. Dissolve 50 gm of mercuric chloride and 35 gm of potassium iodide in 200 ml of water and cool.
  2. Dissolve 50 gm of sodium hydroxide in 50 ml of water and cool.

  Add the cold solution 2 to the cold solution 1 and make up to 100 ml. Allow the mixture to stand and decant the clear supernatant (stable for many months) for use. [Store in dark brown bottles away from light]
- **Ninhydrin solution:** Ninhydrin (0.2% w/v) solution in a mixture of 95 volumes of 1-butanol and 5 volumes of acetic acid (2M).
- **Phosphomolybdic acid reagent:** 20% ethanolic solution of phosphomolybdic acid.
- **Phosphoric acid solution (10% w/v):** Dilute 69 ml of phosphoric acid to 100 ml with water.
- **Picric acid solution:** 1% w/v solution of picric acid in hot water.
- **Picric acid solution (alkaline):** Mix 20 ml of picric acid solution with 10 ml of dilute sodium hydroxide solution. Add sufficient water to produce 100 ml.
- **Potassium dichromate solution:** Dissolve 5 gm of potassium dichromate in 5 ml of water and cool. Add 100 ml of sulphuric acid slowly with stirring (exothermic reaction, hence use fume cupboard).
- **Potassium hydroxide solution:** 5% w/v solution of potassium hydroxide in water.
- **Potassium iodide solution:** 16.6% w/v solution of potassium iodide in water.

- **Potassium mercuric-iodide solution (Mayer's reagent):** Add mercuric chloride (1.36 gm) dissolved in water (60 ml) to a solution of potassium iodide (5 gm) in water (20 ml) mix and add sufficient water to produce 100 ml.
- **Potassium mercuric-iodide solution (Alkaline Nessler's reagent):** To potassium iodide (3.5 gm), add mercuric chloride (1.25 gm) dissolved in water (80 ml), add cold saturated solution of mercuric chloride in water with constant stirring until a slight red precipitate remains. Dissolve NaOH (12 gm) in the solution, add a little more of the cold saturated solution of mercuric chloride and sufficient water to produce 100 ml. Allow to stand and decant the clear liquid.
- **Potassium permanganate solution (1% w/v):** *See Bayer's reagent.*
- **Phenyl hydrazine solution:** Dissolve phenyl hydrazine hydrochloride (65 gm) previously recrystallized from aqueous alcohol in sufficient of a mixture of sulphuric acid (70 ml) and water (30 ml) to produce 100 ml.
- **Pyridine bromide solution:** Dissolve pyridine (10 ml) and sulphuric acid (5.4 ml) in glacial acetic acid (20 ml) keeping the mixture cool. Add bromine (2-6 ml) dissolved in glacial acetic acid (20 ml) and dilute with glacial acetic acid to 100 ml. Pyridine bromide solution must be prepared freshly.
- **Seliwanoff's reagent:** Dissolve 0.05 gm of resorcinol in 100 ml of hydrochloric acid (4 N).
- **Schiff's reagent:** Dissolve 0.2 gm of rosaniline hydrochloride (fuchsine) in 120 ml of hot water. Cool, add 20 ml of solution of sodium hydrogen sulphite (10% w/v) and 2 ml of hydrochloric acid. Dilute the solution to 200 ml with water. [Store at 4 °C and protect from light]
- **Silver nitrate solution:** 1.7% w/v solution of silver nitrate in water.
- **Silver nitrate (ammoniacal):** Dissolve 2.5 gm of silver nitrate in 80 ml of water and add dilute ammonia [until the precipitate first formed dissolves]. Set aside, decant and add water to produce 100 ml.
- **Simon's reagent:**

  ***Solution 1*:** Dissolve 1 gm of sodium nitroprusside in 100 ml of water and add 2 ml of acetaldehyde to the solution with thorough mixing.

  ***Solution 2*:** Freshly prepared 2% sodium carbonate in distilled water.
- **Sodium bicarbonate solution:** 5% w/v solution of sodium bicarbonate in water.
- **Sodium carbonate solution:** 10.6% w/v solution of sodium carbonate in water.
- **Sodium hydroxide solution:** 20% w/v solution of sodium hydroxide in water.
- **Sodium hydroxide solution (dilute):** 5% w/v solution of sodium hydroxide in water.
- **Sodium nitrite solution:** 10% w/v solution of sodium nitrite in water.

- **Sodium nitroprusside reagent:** 1% w/v solution of sodium nitroprusside in water.
- **Sodium sulphide solution***:* Dissolve 10 gm of sodium sulphide in water and produce 30 ml with water.
- **Stannous chloride solution:** Dissolve stannous chloride (30 gm) in hydrochloric acid (100 ml) and add sufficient water to produce 100 ml.
- **Starch indicator (0.5%):** Boil 0.25 gm of starch with 50 ml of distilled water with stirring. Cool and use for the experiments.
- **Steyn reagent:** Prepare a solution of 5 gm sodium bicarbonate and 0.5 gm picric acid in 100 ml of water.
- **Sudan III solution:** Dissolve 50 mg of Sudan III in 25 ml of ethanol/methanol. Warm to dissolve, cool and add 25 ml of glycerin. Mix and filter if any particles persists.
- **Sulphuric acid-formaldehyde reagent:** *See Formaldehyde-sulphuric acid* reagent.
- **Thioacetamide reagent:** Add 1 ml of a mixture of sodium hydroxide (1M, 15 ml), water (5 ml) and glycerin (20 ml) to a solution of thioacetamide (4% w/v, 0.2 ml) in water. Heat on water bath for twenty seconds and coll. Thioacetamide reagent should be prepared immediately before use.
- **Vanillin hydrochloric acid reagent:** Vanillin (1 gm) is dissolved in ethanol (100 ml) and Conc. hydrochloric acid (3 ml) is added.
- **Vanillin sulphuric acid reagent:** Vanillin (3 gm) is dissolved in ethanol (100 ml) and Conc. sulphuric acid (1 ml) added.
- **Tollen's reagent:** To 20 ml of 0.1M silver nitrate add sufficient strong ammonia solution to dissolve the initial brown precipitate to get clear solution.
- **Vanillin reagent:** Dissolve 1 gm of vanillin in 20 ml of sulphuric acid, warming if necessary
- **Wasicky reagent (Van UrK reagent):** Dissolve 2 gm of *p*-Dimethylamino benzaldehyde (*p*-DMAB) to 50 ml of 95% ethanol and 50 ml of concentrated hydrochloric acid. The reagent should be freshly prepared.

INDEX

B

D

E

F

G

H

I

J

K

L

M

N

O

P

Q

R

T

U

V

W

X

Z

9 789352 300792

Printed at Repro India Ltd.